Extremophiles for Sustainable Agriculture and Soil Health Improvement

Anuj Ranjan • Vishnu D. Rajput
Abhishek Chauhan
Evgeniya Valer'evna Prazdnova
Tatiana Minkina • Sajad Majeed Zargar
Editors

Extremophiles for Sustainable Agriculture and Soil Health Improvement

Springer

Editors
Anuj Ranjan
Academy of Biology and Biotechnology
Southern Federal University
Rostov-on-Don, Russia

Abhishek Chauhan
Institute of Environmental Toxicology,
Safety and Management
Amity University
Noida, Uttar Pradesh, India

Tatiana Minkina
Academy of Biology and Biotechnology
Southern Federal University
Rostov-on-Don, Russia

Vishnu D. Rajput
Academy of Biology and Biotechnology
Southern Federal University
Rostov-on-Don, Russia

Evgeniya Valer'evna Prazdnova
Academy of Biology and Biotechnology
Southern Federal University
Rostov-on-Don, Russia

Sajad Majeed Zargar
Proteomics Lab, Division of Plant
Biotechnology
Sher-e-Kashmir University of Agricultural
Sciences and Technology of Kashmir
(SKUAST- Kashmir)
Srinagar, Jammu and Kashmir, India

ISBN 978-3-031-70202-0   ISBN 978-3-031-70203-7   (eBook)
https://doi.org/10.1007/978-3-031-70203-7

Foreword

In the domain of soil microbiology, extremophiles emerge as significant subjects deserving attention of scientific community due to their unique adaptability. Extremophiles, such as thermophiles, psychrophiles, acidophiles, alkaliphiles, barophiles, and halophiles, have adapted well to extreme environments, including manmade ones like polluted land. Their unique physiology and biochemistry play a crucial role in survival. Exploring their adaptation mechanism has provided insights into their ability to produce enzymes and metabolites, which improve crop health and disease management, and secrete plant hormones and bioactive compounds. The book, entitled *Extremophiles for Sustainable Agriculture and Soil Health Improvement*, provides a thorough exploration of extremophiles and their applications in agriculture. In this comprehensive collection, learned contributors converge to describe the complex relationship between extremophile microorganisms and their environment, diving into the depths of microbial ecology and diversity with scientific precision. This collection has five parts and each part has chapters under respective themes as per respective parts.

Part I, "Boundary limits and Adaptation of Extremophiles," the foundational part, routes the underlying principles of microbial ecology. Beginning with an in-depth analysis of microbial taxonomy and community structures, it progresses to examine the dynamics of microbial interactions. It further extends to the role of diversity in shaping ecological processes, providing readers to comprehend the elaborated extremophile microbe relationships.

Part II, "Potential Application of Extremophiles in Agricultural and Improving Soil Health," focuses on the practical applications of extremophiles, from the bioactive compounds of extremophilic bacteria for biocontrol and plant disease management to the utilization of metallotolerant microbes in remediating heavily polluted soil.

Part III, "Biotechnology and Genetic Basis of Extremophile," investigates into capabilities such as nutrient solubilization, cold stress adaptation, and genomic exploration of extremophiles using the biotechnological and genetic aspects. The part also highlights the improvement of saline soil fertility using halotolerant microbes. This part features the genetic pool of extremophiles as a prospective source for developing stress-resilient plants and microbes.

Part IV, "Multi-omics Approach for Exploring Extremophiles," introduces multi-omics technology-based approaches, emphasizing the role of extremophiles in ensuring food and nutritional security. Genomic and metagenomic prospecting of extremophiles emerges as a potent tool for a deeper understanding of their genetic makeup and potential applications and supporting sustainable agriculture.

The final Part V, "Exploiting Extremophiles by Nano Approach", explores the nanotechnological frontier, where extremophiles play a vital role in synthesizing nanoparticles for agricultural applications. This intersection of nanotechnology and extremophiles offers innovative possibilities for enhancing the efficiency and precision of agricultural use by integrating the extremophile as PGPR with various nanomaterials.

In this compilation, contributors provide valuable insights into the novel ground of extremophiles, offering exploration of their potential in sustainable agriculture and soil health improvement. As readers navigate through these pages, they engage in a journey of scientific investigation, where the adaptability of extremophiles becomes a focal point for advancing our understanding of sustainable agricultural practices.

University of Réunion Island, France   Laurent DUFOSSÉ

Preface

Microbes are ubiquitously present on the earth. They have been reported in the hot spring, coldest places, poles of the earth, the deep oceans, volcanoes, deserts, and the thin atmosphere. Adaptation in such extreme environments requires a special ability for survival. Extremophiles such as thermophiles, psychrophiles, acidophiles, alkaliphiles, barophiles, and halophiles have adapted well to their respective environment. Such extreme conditions can also be manmade, for example heavily polluted land laden with varieties of persistent organic pollutants and heavy metals or even radioactive wastes.

Resilience to the extreme environment is the superior attribute of the extremophiles. Their unique physiology and biochemistry play a vital role in survival in extreme conditions. Exploring potential extremophiles' adaptation physiology, biochemistry, and genetics has provided several fundamental insights. They produce a variety of enzymes and metabolites that have been useful in improving crop health and disease management. Many of them are capable of solubilizing nutrients (NPK), secreting plant hormones, bioactive compounds (with antibacterial and antifungal properties) against plant pathogens, etc.

In the present day, the major environmental concern in agriculture practice is the degradation of soil quality and fertility due to the use of synthetic agrochemicals. Polluted soil has a direct association with its adverse effect on crop production, such as reduced yield and assimilation of heavy metals by the crops, thereby entering the food chains and food webs.

The adaptation of extremophiles in man-made extreme conditions (e.g., polluted soil) has been advantageous to those microbes as they were able to convert toxic pollutants into beneficial resources for their metabolism, thereby reducing their ecotoxicity and the loads from the soil. Such adaptation of extremophiles is a piece of evidence that they can be a potential candidate for effective bioremediation of several pollutants in the soil. Harnessing their competency to tackle adverse environmental conditions, for example, highly saline soil, mining areas loaded with metal pollutions, cold stressed plants, presence of synthetic organic pollutants in the soil, etc., could be the potential application of extremophiles. Application of extremophiles can be elaborated further with examples of metallotolerant bacteria that can be very useful for the remediation of soil in mining areas where the soil is loaded with varieties of metal and organic pollutants. Similarly, psychrophilic microbes can help improve the cold stressed plant for the region/countries where agriculture is badly affected by cold weather.

With the advances in microbial technology, scientists have been able to use them in improving or developing crops with beneficial traits from other species. Multi-omics approaches have added another dimension where the extremophiles have been explored at the genomic, transcriptomic, proteomic, and metabolomic levels for their beneficial aspects in agriculture. Further, coupling extremophile potential with nanotechnology has proven far more effective in agriculture, especially in soil amendments. Nanoparticles with varied properties have been studied for improving the bioremediation of pollutants from the soil, plant disease management, synthesis of green NPs for agricultural application, etc.

This book, *Extremophiles for Sustainable Agriculture and Soil Health Improvement*, anticipates bringing recent advances and studies done on extremophile application in the agriculture sector collectively on one single platform as an edited book and contents in the form of chapters for the benefit of readers from the scientific community who exclusively works on soil remediation, nanobioremediation, crop yield, plant physiology and biochemistry, plant disease, and PGPR.

Rostov-on-Don, Russia   Anuj Ranjan
Rostov-on-Don, Russia   Vishnu D. Rajput
Noida, Uttar Pradesh, India   Abhishek Chauhan
Rostov-on-Don, Russia   Evgeniya Valer'evna Prazdnova
Rostov-on-Don, Russia   Tatiana Minkina
Srinagar, Jammu and Kashmir, India   Sajad Majeed Zargar

Acknowledgment

This book, titled *Extremophiles for Sustainable Agriculture and Soil Health Improvement*, is the outcome of collaborative efforts from the entire editorial team. Throughout the process, we recognized the utmost importance of meticulous planning, effective execution, adherence to timelines, and successful project delivery. Our gratitude extends to our respective institutions, with special appreciation for the unwavering support from colleagues, friends, and family members, whose encouragement and motivation were invaluable.

We sincerely acknowledge the support of the Strategic Academic Leadership Program of the Southern Federal University ("Priority 2030"), Rostov-on-Don, Russia.

We express our heartfelt thanks to all the contributing authors whose extensive research and rich experience in extremophilic microbes have significantly enriched the content of this publication. The pleasure of editing this book was derived from the stimulating and cooperative engagement of our contributors.

A thoughtful appreciation goes to the team at Springer for demonstrating confidence in our editorial team and entrusting us with the responsibility of publishing for the global scientific community. Their support is instrumental in bringing this valuable contribution to completion.

Contents

Part I   Boundary Limits and Adaptation of Extremophiles

1 **Extremophile: Occurrence, Ecological Diversity
and Taxonomic Aspects**   3
Chesta Saini, Deepesh Kumar Neelam, Jebi Sudan,
and Sajad Majeed Zargar

2 **Survival Strategies of Extremophiles: Physiology
and Biochemistry** 21
R. Mythrayee and K. Veena Gayathri

3 **Microbial Adaptation in Different Extreme Environmental
Conditions and Its Usefulness in Differently Polluted Soil** 47
Jayati Arora, Arpna Kumari, Anuj Ranjan, Vishnu D. Rajput,
Sudhir Shende, Evgeniya Valer'evna Prazdnova, Saglara
S. Mandzhieva, Svetlana Sushkova, Tatiana Minkina, Abhishek
Chauhan,
Rajpal Srivastav, and Tanu Jindal

4 **Extremophiles Adaptation and Its Utilization in Mitigating
Abiotic Stress in Crops** 63
Adesh Kumar, Monika Shrivastava, and Pallavi Saxena

**Part II   Potential Application of Extremophiles in Agricultural
and Improving Soil Health**

5 **Exploration of Extremophiles: Potential Applications
in Agriculture and Soil Health Improvement Utilizing
Extremophiles** 91
Vinay Mohan Pathak, Nitika Rana, Sumitra Pandey, A. K. Sarkar,
Abhishek Chauhan, Tanu Jindal, Mukta Sharma, Seeta Dewali,
Satpal Singh Bisht, Divya, Akansha Sharma, Monika Yadav,
Balwant Singh Rawat, Somdatt Tyagi, D. P. Uniyal, Pradeep Kumar,
Sevaram Singh, Baljinder Kaur, Jose Maria Cunill,
and Satish Chandra Garkoti

6   **Bioactive Compounds/Metabolites from Extremophiles
    for Biocontrol and Plant Disease Management** 121
    Monika Shrivastava, Adesh Kumar, and Pallavi Saxena

7   **Bioactive Molecules Derived from Extremophilic Fungi
    and Their Agro-Biotechnological Application** 137
    Namita Ashish Singh, Avinash Marwal, Juhi Goyal, and Nitish Rai

8   **Metallotolerant Microbes for Improving the Health
    of Heavily Polluted Soil.** 163
    Sarieh Tarigholizadeh, Roghayeh Heydari, Svetlana Sushkova,
    Saglara Mandzhieva, Sudhir Shende, Vishnu D. Rajput,
    and Tatiana Minkina

9   **Remediation of Soil Organic Pollutants by Microbes
    from Extreme Environment** 199
    Dinoo Gunasekera and Disna Ratnasekera

Part III   **Biotechnology and Genetic Basis of Extremophile**

10 **Functional Insights of Nutrients Solubilizing Extremophiles
    for Potential Agriculture Application** 221
    Bhalerao Bharat, Khaire Pravin, Borase Dhyaneshwar,
    Kamble Bhimrao, Arjun Singh, Murugan Kumar, Aniket Gade,
    and Arunima Mahto

11 **Extremophiles and Their Genetic Aspects of Potential
    Bioactive/Metabolites Beneficial for Promoting
    Plant Health and Soil Fertility** 251
    Bal Krishna, Parkash Verma, Rakesh Kumar, Anil Kumar Singh,
    Priyanka Upadhyay, Ashutosh Kumar, Talekar Nilesh Suryakant,
    Birender Singh, Sudeepa Kumari Jha, and Juli Kumari

12 **Biotechnology and Genomics Exploration of Halotolerant
    Microbes: Application for Improving the Fertility of Saline Soil** 281
    Smita Kumari and Balaram Mohapatra

13 **Biotechnological Insights into Cold-Stressed, Adapted
    Microorganisms for Plant Health and Soil Improvement** 301
    Vishnu Mishra, Jawahar Singh, and Vishal Varshney

14 **Thermophiles and Their Diverse Function in Agricultural
    and Biotechnological Applications.** 317
    Himanshi Aggarwal, Divya Chaudhary, Jaagiriti Tyagi,
    Naveen Chandra Joshi, Sakshi Arora, Vaibhav Mishra,
    and Manoj Kumar

15 **Thermophilic Microbes: Their Role in Plant Growth
    Promotion and Mitigation of Biotic Stress** 337
    Sumit Kumar, Mehjebin Rahman, Mateti Gayithri, Anjali,
    Ali Chenari Bouket, R. Naveenkumar, Anuj Ranjan,
    Vishnu D. Rajput, Tatiana Minkina, and Rupesh Kumar Singh

16 **Biotechnology of Promising Genes from Extremophiles
    to Produce Stress-Resilient Plants and Microbes
    for Sustainable Agriculture** 361
    Manmeet Kaur, Diksha Singla, Kamal Kapoor, Gautam Chhabra,
    Sezai Ercisli, Mehmet Ramazan Bozhuyuk, Shiv K. Yadav,
    and Ravish Choudhary

17 **Insight into Soil Nutrient Management in Agriculture
    by Acidophilus Microbes** 389
    Vaibhav Mishra, Neeraj Shrivastava, Smriti Shukla,
    and Rupesh Kumar Basniwal

Part IV Multi-omics Approach for Exploring Extremophiles

18 **Omics Technology in Food and Nutritional Security
    of Agricultural Crops: Role of Extremophiles** 405
    Tamana Khan, Sabba Khan, Diksha Singh, Aaqif Zaffar,
    Labiba Shah, Rizwan Rashid, Parvaze A. Sofi, Baseerat Afroza,
    and Sajad Majeed Zargar

19 **Genomic and Metagenomic Prospecting of Extremophiles
    to Support Sustainable Development** 425
    Mohit Gururani, Rishika Malhotra, Abhishek Singh,
    Raj Kishor Kapardar, and Rajpal Srivastav

Part V Exploiting Extremophiles by Nano Approach

20 **Nanotechnology and Extremophiles: Agricultural
    Applications and Possibilities.** 441
    Dinoo Gunasekera, Parakkrama Wijerathna, and Disna Ratnasekera

21 **Nanoparticles Synthesis Using Extremophilic Microbes
    and their Potential Agricultural Applications** 455
    Girima Nagda, Nitish Rai, Jaya, Shakshi, Chhavi Bhalothia,
    and Namita Ashish Singh

**Correction to: Biotechnology of Promising Genes
from Extremophiles to Produce Stress-Resilient Plants
and Microbes for Sustainable Agriculture** C1
Manmeet Kaur, Diksha Singla, Kamal Kapoor, Gautam Chhabra,
Sezai Ercisli, Mehmet Ramazan Bozhuyuk, Shiv K. Yadav,
and Ravish Choudhary

Index 485

About the Editors

Anuj Ranjan is a Leading Researcher and Post-Doc Fellow at Southern Federal University, Russia. His area of work includes beneficial microbes for the environment and health, bioactive compounds, and environmental toxicology. His expertise is in genomics, structural bioinformatics, target characterization, molecular docking, and simulation. He has published over 70 publications that include over 45 peer-reviewed articles and 25 book chapters/conference proceedings/ abstracts along with one authored book. He has also reviewed over 55 research/review articles, books, and grants. He has received awards from CSIR, Govt. of India, and the American Chemical Society (2016), and has presented his work at North American Chemical residue workshops (NACRW) (2017) in Naples, Florida, USA. He has also been a scientific member of the 34th Indian Scientific Expedition to Antarctica for Polar Environment Monitoring program of the Government of India. Dr. Ranjan is also an active member of the Society of Environmental Toxicology and Chemistry, USA, and the International Network on Soil Pollution (Food and Agriculture Organization of the United Nations).

Vishnu D. Rajput is the Head of "Soil Health Lab" at Southern Federal University, Rostov-on-Don, Russia. His ongoing research is based on the investigation of effective remediation approaches using biochar/nano-biochar-based sorbents and nanomaterials. With long experience and experimental work, Dr. Rajput comprehensively detailed the state of research in environmental science in regard to "how nanoparticles/heavy metals interact with plants, soil, microbial community, and the larger environment as well as possible remediation technology using nanoparticles/nano-biochar. Recently he is working nano-enhanced remediation of soil and water system. He has published 346 scientific publications, 88 chapters, and 28 books, with an H-Index: 39. He is an internationally recognized reviewer and received an Outstanding Reviewing Certificate by Elsevier and Springer. He is an editorial board member of various high-impacted journals such as *Biochar*. He is holding national and international projects including Mega-Grant and BRICS. Dr. Rajput has received "Certificate for Appreciation 2019" 2021, "Certificate of Honor 2020," Diploma Award 2021 and 2022, and Letter of Gratitude 2022 and 2023 from Southern Federal University, Russia, for outstanding contribution to academic and creative research, and publication activities. He has also received "Highly Qualified Specialist" status from the Russian government. In 2023, he is included in the list of 2% worlds potential scientists by Elsevier BV publishing house.

Abhishek Chauhan has a PhD in Microbiology from Gurukul Kangri University, Haridwar (Uttarakhand), India. He is currently working as a Senior Scientist at Amity University, Noida (Uttar Pradesh), India, and formerly worked as a Scientist "C" and Head at the Department of Microbiology, Shriram Institute for Industrial Research (SRI), Delhi, India. He has strong analytical and supervision ability and has done over 25 Industry sponsored projects. While working at SRI, Delhi, he has developed and validated several microbiological methods related to food, water, drugs, and pharmaceuticals. He has a total of 18 years of rich experience and has guided over 10 postgraduate and 3 PhD on emerging issues of microbiology and biotechnology. He has authored a research book on key aspects of "Antibacterial Activity of Cyanobacteria" and edited a book on *Plants and Microbes: An Innovative Approach*. He has delivered various plenary lectures and invited talk at several national and international symposium / conferences and has been a key speaker of teleconferencing talk on "Food and Diet" under EduSAT network, Vigyan Prasar, DST, Govt. of India. He has been a reviewer for many peer-reviewed journals and has published more than 45 research papers and honored with prestigious awards such as "Young Scientist Award" (2013), STOX-Appreciation Award (2016), NCEEBR-Certificate of Excellence for Oral Presentation, and Best Presentation Award (IIPA-DST Govt. of India) 2012. Dr. Chauhan has also participated in 35th Indian Scientific Expedition to Antarctica and 10th Indian Southern Ocean Expedition, scientific ventures of Ministry of Earth Science, Govt. of India. He has also been certified on various quality certifications and accreditations such as ISO: 17025 (Laboratory Quality Management System), ISO: 17043 (Proficiency Testing), ISO: 22000, ISO:9001, and ISO:14001. He is an active member of the Indian Science Congress Association, Society for Plant Research, and Association of Microbiologists of India.

Evgeniya Valer'evna Prazdnova is the Head of the Laboratory of Molecular Genetics of Microbial Consortia in the D.I. Ivanovsky Academy of Biology and Biotechnology, Southern Federal University, Rostov-on-Don, and Senior Researcher in Agrobiotechnology Center, Don State Technical University. Her scientific interests are: genetics, biochemistry and physiology of microorganisms, genetics and epigenetics of symbiotic interactions of micro- and macroorganisms, antioxidants, antimutagens, gerontoprotectors (geroprotectors), and probiotics. She managed projects such as: RFBR project "Study of the adaptogenic effect of low molecular weight organic cations using bacterial biosensors"; Grant of the President of the Russian Federation for state support of young Russian scientists "Isolation of metabolites of probiotic bacteria exhibiting antimutagenic, SOS-inhibiting and antioxidant activity."

Total publication number—117 (57 in Scopus). h-index: 14 (Scopus), 10 (WOS).

Tatiana Minkina is the Head of Soil Science and Land Evaluation Department of Southern Federal University and the Head of International Master's Degree Educational Program "Management and Estimation of Land Resources" (2015-2022, accreditation by ACQUIN). Her scientific interests include soil science, biogeochemistry of trace elements, environmental soil chemistry, soil monitoring, assessment, and modeling and remediation using physicochemical treatment methods. She was awarded in 2015 with Diploma of the Ministry of Education and Science of the Russian Federation for many years of long-term work for development and improvement of the educational process and significant contribution to the training of highly qualified specialists. Currently, she is handling projects funded by Russian Scientific Foundation, Ministry of Education and Science of the Russian Federation, and Russian Foundation of Basic Research. She is a Member of Expert Group of Russian Academy of Science, the International Committee on Contamination Land, Eurasian Soil Science Societies, the International Committee on Protection of the Environment, and the International Scientific Committee of the International Conferences on Biogeochemistry of Trace Elements. Her total scientific publications are: 757 (389 in English). She is the invited editor of an Open Access Journal by MDPI Water, ISSN 2073-4441 (Impact factor: 2.524). She is an editorial board member of *Geochemistry: Environment, Exploration, Analysis*; ISSN: 1467-7873 (Impact factor: 1.109; 5yr IF: 1.769; SJR: 0.334; SNIP: 0.598), and *Eurasian Journal of Soil Science*; e-ISSN: 2147-4249.

Sajad Majeed Zargar , PhD, is currently a Senior Assistant Professor at Sher-e-Kashmir University of Agricultural Sciences and Technology of Kashmir (SKUAST-Kashmir) in India and a Visiting Professor at the University of Padova, Italy. He was previously a Visiting Professor at the Nara Institute of Science and Technology, Japan. He has worked as an Assistant Professor at SKUAST-Jammu, Baba Ghulam Shah Badshah University, Rajouri (BGSB) in India. He has also worked as a scientist at Advanta India Limited, Hyderabad, India, and TERI (The Energy & Resources Institute), New Delhi, India. Dr. Zargar is the recipient of World Bank-funded NAHEP-ICAR overseas fellowship, CREST overseas fellowship from DBT, India, Goho grant from Govt. of Japan, and Erasmus Fellowship from European Commission. He has received several awards for his work and research. He is also the member and representative of INPPO (International Plant Proteomics Organization). He has chaired a session in the 3rd INPPO World Congress held at University of Padova, Italy, in 2018. He is a member of various International and national scientific societies. His editorial activities and scientific memberships include publishing research and review articles in international journals and as a reviewer. He has been affiliated with several internationally reputed journals and is also a reviewer of reputed journals such as *Journal of Advanced Research*, *Frontiers in Plant Science*, *3 Biotech*, *Scientia Horticulture*, *Methods in Ecology and Evolution*, *Australian Journal of Crop Science*, and many others. He has also edited several books that are published by internationally reputed publishers. Dr. Zargar has been invited to give many lectures at professional meetings and workshops and has received grants for research projects under his supervision. He is presently coordinator / principal investigator of rice genomics and buckwheat genetics projects funded by DBT, New Delhi, India. He has supervised many MSc and PhD students. He has been the mentor of three Post Doc Fellows funded by SERB, New Delhi and DST, New Delhi. He is the in-charge Scientist of the Genomics Lab. and Proteomics Lab. at the Division of Plant Biotechnology, SKAUST-Kashmir, INDIA.

Abbreviations

ABA	Abscisic acid
ABC	ATP-binding cassette
ABRC	*Arabidopsis* Biological Resource Center
ACC	1-Aminocyclopropane-1-carboxylate
ACP	Acyl carrier protein
AFP	Antifreeze proteins
AHL	N-acyl-l-homoserine lactone
AMF	Arbuscular mycorrhiza fungi
ANAMMOX	Anaerobic ammonium oxidation
APX	Ascorbate peroxidase
ARA	Arachidonic acid
ARDRA	Technique amplified ribosomal DNA restriction analysis
ATP	Adenosine triphosphate
CAE	Cold-active enzymes
CAP	Cold adaptation protein
CAT	Catalase
CRISPR	Clustered regularly interspaced short palindromic repeats
CSP	Cold shock proteins
DHA	Docosahexaenoic acid
DNA	Deoxyribonucleic acid
DPG	2,3-Diphosphoglycerate
EAP	Extensible authentication protocol
ED	Entner–Doudoroff
EDF	Extracellular death factor
EDTA	Ethylenediaminetetraacetic acid
EM	Embden–Meyerhof
EMP	Embden–Meyerhof–Parnas
EPA	Eicosapentaenoic acid
EPS	Exopolysaccharides
ER	Eleno reductase

FA	Fanconi anemia pathway
FAO	Food and Agriculture Organization
FIS	Factor for inversion stimulation
FNR	Fumarate and nitrate reductase
GA	Gibberellic acid
GAPDH	Glyceraldehyde 3-phosphate dehydrogenase
GDGT	Glycerol dialkyl glycerol tetraether
GPX	Glutathione peroxidase
GR	Glutathione reductase
GSH	Glutathione
GSK	Glycogen synthase kinase
GWM	Genome-wide Methylation
HCN	Hydrogen cyanide
HGT	Horizontal gene transfer
HM	Heavy metals
H-NS	Histone-like nucleoid structuring
HR	Homologous recombination
HSP	Heat shock proteins
IAA	Indole-3-acetic acid
IC	Inhibitory concentration
IHF	Integration host factor
INDIGO	Integrated Data Warehouse of Microbial Genomes
IPCC	Intergovernmental Panel on Climate Change
IR	Ionizing radiation
IRI	Ice recrystallization inhibition
JNK	C-jun N-terminal kinases
KAS	β-ketoacyl-acyl carrier protein synthase
LC	Liquid chromatography
LC-PUFA	Long-chain polyunsaturated fatty acids
LEO	Lower earth orbit
LPS	Lipopolysaccharide
LSU	Large subunit
LUCA	Last universal common ancestor
MAA	Mycosporine-like amino acids
MDA	Malondialdehyde
MDR	Multidrug-resistant
MIC	Minimum inhibitory concentration
MS	Mass spectrometry
NADH	Nicotinamide adenine dinucleotide
NADP	Nicotinamide adenine dinucleotide phosphate
NF	Nuclear factor
NGS	Next-generation sequencing
NHEJ	Non-homologous end-joining
NPs	Nanoparticles

NRPS	Non-ribosomal peptide synthetases
OECD	Organisation for Economic Co-operation and Development
OP	Organic pollutants
PAEM	Plant-associated extremophilic microbes
PAL	Phenylalanine ammonia-lyase
PAM	Protospacer adjacent motif
PCA	Phenazine-1-carboxylic acid
PCR	Polymerase chain reaction
PEP	Phosphatidyl pyruvate
PFAS	Perfluoroalkyl and polyfluoroalkyl
PGK	Phosphoglycerate kinase
PGL	6-Phosphogluconolactone
PGP	Plant growth promotion
PGPF	Plant growth-promoting fungi
PGPM	Plant growth-promoting microbes
PGPR	Plant growth-promoting rhizobacteria
PKS	Polyketide synthase
PLFA	Phospholipid fatty acid
PMF	Proton motive force
PMI	Precision agriculture management
POD	Peroxidase
POP	Persistent organic pollutants
POX	Phenol peroxidase
PP	Pentose phosphate
PQQ	Pyrroloquinoline-quinone
PS	Phosphate solubilization
PSB	Phosphate solubilizing bacteria
PTS	Phosphotransferase system
QS	Quorum sensing
RNA	Ribonucleic acid
RND	Resistance-nodulation-division
ROS	Reactive oxygen species
SA	Salicylic acid
SDS	Sodium dodecyl sulfate
SIP	Stable isotope probing
SNP	Single-nucleotide polymorphism
SOB	Sulfur-oxidizing bacterium
SOD	Superoxide dismutase
TALENS	Transcription activator-like effector nucleases
TBP	Transcription-binding protein
TCA	Tricarboxylic acid
TEM	Transmission electron microscope
TH	Thermal hysteresis
TILLING	Targeting induced local lesions in genome

TLC	Thin layer chromatography
TLS	Translesion synthesis
USA	United States of America
UV	Ultraviolet
UVR	Ultraviolet radiation
VOC	Volatile organic compounds
XPF	Nucleotide excision repair
ZFN	Zinc finger nucleases

Part I
Boundary Limits and Adaptation of Extremophiles

Chapter 1
Extremophile: Occurrence, Ecological Diversity and Taxonomic Aspects

Chesta Saini, Deepesh Kumar Neelam, Jebi Sudan, and Sajad Majeed Zargar

1.1 Introduction

The interesting microorganisms that live in harsh conditions have attracted scientists' attention for the past few decades (Rampelotto 2013). Such microorganisms are known as extremophiles (from Latin '*extremus*' meaning 'extreme' and Greek '*philia*' meaning 'love'), and this term was given by Bob MacElroy in 1974 (Rothschild 2007). The majority of extremophiles belong to the bacteria, eukarya and archaea domains (Rothschild and Mancinelli 2001). In the middle of the 1960s, extremophile microorganisms attracted attention when *Thermus aquaticus*, a filamentous bacterium that thrives at extreme temperatures, was isolated by T. Brock and his colleagues from 'Yellowstone National Park in the United States' (Ramirez 2021). The morphological, genetic and physiological characteristics of extremophile microorganisms that enable them to tolerate extremely selective environmental circumstances are especially impressive (Kohli et al. 2020). The organisms can be classified as alkaliphiles, acidophiles, thermophiles, halophiles, psychrophiles, piezophiles, xerophiles, radiophiles and endolithic, and certain extremophiles can adapt to several stresses (polyextremophile) (Gupta et al. 2014) based on extreme conditions, including salt concentrations, high temperature, pressure, low temperature, pH, water availability, nutrient concentration, circumstances with high amounts of radiation and also toxic compounds (organic solvents) and hazardous heavy metals (Satyanarayana et al. 2005). Research on extremophiles has advanced to the extent that Portugal hosted the 'First International Congress on Extremophiles' in 1996 and the scientific journal *Extremophiles* was founded there in 1997. Another

C. Saini · D. K. Neelam (✉)
Department of Microbiology, JECRC University, Jaipur, India

J. Sudan · S. M. Zargar
Proteomics Lab, Division of Plant Biotechnology, Sher-e-Kashmir University of Agricultural Sciences and Technology of Kashmir (SKUAST- Kashmir), Srinagar, Jammu and Kashmir, India

© The Author(s), under exclusive license to Springer Nature Switzerland AG 2024
A. Ranjan et al. (eds.), *Extremophiles for Sustainable Agriculture and Soil Health Improvement*, https://doi.org/10.1007/978-3-031-70203-7_1

organization named 'The International Society for Extremophiles' (ISE) was established in 2002 to exchange knowledge and expertise in the quickly expanding field of extremophile research (Gupta et al. 2014).

Agriculture sector is facing significant stress to produce sufficient food for the world's population, but such sectors are seriously threatened by some harsh environmental factors (Rizvi et al. 2021), including drought (Kapoor et al. 2020), salinity (Zahra et al. 2020), and extremely high (Jiang et al. 2020; Alsamir et al. 2021) and low temperatures (Chi et al. 2021; Sanghera 2020). Extremophiles' diversity and ability to adapt to various stresses have resulted in increased interest in their possible applications in a variety of industrial operations, such as food production, biotechnology and the medical and pharmaceutical industries. For the improvement of crop growth and health in sustainable agriculture, extreme conditions have recently become more significant as potential sources of plant-growth-promoting (PGP) agents (Bouri et al. 2022). For example, tomato seedlings grown in extreme salinity conditions using *Achromobacter piechaudii* showed PGP characteristics of ACC deaminase (Yang et al. 2009). Tomato seedling growth under saline conditions was boosted by >66% when *Azospirillum piechaudii*, with PGP properties of ACC, was compared to the control. According to significant studies on PGP microbe-mediated plant resistance to salt stress, the impacts of salt stress can be reduced in diverse plant species through inoculation with bacteria. Most amino acid concentration increases when *Azospirillum* introduced into salt-stressed maize cultivars (cv. 323 and cv. 324). The inoculation of *Azospirillum* significantly changed the selectivity of K^+, $Ca2^+$ and Na^+ particularly in the salt-sensitive cultivar cv. 323. The K^+/ Na^+ ratio of cv. 324 was comparatively higher than that of cv. 323, indicating significantly higher salt tolerance of cv. 324 (Hamdia et al. 2004). The various groups of microorganisms, including bacteria (*Proteobacteria, Firmicutes, Deinococcus-Thermus, Bacteroidetes, Actinobacteria* and *Acidobacteria*) and archaea (*Euryarchaeota*), have been reported to be associated with plants (Verma et al. 2017).

More than 157 halophilic archaea were identified, but only 20 isolates were recognized as 17 unique species of 11 genera, including *Natronoarchaeum, Haloterrigena, Halosarcina, Halolamina, Haloferax, Halococcus, Halobacterium, Natrinema, Natrialba, Halostagnicola* and *Haloarcula*, related to various salt-tolerant plants (*Suaeda nudiflora, Sporobolus, Dichanthium, Cenchrus* and *Abutilon*) utilizing Haloarchaea Phosphate Solubilization (HPS) medium, which exhibit P solubilization. *Natrinema* sp. strain IARI-WRAB2 was determined to be the most effective P solubilizer (134.61 mg/L). These halophilic archaea that solubilize phosphate play an important role in providing P nutrition to the plants that are developing in the highly saline soils (Yadav et al. 2015). Among distinct taxa, *Exiguobacterium, Pseudomonas* and *Bacillus* have the greatest characteristics for promoting plant growth at lower temperatures (Selvakumar et al. 2011; Yadav et al. 2016).

1.2 Extremophiles: Occurrence, Adaptation and Role in Biogeochemical Cycles

Extremophiles can grow in a variety of environments that used to be considered unsuitable for life, including extremely hot niches, salt solutions, ice and acidic and alkaline environments, and certain may even thrive in organic solvents, toxic waste and heavy metal environments (Rampelotto 2013). Extremophiles are evolved relics that have developed molecular, biochemical and cellular adaptation mechanisms. They develop enzymes that can remain stable and function in a variety of extreme conditions. To protect themselves from extremes of solar radiation, temperature, pressure, pH and salinity, these microorganisms also synthesize a wide range of additional metabolites and molecules, like surface-active substances and extremolytes (Thakur et al. 2022). This suggests that their physiological and molecular characteristics should have been developed appropriately to survive under such conditions. They produce proteins, particularly enzymes referred to as extremozymes, which are essential to their survival. Additionally, these extremozymes are utilized in a variety of industry areas (Rao et al. 2022). Extremophiles surviving high radiation levels have been connected to their effective DNA repair methods and capability to synthesize extremozymes and extremolytes, which are protective metabolic products (Singh and Gabani 2011). Extremophiles under high pressure have established methods to maintain metabolic activity by intrinsic and extrinsic adaptations. Intrinsic adaptation includes modifications to the sequences and structure of proteins, and extrinsic adaptation is accomplished by micro-molecular metabolites that might prevent metabolic functioning under stress (Kumar et al. 2018).

Proteins found in halophilic microorganisms adapt to salty environments and maintain their activity by decreasing bulky hydrophobic residues and enhancing the protein surface with acidic or short polar side chain amino groups (Ortega et al. 2015). The halophilic proteins contain several unique peptides that might help stabilize and enhance the performance of proteins in highly halophilic environments (Kumar et al. 2018). Microorganisms are the primary initiators of biogeochemical cycles in the Earth's biosphere (Zheng et al. 2018); thus, it can be concluded that extremophilic microbes are also essential in biogeochemical cycles. Extremophilic organisms and their habitat have shown that most of the archaea, certain bacteria and cyanobacteria perform metabolic activities necessary for the biogeochemical cycles of nitrogen, carbon and sulfur. Bacteria and archaea are two of the main organisms involved in various processes, including methanogenesis, sulfidogenesis and anaerobic ammonium oxidation (ANAMMOX), among others (González and Terrón 2021). For example, ammonia released during nitrogen fixation in saline lakes can be converted to 'nitrate' through 'nitrite' via haloalkaliphilic nitrifiers. A subpopulation of *Nitrosomonas halophila* that is highly tolerant to alkaline conducts the oxidation of ammonia to nitrite in saline soils and lakes, while *Nitrobacter alkalicus* that is moderately tolerant to alkaline conducts the oxidation of nitrite (Sorokin et al. 2014).

1.3   Taxonomic Aspects of Extremophiles

The large variety of microbes, including various new, phylogenetically deep-rooted taxa, thrive and survive in harsh habitats (Shu and Huang 2022). Extremophiles have been discovered at depths of 6.7 km under the Earth's crust and more than 10 km within the sea, at 110 MPa pressures, in extreme acidic (pH 0) and basic (pH 12.8) environments and in temperatures ranging from 122 °C in hydrothermal vents to −20 °C in freezing ocean water (Rampelotto 2013). These distinct and low-complexity environments offer an excellent chance to investigate the function, evolution and structure of natural microbe communities. A huge uncultured diversity of microbes and most archaea under extreme habitats have been revealed by marker gene surveys that have clarified the ecological and pattern factors of these extremophile assemblies (Shu and Huang 2022). Extremophiles are classified into several categories based on the environments in which they thrive, including psychrophiles, alkaliphiles and acidophiles, barophiles, thermophiles and halophiles (Gallo et al. 2021). Extremophilic microbe phylogeny has been connected to issues about the early origin and evolution of life on Earth. Specifically, among the archaeal thermophiles and halophiles, many hyperthermophiles, thermophiles and halophiles seem to have extremely deep lineages (Al'Abri 2011). The surroundings that the last universal common ancestor (LUCA) of existent life may have inhabited are a subject of intense discussion. The research community believes that extremophiles first appeared during the evolution of prokaryotes even though it is still unclear whether the LUCA was an extremophile (hyperthermophile) (Catchpole and Forterre 2019; de la Haba et al. 2022).

The identification and genetic features of the significant novel lineage that significantly increase microbial diversity and modify the structure of the tree of life have been created possible by recent omics investigations that have revealed connections among community activity and external factors. Such attempts have helped us learn a lot more about the ecology, diversity and evolution of the microbes that live in Earth's extreme habitats as well as improved the research of microbiomes and their functions in more complicated environments (Shu and Huang 2022). As a result, understanding the metabolic features and taxonomy of microbial diversity in extreme conditions has become a difficult challenge in the fields of microbial biotechnology and microbiology. This difficulty has been solved to an extent by the development and improvement of next-generation sequencing (NGS) and computer tools for NGS analysis of data. In addition, it has also been discovered that single-cell genomics and metagenomic techniques must be integrated synergistically in order to highly exploit the extremophilic microbial metabolic and genetic diversity. Understanding the extremophilic microbe's biology, such as their metabolic capabilities, molecular methods of adaptation and special genomic traits, like codon reassignments, is believed to result from a synergistic approach (Goyal et al. 2020).

1.3.1 Types of Extremophiles

There are various extremophiles, including the following, which can grow in extreme conditions.

Alkaliphiles

Microorganisms that thrive ideally or extremely well at above pH 9 are referred to be alkaliphiles; however, they cannot multiply or slowly develop at pH 6.5 (Horikoshi 2004). According to the taxonomy that is currently established, alkaliphiles can be split into two separate groups based on pH values for development (Horikoshi and Bull 2011). Especially, those microorganisms that grow slightly less than 7.0 pH (but not lesser than 6.0) have been labelled as facultative alkaliphiles, whereas those microorganisms that grow between the range of pH 7 and 12 have been classed as obligatory alkaliphiles. Additionally, alkaliphiles are also classified as obligate alkaliphiles (true alkaliphiles) that cannot develop or multiply at pH 7.0 or lesser and their ideal growth at greater than pH 7.0, and those microorganism's development ideal pH ranges from 7.0 to lower (but not below 6.0), but maximum growth pH ranges from 8.0 to 8.5 known as alkali-tolerant (Horikoshi and Bull 2011). Alkaliphiles can be extracted from natural habitats like garden soil (Horikoshi 2004). Alkaliphiles are majorly distributed around the subtropical, tropical and intracontinental cryo-arid zones of the Earth (Kevbrin 2020). Alkalinity is produced in these lakes by the chemical reactions of metamorphic volcanic rocks that contain carbonatite or are igneous by CO_2-bearing waters. Water becomes alkaline as a result of the release of several dissolved carbonate species. All soda lakes in hydrologically enclosed basins have the same characteristic of a progressive evaporative concentration of inorganic salts, with sodium carbonates predominating (Deocampo and Renaut 2016). Gram-negative alkaliphiles were discovered to be restricted to the proteobacteria γ3 subdivision, with many of the isolates being connected to the Halomonas/Deleya group. Both the low % G + C and high % G + C divisions of the gram-positive lineage contained gram-positive alkaliphiles, with many of the isolates being connected to the *Bacillus* group and others to *Arthrobacter* sp. Alkaliphilic archaea and members of the genera *Natronococcus* and *Natronobacterium* shared a close relationship (Duckworth et al. 1996). One of the most frequently synthesized enzymes in industry is proteases and protein-degrading enzymes (Fujinami and Fujisawa 2010). Natural detergents contain alkaline enzymes, such as alkaline proteases and alkaline cellulases, which are derived from alkaliphiles. The industrial preparation of cyclodextrin using alkaline cyclomaltodextrin glucanotransferase is another significant use. This enzyme decreased the cost of production and made it possible for cyclodextrin to be used extensively in foods, chemicals and medications. Moreover, it has been suggested that alkali-treated wood pulp may undergo biological bleaching due to the xylanases that are synthesized by alkaliphiles (Horikoshi 2004).

Acidophiles

Between pH values 0 and 11.0, microorganism growth is possible. Microbes that can survive in extremely acidic conditions (pH ≤ 3) belong to all three domains of life: bacteria, archaea and eukarya (Oren 2010). Solfataric fields, acid mine drainage sites, fumaroles, acido-thermal hot springs and coal slags are all favourable environments for acidophiles to grow (Tiquia-Arashiro and Rodrigues 2016). Pressure up to 5.0 MPa, low salinity, a temperature range between 25 °C and 90 °C, low pH values, certain heavy metals and either aerobic or anaerobic conditions are present in these environments (Hallberg et al. 2010; Reeb and Bhattacharya 2010). Certain acidophiles can thrive in warm environments. Moreover, industrialized leaching of copper or other metals from ores is carried out by acidophilic prokaryotes (Oren 2010). Acidophiles use various pH homeostatic processes to restrict the entry of protons through cytoplasmic membranes, protons purging and their impacts. Acidophiles have an extremely impermeable cell membrane that prevents the influx of protons into the cytoplasm, helping to maintain pH (Konings et al. 2002). Since the membrane permeability of the proton monitors the rate at which the proton leaks inward, balance among influx of proton by energetic, transport systems and permeability of proton and rate of outward pumping of proton controls in case cells can sustain a suitable proton motive force (PMF) (Tiquia-Arashiro and Rodrigues 2016). To boost acid resistance, various bacteria of acidophilic insert ω-alicyclic fatty acids into the membrane (Chang and Kang 2004); at neutral pH, these acid-resistant lipids lose their structural integrity (van de Vossenberg et al. 1998). Another strategy includes the transmembrane channels to decrease H^+ influx. *Acidithiobacillus ferrooxidans* increase the production of 'Omp40', a channel with the smallest pore size when the extracellular pH is low. A Donnan potential is produced by the accumulation of 'monovalent cations' in the cytoplasm, which is another type of adaptive mechanism. High intracellular cation concentration creates the positive charge gradient $\Delta\psi$, which inhibits H^+ influx despite favorable concentration gradients. In order to facilitate the development of this Donnan potential, K^+/H^+ antiporters with stoichiometries greater than 1:1 are frequently used. Those antiporters and ATP-dependent H^+ pumps also assist in promoting H^+ efflux and preventing cytoplasmic acidification and are present in acidophiles of all three domains (Enami et al. 2010).

Thermophiles

Over the past two decades, a great deal of study has been done on both thermophilic microorganisms and thermostable enzymes. However, the innovative work of Brock and his collaborators in the 1960s sparked the focus on thermophiles and how their proteins are capable of functioning at higher temperatures (Brock and Freeze 1969). Psychrophiles (<20 °C), mesophiles (mid-temperatures) and thermophiles (high temperatures, >55 °C) are the three main types of microorganisms based on their ideal growth temperatures (Brock 1986). Just a few eukaryotes are believed to have the capacity to survive at this temperature, while several fungi can survive in the

50–55 °C range (Maheshwari et al. 2000). A few years ago, Kristjansson and Stetter proposed a new boundary between the thermophiles and the hyperthermophiles (growth at and above 80 °C), which is now widely accepted (Kristjansson and Stetter 2021). The majority of thermophilic bacteria now defined grow below the hyperthermophilic boundaries (with rare exceptions, such as *Thermotoga* and *Aquifex*), whereas the archaea predominate among hyperthermophilic species (Stetter 1996). Thermophiles and hyperthermophiles can be found around the world, like hot springs, volcanic environments, fumaroles, mud pots, coastal thermal springs, deep-sea hydrothermal vents and geysers. They can also be found in artificial environments like spray dryers, reactors and hot composting facilities (Urbieta et al. 2015). Thermostable enzymes provide strong catalyst substitutes that can endure the frequently aggressive circumstances of industrial processing (Turner et al. 2007). Thermophiles and their bioproducts enable a variety of commercial, medical and agricultural uses, as well as potential remedies for environmental harm and the need for biofuels (Urbieta et al. 2015). Thermophile microorganisms are also utilized in the process of leaching and removal of heavy metals from the trash (Ilyas et al. 2014).

Psychrophiles

Eighty-five per cent portion of the biosphere is continuously exposed to temperatures below 5 °C (Margesin and Miteva 2011). From the Antarctic to the Arctic, high mountains to the depths of the ocean, there are cold ecosystems. Deep sea, followed by permafrost, glaciers, sea ice and snow represents the majority of this low-temperature ecosystem. Cold deserts, cold soils (particularly subsoils), caverns and cold lakes are more examples of cold habitats (Margesin and Miteva 2011). Psychrophiles are cold-loving microorganisms that contribute significantly to the biomass of the Earth and play important roles in the biogeochemical cycles on a global scale (Siddiqui et al. 2013). In 1975, the word "psychrophile" was defined by Morita to refer to microbes that are cold-loving or cold-adapted, and having minimal, an optimal and maximal temperature for development and growth at or below 0 °C, 15 °C and 20 °C (Baross and Morita 1978). The first reports of psychrophiles were in 1884, but the majority of the previous research focused on psychrotrophic bacteria rather than real psychrophiles (Moyer and Morita 2007). There was significant discussion because researchers were not dealing with bacteria that could withstand extremely low temperatures, and as a result, various terms were developed to describe psychrophiles, like thermophobic bacteria, rhigophile, psychrobe, cryophile, Glaciale Bakterien, psychrotolerant, facultative psychrophile, psychrotrophic and psychrocartericus (Morita 1975). In cold environments, the capacity of psychrophiles to proliferate and survive indicates that they have overcome major challenges like decreased membrane fluidity, reduced enzyme activity, altered transport of nutrients and waste products, translation and cell division, decreased rates of transcription, intracellular ice formation, inappropriate protein folding and protein cold denaturation (D'Amico et al. 2006). The scientific community has focused on using psychrophilic enzymes in a wide range of pharmacological and industrial applications due to their exceptional properties (Parvizpour et al. 2021). Psychrophiles enzymes used in the food industries, detergent, specific biotransformation, contact lens cleaning fluids, environmental bioremediations, lactose content of milk and ice-nucleating proteins specially used in artificial snow or ice cream manufacturing and lipids have potential uses in dietary supplements (unsaturated fatty acids) through the Antarctic marine psychrophiles (Russell 1998).

Halophiles

Halophiles are salt-loving microorganisms that thrive in saline habitats. Based on NaCl requirement, halophiles are classified as slight halophiles (2–5%), moderate halophiles (5–20%) and extreme halophiles (20–30%) (DasSarma and Arora 2001; DasSarma and DasSarma 2017). Halophiles mainly include the phototrophic, methanogenic, heterotrophic archaea, lithotrophic, heterotrophic, photosynthetic bacteria and heterotrophic and photosynthetic eukaryotes (DasSarma and DasSarma 2017). Halophiles are widespread throughout the globe in saline habitats like coastal and natural saline brines in the arid, deep-sea, artificial salterns and salt mines (DasSarma and Arora 2001). About 2700 BC (Bass Becking 1931), the first microorganism was discovered in a saline environment. From the 1920s to 1940s, various halophilic bacteria were extracted from several sites like animal hides, anchovies and fish (Gunjal and Badodekar 2022). In silico research aids in understanding the exact/detailed biology of halophiles (Oren 2014), and halophiles' biochemistry, ecology and physiology have been examined (Oren 2015). In silico genetic engineering and post-genomic investigations have established new methods for halophilic bacteria to optimize their growth (Yue et al. 2014). Elazari extracted several halophiles from the Dead Sea, including extreme halophiles such as *Micrococcus morrhuae* and *Halobacterium trapanicum* and moderate halophiles such as *Flavobacterium halmephium*, *Pseudomonas halestrogus* and *Chromohalobacterium marismortui* (Edbeib et al. 2016). *Halobacterium* NRC-1 was the first halophile for which genome sequencing has been examined (Ng et al. 2000). The most frequently seen halophiles are archaea and bacteria like *Salinibacter*, *Halomonas* and *Halobacterium*. One of the most abundant sources of carotenoids found in nature so far is the halophilic eukarya *Dunaliella salina* (Waditee-Sirisattha et al. 2016). Strain MLA3 was extracted from Sambhar Salt Lake, Rajasthan and it was motile, rod-shaped, gram negative bacteria. The best growth was observed in 10% NaCl at pH 8.0 and 30 °C. According to 16S rRNA gene sequence analysis, strain MLA3 was phylogenetically recognized as a member of the *Halovibrio* genus and it was grouped into a clade with the type species *Halovibrio variabilis* DSM 3050 (Neelam et al. 2018). C12A1 was a moderately halo-alkaliphilic bacterial strain discovered from Sambhar Salt Lake in Rajasthan, India. Strain C12A1 was a rod-shaped, motile, gram positive bacterium that produced carotenoids and oval endospores. The best growth was observed at pH 8.0, 37 °C and 10–15% (w/v) NaCl concentration. According to 16S rRNA gene sequence analysis, strain C12A1 exhibited

98.50% and 98.87% similarity to *P. salipiscarius* and *P. halophilus*. However, C12A1 was grouped in the clade of *P. salipiscarius* strains, although it exhibited a unique lineage. As a result, C12A1 was identified as a *Piscibacillus* species (Neelam et al. 2019).

Extremely halophilic archaea are also capable of producing exceptional red pigments known as carotenoid compounds. These coloured pigments have been found to have powerful antioxidant and immune-boosting properties and are likely to protect against premature ageing (Waditee-Sirisattha et al. 2016). Heavy metals, such as arsenic, cadmium, and mercury, are among the compounds present in wastewater. As a result, it is being carried out by halophilic bacteria that can withstand the heavy metal (Yin et al. 2015). The first halophilic archaeon discovered in Thai fish sauce was nampla, a strain that looked similar to *Halobacterium salinarum* (Thongthai et al. 1992). In the leather industry, halophilic proteases are also utilized (DelgadoGarca et al. 2012). Many therapeutic enzymes have been used in cancer treatments, such as L-asparaginase, L-arginase, L-tyrosinase, L-glutaminase, α- and β-glucosidase, and β-galactosidase (Bar 1970). This means they are a viable option for a variety of industrial applications that require high salinity and/or low water activity (Daoud and Ali 2020).

Piezophiles

Microorganisms that thrive in high-pressure environments are termed piezophiles or barophiles (Abe and Horikoshi 2001), and these microorganisms' growth arises under hydrostatic pressure (Prieur 2014). Piezophiles have been adapted to withstand pressures between 40 and 110 MPa (Rampelotto 2010). Piezophiles are classified as piezo-tolerant microorganisms that only tolerate high pressure, piezophile microorganisms require high pressures for better growth and obligate/strict piezophile microorganisms need pressure higher than the atmosphere to grow. Although piezophiles can be found in a variety of environments with high hydrostatic pressure (deep oil reservoirs, deep aquifers, etc.), the deep ocean floor is where they are most commonly found (Prieur 2014). They are primarily found in the genera *Moritella*, *Shewanella*, *Pyrococcus* and *Methanococcus* (Kato and Nogi 2001). Hydration and the tertiary protein structures are affected by pressure. The packing of lipids membrane is constricted by high pressure, which ultimately impacts its fluidity. To get over this, piezophile microorganisms have modified their cell membranes by adding phosphatidylcholine and monounsaturated fatty acids rather than polyunsaturated fatty acids and phosphatidylethanolamine (Coker 2019). They have high multimerization and broad hydrophobic cores (Sarma et al. 2023). *Pyrococcus horikoshii* has a TET3 peptidase (piezophilic protein) that demonstrates the development of a dodecamer, which is essential for creating dense layers surrounding the cells so that water cannot easily permeate them under high pressure (Reed et al. 2013). An enzyme such as chymotrypsin that can function under higher temperatures and pressure has many benefits for applications in biotechnology (Kumar et al. 2011).

Xerophiles

Xerophile microorganisms can thrive in highly dry/arid environments or the presence of very little water activity. Xerophiles live in dry places like the Dry Valleys, Antarctica and other cold and hot deserts. But only a few numbers of specific genera of algae, lichens, fungi, yeast and bacteria can thrive in this habitat (Rothschild and Mancinelli 2001). To deal with low water activity, xerophiles accumulate high intracellular amounts of compatible osmolytes (glycerol, trehalose, betaine or glycine) using identical adaptation to "salt-out" halophiles (Lebre et al. 2017). Xerophiles are frequently radiation-resistant due to the processes to deal with desiccation provide them with oxidative stress and ionizing radiation (Hallsworth 2018). Antioxidants are abundant in xerophiles to protect against protein damage from radiation and desiccation (Schmid et al. 2020). Manganese complexes are used as antioxidants by *Deinococcus radiodurans* (Daly 2009). Stored grains, dry foods, spices, oilseeds and nuts are believed to spoil because of xerophiles (Gupta et al. 2014).

Radiophiles

Radiophiles thrive in habitats with higher levels of radiation (X-rays, gamma and UV) and oxidative stress, and they can repair significant DNA damage. Radiophile microorganisms are found in several microbiological species and groups, such as bacteria from the genera *Kineococcus*, *Rubrobacter*, *Bacillus* and *Deinococcus*, as well as members of the family *Geodermatophilaceae* and cyanobacteria, including the genera *Chroococcidiopsis* and *Nostoc* (Gtari et al. 2012; Gabani and Singh 2013). In 1965, Anderson reported discovering the first strain of *Deinococcus radiodurans* in sterilized X-ray cans. This bacterium was gram-positive and nonsporulated and had a red pigment colony. Excision repair and recombination repair of DNA are the two types of repair mechanisms found in *Deinococcus radiodurans* (Shukla et al. 2020). The radiation-resistant carotenoid pigment was derived from *Deinococcus radiodurans*, and it may have significant biotechnological applications (Lemee et al. 1997). Bacterioruberin synthesized by radiation-resistant microorganisms (*Rubrobacter* and *Halobacterium*) has been proposed to have use in preventing skin cancer in humans because it helps repair DNA strands damaged by ionized UV radiation (Singh and Gabani 2011). Many UV radiation–protective substances, including melanin, bacterioruberin, ectoines, mycosporine-like amino acids (MAAs) and scytonemin, have been identified from UV-resistant extremophilic bacteria (Rastogi and Incharoensakdi 2014). Deinoxanthin, a carotenoid that was isolated from the radio-resistant bacteria *D. radiodurans*, caused apoptosis in cancer cells, indicating that it might be effective as a chemopreventive agent (Choi et al. 2014; Raddadi et al. 2015). An exceptionally radiation-resistant thermophilic bacterium identified as *Deinococcus geothermalis* is being developed for in situ bioremediation of radioactive wastes, and it is linked to the mesophile *Deinococcus radiodurans*. *D. geothermalis* is transformable with plasmids developed for

D. radiodurans and created a Hg(II)-resistant. *D. geothermalis* strain can reduce Hg(II) at high temperatures. *D. geothermalis* can also reduce U(VI), Cr(VI) and Fe(III)-nitrilotriacetic acid, and these properties supported the potential invention of the thermophilic radiophile for bioremediation of radioactive combined waste habitats at higher temperatures (>55 °C) (Brim et al. 2003).

Endoliths

Endolithic microorganisms are those that colonise inside rocks (De Los Ríos et al. 2014). Endoliths are divided into four categories: (i) cryptoendoliths, which are organisms that live in naturally occurring structure cavities; (ii) chasmoendoliths, which are growing in the fissures and cracks of the rock; (iii) euendoliths, which actively pass through rocks to create new tunnels and cavities and speed up biogenic weathering of rocks; and (iv) autoendoliths, which are organisms that aid in the mineral deposition on rocks (Sajjad et al. 2022). Microorganisms most frequently colonize this endolithic environment in cold and hot deserts around the globe. Warmer and moderate temperatures, radiation, moisture and longer exposure to liquid water are present at the surface of the Earth (Warren-Rhodes et al. 2006). Various locations around the world have reported finding endolithic organisms. There have been finds in warmer, arid and hyper-arid deserts like the Atacama in Chile, the Mojave and Sonora in the United States, the Negev in Israel, the Gobi in China and Mongolia, the Namib in Namibia and Angola, the Depression of Turpan in China and the Al-Jafr basin in Jordan. There have also been finds in cold deserts like the Antarctic and Arctic, as well as in ocean trenches rocks and deep subsoil (Inagaki et al. 2002). Those communities of microbes usually consist of fungi, cyanobacteria, heterotrophic bacteria, lichen, green and red algae and non-photoautotrophic bacteria (Omelon 2016). Endolithic organisms exhibit a diverse range of metabolic processes; several of these groups have been found with genes involved in iron uptake, carbon fixation and sulfur metabolism. Furthermore, it is yet unknown if they metabolise them through the rocks around them or if they first secrete an acid to dissolve them. Genes involved in nitrogen fixation have been discovered in the endolithic population (Meslier and DiRuggiero 2019).

Polyextremophiles

Polyextremophiles are extremophiles that can survive in various extreme environmental situations (Fig. 1.1) (Chela-Flores 2013). Chela-Flores (2013), have developed characteristics that enable them to survive in hazardous conditions, like Deep Lake in Antarctica, where temperatures can drop to −20 °C, and only remain liquid due to extremely high salt concentrations (Karan et al. 2020). A polyextremophile, *Nesterenkonia* sp. AN1, was discovered in the soil of the Antarctic desert. The adaptations found in *Nesterenkonia* sp. AN1 include changes in membrane fluidity and responses to oxidative and osmotic stress carried by cold exposure (Aliyu et al. 2016).

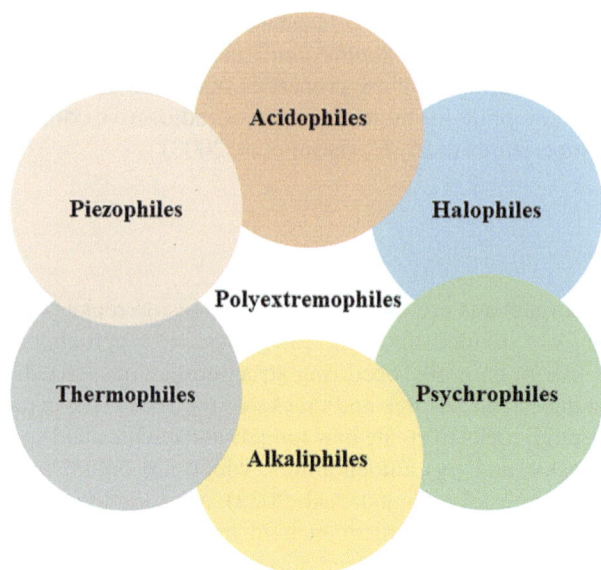

Fig. 1.1 Polyextremophiles: A collection of various extremophiles. The following image shows the severe environments in which microbes survive and overlaps are instances of possible polyextremophiles

Using statistical methods, the synthesis of xylanase (cellulase-free) and thermo-alkali-stable from the polyextremophilic new *Bacillus halodurans* TSEV1 (MTCC 10962) was optimized through batch fermentation with optimum pH 9 at higher (80 °C) temperature. The production of xylanase was increased 7.35-fold as a result of the optimization of fermentation factors. When this xylanase was used to pre-bleach wheat straw kraft pulp resulted in a 5.85% increase in brightness and 14.6% decrease in kappa number. Thus, it is convenient to use the xylanase of TSEV1 in the development of an environmentally friendly bleaching method for paper pulp (Kumar and Satyanarayana 2012). Polyextremophiles are useful in a wide variety of circumstances. Halopsychrophiles and halothermophiles are potential resources of beneficial enzymes (Elleuche et al. 2014). Polyextremophiles' halophilic amylase properties may make them effective catalysts under higher saline and alkaline pH in activities like starch hydrolysis and applications like the food industry, bioremediation and detergent manufacturing (Rekadwad and Khobragade 2017; Sysoev et al. 2021).

1.4   Conclusion

Extremophiles have emerged as unique organisms that thrive in harsh conditions. This chapter has discussed about various extremophiles their diversity and taxonomical aspects. These organisms produce special biomolecules and metabolites that have novel applications in the food, chemical and pharmaceutical industries. New developments in the production and cultivation of extremophiles, their genetic modification and cloning in heterologous microorganisms could expand the range of applications in various industries, including agriculture.

References

Abe F, Horikoshi K (2001) The biotechnological potential of piezophiles. Trends Biotechnol 19:102–108

Al'Abri K (2011) Use of molecular approaches to study the occurrence of extremophiles and extremodures in non-extreme environments. Mol Biol Biotechnol, pp 1–402

Alsamir M, Mahmood T, Trethowan R, Ahmad N (2021) An overview of heat stress in tomato (Solanum lycopersicum L.). Saudi J Biol Sci 28:1654–1663

Aliyu H, De Maayer P, Cowan D (2016) The genome of the Antarctic polyextremophile Nesterenkonia sp. AN1 reveals adaptive strategies for survival under multiple stress conditions. FEMS Microbiol Ecol 92:1–11

Bar D (1970) Enzymes used as drugs. Lille Medical: Journal de la Faculte de Medecine et de Pharmacie de L'universite de Lille 15:827–847

Baross JA, Morita RY (1978) Microbial life at low temperatures: ecological aspects. Microbial Life in Extreme Environments, pp 9–71

Bouri M, Mehnaz S, Sahin F (2022) Extreme Environments as Potential Sources for PGPR. In: Secondary Metabolites and Volatiles of PGPR in Plant-Growth Promotion. Springer, pp 249–276

Brim H, Venkateswaran A, Kostandarithes HM et al (2003) Engineering Deinococcus geothermalis for bioremediation of high-temperature radioactive waste environments. Appl Environ Microbiol 69:4575–4582

Brock TD (1986) Introduction: an overview of the thermophiles. Thermophiles: General, Molecular, and Applied Microbiology, pp 1–16

Brock TD, Freeze H (1969) Thermus aquaticus gen. n. and sp. n., a nonsporulating extreme thermophile. J Bacteriol 98:289–297

Catchpole RJ, Forterre P (2019) The evolution of reverse gyrase suggests a nonhyperthermophilic last universal common ancestor. Mol Biol Evol 36:2737–2747

Chang S-S, Kang D-H (2004) Alicyclobacillus spp. in the fruit juice industry: history, characteristics, and current isolation/detection procedures. Crit Rev Microbiol 30:55–74

Chela-Flores J (2013) Polyextremophiles: Summary and conclusions. In: Polyextremophiles: Life Under Multiple Forms of Stress. Springer, pp 611–615

Chi YX, Yang L, Zhao CJ et al (2021) Effects of soaking seeds in exogenous vitamins on active oxygen metabolism and seedling growth under low-temperature stress. Saudi J Biol Sci 28:3254–3261

Choi Y-J, Hur J-M, Lim S et al (2014) Induction of apoptosis by deinoxanthin in human cancer cells. Anticancer Res 34:1829–1835

Coker JA (2019) Recent advances in understanding extremophiles. F1000Research 8:1–7

D'Amico S, Collins T, Marx J et al (2006) Psychrophilic microorganisms: challenges for life. EMBO Rep 7:385–389

Daoud L, Ali MB (2020) Halophilic microorganisms: interesting group of extremophiles with important applications in biotechnology and environment. In: Physiological and biotechnological aspects of extremophiles. Elsevier, pp 51–64

Daly MJ (2009) A new perspective on radiation resistance based on Deinococcus radiodurans. Nat Rev Microbiol 7:237–245

DasSarma S, Arora P (2001) Halophiles. eLS 8:458-466

DasSarma S, DasSarma P (2017) Halophiles. eLS, pp 1–13

De Los Ríos A, Wierzchos J, Ascaso C (2014) The lithic microbial ecosystems of Antarctica's McMurdo Dry Valleys. Antarct Sci 26:459–477

de la Haba RR, Antunes A, Hedlund BP (2022) Extremophiles: Microbial genomics and taxogenomics. Front Microbiol 13:1–7

Delgado-Garcia M, Valdivia-Urdiales B, Aguilar-Gonzalez CN, Contreras-Esquivel JC, Rodriguez-Herrera R (2012) Halophilic hydrolases as a new tool for the biotechnological industries. J Sci Food Agric 92:2575–2580

Deocampo DM, Renaut RW (2016) Geochemistry of African soda lakes. Soda Lakes of East Africa, pp 77–93

Duckworth AW, Grant WD, Jones BE, Van Steenbergen R (1996) Phylogenetic diversity of soda lake alkaliphiles. FEMS Microbiol Ecol 19:181–191

Edbeib MF, Wahab RA, Huyop F (2016) Halophiles: biology, adaptation, and their role in decontamination of hypersaline environments. World J Microbiol Biotechnol 32:1–23

Elleuche S, Schröder C, Sahm K, Antranikian G (2014) Extremozymes—biocatalysts with unique properties from extremophilic microorganisms. Curr Opin Biotechnol 29:116–123

Enami I, Adachi H, Shen J-R (2010) Mechanisms of acido-tolerance and characteristics of photosystems in an acidophilic and thermophilic red alga, *Cyanidium caldarium*. In: Red algae genomic age, pp 373–389

Fujinami S, Fujisawa M (2010) Industrial applications of alkaliphiles and their enzymes–past, present and future. Environ Technol 31:845–856

Gabani P, Singh OV (2013) Radiation-resistant extremophiles and their potential in biotechnology and therapeutics. Appl Microbiol Biotechnol 97:993–1004

Gallo G, Puopolo R, Carbonaro M et al (2021) Extremophiles, a nifty tool to face environmental pollution: From exploitation of metabolism to genome engineering. Int J Environ Res Public Health 18:1–24

González AG, Terrón RP (2021) Importance of extremophilic microorganisms in biogeochemical cycles. GSC Adv Res Rev 9:82–93

Goyal D, Swaroop S, Pandey J (2020) Harnessing the genetic diversity and metabolic potential of extremophilic microorganisms through the integration of metagenomics and single-cell genomics. In: Extremophilic microbes and metabolites-diversity, bioprospecting and biotechnological applications. IntechOpen, pp 1–20

Gtari M, Essoussi I, Maaoui R et al (2012) Contrasted resistance of stone-dwelling Geodermatophilaceae species to stresses known to give rise to reactive oxygen species. FEMS Microbiol Ecol 80:566–577

Gupta GN, Srivastava S, Khare SK, Prakash V (2014) Extremophiles: an overview of microorganism from extreme environment. Int J Agric Environ Biotechnol 7:371–380

Gunjal AB, Badodekar NP (2022) Halophiles. In Physiology, Genomics, and Biotechnological Applications of Extremophiles, pp 13–34

Hallberg KB, González-Toril E, Johnson DB (2010) Acidithiobacillus ferrivorans, sp. nov.; facultatively anaerobic, psychrotolerant iron-, and sulfur-oxidizing acidophiles isolated from metal mine-impacted environments. Extremophiles 14:9–19

Hallsworth JE (2018) Stress-free microbes lack vitality. Fungal Biol 122:379–385

Hamdia MAE-S, Shaddad MAK, Doaa MM (2004) Mechanisms of salt tolerance and interactive effects of Azospirillum brasilense inoculation on maize cultivars grown under salt stress conditions. Plant Growth Regul 44:165–174

Horikoshi K (2004) Alkaliphiles. Proc Japan Acad Ser B 80:166–178

Horikoshi K, Bull AT (2011) Prologue: Definition, categories, distribution, origin and evolution, pioneering studies and emerging fields. K Horikoshi, G Antranikan, Bull, FT Robb, KO Stetter (Eds), Extremophiles Handbook, pp 4–15

Ilyas S, Lee J, Kim B (2014) Bioremoval of heavy metals from recycling industry electronic waste by a consortium of moderate thermophiles: process development and optimization. J Clean Prod 70:194–202

Inagaki F, Takai K, Komatsu T et al (2002) Profile of microbial community structure and presence of endolithic microorganisms inside a deep-sea rock. Geomicrobiol J 19:535–552

Jiang Y, Lindsay DL, Davis AR et al (2020) Impact of heat stress on pod-based yield components in field pea (Pisum sativum L.). J Agron Crop Sci 206:76–89

Kapoor D, Bhardwaj S, Landi M et al (2020) The impact of drought in plant metabolism: How to exploit tolerance mechanisms to increase crop production. Appl Sci 10:1–19

Karan R, Mathew S, Muhammad R et al (2020) Understanding high-salt and cold adaptation of a polyextremophilic enzyme. Microorganisms 8:1–19

Kato C, Nogi Y (2001) Correlation between phylogenetic structure and function: examples from deep-sea Shewanella. FEMS Microbiol Ecol 35:223–230

Kevbrin V V (2020) Isolation and cultivation of alkaliphiles. Alkaliphiles Biotechnol:53–84

Kohli I, Joshi NC, Mohapatra S, Varma A (2020) Extremophile–an adaptive strategy for extreme conditions and applications. Curr Genomics 21:96–110

Konings WN, Albers S-V, Koning S, Driessen AJM (2002) The cell membrane plays a crucial role in survival of bacteria and archaea in extreme environments. Antonie Van Leeuwenhoek 81:61–72

Kristjansson JK, Stetter KO (2021) Thermophilic bacteria. In: Thermophilic bacteria. CRC Press, pp 1–18

Kumar V, Satyanarayana T (2012) Thermo-alkali-stable xylanase of a novel polyextremophilic Bacillus halodurans TSEV1 and its application in biobleaching. Int Biodeterior Biodegradation 75:138–145

Kumar L, Awasthi G, Singh B (2011) Extremophiles: a novel source of industrially important enzymes. Biotechnology 10:121–135

Kumar A, Alam A, Tripathi D et al (2018) Protein adaptations in extremophiles: an insight into extremophilic connection of mycobacterial proteome. In: Seminars in cell & developmental biology. Elsevier, pp 147–157

Lebre PH, De Maayer P, Cowan DA (2017) Xerotolerant bacteria: surviving through a dry spell. Nat Rev Microbiol 15:285–296

Lemee L, Peuchant E, Clerc M, Brunner M, Pfander H (1997) Deinoxanthin: a new carotenoid isolated from Deinococcus radiodurans. Tetrahedron 53:919–926

Maheshwari R, Bharadwaj G, Bhat MK (2000) Thermophilic fungi: their physiology and enzymes. Microbiol Mol Biol Rev 64:461–488

Margesin R, Miteva V (2011) Diversity and ecology of psychrophilic microorganisms. Res Microbiol 162:346–361

Meslier V, DiRuggiero J (2019) Endolithic microbial communities as model systems for ecology and astrobiology. In: Model ecosystems in extreme environments. Elsevier, pp 145–168

Morita RY (1975) Psychrophilic bacteria. Bacteriol Rev 39(2):144–167

Moyer CL, Morita RY (2007) Psychrophiles and psychrotrophs. Encycl Life Sci 1:1-6

Neelam DK, Agrawal A, Tomer AK, Dadheech PK (2018) Characterization and potential applications of Halovibrio variabilis MLA3 (A heterotrophic bacterium) isolated from a saline-alkaline lake of Rajasthan, India. Int J Pharm Biol Sci 8:612–622

Neelam DK, Agrawal A, Tomer AK et al (2019) A Piscibacillus sp. isolated from a soda lake exhibits anticancer activity against breast cancer MDA-MB-231 cells. Microorganisms 7:34–50

Ng WV, Kennedy SP, Mahairas GG et al (2000) Genome sequence of Halobacterium species NRC-1. Proc Natl Acad Sci 97:12176–12181

Omelon CR (2016) Endolithic microorganisms and their habitats. Their World: A Diversity of Microbial Environments. Springer, pp 171–201

Oren A (2015) Halophilic microbial communities and their environments. Curr Opin Biotechnol 33:119–124

Ortega G, Diercks T, Millet O (2015) Halophilic protein adaptation results from synergistic residue-ion interactions in the folded and unfolded states. Chem Biol 22:1597–1607

Oren A (2010) Acidophiles. eLS, pp 1–14

Oren A (2014) Taxonomy of halophilic Archaea: current status and future challenges. Extremophiles 18:825–834

Parvizpour S, Hussin N, Shamsir MS, Razmara J (2021) Psychrophilic enzymes: structural adaptation, pharmaceutical and industrial applications. Appl Microbiol Biotechnol 105:899–907

Prieur D (2014) Piezophile. Encyclopaedia of Astrobiology. Springer, pp 1–2

Raddadi N, Cherif A, Daffonchio D et al (2015) Biotechnological applications of extremophiles, extremozymes and extremolytes. Appl Microbiol Biotechnol 99:7907–7913

Ramirez J (2021) Extremophiles: Defining the physical limits at which life can exist. Extremophiles

Rampelotto PH (2010) Resistance of microorganisms to extreme environmental conditions and its contribution to astrobiology. Sustain 2:1602–1623

Rampelotto PH (2013) Extremophiles and extreme environments. Life 3:482–485

Rao AS, Nair A, Nivetha K et al (2022) Molecular adaptations in proteins and enzymes produced by extremophilic microorganisms. In: Extremozymes and their industrial applications. Elsevier, pp 205–230

Rastogi RP, Incharoensakdi A (2014) Characterization of UV-screening compounds, mycosporine-like amino acids, and scytonemin in the cyanobacterium Lyngbya sp. CU2555. FEMS Microbiol Ecol 87:244–256

Reeb V, Bhattacharya D (2010) The thermo-acidophilic cyanidiophyceae (Cyanidiales). In: Red algae genomic age 409–426

Reed CJ, Lewis H, Trejo E et al (2013) Protein adaptations in archaeal extremophiles. Archaea 2013:1–14

Rekadwad B, Khobragade C (2017) Marine polyextremophiles and their biotechnological applications. In: Microbial Applications Vol. 1: Bioremediation and Bioenergy, pp 319–331

Rizvi A, Ahmed B, Khan MS et al (2021) Psychrophilic bacterial phosphate-biofertilizers: a novel extremophile for sustainable crop production under cold environment. Microorganisms 9:1–28

Rothschild L (2007) Extremophiles: defining the envelope for the search for life in the universe. Planet Syst Orig Life 3:113–134

Rothschild LJ, Mancinelli RL (2001) Life in extreme environments. Nature 409:1092–1101

Russell NJ (1998) Molecular adaptations in psychrophilic bacteria: potential for biotechnological applications. Biotechnol Extrem:1–21

Sanghera GS (2020) Sugarcane disorders associated with temperature extremes and mitigation strategies. East African Scholars J Agri Life Sci 3:101–114

Sarma J, Sengupta A, Laskar MK, Sengupta S, Tenguria S, Kumar A (2023) Microbial adaptations in extreme environmental conditions. In Bacterial Survival in the Hostile Environment. Academic Press, pp 193–206

Sajjad W, Ilahi N, Kang S et al (2022) Endolithic microbes of rocks, their community, function and survival strategies. Int Biodeterior Biodegradation 169:105387

Satyanarayana T, Raghukumar C, Shivaji S (2005) Extremophilic microbes: diversity and perspectives. Curr Sci:78–90

Schmid AK, Allers T, DiRuggiero J (2020) SnapShot: microbial extremophiles. Cell 180:818

Selvakumar G, Joshi P, Suyal P et al (2011) Pseudomonas lurida M2RH3 (MTCC 9245), a psychrotolerant bacterium from the Uttarakhand Himalayas, solubilizes phosphate and promotes wheat seedling growth. World J Microbiol Biotechnol 27:1129–1135

Shu W-S, Huang L-N (2022) Microbial diversity in extreme environments. Nat Rev Microbiol 20:219–235

Shukla PJ, Bhatt VD, Suriya J, Mootapally C (2020) Marine extremophiles: adaptations and biotechnological applications. Encycl Mar Biotechnol 3:1753–1771

Singh OV, Gabani P (2011) Extremophiles: radiation resistance microbial reserves and therapeutic implications. J Appl Microbiol 110:851–861

Sorokin DY, Berben T, Melton ED et al (2014) Microbial diversity and biogeochemical cycling in soda lakes. Extremophiles 18:791–809

Siddiqui KS, Williams TJ, Wilkins D, Yau S, Allen MA, Brown MV, Lauro FM, Cavicchioli R (2013) Psychrophiles. Annu Rev Earth Planet Sci 41:87–115

Stetter KO (1996) Hyperthermophilic procaryotes. FEMS Microbiol Rev 18:149–158

Sysoev M, Grotzinger SW, Renn D, Eppinger J, Rueping M, Karan R (2021) Bioprospecting of novel extremozymes from prokaryotes-the advent of culture-independent methods. Front Microbiol 12:1–16

Thakur N, Singh SP, Zhang C (2022) Microorganisms under extreme environments and their applications. Curr Res Microb Sci 3:1–2

Thongthai C, McGenity TJ, Suntinanalert P, Grant WD (1992) Isolation and characterization of an extremely halophilic archaeobacterium from traditionally fermented Thai fish sauce (nam pla). Lett Appl Microbiol 14:111–114

Tiquia-Arashiro S, Rodrigues DF et al (2016) Alkaliphiles and acidophiles in nanotechnology. Extremophiles: applications in nanotechnology. Springer, pp 129–162

Turner P, Mamo G, Karlsson EN (2007) Potential and utilization of thermophiles and thermostable enzymes in biorefining. Microb Cell Fact 6:1–23

Urbieta MS, Donati ER, Chan K-G et al (2015) Thermophiles in the genomic era: Biodiversity, science, and applications. Biotechnol Adv 33:633–647

van de Vossenberg JLCM, Driessen AJM, Zillig W, Konings WN (1998) Bioenergetics and cytoplasmic membrane stability of the extremely acidophilic, thermophilic archaeon Picrophilus oshimae. Extremophiles 2:67–74

Verma P, Yadav AN, Kumar V et al (2017) Beneficial plant-microbes interactions: biodiversity of microbes from diverse extreme environments and its impact for crop improvement. Plant-microbe Interact agro-ecological Perspect Vol 2 Microb Interact agro-ecological impacts, pp 543–580

Waditee-Sirisattha R, Kageyama H, Takabe T (2016) Halophilic microorganism resources and their applications in industrial and environmental biotechnology. AIMS Microbiol 2:42–54

Warren-Rhodes KA, Rhodes KL, Pointing SB et al (2006) Hypolithic cyanobacteria, dry limit of photosynthesis, and microbial ecology in the hyperarid Atacama Desert. Microb Ecol 52:389–398

Yadav AN, Sharma D, Gulati S et al (2015) Haloarchaea endowed with phosphorus solubilization attribute implicated in phosphorus cycle. Sci Rep 5:1–10

Yadav AN, Sachan SG, Verma P et al (2016) Cold active hydrolytic enzymes production by psychrotrophic Bacilli isolated from three sub-glacial lakes of NW Indian Himalayas. J Basic Microbiol 56:294–307

Yang J, Kloepper JW, Ryu C-M (2009) Rhizosphere bacteria help plants tolerate abiotic stress. Trends Plant Sci 14:1–4

Yin J, Chen J-C, Wu Q, Chen G-Q (2015) Halophiles, coming stars for industrial biotechnology. Biotechnol Adv 33:1433–1442

Yue H, Ling C, Yang T et al (2014) A seawater-based open and continuous process for polyhydroxyalkanoates production by recombinant Halomonas campaniensis LS21 grown in mixed substrates. Biotechnol Biofuels 7:1–12

Zahra N, Raza ZA, Mahmood S (2020) Effect of salinity stress on various growth and physiological attributes of two contrasting maize genotypes. Brazilian Arch Biol Technol 63:1–10

Zheng B, Zhu Y, Sardans J et al (2018) QMEC: a tool for high-throughput quantitative assessment of microbial functional potential in C, N, P, and S biogeochemical cycling. Sci China Life Sci 61:1451–1462

Chapter 2
Survival Strategies of Extremophiles: Physiology and Biochemistry

R. Mythrayee and K. Veena Gayathri

2.1 Introduction

Optimal parameters that can be observed and measured establish the boundaries that support life. However, studies over the last century have shown that organisms have expanded these boundaries in several ways. The scientific community had investigated many severe conditions in the preceding decades in search of microbes that might survive there. There is a unique category of microorganisms known as extremophiles that can tolerate and thrive in a variety of extremes, including radiation, severe pH, crushing pressure, excessive amounts of salt, enormous temperature fluctuations and metallic toxicity (Macelroy 1974). Extremophiles are living microorganisms that flourish in conditions that might be referred to as 'extreme' because of their exceptional living circumstances. Unicellular species like bacteria and archaea comprise the majority of extremophiles (Rampelotto 2013). A distinctive richness of organisms can adapt to various environmental stresses in these harsh conditions. These extremophiles can survive and even thrive in one or more harsh ecosystems, thanks to their adaptive features (Merino et al. 2019).

Archaea is the main phylum that endures adverse circumstances. Individuals of this category are usually quite proficient in adapting to a range of extreme environments, often setting records for extremophiles, despite the fact they are often less adaptive than bacteria and eukaryotes. *Methanopyrus kandleri* strain 116, an archaeal, flourishes at temperatures as high as 122 °C (252 °F), whereas *Picrophilus torridus* from the genus *Picrophilus* belongs to the acidophilic species reported so far and can survive at a pH level of 0.06. The bacteria with the finest adaptation to many harsh environments are the cyanobacteria. From continental hot springs to the Antarctic ice, they frequently establish microbial beds with different bacteria.

R. Mythrayee · K. V. Gayathri (✉)
Department of Biotechnology, Stella Maris College (Autonomous), Chennai, India
e-mail: veenagayathri@stellamariscollege.edu.in

© The Author(s), under exclusive license to Springer Nature Switzerland AG 2024
A. Ranjan et al. (eds.), *Extremophiles for Sustainable Agriculture and Soil Health Improvement*, https://doi.org/10.1007/978-3-031-70203-7_2

Moreover, cyanobacteria can endure extreme metal concentrations, develop endolithic colonies in desert settings, thrive in highly salted and alkaline lakes and withstand xerophilic circumstances (i.e., minimal water accessibility) (Rampelotto 2013).

Extremophiles are categorized according to the types of habitats in which they may survive and reproduce. This includes barophiles, a species that survive under intense atmospheric pressure. Those that thrive at extremely high and low temperatures are referred to as hyperthermophiles and psychrophiles, respectively; acidophiles and alkaliphiles survive in extreme pH levels; organisms that endure high salt environments called halophiles; and organisms that thrive in and take low water activity are termed as xerophiles and radioresistant organisms (Merino et al. 2019). Some extremophiles, called polyextremophiles, have adapted to various extreme environments (Table 2.1). Thermoacidophiles are a typical example of polyextremophiles since they can endure extremely high temperatures and require an acidic pH for survival. *Sulfolobus acidocaldarius*, a polyextremophilic archaeon, survives at pH 3.0 and 80 °C. *Paenibacillus* and *Bacillus* spp. are microorganisms that can endure an array of pH and high temperatures. They are discovered in hot springs located in Assam, India, where the temperature ranges from 20 to 80 °C, and the pH is between 5.0 and 14.0 (Rastädter et al. 2021). Table 2.2 shows a few examples of polyextremophiles and the habitat in which they survive.

Extremophiles might be a vital clue to the beginning of life on Earth because they have inhabited such habitats since the beginning of time. They constitute an

Table 2.1 Examples of polyextremophiles and their requirements

Environmental parameter	Type	Range
Temperature	Psychrophile	<20 °C
	Mesophile	20°–45 °C
	Thermophile	45°–80 °C
	Hyperthermophile	>80 °C
pH	Hyperacidophile	pH < 3.0
Salinity	Acidophile	pH < 5.0
	Neutrophile	pH = 5.0–9.0
	Alkaliphiles	pH > 9.0
	Hyperalkaliphiles	pH > 11.0
	Non-halophile	<1.2%
	Halotolerant	1.2–2.9%
	Halophile	>8.8%
	Extreme halophiles	>14.6%
Pressure	Piezotolerant/barotolerant	0.1–10 MPa
Water activity	Piezophile/Barophile	10–50 MPa
	Hyperpiezophile/hyperbarophile	>50 MPa
	Xerophiles	<0.7
Chemical extremes	Metalotolerant	High concentration of metals
	Toxitolerant	High concentration of toxins
Polyextremophile	Tolerance for multiple parameters	

Table 2.2 Examples of polyextremophiles and their requirements

Type of adaptation	Microbes	Habitat	Requirements	Reference
Acidothermophile	*Acidianus infernus*	Solfatara crater	65°–96 °C pH: 2.0 Salinity: 0.2	Segerer et al. (1986)
Piezopsychrophile	*Colwellia piezophila*	Deep sea	4°–15 °C 40–80 MPa	Nogi et al. (2004)
Alkalithermophile	*Anoxybacillus pushchinensis*	Manure	37°–66 °C pH: 8.0–10.5	Pikuta et al. (2000)
Hyperalkaliphile	*Halomonas campisalis*	Soda lake	4°–50 °C pH: 6.0–12.0 Salinity: 1.1–26.3	Aston and Peyton (2007)
Alkaliphile, piezotolerant, halotolerant	*Oceanobacillus iheyensis*	Deep sea (mud)	15°–42 °C pH: 6.5–10.0 0.1–30 MPa Salinity: 0–21	Lu et al. (2001)
Hyperthermophile	*Geothermobacterium ferrireducens*	Obsidian pool, Yellowstone National Park	65°–100 °C	Kashefi et al. (2002)
Xerotolerant	*Deinococcus geothermalis*	Hot spring	30°–55 °C pH: 5.0–8.0	Frösler et al. (2017)

essential topic of study in many fields, from studies exploring the genesis of life to research on biological acclimatization in challenging environments. They have distinctive molecular pathways that enable them to exhibit great potential as extremozymes and bioactive substances. Due to their potential applications in biotechnology, these compounds are of tremendous economic interest in fields like medicine, industries and agriculture (Macelroy 1974).

Enzymes of microorganisms act as biocatalysts to synthesize various products. There is an increasing need for biocatalysts that can endure harsh industrial process conditions (Adams et al. 1995). Despite their benefits, the enzymes currently used are obtained from mesophilic microbes, and their limited resilience at extreme pH, temperatures and ionic strength restricts the application of these enzymes. However, extremophiles are an abundant resource of extremozymes, demonstrating great stability under adverse environments. As a result, microorganisms that survive in harsh conditions have drawn a lot of interest. Thus, the application of extremophiles and extremozymes in biocatalysis is quickly evolving from a research topic to a commercially feasible technology. Each type of extremophile has significant features that can provide the prospect of industrial application in many sectors, including agriculture. Thus, the production of commercial and scientific uses of extremozymes and extremophiles is receiving significant monetary investment globally (Van den Burg 2003).

Proteins derived from extremophiles are currently being utilized in various industries, from molecular biology reagents to common laundry detergents (Terpe

2013). Owing to their distinct mechanisms of metabolism and resilience to severe environmental factors, they are employed in the bioremediation of polluted sites. Furthermore, researchers are witnessing significant scope in exploring the hostile environments for extremophiles with antimicrobial properties, unique metabolites and enzymes (Macelroy 1974).

Extremophiles' ability to adapt has allowed them to survive in harsh and extreme environments. These microorganisms utilize various strategies and develop multiple molecular tactics to survive in harsh conditions. Extremophiles' adaptation strategies could help researchers determine the application strategies and ways to modify their genes and proteins to enhance soil health and promote sustainable agriculture (Mallik and Kundu 2015).

Extremophiles are good candidates to explore the functional relevance of extremophiles to improve soil health and agricultural practices due to their usage of defence and survival mechanisms and other specialized processes that allow them to survive in different environments. To understand its advantages in sustainable agriculture, a further understanding of the physiology and biochemical understanding is necessary. In this chapter, we intend to discuss various survival strategies of different extremophiles in their respective harsh habitats and adaptation approaches.

2.2   Survival Strategies of Extremophiles

Extremophiles outperform their mesophilic competitors and offer new opportunities for biocatalysis and biotransformation due to their unique catalysis properties and stability under various challenging conditions, including salt, pH, organic solvents and temperature. They are equipped to endure and flourish in environments regarded as hostile or incapable of supporting life. Factors like pressure, temperature, solubility, pH and salinity have altered a significant fraction of organisms' living conditions (Kumar and Singh 2012). The strategies by which various organisms cope with extreme conditions offer a distinct viewpoint on essential elements of cellular processes, such as the genetic blueprints for building macromolecules that are capable of withstanding several extreme parameters and the biochemical limitations to macromolecular stability. These microbes have a broad and flexible metabolic diversity with exceptional physiological abilities to inhabit hostile environments (Rampelotto 2013). They adopt various methods to prosper, and some of the survival and defence mechanisms of selected extremophiles are discussed subsequently.

2.2.1   Psychrophiles

Microorganisms that are either psychrophilic (loving cold) or psychrotolerant (loving cold) are frequently located in extremely low-temperature regions on Earth, such as the polar regions, mountaintops, ice sheets, the seafloor, shallow subterranean systems (cave systems), upper atmosphere, freezers and the surfaces of cold-adapted plants and animals, where temperatures rarely exceed 5 °C (Deming 2002). Several psychrophiles inhabit ecosystems with several stressors, like the deep sea with low temperatures with elevated pressure (piezo-psychrophiles) or the sea ice, which has low temperatures and significant levels of salt (halo-psychrophiles) (Cavicchioli et al. 2002).

In extreme cold ecosystems, a number of cold-adapted microorganisms thrive, exhibiting extraordinary adaptability to extreme cold. Among these are gram-negative bacteria such as *Psychroflexus*, *Psychrobacter*, *Pseudoalteromonas*, *Polaromonas* and *Polaribacter*. Some of the gram-positive bacteria also exhibit adaptations, such as *Bacillus*, *Arthrobacter* and *Micrococcus*. Additionally, archaea species like *Methanogenium*, *Methanococcoides* and *Halorubrum* flourish in these harsh environments along with yeast such as *Candida* and *Cryptococcus*, as well as fungi like *Penicillium* and *Cladosporium* which are also abundant (Cavicchioli et al. 2002). These psychrophilic organisms demonstrate proficiency in breaking down various polymeric substances, with the help of extremozymes: cellulase, amylase, pectinases, xylanase, chitinase, lipase and protease as highlighted by Demirjian et al. (2001).

Psychrophilic microorganisms employ various tactics, such as structural and enzyme adaptations, to acclimatize themselves to lower temperatures. Figure 2.1 shows the methods adapted by psychrophiles to maintain their structural stability. The lipid bilayer that constitutes the cell membrane is altered, and genes that synthesize long-chain fatty acids are activated (D'Amico et al. 2001). These fatty acids (KAS-II and KAS-III), which are highly branched, long-chained and have cis-isomerization, aid in preventing the production of ice crystals in cell membrane fluids (Goordial et al. 2016). To maintain cell membrane fluidity in marine microorganisms, long-chain polyunsaturated fatty acids (LC-PUFAs) such as docosahexaenoic (DHA), eicosapentaenoic (EPA) and arachidonic (ARA) acids play a crucial role (Yoshida et al. 2016). Carotenoids, a membrane pigment, are necessary for stabilizing membranes and helping to protect organisms against freeze–thaw cycles (Dieser et al. 2010). Large amounts of psychrophiles produce non-toxic, organic, low molecular mass osmolytes to maintain the osmotic pressure, which prevents water loss and cell shrinkage (Fonseca et al. 2016).

The synthesis of antifreeze proteins (AFPs) helps the organism preserve its internal fluidity using two different approaches ice recrystallization inhibition (IRI) and thermal hysteresis (TH). In the IRI, AFPs function to inhibit the formation of ice crystals, whereas in TH, the same decreases the freezing point of a solution without impacting its melting point (Kim et al. 2017). Exopolysaccharide (EPS) secretion creates a habitable semi-liquid environment and a barrier against ice formation around the cell (Garcia-Lopez et al. 2021).

Temperature sensing has been additionally related to the differential phosphorylation of lipopolysaccharide (LPS) and membrane proteins. The significance of LPS in cold adaptation is illustrated by investigating how variations in low temperatures affect the composition of LPS. Sea-ice microbes produce polymeric compounds that act as cryogenic substances for both the microbes and enzymes. Psychrophiles

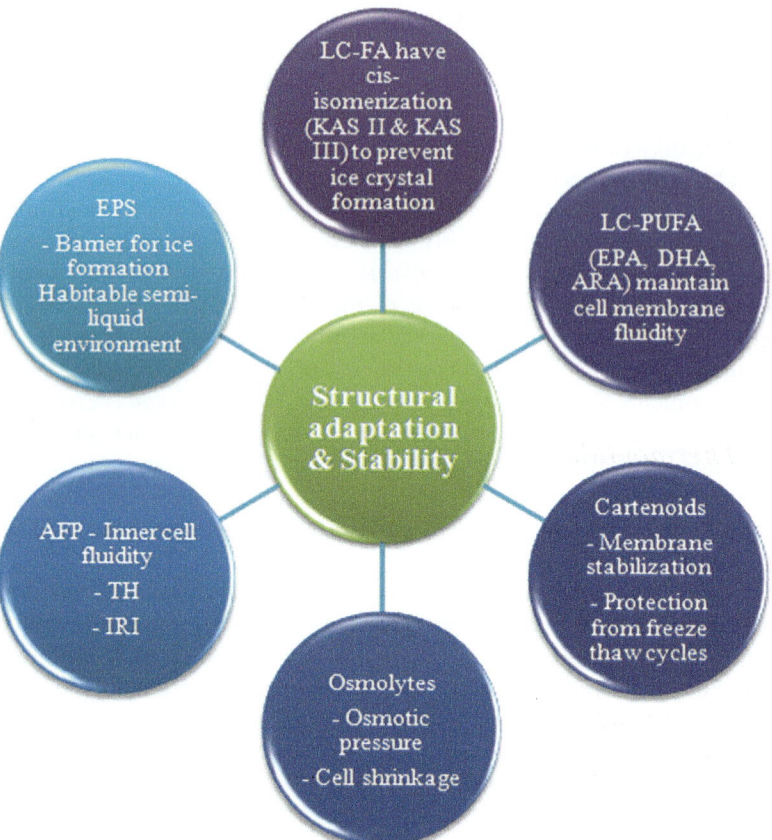

Fig. 2.1 Adaptation strategies employed by psychrophiles to maintain their structure and stability in extreme cold conditions

like *Bacillus Arthrobacter globiformis*, *Geobacter psychrophilus*, *Trichosporon pullulans*, and *Aquaspirillum arcticum* produce heat/cold shock proteins and cold acclimation peptides that serve as transcriptional boosters and RNA-binding molecules when subjected to abrupt fluctuations in external temperature (cold shock) (Berger et al. 1996).

Nucleic acid chaperones and protein chaperones in the cells are expressed more and persistently in psychrophilic microbes to confer genomic and protein stability (Lim et al. 2000). Psychrophilic proteins exhibit fewer H-bonds and ionic interactions than mesophilic proteins and also have more charge bearing functional groups and fewer hydrophobic groups on their surface and longer surface loops. These alterations cause psychrophilic proteins to lose their firmness at low temperatures and attain more structural flexibility for improved catalytic performance. Psychrophilic enzymes exhibit significant activity even at low temperatures, have a lower optimal temperature for activity compared to other enzymes and are significantly sensitive to heat, which are adaptations specific to their cold-loving nature (Georlette et al. 2004). These essential adaptations enable microorganisms to survive in hostile, cold environments.

From an industrial perspective, psychrophiles' cold-active enzymes are significant. A deep-sea resident bacterium, *Exiguobacterium oxidotolerans*, produces B-glycosidase that retains around 61% of its maximum activity at 10 °C, operating within a pH range of 6.6–9.0. Cold-adapted α-amylase extracted from *Bacillus* spp. is widely employed in food industries. EPSs derived from the psychrophilic *Colwellia psychrerythraea* marine bacterium exhibit antifreeze characteristics. It is also utilized as a stabilizer, thickener, gelling agent, pre-biotic, antioxidant and anti-tumoral (Deosthali and Dilipkumar Sajwani 2022).

2.2.2 Thermophiles

Thermophiles can be broadly categorized into moderate (50–60 °C), extreme (60–80 °C) and hyperthermophiles (80–110 °C). The taxa *Bacillus*, *Clostridium*, *Thermoanaerobacter*, *Thermus* and *Fervidobacterium* have a large distribution of extreme thermophiles (Vieille and Zeikus 2001). However, the majority of hyperthermophiles belong to archaea, a domain with four primary phyla: Euryarchaeota (*Pyrococcus*, *Thermococcus*, *Archaeoglobus*, *Ferroglobus* and *Methanopyrus*), Crenarchaeota (*Pyrodictium*, *Sulfolobus*, *Thermoproteus*, *Pyrolobus*, *Acidianus* and *Thermofilum*), Nanoarchaeota (*Nanoarchaeum equitans*) and *Korarchaeota* (Huber et al. 2002).

Thermophilic enzymes, out of all extremozymes, have garnered the most interest over the past four decades (Haki and Rakshit 2003). They perform exceptionally well in abrasive industrial operations. For protein engineering, these enzymes are frequently employed as prototypes for studying thermal stability and thermal activity (Sterner and Liebl 2001). Structural characteristics of thermophilic extremozymes have also garnered a lot of interest. To elucidate the mechanisms underlying thermostability, various three-dimensional components were formed and contrasted with their mesophilic counterparts (Kumar and Nussinov 2001a). Enzyme's structural characteristics, like distinctive salt bridges, extensive hydrogen bonding and hydrophobic interactions, serve as stabilizing factors that permit their thermophilicity (Sterner and Liebl 2001).

To survive in higher temperatures, thermophilic organisms must alter their cellular membranes, protein assembly and packaging techniques and routes of metabolism and generate HSPs, as shown in Fig. 2.2. Membrane phospholipids have proved crucial in protecting cell internals from their harsh environment. Lipid monolayers found in thermophilic archaea are composed of glycerol dialkyl glycerol tetraethers (GDGTs), which reduce the membrane's permeability to solutes. At high temperatures, C40 isoprenoid cyclization in *Thermoplasma* and *Sulfolobus* strains increased, resulting in tight packaging of lipids and lowered cell membrane fluidity (Sahonero-Canavesi et al. 2022).

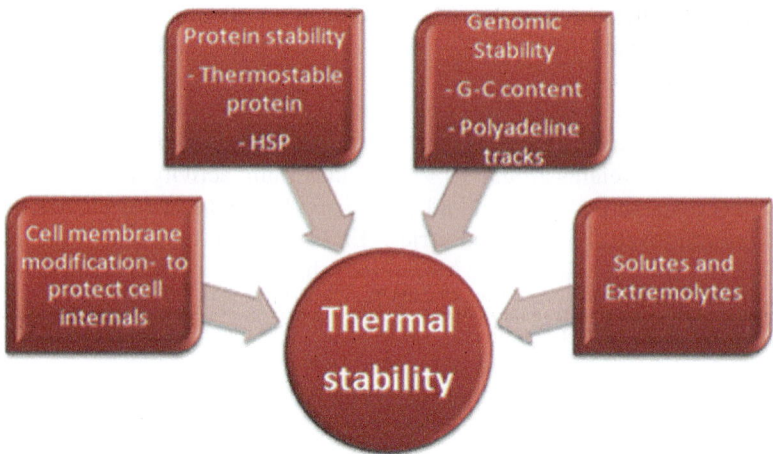

Fig. 2.2 Techniques adapted by thermophiles to survive in high temperatures

Thermostable proteins greatly enhance the ability of organisms to combat the adverse effects of high temperatures. The decline in amino acids like histamine, threonine and glutamine which are thermolabile and a spike in ionized residues like negatively charged glutamic acid suggest that ionic bonds aid protein stabilization at higher temperatures (Friedman et al. 2004). The increase also influences protein stability in intra-helical salt bridges (Kumar and Nussinov 2001b). Chaperonins are a special class of thermostable proteins produced by thermophiles that are resistant to denaturation and proteolysis. Chaperones known as HSPs aid in protein folding, refolding and unfolding and refolding of misfolded proteins. Heat-shock chaperones can be categorized into five main conserved classes: Hsp100s, Hsp90s, Hsp70s, Hsp60s and sHsps (Netzer and Hartl 1998).

Thermophiles generate and conserve suitable solutes to enable them to withstand environmental stresses such as desiccation and high-temperature fluctuations. The thermophile *Methanobacterium thermoautotrophicum* has a substance called cyclic 2,3-diphosphoglycerate (cDPG), which serves as a molecule for phosphate and energy storage and shields plasmid DNA from the oxidative damage caused by hydroxyl radicals (Kanodia and Roberts 1983). Various other extrapolates have been shown to promote thermostability, including ectoin, mannosyl-glyceramide, di-myo-inositol 1,1-phosphate, proline, mannitol and glucosyl-glycerate (Becker and Wittmann 2020). An elevated G + C composition in the stem region confers the thermostability of structural RNAs. Additionally, single-stranded RNA gains thermostability from polyadenine tracts in mRNA (Galtier and Lobry 1997).

Thermophilic enzymes are currently being utilized in various commercial applications. Quorum-sensing inhibitors and antibacterial substances are extracted from cyanobacterial mats near hot springs in Oman (Charlesworth and Burns 2016). Pretaq, dipeptides and DNA-processing enzymes are all synthesized using thermolysin. The most stable prolidase, isolated from *Pyrococcus furiosus*, displays no indication of functionality decline even after 12 h at temperatures exceeding 100 °C. AgaP4383, a unique thermostable agarase which is resistant to pH change that was isolated from *Flammeovirga pacifica* WPAGA1, demonstrates no reduction of endolytic efficacy against agar at 50 °C for 10 hours. *Thermothelomyces thermophila*'s laccase-like multicopper oxidase performs oxidative cyclization of 2′,3,4-trihydroxychalcone; aside from that, it is also utilized in bioremediation, biosensor, dye decolourization and so on (Deosthali and Dilipkumar Sajwani 2022).

2.2.3 Radiation-Resistant Organisms

Microorganisms that have a high tolerance for increased levels of UV and ionizing radiation are termed radiophiles (radiation-resistant organisms). *Deinococcus radiodurans*, *Deinococcus radiophilus* (Daly 2000), *Thermococcus radiotolerans* and *Thermococcus marinus* spp. (Jolivet et al. 2004) are a few examples of radiophiles. Radiation primarily affects DNA and RNA. To tolerate radiation, different mechanisms are employed at the transcriptional and proteomic levels. *Deinococcus radiodurans*, a unique microbe studied extensively for its resilience to chemicals, dehydration, radiation and oxidative stress, is recognized as a polyextremophile and is highly resistant to radiation which makes it a focus of research on radiation adaptation techniques (Sandigursky et al. 2004). Cellular cleansing is crucial before DNA repair to eliminate products created by radiation's harmful effects. This prevents the incorporation of fragmented bases during DNA synthesis and lowers the degree of mutagenesis by eliminating damaged DNA fragments (Battista 1997).

The sun's ultraviolet rays induce the formation of dimers within DNA strands, altering the molecular structure of DNA (Roy 2017). Extremophiles possess the capacity to repair DNA damaged by radiation, enabling their survival in environments with higher UV and ionizing radiation. *Dictyostelium discoideum* displays the translesion synthesis (TLS), Fanconi anaemia (FA) pathway and nucleotide excision repair (XPF) and provides its ability to withstand high levels of radiation and nucleotide cross-linking agents (Kumar et al. 2018).

DNA repair mechanism in *D. radiodurans* (Fig. 2.3) is induced by genes, proteins and enzymes specific to particular pathways. Three distinct survival routes were discovered through homologous recombination, which induces gene induction. The uvrABC system protein A, which serves excision repair in DNA damage, is revealed by the UVR-induced gene *uvrA*. RecQ regulates DNA helicase to control the type and quantity of DNA damage required for RecQ to function. DNA repair proteins, DNA metabolism, replication and SOS inducibility functionally represent recO and recF when the RecE pathway modulates them. RecR's modification while functioning with recF and recO could endure DNA recombination and interstrand cross links (Singh and Gabani 2011).

Reactive oxygen species can be eliminated enzymatically with the aid of catalases like KatA and superoxide dismutases like SodA. In *D. radiodurans*, pyrroloquinoline quinone (PQQ) acts as an antioxidant and confers resistance to γ-radiation

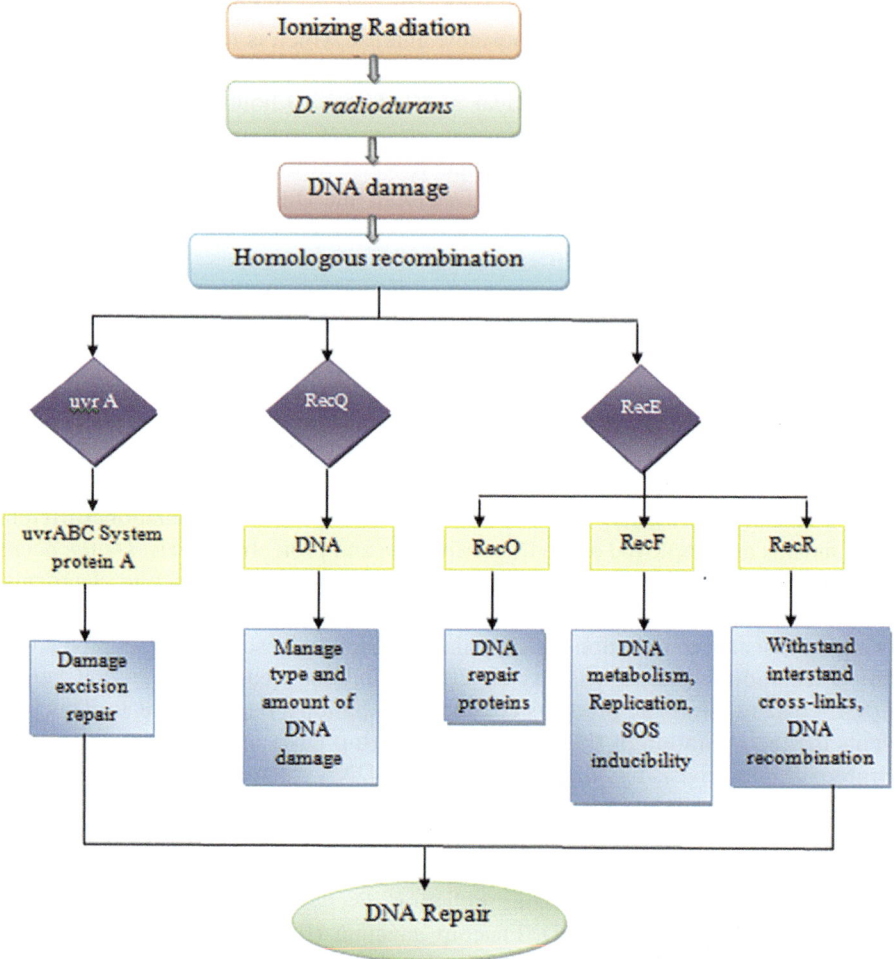

Fig. 2.3 DNA repair mechanism in radiation-resistant bacteria, *Deinococcus radiodurans*, as part of its survival strategy against radiation

(Jung et al. 2017). PprA and DrRecA have also been determined to be essential in gamma radiation resistance. Guanine quadruplex (G4) secondary DNA and RNA structures confer radio-resistance in *D. radiodurans* (Mishra et al. 2019).

Radiophiles find their applications in radioactive waste disposal, biomining and environmental remediation. *Sphingomonas*, *D. radiodurans*, *Bacillus*, *C. crescentus* and *H. noricense* produce phosphatases that are used to biomineralize uranium. Mycosporine-like amino acids (MAA) absorb UV rays and antioxidants; hence, they are used in sunscreens. By mending DNA strands harmed by ionizing radiation, bacterioruberin generated by *Rubrobacter radiotolerans* can prevent screen cancer. It has been reported that *Shewanella putrefaciens* and *Geobacter sulfurreducens* can transform soluble uranium into insoluble uranium (Deosthali and Dilipkumar Sajwani 2022).

2.2.4 Xerophiles

Life can exist in the driest places on Earth, including deserts and dry stones. Yet unique microorganisms belonging to fungi, lichens and algae can thrive in such arid conditions. These unique organisms are termed xerophiles (Madigan and Marrs 1997).

Biofilm development, enhanced production of extracellular DNA (eDNA) and EPS increased cytoplasmic Mn^{2+}, import and synthesis of suitable osmolytes, HSP production and aggregation of proteins are survival mechanisms used by Xerophiles, as shown in Fig. 2.4. Colanic acid is produced as EPS by some *Escherichia coli* K-12 mucoid strains. In contrast to their non-mucoid strain, which demonstrated 0.7–5% survival, this conferred resistance against desiccation, resulting in up to 35% survival (Ophir and Gutnick 1994). Methanochondroitin, an EPS produced by *Methanosarcina barkeri*, provides desiccation resistance by maintaining water levels inside the cell (Anderson et al. 2012).

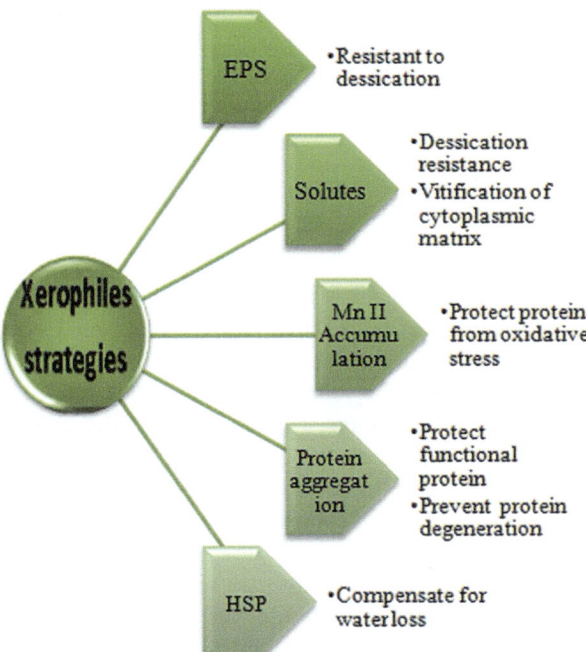

Fig. 2.4 Adaptation strategies utilized by xerophilic organisms in dry conditions to survive better

By enabling vitrification of the cytoplasmic matrix, the accumulation of solutes or osmoprotectants contributes significantly to desiccation resistance. *Acinetobacter baumannii*, a nosocomial infectious agent, was demonstrated to gather exogenous trehalose in response to desiccation (Zeidler and Müller 2019). When subjected to desiccation using NaCl, *Rhizobia* accumulated solutes such as K^+ ions, betaine, trehalose and sucrose (Vriezen et al. 2007).

While *D. radiodurans* was subjected to desiccation stress, Mn^{2+} accumulated, shielding the protein structures against oxidative damage induced by desiccation (Jin et al. 2019). *Acinetobacter baumannii* showed protein aggregation during desiccation, protecting functional proteins from desiccation (Wang et al. 2020). In yeast faced with desiccation, pyruvate kinase exhibited reversible accumulation, which inhibited protein degradation and guaranteed that the cell cycle restarted after distress. To compensate for water loss, HSPs create hydrogen bonds that connect to different molecules (Saad et al. 2017). HSPs were found to be upregulated in the cyanobacterium *Anabaena* sp. 7120 and *D. radiodurans* in response to desiccation (García 2011).

Desiccation-resistant bacteria *Azospirillum brasilense* Sp7, *Sphingomonas* sp. OF178 and *Acinetobacter* sp. EMM02 enhance maize plant development. *Pseudomonas fluorescens*, *Pseudomonas migulae* and *Enterobacter hormaechei* enhanced soil moisture and soil/root ratio, which improved plant growth. Non-ribosomal peptide synthase gene clusters present in *Lechevalieria* and *Amycolatopsis* species produce antitumoral drugs. For the bioleaching of copper, organisms such as *Acidithiobacillus thiooxidans* strain Licanantay DSM 17318 and *Acidithiobacillus ferrooxidans* strain Wenelen DSM 16786 were patented (Deosthali and Dilipkumar Sajwani 2022).

2.2.5 Halophiles

For survival, halophilic microorganisms need extremely high salt (NaCl) concentrations. Salterns and hypersaline lakes, such as the Great Salt Lake, Dead Sea and Solar Lakes, are home to them in Africa, Europe and the USA. Salt lakes are also discovered even in Antarctica. They accumulate quantities of KCl or NaCl that are in equilibrium with their surroundings. Therefore, halophilic proteins have to withstand extremely saline environments (up to 5 M NaCl and 4 M KCl) (Oren 2002). *Salinibacter ruber* is a bacterium, whereas halophiles such as *Haloarcula*, *Halobacterium*, *Natronobacterium*, *Haloferax*, *Halococcus* and *Natronococcus* are members of the archaea (Oren 2002).

Halophiles survive in high-concentration salt niches, and this higher salt concentration outside cells causes osmosis. Osmotic balance is maintained primarily by the cellular membranes. Compared to non-halophiles, halophilic organisms' membrane proteins include an increased content of polar and acidic amino acids. Halophiles lack the tetraethers that link the two halves of a bilayer together, allowing membranes to remain fluid even in highly salinized environments. The primary element of the lipid bilayer is phosphatidylglycerol phosphate, which is replaced by sulphate in some Halobacterium species. Therefore, a significant negative charge concentration acts as an adjustment strategy to deal with both internal K^+ concentration and high external Na^+ concentrations (Russell 1989).

Two mechanisms, the organic osmolyte and salt-in-cytoplasm methods, maintain the osmotic pressure within their cytoplasm. Halophiles build up K^+ ions and Cl^- ions as counter ions in their cytoplasm as part of the salt-in-cytoplasm approach, which is later replaced by Na^+ ions in the stationary phase. Enzymes are made up of acidic charged amino acids to help them adapt to these high ionic conditions. Consequently, this system keeps the cell hydrated. With this latter method, organisms sustain osmotic equilibrium by either producing suitable organic solutes on their own or absorbing them from their surroundings. These organic solutes may consist of sugars, amino acids or polyols. These organic solutes stabilize proteins and cells in addition to preserving osmotic balance (Kunte 2013).

Archaeal halophiles provide the primary supply of highly halophilic enzymes. Halophiles from the genera *Haloferax*, *Halobacterium*, *Marinococcus*, *Halorhabdus*, *Micrococcus* and *Acinetobacter* have been found to produce halophilic enzymes like xylanases and amylases (Mevarech et al. 2000).

Halophilic enzymes' properties are significantly different from their non-halophilic counterparts even though they are capable of performing identical enzymatic functions. These variations include a notable preponderance of acidic over basic amino residues and the requirement for considerable amounts of salt (1–4 M NaCl) for activation and stabilization (Madern et al. 2000). When exposed to higher salt levels, non-halophilic proteins tend to clump and become rigid. In contrast, halophilic proteins become soluble and more flexible because of their high negative surface charge. Firmly bonded water dipoles are responsible for neutralizing this high surface charge (Danson and Hough 1997). On the contrary, for halophilic enzyme's stability, a higher salt concentration is needed which results from a low-affinity salt binding to particular sites on the folded polypeptide chain, which stabilizes the active conformation of the protein (Mevarech et al. 2000).

Halophilic Antarctic archaea *Halorubrum lacusprofundi*'s amino acid sequences were examined by DasSarma et al. Numerous polar and acidic amino acids were found along with the substitution of smaller for more extensive amino acids, resulting in protein misfolding, which provided the flexibility needed to function under duress (DasSarma et al. 2013). Halophiles exhibit various strategies to cope with stress from high salinity, including the overexpression of HSPs, DNA repair modules, stress proteins, proteasome subunits, UVR systems and biofilm development.

Halophiles have been proposed as a viable bioremediation system for industrial wastewater, especially contaminated with pesticides and herbicides, for instance, using the halophilic community to degrade phenolic compounds in saline conditions. Food and cosmetic industries often utilize EPS produced by halophiles as gelling, emulsifying and stabilizing agents. It also aids in bioremediation of HMs, and *Halomonas* can concentrate lead and cadmium effectively. Tiny antimicrobial peptides called halocins derived from *Haloarchaea* can effectively treat cardiovascular conditions and prevent spoilage in the textile industry (Charlesworth and Burns 2016).

2.2.6   *Acidophiles*

True acidophiles like the archaea *Picrophilus oshimae* and *Picrophilus torridus* identified in sulphuric acid-soaked Japanese soils thrive at pH 0.7 and a temperature of 60 °C and produce enzymes that hydrolyse starch, including pullulanases, glucoamylases, glucosidases and amylases. Acidothermophiles flourish in acidic and elevated-temperature environments. For instance, *Sulfolobus solfataricus*, an acidothermophile, may survive at pH 3.0 and 80 °C (Serour and Antranikian 2002).

When compared to other acidophiles, *Picrophilus oshimae* sustains an internal pH value of 4.6 (Schleper et al. 1995). Thermophilic acidophiles, such as *Leptospirillum* thermoferrooxidans and *Acidiphilum* species, inhabit acid mine drains. The acid tolerance of rusticyanin, an acid-stable electron carrier found in *T. ferrooxidans*, is attributed to a high degree of intrinsic secondary structure and the hydrophobic environment within the cell where it is located (Morozkina et al. 2010).

Acidophiles that survive in highly acidic conditions adopt various pathways to survive at minimal pH. The impenetrability of the membrane to H^+ is one adaptation that helps acidophiles maintain their internal pH. Other mechanisms include the breakdown of organic acids, active proton extrusion and buffering of cytoplasm (Fig. 2.5) (Baker-Austin and Dopson 2007).

Acidophiles' hard, impermeable cell membrane limits protons' ability to enter the cytoplasm. Another strategy for maintaining pH homeostasis is controlling the cell membrane's porosity and thickness. Amaro et al. (1991) described the acidophile's exterior membrane porin. They disclosed a sizeable exterior loop that controls the cell's pore size and ion selectivity (Amaro et al. 1991).

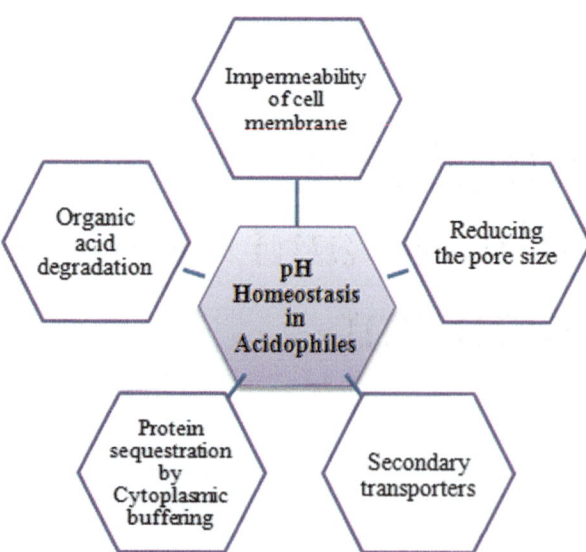

Fig. 2.5   pH homeostasis in acidophiles to maintain their internal pH

In acidophiles, a reversed membrane potential is produced by secondary transporters such as Na^+/K^+ transporters to prevent H^+ from flowing inward. Acidophiles utilize their capacity for buffering to encapsulate and remove H^+ as part of their defence strategy against H^+ permeation into the cells. The H^+ sequestration is facilitated by the presence of cytoplasmic buffer molecules in cells, namely, histidine, arginine and lysine (Chen 2021). Another mechanism is uncoupling the organic acids. Protons in the cytoplasm split apart, causing a process known as cytoplasmic protonation (Chen 2021). The microorganisms' intracellular enzymes do not require adaptation to function in extreme environments; however, their extracellular enzyme must operate at low pH. Acidophiles are valued because of their capacity to bioleach various metal ions, including HMs that are harmful to most organisms (Charlesworth and Burns 2016). They are essential in metal biomining from low-grade ores.

2.2.7 Alkaliphiles

Alkaliphiles require a pH of 8.0 or greater for optimal growth and thrive best at around pH 10. Haloalkaliphiles, on the other hand, oblige both high salinity, reaching up to 33%, and an alkaline pH exceeding 8.0. Alkalithermophilic microorganisms thrive in two severe circumstances: elevated temperatures (50–85 °C) and pH levels of 8.0 or above (Eichler 2001). Alkaline hot springs, alkaline steam vents and alkaline lakes in Egypt, Africa and Israel are examples of alkaliphilic and thermobiotic environments from which alkalithermophiles and alkaliphiles have primarily been extracted (Sorokin et al. 2011; Mesbah and Wiegel 2012). However, diverse alkalithermophiles were further identified in mesobiotic settings that range from mildly acidic to neutral, as well as sporadically in sewage and acidic soil samples.

Alkalophilic cyanobacteria dominate microbial communities below pH 10.0. Among these microorganisms, there is a wide diversity of aerobic species, including *Bacillus*, *Micrococcus* and *Pseudomonas*, as well as anaerobic species like *Clostridium thermohydrosulfurican*. Thermophilic bacteria such as *Anaerobranca horikoshii* and *Clostridium paradoxum*, along with hyperthermophilic archaebacteria like *Thermococcus alcaliphilus*, are also discovered within these genera. Additionally, haloalkaliphilic microorganisms include archaebacteria, cyanobacteria such as *Synechocystis* and *Nostoc*, methanogens like *Methanosalsus zhilinaeae* and sulfur-oxidizing microorganisms (Zhilina et al. 2005).

All alkaliphiles regulate a lower cellular pH than their surrounding environment due to intracellular pH homeostasis. To sustain the reversed pH gradient, metabolic energy is essential. Alkaliphiles actively uptake solutes from their surroundings. Aerobic alkaliphiles employ the Na^+/H^+ antiporter alongside H^+-linked respiration to regulate intracellular pH. Additionally, alkaliphiles serve as the primary hosts for the cation/proton antiporter referred to as the Mrp antiporter (Krulwich et al. 2011). Alkaliphiles possess a higher amount of anionic charged residues in their cell walls, primarily glutamic and glucuronic acids. Due to interactions with cations like H^+, this extremely negatively charged cell wall structure slows down the rate at which H^+ leaves the cell surface as a result of the environment's alkaline bulk phase equilibrating. This has a significant impact on the bioenergetics and pH homeostasis of alkaliphiles (Mamo 2020).

When reactions are conducted at higher temperatures and an alkaline pH, thermoalkaliphilic enzymes have significant biocatalytic potential (Horikoshi 1999). Alkaliphilic proteins exhibit a lower proportion of essential amino acids such as lysine and arginine while containing a higher concentration of acidic amino acids like aspartate and glutamate. For instance, the SecY protein discovered in *Bacillus halodurans* C-125 and the c-binding domain of cytochrome oxidase subunit II identified in *Bacillus pseudofirmus* OF4 serve as examples (Krulwich 2006). Survival strategies aside from the ones mentioned earlier still apply because many alkaliphiles happen to be thermophilic or halophilic.

In industries, alkaliphiles are mainly employed in the textile and detergent sectors. They are essential to biogeocycling. They can also degrade xenobiotic compounds. Various *Bacillus* strains, including *Bacillus* LP-Ya, *Bacillus cohnii* D-6, *Bacillus* NP-1 and *Bacillus* KSM-9860, produce alkaline protease, which is essential in detergent, food and tanning industries. From the alkalophilic *Nocardiopsis* species, the antibiotic naphthospironone A was derived and exhibited antibiotic and cytotoxic action (Deosthali and Dilipkumar Sajwani 2022).

2.2.8 Piezophiles

Piezophiles are organisms that thrive best in environments with high pressure. Piezophiles, comprising several types of thermophiles and hyperthermophiles, reside in the world's oceans, with an average pressure of 38 MPa (Abe and Horikoshi 2001). They mostly belong to the genera *Colwellia*, *Shewanella*, *Moritella*, *Pyrococcus*, *Methanococcus* and *Thermus* (Yano and Poulos 2003).

Piezophilic enzymes stable under high pressure have been discovered across various microorganisms, with most of them being either thermophilic or psychrophilic. These enzymes display optimal development conditions above one atmosphere (Yano and Poulos 2003). Microbes residing in deep waters typically encounter pressures not exceeding 120 MPa, and protein denaturation requires pressures exceeding 400 MPa (Gross and Jaenicke 1994). It is presumed that pressure does not significantly influence the structure and function of piezophilic proteins, and they require no particular pressure-related adaptations (Van den Burg 2003).

Piezolytes, molecules that stabilize cell proteins against high pressure, are a crucial component of microorganisms when exposed to high hydrostatic pressures. Trimethylamine, an N-compound that is prevalent in the ocean, is absorbed by piezotolerant bacteria, such as *Myroides profundi*, and converted to trimethylamine oxide. In hydrostatic conditions, trimethylamine oxide serves as a piezolyte and aids in cell growth and survival. Piezophiles synthesize a lot of PUFA, which serves to stabilize the cell membrane against pressure. Antioxidant proteins are another strategy for ensuring survival (Gross and Lehle 1993).

In a study investigating gene expression in psychrophilic piezophiles *Photobacterium profundum* SS9 under higher pressure, researchers found that the OmpH protein facilitates nutrient transport into the cell. Notably, the highest expression of this protein appeared at the pressure optimal for microbe growth, which was recorded at 28 MPa. Interestingly, under normal pressure conditions (0.1 MPa), the production of the OmpH protein was absent, confirming its role in adaptation (Welch and Bartlett 1996). Furthermore, it was shown that the rpoE-like locus of the SS9 strain is essential under both low-growth temperature and high-pressure conditions and controls the expression of the OmpH protein (Chi and Bartlett 1993). Later, various benthic bacteria were shown to harbour this conservative operon (Li et al. 1998). Furthermore, it has been anticipated that the transmembrane proteins ToxS and ToxR are controlling gene expression.

High-temperature-tolerant and high-pressure-tolerant enzymes provide many advantages in the field of biotechnology. Piezophile *Photobacterium profundum* can be utilized to synthesize toxic components at elevated pressure and low temperature, where there would be little harm to human health while yet allowing for the generation of the toxin. Certain piezophilic enzymes are essential to taste like proteolytic enzymes required for cheese ripening (Charlesworth and Burns 2016). A table elaborating on the application of extremophiles with examples, their adaptation requirements and the application of extremozymes produced by them (Table 2.3).

2.2.9 Metallophiles

Metallophiles are microorganisms that may flourish in an environment with high metal concentrations. These organisms, which include various species of the genus *Ralstonia*, inhabit heavy-metal-rich commercial sediments, soils and wastes. Due to intense anthropogenic activity or biotopes, metal-resistant *Ralstonia* has successfully adapted to these hostile conditions. *Ralstonia metallidurans* can withstand deadly heavy metal (HM) concentrations in millimolar amounts. It was initially found in 1976 in the severely contaminated sludge of a zinc decantation tank in Belgium that held several HMs. *Ralstonia* often has one or two large megaplasmids with genes conferring resistance metals such as Cd, Zn, Co, Cu, Pb, Hg, Cr and Ni (Mergeay et al. 2003).

These microbes, such as *Ralstonia*, may have applications in the bioremediation of HMs. Their enzymes may effectively immobilize some HMs besides serving as biosorbents. By reducing these elements to a decreased redox state, they are able to produce metal compounds with less bioactivity. They display various enzymatic processes that oxidize, reduce, methylate and alkylate specific metal species. In bioremediation, additional biological processes leading to less toxic metallic

Table 2.3 Extremophilic enzymes produced by extremozymes

Type	Organism	Enzyme	pH	Temperature	Applications	Reference
Hyperthermophile	*Methanococcus jannaschii*	α-Amylase	5.0–8.0	120 °C	Starch industry	Li et al. (2002)
	Sulfolobus solfataricus	α-Glucosidase	4.5	120 °C	Hydrolysis of starch	Kim et al. (2004)
Psychrophile	*Psychrobacter* sp. Ant300	Esterase	7.0–9.0	5°–25 °C	Food and paper industry	Suzuki et al. (2001)
	Clostridium strain PXYL1	Xylanase, cellulase	5.0–6.0	20 °C	Paper and pulp, textile industry	Akila and Chandra (2003)
Alkaliphiles	*Bacillus* isolate KSM-K38	α-Amylase	8.0–9.0	55°–60 °C	Detergent formulations and starch industry	Hagihara et al. (2001)
	Nesterenkonia sp. AL-20	Alkaline protease	10.0	74 °C	Detergents, leather industry and food and feed industry	Soto-Padilla et al. (2018)
Halophiles	*Halothermothrix orenii*	α-Amylase	7.5	65 °C	Textile industry, sweetener in food industry	Bhattacharya and Pletschke (2014)
	Halorhabdus utahensis	β-Xylanase, β-xylosidase	–	55°–70 °C	Hydrolysing glycosidic bonds	Werner et al. (2014)

substances have been used in conjunction with enzymatic alterations, resulting in metal precipitation and immobilization (Valls and De Lorenzo 2002).

2.3 Survival Mechanisms

Extremophiles tend to adapt mechanisms to ensure their survivability and fitness in harsh and hostile physiological environments. Extremophiles and the enzymes they generate are capable of withstanding both abiotic and biotic stresses, and they were engaged in adaptations and processes to maintain the stability of their existence and the enzymes. These extremozymes are remarkable biocatalysts and present novel prospects for biocatalysis and biotransformation. The following are some of the coping techniques that these extremophiles have developed.

2.3.1 Evolutionary Diversity

Extremophiles are very interesting to basic and applied microbiologists because of their biochemical characteristics. According to research, extremophiles possess essential genes preserved throughout evolution. These essential genes are crucial for producing the beneficial products that allow the organisms to endure adverse environmental circumstances. Extremophiles' ability to adapt to their environment depends on many vital genes, including *dnaKSf, dnaKCp, pyr-GCp, ftsZCp, hemCCp, murGCp, ligAPh, ligASf, ligACp, fmtCp, tyrSCp* and *cmkCp*. This strategy, called 'evolutionary diversity', may be exploited to create temperature-sensitive, persistent bacterial infections in the future by leveraging the traits of the key genes (Duplantis et al. 2010).

2.3.2 Increased Catalytic Activity

The metabolic fluxes between psychrophilic microorganisms are another extremophilic mechanism. These organisms flourish in extremely cold environments, generate enzymes that have high catalytic efficiency in the cold and adapt to the decline in the rate of chemical reactions caused by low temperatures. According to some theories, cold-adapted DNA ligase exhibits unique properties at the active sites, including enhanced performance at moderate and low temperatures, strong structural versatility and general molecular structure instability, which may have implications for biotechnology applications. In contrast, mesophiles and thermophiles exhibit low-temperature activity reduction, strong stability and decreased flexibility (Georlette et al. 2003). It is necessary to investigate further the significance of enhanced catalytic capacity in psychrophilic organisms resulting from the balance

between cold temperature and reduced thermal energy, leading to particular alterations in the molecular structure.

2.3.3   Amino Acid Accumulation

Some extremophile adaptation strategies involve the ability of the organisms to
modulate their cytoplasmic activity under neutral pH conditions. Halophiles adjust
to the environment by controlling the salinity level in the cytoplasm and the cytoplasmic protein formation of anionic amino acid molecules on the cell surfaces.
This attribute enhances their reliability and performance in inorganic solvents.
Additionally, halophiles frequently gather larger concentrations of neutral organic
compounds with low molecular weight to lower their osmotic pressure (Hendry 2006).

2.3.4   Aggregation Resistance Strategies

An organism needs to maintain its functional state even while under tremendous
stress to maintain the metabolic flux and cellular processes. It is conceivable to
resolve the reaction of the protein areas that are prone to aggregation by comprehending the aggregation resistance tactics of thermophilic proteins. Thermophiles
generate proteins that aid in tackling the accumulation of proteins, which lowers the
organism's functional status (Kufner and Lipps 2013). Comparing analysis of 373
protein families, Thangakani et al. (2012) discovered that thermophilic proteins utilized the aggregation resistance techniques more effectively compared to their
mesophilic counterparts. Thermophiles build up osmolyte molecules to stabilize
their proteins and nucleic acids, which helps them survive in harsh environments
(Thangakani et al. 2012).

2.3.5   Activation of the Nuclear Factor

Another method for extremophiles to thrive in hostile environments is through the
capacity of HSPs to suppress the gene expression of pro-inflammatory cytokines.
Buommino et al. (2005) specified that activation of kappa-B (NF-kappaB) is necessary for pro-inflammatory cytokines transcription. Specific HSPs are activated by
ectoine, which halophiles generate (Buommino et al. 2005). HSPs are recognized
for their function in protecting the cellular structures and cell and tissue repair in
response to stresses and trauma they may experience in harsh environments
(Morimoto and Gabriella Santoro 1998).

2.4   Conclusion

The adaptability strategies and functional diversity of extremophiles, alongside their enzymes, offer significant potential for enhancing soil health and promoting sustainable agriculture. Their ability to thrive in harsh environments presents a promising alternative to excessive reliance on agrochemicals and synthetic fertilizers. Extremophiles' resilience and metabolic activity under extreme conditions make them valuable assets in agricultural practices, contributing to reduced environmental impact and improved sustainability. Moreover, the unique biomolecules and metabolic pathways of extremophiles hold promise for unlocking new insights into life's complexities and for developing innovative solutions for sustainable environmental applications in agriculture.

References

Abe F, Horikoshi K (2001) The biotechnological potential of piezophiles. Trends Biotechnol 19

Adams MWW, Perler FB, Kelly RM (1995) Extremozymes: expanding the limits of biocatalysis. Bio/Technology 13

Akila G, Chandra TS (2003) A novel cold-tolerant clostridium strain PXYL1 isolated from a psychrophilic cattle manure digester that secretes thermolabile xylanase and cellulase. FEMS Microbiol Lett 219. https://doi.org/10.1016/S0378-1097(02)01196-5

Amaro AM, Chamorro D, Seeger M et al (1991) Effect of external pH perturbations on in vivo protein synthesis by the acidophilic bacterium Thiobacillus ferrooxidans. J Bacteriol 173. https://doi.org/10.1128/jb.173.2.910-915.1991

Anderson KL, Apolinario EE, Sowers KR (2012) Desiccation as a long-term survival mechanism for the archaeon Methanosarcina barkeri. Appl Environ Microbiol 78. https://doi.org/10.1128/AEM.06964-11

Aston JE, Peyton BM (2007) Response of Halomonas campisalis to saline stress: changes in growth kinetics, compatible solute production and membrane phospholipid fatty acid composition. FEMS Microbiol Lett 274. https://doi.org/10.1111/j.1574-6968.2007.00851.x

Baker-Austin C, Dopson M (2007) Life in acid: pH homeostasis in acidophiles. Trends Microbiol 15

Battista JR (1997) Against all odds: the survival strategies of Deinococcus radiodurans. Ann Rev Microbiol 51

Becker J, Wittmann C (2020) Microbial production of extremolytes—high-value active ingredients for nutrition, health care, and Well-being. Curr Opin Biotechnol 65

Berger F, Morellet N, Menu F, Potier P (1996) Cold shock and cold acclimation proteins in the psychrotrophic bacterium Arthrobacter globiformis SI55. J Bacteriol 178. https://doi.org/10.1128/jb.178.11.2999-3007.1996

Bhattacharya A, Pletschke BI (2014) Review of the enzymatic machinery of Halothermothrix orenii with special reference to industrial applications. Enzym Microb Technol 55

Buommino E, Schiraldi C, Baroni A et al (2005) Ectoine from halophilic microorganisms induces the expression of hsp70 and hsp70B′ in human keratinocytes modulating the proinflammatory response. Cell Stress Chaperones 10. https://doi.org/10.1379/CSC-101R.1

Cavicchioli R, Siddiqui KS, Andrews D, Sowers KR (2002) Low-temperature extremophiles and their applications. Curr Opin Biotechnol 13

Charlesworth J, Burns BP (2016) Extremophilic adaptations and biotechnological applications in diverse environments. AIMS Microbiol 2

Chen X (2021) Thriving at low pH: adaptation mechanisms of Acidophiles. In: Acidophiles—Fundamentals and Applications

Chi E, Bartlett DH (1993) Use of a reporter gene to follow high-pressure signal transduction in the deep-sea bacterium Photobacterium sp. strain SS9. J Bacteriol 175. https://doi.org/10.1128/jb.175.23.7533-7540.1993

D'Amico S, Claverie P, Collins T et al (2001) Chapter 3 cold-adapted enzymes: an unachieved symphony. Cell and molecular response to stress 2. https://doi.org/10.1016/S1568-1254(01)80005-X

Daly MJ (2000) Engineering radiation-resistant bacteria for environmental biotechnology. Curr Opin Biotechnol 11

Danson MJ, Hough DW (1997) The structural basis of protein halophilicity. Comparative Biochemistry and Physiology - A Physiology, In

DasSarma S, Capes MD, Karan R, DasSarma P (2013) Amino acid substitutions in cold-adapted proteins from Halorubrum lacusprofundi, an extremely halophilic microbe from Antarctica. PLoS One 8. https://doi.org/10.1371/journal.pone.0058587

Deming JW (2002) Psychrophiles and polar regions. Curr Opin Microbiol 5

Demirjian DC, Morís-Varas F, Cassidy CS (2001) Enzymes from extremophiles. Curr Opin Chem Biol 5

Deosthali C, Dilipkumar Sajwani N (2022) Extremophiles: applications and adaptive strategies. International Journal of Science & Engineering Development Research 7:378–390

Dieser M, Greenwood M, Foreman CM (2010) Carotenoid pigmentation in Antarctic heterotrophic bacteria as a strategy to withstand environmental stresses. Arct Antarct Alp Res 42. https://doi.org/10.1657/1938-4246-42.4.396

Duplantis BN, Osusky M, Schmerk CL et al (2010) Essential genes from Arctic bacteria used to construct stable, temperature-sensitive bacterial vaccines. Proc Natl Acad Sci USA 107. https://doi.org/10.1073/pnas.1004119107

Eichler J (2001) Biotechnological uses of archaeal extremozymes. Biotechnol Adv 19. https://doi.org/10.1016/S0734-9750(01)00061-1

Fonseca F, Meneghel J, Cenard S, et al (2016) Determination of intracellular vitrification temperatures for unicellular microorganisms under conditions relevant for cryopreservation. PLoS One 11. https://doi.org/10.1371/journal.pone.0152939

Friedman R, Drake JW, Hughes AL (2004) Genome-wide patterns of nucleotide substitution reveal stringent functional constraints on the protein sequences of thermophiles. Genetics 167. https://doi.org/10.1534/genetics.104.026344

Frösler J, Panitz C, Wingender J et al (2017) Survival of Deinococcus geothermalis in biofilms under desiccation and simulated space and Martian conditions. Astrobiology 17. https://doi.org/10.1089/ast.2015.1431

Galtier N, Lobry JR (1997) Relationships between genomic G+C content, RNA secondary structures, and optimal growth temperature in prokaryotes. J Mol Evol 44. https://doi.org/10.1007/PL00006186

García AH (2011) Anhydrobiosis in bacteria: from physiology to applications. J Biosci 36

Garcia-Lopez E, Alcazar P, Cid C (2021) Identification of biomolecules involved in the adaptation to the environment of cold-loving microorganisms and metabolic pathways for their production. Biomol Ther 11. https://doi.org/10.3390/biom11081155

Georlette D, Damien B, Blaise V et al (2003) Structural and functional adaptations to extreme temperatures in psychrophilic, mesophilic, and thermophilic DNA ligases. J Biol Chem 278. https://doi.org/10.1074/jbc.M305142200

Georlette D, Blaise V, Collins T et al (2004) Some like it cold: biocatalysis at low temperatures. FEMS Microbiol Rev 28

Goordial J, Raymond-Bouchard I, Zolotarov Y et al (2016) Cold adaptive traits revealed by comparative genomic analysis of the eurypsychrophile Rhodococcus sp. JG3 isolated from high elevation McMurdo Dry Valley permafrost, Antarctica. FEMS Microbiol Ecol 92. https://doi.org/10.1093/femsec/fiv154

Gross M, Jaenicke R (1994) Proteins under pressure: the influence of high hydrostatic pressure on structure, function and assembly of proteins and protein complexes. Eur J Biochem 221

Gross M, Lehle K (1993) Pressure induced dissociation of ribosomes and elongation cycle intermediates. European Journal of Biochem 218:468

Hagihara H, Igarashi K, Hayashi Y et al (2001) Novel α-amylase that is highly resistant to chelating reagents and chemical oxidants from the Alkaliphilic bacillus isolate KSM-K38. Appl Environ Microbiol 67. https://doi.org/10.1128/AEM.67.4.1744-1750.2001

Haki GD, Rakshit SK (2003) Developments in industrially important thermostable enzymes: a review. Bioresour Technol 89

Hendry P (2006) Extremophiles: There's more to life. Environ Chem 3

Horikoshi K (1999) Alkaliphiles: some applications of their products for biotechnology. Microbiol Mol Biol Rev 63. https://doi.org/10.1128/mmbr.63.4.735-750.1999

Huber H, Hohn MJ, Rachel R et al (2002) A new phylum of archaea represented by a nanosized hyperthermophilic symbiont. Nature 417. https://doi.org/10.1038/417063a

Jin M, Xiao A, Zhu L et al (2019) The diversity and commonalities of the radiation-resistance mechanisms of Deinococcus and its up-to-date applications. AMB Express 9

Jolivet E, Corre E, L'Haridon S et al (2004) Thermococcus marinus sp. nov. and Thermococcus radiotolerans sp. nov., two hyperthermophilic archaea from deep-sea hydrothermal vents that resist ionizing radiation. Extremophiles 8. https://doi.org/10.1007/s00792-004-0380-9

Jung KW, Lim S, Bahn YS (2017) Microbial radiation-resistance mechanisms. J Microbiol 55

Kanodia S, Roberts MF (1983) Methanophosphagen: unique cyclic pyrophosphate isolated from Methanobacterium thermoautotrophicum. Proc Natl Acad Sci 80. https://doi.org/10.1073/pnas.80.17.5217

Kashefi K, Holmes DE, Reysenbach AL, Lovley DR (2002) Use of Fe(III) as an electron acceptor to recover previously uncultured hyperthermophiles: Isolation and characterization of Geothermobacterium ferrireducens gen. nov., sp. nov. Appl Environ Microbiol 68. https://doi.org/10.1128/AEM.68.4.1735-1742.2002

Kim MS, Park JT, Kim YW et al (2004) Properties of a novel thermostable glucoamylase from the hyperthermophilic archaeon sulfolobus solfataricus in relation to starch processing. Appl Environ Microbiol 70. https://doi.org/10.1128/AEM.70.7.3933-3940.2004

Kim HJ, Lee JH, Hur YB et al (2017) Marine antifreeze proteins: structure, function, and application to cryopreservation as a potential cryoprotectant. Mar Drugs 15. https://doi.org/10.3390/md15020027

Krulwich TA (2006) Alkaliphilic Prokaryotes. In: The Prokaryotes

Krulwich TA, Liu J, Morino M et al (2011) Adaptive mechanisms of extreme Alkaliphiles. In: Extremophiles Handbook

Kufner K, Lipps G (2013) Construction of a chimeric thermoacidophilic beta-endoglucanase. BMC Biochem 14. https://doi.org/10.1186/1471-2091-14-11

Kumar S, Nussinov R (2001a) How do thermophilic proteins deal with heat? Cell Mol Life Sci 58

Kumar S, Nussinov R (2001b) Fluctuations in ion pairs and their stabilities in proteins. Proteins Struct Funct Genet 43. https://doi.org/10.1002/prot.1056

Kumar R, Singh A (2012) Smart therapeutics from extremophiles: unexplored applications and technological challenges. In: Extremophiles: Sustainable Resources and Biotechnological Implications

Kumar A, Alam A, Tripathi D et al (2018) Protein adaptations in extremophiles: an insight into extremophilic connection of mycobacterial proteome. Semin Cell Dev Biol 84

Kunte HJ (2013) Osmoregulation in halophilic bacteria. Extremophiles II

Li L, Kato C, Nogi Y, Horikoshi K (1998) Distribution of the pressure-regulated operons in deep-sea bacteria. FEMS Microbiol Lett 159. https://doi.org/10.1016/S0378-1097(97)00560-0

Li M, Kim JW, Peeples TL (2002) Amylase partitioning and extractive bioconversion of starch using thermoseparating aqueous two-phase systems. J Biotechnol 93. https://doi.org/10.1016/S0168-1656(01)00382-0

Lim J, Thomas T, Cavicchioli R (2000) Low temperature regulated DEAD-box RNA helicase from the Antarctic archaeon, Methanococcoides burtonii. J Mol Biol 297. https://doi.org/10.1006/jmbi.2000.3585

Lu J, Nogi Y, Takami H (2001) Oceanobacillus iheyensis gen. Nov., sp. nov., a deep-sea extremely halotolerant and alkaliphilic species isolated from a depth of 1050 m on the Iheya Ridge. FEMS Microbiol Lett 205. https://doi.org/10.1016/S0378-1097(01)00493-1

Macelroy RD (1974) Some comments on the evolution of extremophiles. In: BioSystems

Madern D, Ebel C, Zaccai G (2000) Halophilic adaptation of enzymes. Extremophiles 4

Madigan M, Marrs B (1997) Extremophiles. In: 4th edn. Scientific American, pp 82–87

Mallik S, Kundu S (2015) Molecular interactions within the halophilic, thermophilic, and mesophilic prokaryotic ribosomal complexes: clues to environmental adaptation. J Biomol Struct Dyn 33. https://doi.org/10.1080/07391102.2014.900457

Mamo G (2020) Challenges and adaptations of life in alkaline habitats. In: Advances in Biochemical Engineering/Biotechnology

Mergeay M, Monchy S, Vallaeys T et al (2003) Ralstonia metallidurans, a bacterium specifically adapted to toxic metals: towards a catalogue of metal-responsive genes. FEMS Microbiol Rev 27

Merino N, Aronson HS, Bojanova DP et al (2019) Living at the extremes: extremophiles and the limits of life in a planetary context. Front Microbiol 10. https://doi.org/10.3389/fmicb.2019.00780

Mesbah NM, Wiegel J (2012) Life under multiple extreme conditions: diversity and physiology of the halophilic alkalithermophiles. Appl Environ Microbiol 78

Mevarech M, Frolow F, Gloss LM (2000) Halophilic enzymes: proteins with a grain of salt. In: Biophysical Chemistry

Mishra S, Chaudhary R, Singh S et al (2019) Guanine quadruplex DNA regulates gamma radiation response of genome functions in the radioresistant bacterium deinococcus radiodurans. J Bacteriol 201. https://doi.org/10.1128/JB.00154-19

Morimoto RI, Gabriella Santoro M (1998) Stress-inducible responses and heat shock proteins: new pharmacologic targets for cytoprotection. Nat Biotechnol 16

Morozkina EV, Slutskaya ES, Fedorova TV et al (2010) Extremophilic microorganisms: biochemical adaptation and biotechnological application (review). Appl Biochem Microbiol 46

Netzer WJ, Hartl FU (1998) Protein folding in the cytosol: chaperonin-dependent and -independent mechanisms. Trends Biochem Sci 23

Nogi Y, Hosoya S, Kato C, Horikoshi K (2004) Colwellia piezophila sp. nov., a novel piezophilic species from deep-sea sediments of the Japan trench. Int J Syst Evol Microbiol 54. https://doi.org/10.1099/ijs.0.03049-0

Ophir T, Gutnick DL (1994) A role for exopolysaccharides in the protection of microorganisms from desiccation. Appl Environ Microbiol 60

Oren A (2002) Molecular ecology of extremely halophilic archaea and bacteria. FEMS Microbiol Ecol 39

Pikuta E, Lysenko A, Chuvilskaya N et al (2000) Anoxybacillus pushchinensis gen. Nov., sp. nov., a novel anaerobic, alkaliphilic, moderately thermophilic bacterium from manure, and description of Anoxybacillus flavithermus comb. nov. Int J Syst Evol Microbiol 50. https://doi.org/10.1099/00207713-50-6-2109

Rampelotto PH (2013) Extremophiles and extreme environments. Life 3

Rastädter K, Wurm DJ, Spadiut O, Quehenberger J (2021) Physiological characterization of sulfolobus acidocaldarius in a controlled bioreactor environment. Int J Environ Res Public Health 18. https://doi.org/10.3390/ijerph18115532

Roy S (2017) Impact of UV radiation on genome stability and human health. In: Advances in Experimental Medicine and Biology

Russell NJ (1989) Adaptive modifications in membranes of halotolerant and halophilic microorganisms. J Bioenerg Biomembr 21. https://doi.org/10.1007/BF00762214

Saad S, Cereghetti G, Feng Y et al (2017) Reversible protein aggregation is a protective mechanism to ensure cell cycle restart after stress. Nat Cell Biol 19. https://doi.org/10.1038/ncb3600

Sahonero-Canavesi DX, Villanueva L, Bale NJ et al (2022) Changes in the distribution of membrane lipids during growth of Thermotoga maritima at different temperatures: indications for the potential mechanism of biosynthesis of ether-bound diabolic acid (membrane-spanning) lipids. Appl Environ Microbiol 88. https://doi.org/10.1128/AEM.01763-21

Sandigursky M, Sandigursky S, Sonati P et al (2004) Multiple uracil-DNA glycosylase activities in Deinococcus radiodurans. DNA Repair (Amst) 3. https://doi.org/10.1016/j.dnarep.2003.10.011

Schleper C, Piihler G, Kühlmorgen B, Zillig W (1995) Life at extremely low pH. Nature 375

Segerer A, Neuner A, Kristjansson JK, Stetter KO (1986) Acidianus infernus gen. Nov., sp. nov., and Acidianus brierleyi comb. nov.: Facultatively aerobic, extremely acidophilic thermophilic sulfur-metabolizing archaebacteria. Int J Syst Bacteriol 36. https://doi.org/10.1099/00207713-36-4-559

Serour E, Antranikian G (2002) Novel thermoactive glucoamylases from thermoacidophilic Archea thermoplasma acidophilum, picrophilus torridus and picrophilus oshimae. Anton Leeuw Int J Gen Mol Microbiol 81. https://doi.org/10.1023/A:1020525525490

Singh OV, Gabani P (2011) Extremophiles: radiation resistance microbial reserves and therapeutic implications. J Appl Microbiol 110

Sorokin DY, Kuenen JG, Muyzer G (2011) The microbial sulfur cycle at extremely haloalkaline conditions of soda lakes. Front Microbiol 2. https://doi.org/10.3389/fmicb.2011.00044

Soto-Padilla MY, Gortáres-Moroyoqui P, Cira-Chávez LA, Estrada-Alvarado MI (2018) Biochemical and molecular characterization of a native haloalkaliphilic tolerant strain from the Texcoco lake. Pol J Microbiol 67. https://doi.org/10.21307/pjm-2018-047

Sterner R, Liebl W (2001) Thermophilic adaptation of proteins. Crit Rev Biochem Mol Biol 36

Suzuki T, Nakayama T, Kurihara T et al (2001) Cold-active lipolytic activity of psychrotrophic Acinetobacter sp. strain no. 6. J Biosci Bioeng 92. https://doi.org/10.1263/jbb.92.144

Terpe K (2013) Overview of thermostable DNA polymerases for classical PCR applications: from molecular and biochemical fundamentals to commercial systems. Appl Microbiol Biotechnol 97

Thangakani AM, Kumar S, Velmurugan D, Gromiha MSM (2012) How do thermophilic proteins resist aggregation? Proteins: structure. Function and Bioinformatics 80. https://doi.org/10.1002/prot.24002

Valls M, De Lorenzo V (2002) Exploiting the genetic and biochemical capacities of bacteria for the remediation of heavy metal pollution. FEMS Microbiol Rev 26

Van den Burg B (2003) Extremophiles as a source for novel enzymes. Curr Opin Microbiol 6

Vieille C, Zeikus GJ (2001) Hyperthermophilic enzymes: sources, uses, and molecular mechanisms for thermostability. Microbiol Mol Biol Rev 65. https://doi.org/10.1128/mmbr.65.1.1-43.2001

Vriezen JAC, De Bruijn FJ, Nüsslein K (2007) Responses of rhizobia to desiccation in relation to osmotic stress, oxygen, and temperature. Appl Environ Microbiol 73

Wang X, Cole CG, Dupai CD, Davies BW (2020) Protein aggregation is associated with Acinetobacter baumannii desiccation tolerance. Microorganisms 8. https://doi.org/10.3390/microorganisms8030343

Welch TJ, Bartlett DH (1996) Isolation and characterization of the structural gene for OmpL, a pressure-regulated porin-like protein from the deep-sea bacterium Photobacterium species strain SS9. J Bacteriol 178. https://doi.org/10.1128/jb.178.16.5027-5031.1996

Werner J, Ferrer M, Michel G et al (2014) Halorhabdus tiamatea: Proteogenomics and glycosidase activity measurements identify the first cultivated euryarchaeon from a deep-sea anoxic brine lake as potential polysaccharide degrader. Environ Microbiol 16. https://doi.org/10.1111/1462-2920.12393

Yano JK, Poulos TL (2003) New understandings of thermostable and piezostable enzymes. Curr Opin Biotechnol 14

Yoshida K, Hashimoto M, Hori R et al (2016) Bacterial long-chain polyunsaturated fatty acids: their biosynthetic genes, functions, and practical use. Mar Drugs 14

Zeidler S, Müller V (2019) The role of compatible solutes in desiccation resistance of Acinetobacter baumannii. Microbiology 8. https://doi.org/10.1002/mbo3.740

Zhilina TN, Zavarzina DG, Kolganova T V., et al (2005) "Candidatus Contubernalis alkalaceticum," an obligately syntrophic alkaliphilic bacterium capable of anaerobic acetate oxidation in a coculture with Desulfonatronum cooperativum. Mikrobiologiya 74:

Chapter 3
Microbial Adaptation in Different Extreme Environmental Conditions and Its Usefulness in Differently Polluted Soil

Jayati Arora, Arpna Kumari, Anuj Ranjan, Vishnu D. Rajput,
Sudhir Shende, Evgeniya Valer'evna Prazdnova, Saglara S. Mandzhieva,
Svetlana Sushkova, Tatiana Minkina, Abhishek Chauhan, Rajpal Srivastav,
and Tanu Jindal

3.1 Introduction

Microorganisms, present in abundance throughout the Earth, display a wide-ranging and pervasive presence. Their quantity and diversity differ from one environment to another. It is noteworthy that as of now, we have been able to culture a mere 1% of the entire spectrum of microorganisms found on our planet, using precise growth mediums and conditions. Nonetheless, there is a strong belief that numerous untapped habitats are teeming with a rich tapestry of life forms, offering tremendous opportunities for exploration and discovery (Shukla and Singh 2020).

Jayati Arora and Arpna Kumari contributed equally with all other contributors.

J. Arora
Amity Institute of Environmental Sciences, Amity University, Noida, Uttar Pradesh, India

A. Kumari (✉)
Department of Applied Biological Chemistry, Graduate School of Agricultural and Life Sciences, The University of Tokyo, Tokyo, Japan
e-mail: arpnakumari09@g.ecc.u-tokyo.ac.jp

A. Ranjan · V. D. Rajput · S. Shende · E. V. Prazdnova · S. S. Mandzhieva · S. Sushkova ·
T. Minkina
Academy of Biology and Biotechnology, Southern Federal University, Rostov-on-Don, Russia

A. Chauhan · T. Jindal
Institute of Environmental Toxicology, Safety and Management, Amity University, Noida, Uttar Pradesh, India

R. Srivastav
Amity Institute of Biotechnology, Amity University Uttar Pradesh, Noida, India

© The Author(s), under exclusive license to Springer Nature Switzerland AG 2024
A. Ranjan et al. (eds.), *Extremophiles for Sustainable Agriculture and Soil
Health Improvement*, https://doi.org/10.1007/978-3-031-70203-7_3

Extremophilic microorganisms, a relatively understudied group, exhibit remarkable adaptability to environments conventionally deemed inhospitable for life. These microorganisms call harsh habitats home, characterized by extreme fluctuations in factors like temperature, salinity, nutrient availability, pressure, and various other environmental variables.

The term 'extremophile', coined by MacElroy in 1974, borrows from the Latin word 'extremus', signifying 'being on the outside'. It serves to describe organisms capable of thriving in conditions that would typically prove fatal to most eukaryotic cells (Pakchung et al. 2006). Extremophiles are life forms that flourish in incredibly hostile, and sometimes even deadly, settings. These settings encompass scorching hot domains, frigid polar regions, and saline lakes. Astonishingly, certain extremophiles exhibit the ability to persist in the presence of toxic waste, organic compounds, and heavy metal pollutants that would spell doom for other life forms. These remarkable organisms can be found in diverse locations such as hydrothermal vents, the depths of the sea, and volcanic environments.

Despite their status as the most abundant organisms on Earth, extremophiles have received relatively limited attention in scientific study. Extremophiles can be broadly categorized into two main groups: those exclusively adapted to extreme conditions and those with the capacity to endure specific types of extreme environments (Basak et al. 2020). These exceptional life forms encompass representatives from all three domains of life: bacteria, eukarya, and archaea. Among these, archaea exhibit remarkable resilience in extreme conditions albeit with somewhat less adaptability compared to bacteria. Cyanobacteria, a type of bacteria, have particularly well-developed adaptations for thriving in harsh environments. Fungi, belonging to the eukaryotic domain, demonstrate remarkable versatility in withstanding challenging conditions. A remarkable example of a eukaryotic polyextremophile is the tardigrade, a microscopic invertebrate. Tardigrades display an astonishing ability to endure temperatures spanning from $-272\ °C$ to $151\ °C$, withstand pressures up to 6000 times atmospheric pressure, survive extreme desiccation, and even tolerate exposure to X-rays and gamma rays (Rampelotto 2013).

A thorough grasp of phylogenetics and microbial metabolism serves as a valuable foundation for delving into extremophiles and the invaluable insights they provide. To attain a more holistic understanding of extremophiles and their relevance, it is crucial to incorporate additional layers of knowledge. These encompass fields like geochemistry, the stability of biomolecules, small molecule chemistry, biotechnology, and the evolutionary aspects of extremophiles. By synergizing these diverse areas of study, we can broaden our knowledge and pave the way for a more profound exploration of extremophiles and their profound significance.

3.2   Extreme Environmental Conditions and Microbial Life Forms

Extremophiles have captured the attention of researchers worldwide, offering a glimpse into the remarkable adaptations and unique biochemical processes that support life in the environmental extremes. The exploration of different extremophile types and their specific adaptations holds immense significance for varied scientific fields, including astrobiology, biotechnology, and environmental science.

Extremophiles can be classified based on their specific requirements for thriving in distinct environmental niches. These niches encompass extreme temperature ranges (both high and low), pH levels (either highly acidic or strongly alkaline), high-pressure conditions, and elevated salinity levels, among others. Furthermore, there exist extremophiles that flourish under unconventional extreme circumstances, such as desiccation (extremely low water availability), or those capable of enduring excessive ionizing radiation, for example, *Deinococcus radiodurans* (Makarova et al. 2001).

Some extremophiles exhibit multiple extreme traits, earning them the classification of polyextremophiles. An excellent example of this is the bacterium *Shewanella benthica*, which exhibits adaptations as both a barophile (suited for high-pressure environments) and a psychrophile (adapted to cold temperatures). This diverse categorization system enhances our understanding of the spectrum of extremophilic adaptations and their capacity to not only survive but also thrive in diverse extreme environments (Pakchung et al. 2006). Microorganisms are well-adapted to thrive optimally within their native ecological settings, a fact underscored by their specific physiological traits, including temperature, pH, and salinity preferences. When these ideal conditions are significantly disrupted, it places stress on the microorganism. The organism's fate, whether it survives, stops growing, or undergoes a prolonged lag phase with reduced biomass yield, is determined by the degree and duration of the environmental change.

Bacteria, in particular, exhibit remarkable resilience to minor fluctuations in environmental factors, capable of adapting swiftly over short time spans, spanning from mere minutes to a few days. This adaptability is achieved through either tolerating the stress and deploying survival mechanisms or actively resisting the unfavourable conditions. Microorganisms have the potential to push their tolerance limits further if they are given ample opportunities to sense and adjust to the deteriorating environment (Russell et al. 1995).

3.2.1   High- and Low-Temperature Environments

Temperature plays a crucial role in shaping the activities and evolution of organisms and is easily measurable. Not all temperature ranges are equally suitable for the growth and reproduction of life. High-temperature environments, particularly those linked to volcanic activity like hot springs, are of particular interest as they show-case the extreme conditions to which evolution has adapted. These natural habitats have likely existed for most of the Earth's evolutionary history, making them valu-able for scientific study (Brock 2012).

In the early stages of Earth's history, the predominant life forms were primarily thermophilic anaerobes. These organisms relied on a chemoheterotrophic metabolic process to thrive in an environment characterized by hydrothermal sources. Thermophilic microorganisms can be categorized into three main groups based on their temperature preferences: moderately thermophilic, with an optimal growth temperature range of 50–60 °C; thermophilic, with an optimal growth temperature exceeding 70 °C; and hyperthermophilic, with an optimal growth temperature sur-passing 80 °C (Brock 2012). These remarkable organisms have successfully adapted to a variety of environments, including mud pots, hot springs, deep-sea hydrother-mal vents, geothermal waters, volcanic regions, deep mining areas, and even human-made settings such as incinerator units and anaerobic reactors (Orellana et al. 2018).

Organisms inhabiting the Antarctic, Arctic, glacial regions, and cold deep oceans have developed adaptations to thrive in consistently cold environments (Deming 2002). The deep sea floor, encompassing the largest low-temperature ecosystem on Earth, is especially significant due to the extensive coverage of oceans across the planet. Within this global-scale ecosystem, psychrophiles thrive in extreme condi-tions, with a consistent temperature of 4 °C below 1000 meters deep. These organ-isms are true extremophiles, as they have adapted to not only low temperatures but also other challenging environmental factors (Feller and Gerday 2003).

3.2.2 High Salinity and Halophilic Adaptation

Aqueous environments can have varying ion concentrations, ranging from nearly zero to saturation. When ion concentration increases, certain ions precipitate before others. The most soluble ions are Na^+, K^+, and Mg^{2+} (cations) and Cl^-, SO_4^{2-}, and HCO_3^- (or CO_3^{2-}, depending on pH) (anions). Consequently, these ions are found in the highest concentrations in seawater and saline lakes. Marine habitats exhibit high species diversity and are not extreme environments. In contrast, hypersaline waters lack fish and have low invertebrate diversity, making them unmistakably extreme environments (Brock 2012).

Halophiles, categorized as organisms that need >2.5 M NaCl for optimal growth, can be found in archaea, bacteria, and eukarya. In contrast, halotolerant microor-ganisms can tolerate up to 15% salinity without being reliant on salt for growth. The metabolic diversity of halophiles is extensive and includes oxygenic and anoxy-genic phototrophs, aerobic heterotrophs, fermenters, denitrifiers, sulfate reducers, and methanogens. These adaptations enable halophiles to thrive in high-salt envi-ronments and perform various energy-generating processes (Oren 2002).

3.2.3 Acidic and Alkaline Environments

In environments where most organisms struggle to survive, some thrive and multiply extensively. Surprisingly, acid environments host a diverse range of organisms, including bacteria, algae, protozoa, and invertebrates. This diversity is more understandable when we consider that acidic environments, often created by sulfuric acid, are widespread on Earth. Acidity is more common than alkalinity in nature because it develops in oxygen-rich environments, and the biosphere is predominantly aerobic. Sulfuric acid is formed through the oxidation of sulfides like hydrogen sulfide and pyrite, which are prevalent in volcanic areas, geothermal areas, bogs, swamps, and the sea. When oxygen enters the picture, sulfides rapidly oxidize, either spontaneously or through the action of specialized sulfur-oxidizing bacteria, leading to sulfuric acid production (Brock 2012).

The majority of microorganisms have an ideal pH range for growth, typically spanning approximately 1 pH unit above and below their optimal value. Acidophilic microorganisms thrive at pH values below 2, while alkaliphilic microorganisms excel at pH values above 10 (Rampelotto 2013).

Some acidophilic microorganisms possess a unique ability to maintain a neutral pH inside their cells, negating the need for internal enzymes to adapt to extremely low pH conditions. Research shows that the internal pH of acidophiles typically falls within the range of 5–7. However, it is important to note that the extracellular proteins of these organisms are typically adapted to function optimally in highly acidic environments, with their peak functionality occurring at low pH levels (Rainey and Oren 2006).

Alkaliphilic microorganisms require a growth medium with a pH higher than 8.0 and cannot thrive at neutral pH levels. Among them, cyanobacteria are exceptionally alkaliphilic, thriving in environments with pH values ranging from 12 to 13. Conversely, alkalitolerant organisms grow optimally at neutral pH but can also survive in pH levels between 10 and 11 (Pikuta et al. 2007). Alkaliphiles inhabit environments like soda lakes, carbonate-rich springs, and alkaline soils. They have mechanisms to regulate their cytoplasmic pH, keeping it at least 2 pH units lower than the surrounding environment, which typically has a pH value of around 10 (Bull 2004).

3.2.4 Microbial Communities in Other Extreme Environments

In the profound depths of the ocean floor, where the average depth plunges to approximately 3800 m, the pressure escalates to about 38 MPa, a staggering 375 times greater than the atmospheric pressure at sea level (0.1013 MPa). Barophiles, also referred to as piezophiles, have adeptly adapted to flourish in these conditions of high hydrostatic pressure, making homes in deep-sea trenches and the digestive systems of bottom-dwelling animals. These microorganisms classified as barophilic demonstrate optimal growth at pressures surpassing 40 MPa. Some, known as hyperbarophiles, thrive in environments with pressures as high as 60–70 MPa, while barotolerant bacteria can achieve optimal growth at pressures below 40 MPa, adapting well even under atmospheric pressure (Pikuta et al. 2007).

Collected samples from the Mariana Trench, the deepest part of the ocean floor at 10,898 m, revealed a rich variety of organisms, including notable species like *Shewanella* and a recently identified bacterium named *Moritella yayanosii* sp. (nov.). Intriguingly, *Moritella yayanosii* sp. (nov.) exhibited its best growth at 80 MPa but could not thrive at 50 MPa. This observation exemplifies the concept of barophily, where organisms that are not adapted to high pressure exhibit abnormal morphologies and eventually perish when subjected to elevated pressures (Kato et al. 1998).

In contrast to typical bacteria and fungi, which struggle to thrive at low water activity levels, microorganisms found in extreme environments, known as xerophiles, showcase the remarkable ability to survive well below these limits. Xerotolerant organisms, capable of enduring dry conditions without relying on low water activity, have evolved various adaptive strategies to counteract the detrimental effects of desiccation on their morphology, physiology, and biochemistry. These strategies include entering dormancy, forming extracellular polysaccharides and biofilms, increasing cyclopropane fatty acids in cell membranes, and enhancing glycine production to counteract desiccation and prevent denaturation. Examples of xerophiles encompass *Aspergillus penicillioides*, *Cereus jamacaru*, *Deinococcus radiodurans*, *Aphanothece halophytica*, *Anabaena*, *Bradyrhizobium japonicum*, and *Saccharomyces bailii* (Lebre et al. 2017).

3.3   Significance of Extreme Environmental Conditions and Microbial Adaptation

The discovery of varied types of microorganisms thriving in extremes of climatic conditions, ranging from the glacial polar waters to the boiling waters of hydrothermal vents, at the base of Mariana Trench (Kato et al. 1998), at extremely high levels of ultraviolet radiation (DeVeaux et al. 2007) and in NaCl- and KCl-saturated lakes, has been the driving force towards comprehending the evolutionary adaptive strategies in extremophiles. Biomolecular adaptations in extremophiles are diversified, including those in proteins, membranes, nucleic acids, and other smaller molecules. In comparison to a non-extremophilic microorganism, proteins in extremophiles exhibit relative changes in terms of the occurrence of amino acids and their residues which are responsible for various properties of the protein like stability and conformational flexibility under environmental stress-like conditions. The structures of membranes are altered by temperature. Nucleic acid integrity in extremophiles can be restored by minor structural modifications to the RNA nucleobases. In conditions of high salt and micronutrient deficiency, a range of osmolytes and minute molecules are produced or sequestered by these microbes as an adaptive mechanism (Pakchung et al. 2006).

In the last 10 years, there has been a growing body of research centred on understanding how microorganisms thriving in extreme environments manage to withstand stress (Zhou et al. 2011). Several studies have revealed that genome plasticity, encompassing factors like codon bias, nucleotide skew, and horizontal gene transfers (HGTs), plays a crucial role in facilitating evolutionary adaptations to extreme conditions (Zeldovich et al. 2007). Nevertheless, researchers have not yet identified consistent, overarching patterns when it comes to the adaptive mechanisms of microorganisms inhabiting these challenging settings. This complexity likely arises from the diverse array of selective pressures experienced in extreme environments. For most microorganisms, adapting to such stressful conditions is a highly intricate and dynamic process, shaped by the interplay of various evolutionary forces (Tyson et al. 2004; Li et al. 2014).

Extremophiles, due to their unique adaptability, hold great ecological significance and are becoming increasingly valuable in biotechnological research and industrial applications. Their ability to maintain cell membrane stability and fluidity in extreme environmental conditions helps in the preservation of their genetic material. This adaptation translates into their possession of unique genes. These exceptional characteristics portray extremophiles as invaluable resources for the production of crucial biomolecules that remain stable in conditions of high or low temperatures, extreme pH levels, and very high pressure and even in the presence of hazardous pollutants (Arora and Panosyan 2019). Extremophiles and their metabolic products serve as a source of robust enzymes capable of functioning in various extreme conditions. Due to these reasons, they find application in biodegradation and bioremediation processes in challenging environments. These microbes also play a pivotal role in the production of biofuels and bioenergy, and they contribute as a source of specialized pigments for solar cells designed to operate in extreme conditions (Martínez et al. 2022).

The potential of extremophilic microbes is vast, and they are poised to make a substantial contribution to the realization of sustainability goals and the development of a bio-based economy in the future.

Extremophiles, crucial subjects of study for various therapeutic and medical applications, exhibit intriguing mechanisms for survival in extreme environments. The focus of research lies in comprehending how these extremophiles develop defensive strategies to endure harsh conditions and how their metabolisms play a vital role in these survival processes.

Studying the adaptation mechanisms of extremophiles provides valuable insights for researchers seeking to understand the molecular elements—such as proteins and genes—involved in their survival. This knowledge serves as a foundation for exploring how these molecular components can be modified and harnessed for therapeutic implications (Babu et al. 2015).

Horizontal gene transfer or HGT refers to the exchange of genetic material between organisms that lack a parent–offspring relationship. Widely acknowledged as a mechanism for adaptation in bacteria and archaea, HGT plays a significant role in microbial evolution. While microbial antibiotic resistance and pathogenicity are commonly linked to HGT, its implications extend beyond disease-causing organisms. For a transferred gene to persist in the recipient lineage over extended periods, it typically must confer a selective advantage, either to itself (as a selfish genetic element) or to the recipient. Early research on HGT primarily concentrated on genes that offered such advantages (Soucy et al. 2015).

HGT encompasses several well-recognized modes, including bacterial transformation, bacterial conjugation, and transduction by viruses such as bacteriophages. These mechanisms, considered outcomes of Darwinian evolution, play crucial roles in the acquisition of new genotypes or phenotypes by existing entities. The five theoretical mechanisms facilitating this process are bacterial transformation, bacterial conjugation, transduction, the utilization of gene transfer agents, and membrane vesicle transfers (Jheeta 2020).

3.4 Role of Microorganisms in Soil Health

The rapid projected growth in the global population, expected to reach 8.9 billion by 2050, is driving increased demands for agricultural products (Lichtfouse et al. 2009). Meeting such demands through the intensive use of synthetic fertilizers and pesticides has resulted in land degradation and environmental pollution across various agroecosystems and has had detrimental impacts on human health, wildlife, and aquatic ecosystems (Devarinti 2016).

In order to address these challenges, sustainable agriculture has emerged as a holistic and eco-friendly alternative which is characterized by an integrated approach that offers solutions to most of the practical issues related to food production (Lal 2008). Sustainable agriculture combines principles from biology, physics, chemistry, and ecology to develop innovative practices that are environment-friendly (Lichtfouse et al. 2009; Tahat et al. 2020).

The application of beneficial microorganisms offers a promising alternative to the use of harmful chemical fertilizers and pesticides. Microbes play a pivotal role in enhancing crop productivity and soil management. Soil microbes associated with plants are found to be extremely beneficial in promoting plant growth and development, including essential functions like nutrient cycling and overall crop yield. The dynamics of soil microbial communities are a critical determinant of soil productivity (Yan et al. 2015).

The interaction between plants and microbes is a key factor influencing ecosystem function, with the nature of these interactions being highly variable and dependent on nutrient availability. Plant-growth-promoting microbes, utilized to boost plant growth through various mechanisms, including plant growth regulation and nitrogen fixation, have gained special attention in recent times (Ahmad et al. 2008). Moreover, these microbes have demonstrated their ability to stimulate plant growth even under stressful conditions such as drought and salinity, by regulating processes like cell division, enlargement, and differentiation (Kumar and Verma 2019).

However, it is essential to recognize that these interactions within plants involve complex factors at multiple levels, spanning genetics, physiology, ecology, and morphology (Islam et al. 2014). To address the challenges posed by conventional agricultural practices, a major shift towards biological control is imperative. The use of microbes for soil health and plant growth promotion is both sustainable and cost-effective, paving the way for a more resilient agriculture system in the future benefiting both the environment and farmers (Kumar and Verma 2019).

3.4.1 Role of Microorganisms in Polluted Soil Remediation

Microorganisms play a crucial role in converting toxic elements into less harmful forms, such as water and carbon dioxide, through a process known as mineralization (Kumar et al. 2022). This bioremediation process involves the use of various microorganisms, including bacteria, fungi, and algae. Due to their diverse distribution, microbes possess exceptionally impressive metabolic activities and can grow in diverse environmental conditions. Furthermore, being ubiquitous in nature exhibits versatility in utilizing diverse substrates as carbon sources. This nutritional adaptability of microorganisms is a valuable asset in leveraging their capacity for the biodegradation of pollutants, a process known as bioremediation. This process capitalizes on the unique ability of specific microorganisms to convert, modify, and utilize toxic pollutants, thereby deriving energy and promoting biomass production in the course of the remediation process. Consequently, they are found in unconventional environments where they can absorb a wide array of pollutants. The adaptability of microorganisms to survive in challenging environments, such as acidic conditions for acidophiles, cold climates for psychrophiles, and saline regions for halophiles, enhances their efficiency in bioremediation processes (Ayilara and Babalola 2023).

3.4.2 Soil Pollution and Its Impact

Over recent decades, soil pollution has escalated, posing potential risks to both human and ecological health. This environmental issue primarily stems from human activities, leading to the accumulation of contaminants in soils that may reach concerning levels. Soil pollutants encompass a diverse array of substances, both organic and inorganic, originating from anthropogenic activities or occurring naturally. Monitoring soil quality can be challenging due to the absence of well-defined variables and indicators. The determination of soil contamination hinges on its suitability for the intended use; soil is considered contaminated when it fails to fulfil its expected functions for a specific land use due to the presence of pollutants. The mere detection of anthropogenic compounds or elevated levels above the background does not necessarily indicate pollution; rather, it is essential to assess the degree of contamination and determine whether the associated risks are acceptable or not (Cachada et al. 2018).

The widespread contamination of soil due to human activities, particularly the expansion of industries and intensified agriculture, poses a significant global challenge. This has led to the increased release of diverse xenobiotics into the environment, causing harm to humans, livestock, wildlife, crops, and native plants and disrupting ecological balance. Scientists worldwide are addressing this issue through various means, such as physical, chemical, and thermal processes like excavating and transporting contaminated soil. Unfortunately, these methods are often expensive, labour-intensive, and not always guaranteeing complete pollutant removal or destruction and can result in abrupt changes to the physical, chemical, and biological characteristics of treated soil (Das and Adholeya 2012).

Soil possesses the capability to absorb diverse pollutants, including heavy metals, pesticides, and polycyclic aromatic hydrocarbons, functioning as a natural pollution absorber. This absorption, however, contributes to the contamination of the food chain, posing potential threats to human health. Pesticides emerge as the most prevalent contaminants among all xenobiotics present in soil. The prevalence of pesticides is influenced by rapid population growth in the last five decades and an escalating demand for high-quality food (Riffaldi et al. 2006). In response to these challenges, agricultural producers often adopt practices involving extensive pesticide use, leading to their heightened accumulation in the surface layers of soil and prolonged persistence in the environment. As projected by Oberemok et al. (2015), the use of pesticides in agriculture is anticipated to be 2.7 times greater in 2050 than in 2000, raising concerns about an escalating risk to human health for future generations.

3.4.3 Polluted Soil Conditions and Strategies Utilizing Extremophile Adaptation

Researching the soil matrix presents challenges due to its intricate nature, encompassing various overlapping environmental parameters such as temperature, pH, granulometric composition, oxido-reductive potential, the presence of metal cations, and other ions, along with occasional organic substances. These factors collectively influence the activities of microorganisms and soil enzymes. In the soil environment, the effective degradation of contaminants hinges on the ability of organisms not only to metabolize and degrade them but also to survive in the contaminated surroundings. The degradation rate is influenced by several factors, including the physico-chemical properties of the residue, soil characteristics, climatic conditions (temperature, rainfall, and humidity), and the abundance of microorganisms capable of decomposing the specific contaminant (Wołejko et al. 2020).

Microorganisms play a pivotal role in the soil's buffering capacity. The high genetic adaptability of soil microbial populations enables them to swiftly adapt to alterations in the soil environment. Over time, these microorganisms have evolved and continue to develop various strategies to manage the presence of toxic substances in soils (Puglisi et al. 2012).

3.5 Bioremediation of Pollutants

In recent decades, bioremediation, which involves harnessing the catabolic capabilities of microorganisms, particularly bacteria, to break down various contaminants, has emerged as a promising approach for reducing contaminant levels and restoring impacted areas. Numerous studies on soils from various locations have been published, providing evidence of the effectiveness of diverse bioremediation strategies in mitigating soil contamination (Das and Adholeya 2012).

In a recent investigation conducted by Marques (2018), the study delved into the extensive potential of extremophiles to serve as micro-factories for environmental pollution clean-up. The research focused on elucidating the metabolic and genetic mechanisms offered by these microorganisms as controlled services. Another recent comprehensive review explored polyextremophilic microorganisms sourced from various environments such as deserts, salt flats, ice fields, geothermal springs, and diverse regions in Chile, encompassing the Atacama Desert, Altiplano, Central Chile, Patagonia, and Antarctica. They extensively examined the molecular and physiological capabilities of numerous isolates, underscoring their significant potential for application in bioremediation processes (Kaushik et al. 2021).

Researchers are exploring extremophilic microorganisms, to find new ways to break down recalcitrant pollutants. Unlocking the potential of these microorganisms could lead to the development of eco-friendly technologies for cleaning up contaminated areas, ensuring a healthier environment. Further study and use of extremophilic bacteria may bring about breakthroughs in our efforts to make the planet cleaner and safer.

A study showcased the potential of local acid-loving microalgae biofilms to clean up heavy metals in phosphorus-rich water from a nickel refinery. The Chlorella-like microalgae and their biofilms were able to remove up to 25% of the total metals in the water. Additionally, researchers explored the use of *Galdieria sulphuraria* for recovering metals, with a focus on rare earth metals. Surprisingly, *G. sulphuraria* efficiently recovered lanthanoid ions, even at very low concentrations of just 0.5 mgL^{-1}, achieving over 90% degradation (Minoda et al. 2015).

In polar regions, a diverse array of microorganisms exhibits the capability to break down hydrocarbons, showcasing significant diversity in their genetic make-up. Research indicates that products designed for bioaugmentation are more effective in improving the biodegradation of oil-contaminated waters and soils compared to approaches solely relying on nutrient treatments. However, the effectiveness of these bioaugmentation products is strongly influenced by the composition of the biostimulation consortium. In certain cases, it has been observed that introducing fertilizers, such as Inipol EAP-22, to the native strain proves to be a more potent method for enhancing biodegradation. This complexity underscores the need to customize bioremediation strategies to specific environmental conditions for optimal outcomes (Brakstad 2008).

With their remarkable survival skills, microbes from extreme environments suggest that their enzymes are well-suited to operate optimally under such challenging conditions. Recent findings on enzymes isolated from extremophiles support this idea, revealing distinct characteristics such as exceptional thermal stability, resilience against chemical denaturants like detergents, chaotropic agents, organic solvents, and the ability to endure extremes of pH.

Extremozymes, along with whole microbial cells, are enzymes that have evolved molecular mechanisms to thrive in extreme physicochemical conditions. They hold tremendous potential as industrial biocatalysts, capable of performing under severe operational conditions that would typically deactivate normal enzymes. These extremozymes not only find applications in industrial settings but also serve as a blueprint for designing and engineering proteins with unique properties, valuable across a wide range of industrial applications (Gupta et al. 2014).

3.6 Phytoremediation and Microbial Plant Interactions

Phytoremediation is a recently developed method that involves the use of specific plants, termed hyperaccumulators, to remove or neutralize hazardous toxicants from the environment for clean-up purposes. These hyperaccumulators thrive in soil with elevated concentrations of metals (Mishra et al. 2020). Phytoremediation is recognized as an effective, nonintrusive, in situ, visually appealing, cost-effective, and socially accepted technology for remediating polluted soil. This approach can address pollutants through various forms, including phytostabilization, rhizofiltration, phytoextraction, and phytovolatilization (Alkorta and Garbisu 2001).

Various plant species have proven successful in remediating harmful chemicals from polluted sites. To be effective phytoremediators, these plant species must exhibit specific properties such as rapid growth, substantial biomass, resilience, and tolerance to toxic chemicals. Phytoremediator plant species can accumulate and withstand significantly higher concentrations of toxic pollutants without displaying visible phytotoxic symptoms compared to other plants. The efficacy of different phytoremediator plant species in absorbing contaminants from the soil varies based on factors like biomass, transpiration rate, enzyme abundance, and the presence of associated microbes (Prakash et al. 2019).

Plant–microbe interactions not only contribute to remediating environmental pollution but also play a crucial role in fostering sustainable agricultural development. With the current challenge of reducing pesticide and chemical fertilizer usage in crop production, the environmentally friendly approach of utilizing plant-growth-promoting rhizobacteria (PGPR) has proven effective. PGPR facilitates increased crop yields through direct and indirect mechanisms of plant growth promotion.

These mechanisms involve nutrient and hormonal regulation, inducing resistance against phytopathogens (Mishra et al. 2020).

Utilizing plant-growth-promoting microorganisms in phytoremediation offers several advantages over chemical amendments. The microbial metabolites produced in the rhizosphere in situ are biodegradable and exhibit lower toxicity, enhancing the overall environmental sustainability of the remediation process (Prakash et al. 2019).

In the rhizosphere, a dynamic ecosystem around plant roots, root exudates, and microorganisms are integral components that influence the bioavailability of metals and nutrients. Root exudates serve as an abundant energy and nutrient source for microbes, fostering a reciprocal relationship where microbes, in turn, stimulate further exudation from plant roots. In the co-evolutionary process, plants and their associated microbes establish relationships that can be either beneficial or detrimental, significantly impacting both partners. Root exudates play a role in enhancing the mobility of metals and nutrients through processes such as acidification, formation of organic/amino acid–metal/mineral complexes, electron transfer via redox reactions, and stimulation of rhizosphere microbial activity. These interactions contribute to improving the efficiency of phytoremediation (Ma et al. 2016).

The symbiotic relationships between plants and microbes in the rhizosphere are increasingly recognized as crucial contributors to improving the efficiency of phytoremediation. The incorporation of PGPR has been shown to enhance the removal of organic pollutants, such as polycyclic aromatic hydrocarbons and creosote. This enhancement is likely attributed to the promotion of plant germination, stimulation of accelerated plant growth, and the accumulation of increased root biomass (Huang et al. 2004; Divya and Deepak Kumar 2011).

3.7 Conclusion

The study of microbial adaptation in extreme environmental conditions reveals the resilience and versatility of microorganisms. These adaptations, ranging from extreme temperatures and pH levels to high salinity and heavy metal contamination, contribute to the degradation and detoxification of pollutants, promoting sustainable and eco-friendly solutions. The insights gained from studying microbial adaptation can inform innovative biotechnological applications, enabling the development of tailored solutions for specific types of soil pollution. This knowledge can help address the challenges posed by polluted soils and unlock practical solutions for sustainable environmental management.

Acknowledgement Tatiana Minkina would like to acknowledge the financial support of the Ministry of Science and Higher Education of the Russian Federation (no. FENW-2023-0008).

References

Ahmad F, Ahmad I, Khan MS (2008) Screening of free-living rhizospheric bacteria for their multiple plant growth promoting activities. Microbiol Res 163:173–181

Alkorta I, Garbisu C (2001) Phytoremediation of organic contaminants in soils. Bioresour Technol 79:273–276

Arora NK, Panosyan H (2019) Extremophiles: applications and roles in environmental sustainability. Environ Sustain 2:217–218

Ayilara MS, Babalola OO (2023) Bioremediation of environmental wastes: the role of microorganisms. Front Agron 5:1183691

Babu P, Chandel AK, Singh OV (2015) Survival mechanisms of extremophiles. In: Babu P, Chandel AK, Singh OV (Eds.) Extremophiles and Their Applications in Medical Processes (pp. 9–23). Springer International Publishing. https://doi.org/10.1007/978-3-319-12808-5_2

Basak P, Biswas A, Bhattacharyya M (2020) Exploration of extremophiles genomes through gene study for hidden biotechnological and future potential. In: Physiological and biotechnological aspects of extremophiles. Academic Press, p 315

Brakstad OG (2008) Natural and stimulated biodegradation of petroleum in cold marine environments. In: Psychrophiles: from biodiversity to biotechnology. Springer, pp 389–407

Brock TD (2012) Thermophilic microorganisms and life at high temperatures. Springer Science & Business Media

Bull AT (2004) Microbial diversity and bioprospecting. ASM Press, Washington, D.C.

Cachada A, Rocha-Santos T, Duarte AC (2018) Soil and pollution: an introduction to the main issues. In: Soil pollution. Elsevier, pp 1–28

Das M, Adholeya A (2012) Role of microorganisms in remediation of contaminated soil. In: Satyanarayana T, Johri BN (Eds.) Microorganisms in Environmental Management: Microbes and Environment (pp. 81–111). Springer Netherlands. https://doi.org/10.1007/978-94-007-2229-3_4

Deming JW (2002) Psychrophiles and polar regions. Curr Opin Microbiol 5:301

Devarinti SR (2016) Natural farming: eco-friendly and sustainable. Agrotechnology 5:147

DeVeaux LC, Müller JA, Smith J et al (2007) Extremely radiation-resistant mutants of a halophilic archaeon with increased single-stranded DNA-binding protein (RPA) gene expression. Radiat Res 168:507–514

Divya B, Deepak Kumar M (2011) Plant–microbe interaction with enhanced bioremediation. Res J Biotechnol 6:4

Feller G, Gerday C (2003) Psychrophilic enzymes: hot topics in cold adaptation. Nat Rev Microbiol 1:200–208

Gupta GN, Srivastava S, Khare SK, Prakash V (2014) Extremophiles: an overview of microorganism from extreme environment. Int J Agric Environ Biotechnol 7(2):371–380

Huang X-D, El-Alawi Y, Penrose DM et al (2004) Responses of three grass species to creosote during phytoremediation. Environ Pollut 130:453–463

Islam F, Yasmeen T, Ali Q et al (2014) Influence of Pseudomonas aeruginosa as PGPR on oxidative stress tolerance in wheat under Zn stress. Ecotoxicol Environ Saf 104:285–293

Jheeta S (2020) Extremophiles and horizontal gene transfer: Clues to the emergence of life. In: Seckbach J, Stan-Lotter H (Eds.) Extremophiles as Astrobiological Models (1st ed., pp. 329–358). Wiley. https://doi.org/10.1002/9781119593096.ch16

Kato C, Li L, Nogi Y et al (1998) Extremely barophilic bacteria isolated from the Mariana Trench, challenger deep, at a depth of 11,000 meters. Appl Environ Microbiol 64:1510–1513

Kaushik S, Alatawi A, Djiwanti SR et al (2021) Potential of extremophiles for bioremediation. In: Microbial rejuvenation of polluted environment: volume 1. Springer, Singapore, pp 293–328

Kumar A, Verma JP (2019) The role of microbes to improve crop productivity and soil health. In: Ecological wisdom inspired restoration engineering. Springer, pp 249–265

Kumar G, Lal S, Soni SK et al (2022) Mechanism and kinetics of chlorpyrifos co-metabolism by using environment restoring microbes isolated from rhizosphere of horticultural crops under subtropics. Front Microbiol 13:891870

Lal R (2008) Soils and sustainable agriculture. A review. Agron Sustain Dev 28:57–64

Lebre PH, De Maayer P, Cowan DA (2017) Xerotolerant bacteria: surviving through a dry spell. Nat Rev Microbiol 15:285–296

Li S-J, Hua Z-S, Huang L-N et al (2014) Microbial communities evolve faster in extreme environments. Sci Rep 4:6205

Lichtfouse E, Navarrete M, Debaeke P et al (2009) Agronomy for sustainable agriculture: a review. Sustain Agric:1–7

Ma Y, Oliveira RS, Freitas H, Zhang C (2016) Biochemical and molecular mechanisms of plant-microbe-metal interactions: relevance for phytoremediation. Front Plant Sci 7:918

Makarova KS, Aravind L, Wolf YI et al (2001) Genome of the extremely radiation-resistant bacterium Deinococcus radiodurans viewed from the perspective of comparative genomics. Microbiol Mol Biol Rev 65:44–79

Marques CR (2018) Extremophilic microfactories: applications in metal and radionuclide bioremediation. Front Microbiol 9:1191

Martínez GM, Pire C, Martínez-Espinosa RM (2022) Hypersaline environments as natural sources of microbes with potential applications in biotechnology: the case of solar evaporation systems to produce salt in Alicante County (Spain). Curr Res Microb Sci 3:100136

Minoda A, Sawada H, Suzuki S et al (2015) Recovery of rare earth elements from the sulfothermophilic red alga Galdieria sulphuraria using aqueous acid. Appl Microbiol Biotechnol 99:1513–1519

Mishra A, Mishra SP, Arshi A, Agarwal A, Dwivedi SK (2020) Plant-microbe interactions for bioremediation and phytoremediation of environmental pollutants and agro-ecosystem development. In: Bharagava RN, Saxena G (Eds.) Bioremediation of Industrial Waste for Environmental Safety: Volume II: Biological Agents and Methods for Industrial Waste Management (pp. 415–436). Springer. https://doi.org/10.1007/978-981-13-3426-9_17

Oberemok VV, Laikova KV, Gninenko YI et al (2015) A short history of insecticides. J Plant Prot Res 55:221

Oren A (2002) Diversity of halophilic microorganisms: environments, phylogeny, physiology, and applications. J Ind Microbiol Biotechnol 28:56–63

Orellana R, Macaya C, Bravo G, Dorochesi F, Cumsille A, Valencia R, Rojas C, Seeger M (2018) Living at the frontiers of life: Extremophiles in Chile and their potential for bioremediation. Front Microbiol 9. https://doi.org/10.3389/fmicb.2018.02309

Pakchung AAH, Simpson PJL, Codd R (2006) Life on earth. Extremophiles continue to move the goal posts. Environ Chem 3:77–93

Pikuta EV, Hoover RB, Tang J (2007) Microbial extremophiles at the limits of life.? Crit Rev Microbiol 33:183–209

Prakash G, Soni R, Mishra R, Sharma S (2019) Role of plant-microbe interaction in phytoremediation. In: In vitro plant breeding towards novel agronomic traits: biotic and abiotic stress tolerance. Springer, pp 83–118

Puglisi E, Hamon R, Vasileiadis S et al (2012) Adaptation of soil microorganisms to trace element contamination: a review of mechanisms, methodologies, and consequences for risk assessment and remediation. Crit Rev Environ Sci Technol 42:2435–2470

Rainey FA, Oren A (2006) 1 extremophile microorganisms and the methods to handle them. In: Methods in microbiology. Elsevier, pp 1–25

Rampelotto PH (2013) Extremophiles and Extreme Environments. J Life 3:482–485. https://doi.org/10.3390/3030482

Riffaldi R, Levi-Minzi R, Cardelli R et al (2006) Soil biological activities in monitoring the bioremediation of diesel oil-contaminated soil. Water Air Soil Pollut 170:3–15

Russell NJ, Evans RI, Ter Steeg PF et al (1995) Membranes as a target for stress adaptation. Int J Food Microbiol 28:255–261

Shukla AK, Singh AK (2020) Exploitation of potential extremophiles for bioremediation of xeno-biotics compounds: a biotechnological approach. Curr Genomics 21:161–167

Soucy SM, Huang J, Gogarten JP (2015) Horizontal gene transfer: building the web of life. Nat Rev Genet 16:472–482

Tahat MM, Alananbeh KM, Othman YA, Leskovar DI (2020) Soil health and sustainable agriculture. Sustainability 12:4859

Tyson GW, Chapman J, Hugenholtz P et al (2004) Community structure and metabolism through reconstruction of microbial genomes from the environment. Nature 428:37–43

Wołejko E, Jabłońska-Trypuć A, Wydro U et al (2020) Soil biological activity as an indicator of soil pollution with pesticides–a review. Appl Soil Ecol 147:103356

Yan N, Marschner P, Cao W et al (2015) Influence of salinity and water content on soil microorganisms. Int Soil Water Conserv Res 3:316–323

Zeldovich KB, Berezovsky IN, Shakhnovich EI (2007) Protein and DNA sequence determinants of thermophilic adaptation? PLoS Comput Biol 3:1

Zhou J, He Q, Hemme CL et al (2011) How sulphate-reducing microorganisms cope with stress: lessons from systems biology. Nat Rev Microbiol 9:452–466

Chapter 4
Extremophiles Adaptation and Its Utilization in Mitigating Abiotic Stress in Crops

Adesh Kumar, Monika Shrivastava, and Pallavi Saxena

4.1 Introduction

The world population, currently estimated at approximately 7.8 billion people, is projected to reach 9.7 billion by 2050. This rapid population growth has resulted in an increased demand for food products, placing immense pressure on global food production systems (Mesa-Marín et al. 2020). However, the productivity of crop production per unit of cultivated land has been struggling to keep up with the predicted food demand due to various factors (Elferink and Schierhorn 2016). Climate change, loss of soil structure, nutrient degradation, drought, and soil salinity have all contributed to decreased crop yields (Khan et al. 2019; Mukhtar et al. 2019). Additionally, according to the Food and Agricultural Organization (FAO), there is a projected loss of 50% of total land mass worldwide by the year 2050 (FAO 2008). This alarming trend highlights the urgent need for sustainable agricultural practices and innovative solutions to ensure food security for the growing global population.

Microbial communities exhibit remarkable adaptability, allowing them to thrive in a diverse range of environments, from the ordinary to the most extreme conditions characterized by factors such as temperature, salinity, water scarcity, and pH (Ali et al. 2014). In order to survive and persist in such harsh surroundings, these stress-adaptive microbes have evolved various mechanisms and properties. They possess the ability to produce bioactive compounds and secondary metabolites, which aid in their survival and proliferation in the face of adversity

A. Kumar · M. Shrivastava
Jawaharlal Nehru University, Aruna Asaf Ali Marg, New Delhi, India

P. Saxena (✉)
Academy of Biology and Biotechnology, Southern Federal University,
Rostov-on-Don, Russia

Adjunct Faculty, Centre for Research Impact & Outcome, Chitkara University, Punjab, India

© The Author(s), under exclusive license to Springer Nature Switzerland AG 2024
A. Ranjan et al. (eds.), *Extremophiles for Sustainable Agriculture and Soil Health Improvement*, https://doi.org/10.1007/978-3-031-70203-7_4

(Toral et al. 2021). Extremophiles, a subset of microorganisms, have adapted to inhabit some of the most hostile regions on Earth. They can tolerate high salinity levels (2–5 M NaCl) as observed in halophiles, as well as extreme pH values, such as acidophiles (pH < 4) and alkaliphiles (pH > 9). Furthermore, extremophiles are capable of thriving in a wide temperature range, from freezing cold conditions (−20 °C to 20 °C) inhabited by psychrophiles and psychrotrophic organisms to scorching hot temperatures (60 °C to 115 °C) favored by thermophiles and hyper-thermophiles (Saxena et al. 2016; Sahay et al. 2017). While archaea represent the true extremophiles, bacteria and eukaryotes also possess extremophilic members within their domains. The extremophilic microbiomes encompass various phyla, including *Euryarchaeota*, *Crenarchaeota*, *Firmicutes*, *Proteobacteria*, *Actinobacteria*, *Deinococcus-Thermus*, *Bacteroidetes*, *Basidiomycota*, and *Ascomycota* (Yadav et al. 2015a).

Microbial communities inhabiting extreme and harsh habitats have gained significant attention due to their potential applications in various fields, including white and green biotechnology, medicine, and food production and processing industries (Yadav et al. 2019). The polyextremophilic microbiomes possess the unique ability to thrive and proliferate under the simultaneous presence of multiple stressors and hostile environmental factors. This makes them intriguing subjects of study for their potential contributions to the development of novel biotechnological solutions and the understanding of microbial adaptation strategies in extreme habitats. Extremophilic microbiomes have emerged as crucial players in the circulation of plant nutrients, offering a sustainable solution to enhance agricultural productivity while reducing reliance on chemical fertilizers (Kour et al. 2020). These bioinoculants, when used in agroecological ecosystems, form beneficial associations with plants, thereby improving yields and nutrient status (Verma et al. 2019). To overcome the limitations caused by different abiotic stresses, it is essential to understand that stress in nature is a multifaceted phenomenon influenced by multiple interacting factors rather than isolated incidents (Mahajan and Tuteja 2005). By harnessing the homeostasis maintenance potential of extremophilic microbiomes and addressing both abiotic and biotic stresses, researchers and farmers can strive toward sustainable agricultural practices that promote increased productivity and environmental resilience. The advent of cutting-edge technologies like metagenomics and next-generation sequencing has revolutionized our understanding of microorganisms inhabiting extreme environments, offering valuable insights into their intricate functions and mechanisms of homeostasis (Canganella and Wiegel 2011). These advancements have empowered researchers to delve deeper into the remarkable abilities of microorganisms to thrive and adapt in harsh surroundings, shedding light on their resilient nature and the mechanisms that enable their survival.

4.2 Effects of Abiotic Stress on Crops

Environmental stresses, including abiotic factors such as adverse environmental conditions (Fig. 4.1), pose significant limitations to agricultural production (Boyer 1982). These stresses can lead to substantial crop yield losses, ranging from 50% to 70%. As the agricultural sector looks ahead, climate change emerges as a critical challenge. Rising temperatures, in particular, are expected to decrease water quantity and quality, thereby impacting both human livelihoods and agricultural yields. Freshwater, a vital resource for ecosystems and humanity, faces increasing pressure, with agriculture being the largest water use sector globally, accounting for approximately 70% of water withdrawal (FAO 2008). To address the implications of climate change, sustainable water resource management becomes imperative. However, certain agricultural regions may experience excessive rainfall, causing flooding and adversely affecting crop development and production. In such areas, it

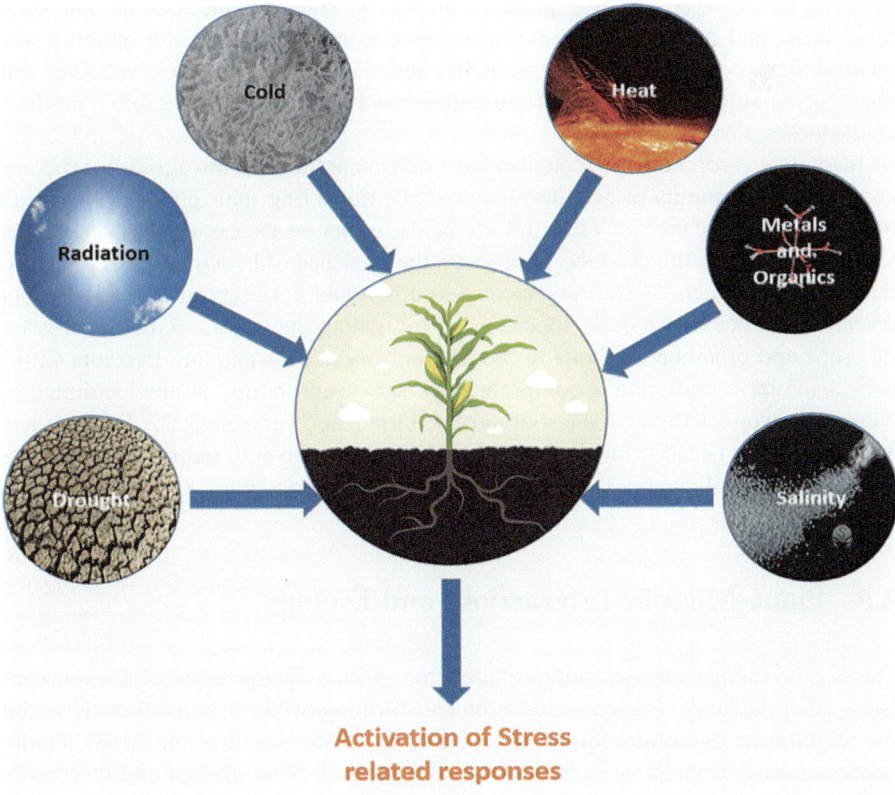

Fig. 4.1 Different abiotic stresses experienced by the crop plants under environmental conditions. These extreme conditions include drought, irradiation, extremely low and high temperatures, heavy metals, and salinity which pose a huge loss in crop development and yields

becomes crucial to carefully select appropriate crops and implement effective agronomic management strategies to regulate soil water content (FAO 2008). By considering these factors, the agricultural sector can strive toward mitigating the impacts of environmental stresses and ensuring more resilient and productive systems. Low temperatures can pose a significant challenge for several macro-thermal species cultivated during the spring, as well as for fruit trees that undergo their blooming phase early in the season, such as peaches or almonds. These species are vulnerable to cold stress, which can result in chilling injury and a detrimental impact on flowers or leaves, ultimately leading to substantial reductions in crop yield.

The level of irradiance, whether high or low, plays a significant role in influencing both the biomass and quality of crops (Fig. 4.1). This is because the processes of photosynthesis and respiration, which are crucial for plant growth and development, are intricately linked to factors such as temperature, light intensity, and light quality. When irradiance is high, plants receive ample energy from sunlight, leading to increased photosynthetic activity and ultimately higher biomass production. Additionally, high irradiance levels promote the synthesis of essential compounds, resulting in improved produce quality. Conversely, low irradiance can hamper photosynthesis and metabolic processes, negatively impacting both the quantity and quality of crop yields. Thus, understanding and managing irradiance levels is essential for optimizing crop productivity and ensuring desirable produce characteristics.

Inadequate crop nutrition can also have detrimental effects on agricultural crops and ornamental plants in Mediterranean areas, impacting their growth, yield, and quality. The imbalance of essential elements, either in excess or deficiency, can significantly affect the produce. Moreover, the presence of heavy metals or xenobiotic compounds acts as limiting factors in urban and agricultural regions. Recent research has focused on understanding and mitigating the impact of abiotic stresses on crops and ornamental plants in the Mediterranean. Mariani and Ferrante (2017 and 2018) have compiled a comprehensive set of agronomic strategies aimed at enhancing crop tolerance in the short term (Mariani and Ferrante 2017; Ferrante and Mariani 2018). These strategies provide valuable insights into improving crop productivity and resilience in the face of challenging environmental conditions.

4.3   Plant–Microbe Interactions and Ecology

Plants exist in intricate symbiotic relationships with a diverse array of microorganisms, such as fungi, bacteria, and archaea, forming what is known as the plant microbiome or phytomicrobiome (Knack et al. 2015; Smith et al. 2017). Plant–microbe associations play a crucial role in promoting plant growth and increasing productivity while providing tolerance to various abiotic stresses. These associations occur in different forms, such as rhizospheric, epiphytic, and endophytic, and are found in extreme environments characterized by high salinity, drought, acidity, alkalinity, and extreme temperatures (Mendes et al. 2013). Extremophilic microorganisms within these associations contribute directly to plant growth by fixing atmospheric nitrogen; solubilizing essential minerals like zinc (Zn), phosphorus (P), and potassium (K); and synthesizing plant hormones such as auxins, gibberellins, and cytokinins (Kour et al. 2020). Additionally, they produce enzymes that facilitate crucial biochemical reactions. Indirectly, extremophiles contribute to plant growth by producing compounds like ammonia (NH_4) and hydrogen cyanide (HCN), as well as various secondary metabolites and hydrolytic enzymes (Robert 2022). These substances act as growth-promoting agents and can also exhibit pesticidal properties against diverse plant pathogens (Elnahal et al. 2022). Overall, the intricate interactions between plants and extremophiles offer numerous benefits for plant growth, productivity, and stress tolerance, making them a vital component of sustainable agriculture. These microorganisms colonize various plant tissues, including the roots, stems, leaves, flowers, and fruits, and their interactions with plants have evolved over time (Baltrus 2017). The ability of plants and microorganisms to communicate before physical contact is established is crucial, as it allows the partners to maximize the benefits derived from each other without causing harm. Different plant parts harbor distinct microbial communities, such as the rhizomicrobiome in roots, caulomicrobiome in stems, and phyllomicrobiome in leaves, while microorganisms residing on the external surfaces of plants are termed epiphytes (Chagas et al. 2018). The rhizosphere, the region surrounding the roots, exhibits higher microbial abundance, activity, and diversity compared to the phyllosphere, mainly due to the nutrient-rich compounds released by the roots (Meharg and Killham 1990; Beneduzi et al. 2012). Recent studies have identified beneficial microorganisms that assist plants, including crops, in coping with various stressors such as nutrient imbalances, salinity, and drought, highlighting the potential for more sustainable agriculture based on the understanding of specific stresses and the role microorganisms can play in their management (Mylona et al. 1995; Yazdani et al. 2009; Subramanian et al. 2016). However, there remains much to learn about the behavior of microorganisms at different pH levels, as soil acidity and alkalinity have direct implications for microbial population dynamics in soil and plant ecosystems (Biswas et al. 2007; Zhalnina et al. 2015).

4.4 Extremophiles Adaptation Strategies and Its Utilization for Abiotic Stress Alleviation

4.4.1 pH Value

Extremophiles, remarkable organisms capable of thriving in extreme environments, exhibit remarkable resistance to pH values. As highlighted by Dhakar and Pandey (2016), these microorganisms can withstand pH levels ranging from as low as 0.1 to as high as 12. Such pH tolerance enables them to inhabit diverse habitats, from sulfidic mines and acidic stomachs to alkaline soda lakes (Dhakar and Pandey

2016). Extremophiles are commonly categorized into acidophiles, which thrive in highly acidic conditions, and alkaliphiles, which flourish in highly alkaline environments (Krulwich et al. 2011). These findings shed light on the extraordinary adaptability of extremophiles and their ability to survive and thrive in seemingly inhospitable pH conditions.

Acidophiles Acidophiles are organisms that thrive in environments with a pH value of less than 3. These remarkable organisms possess the ability to maintain their internal pH within the neutral range, regardless of the acidity of their external environment (Dhakar and Pandey 2016). A diverse group of archaea and bacteria, including *Euryarchaeota, Ferroplasma, Acidobacter, Acidohalobacteria, Leptospirillum, Sulfobacillus, Acidibacillus, Desulfurococcus, Metallosphaera, Pyrococcus, Acidianus, Sulfolobus,* and *Picrophilus*, have adapted to thrive in acidic conditions (Salwan and Sharma 2020; Tripathi et al. 2021). These acidophiles can be found in various natural environments such as marine volcanic vents, acidic sulfur springs, solfataric fields, sulfuric pools, geysers, and areas with acid mine drainage or acid rock drainage. Additionally, they can be found in artificial environments associated with human activities like metal ore and coal mining (Salwan and Sharma 2020). To ensure their survival, acidophiles have developed numerous mechanisms that enable them to tolerate and thrive in acidic conditions. It is worth noting that prokaryotic and eukaryotic cells exhibit different cellular structures, which contribute to the wide range of acid-tolerant mechanisms observed in these organisms (Guan and Liu 2020).

Acidophiles employ various strategies to maintain intracellular pH homeostasis and combat the influx of protons. One such adaptation is the impenetrability of their cell membranes, which limits the entry of protons into the cytoplasm. This membrane characteristic, coupled with the presence of tetraether lipids in the cell membrane, contributes to acid pH tolerance in archaea (Mirete et al. 2017). Acidophiles also utilize techniques such as reducing pore size in membrane channels, generating a positive inside transmembrane potential, and increasing the expression of secondary transporters to minimize proton permeability and prevent proton leakage into the cells (Baker-Austin and Dopson 2007a; Chen 2021). Additionally, acidophiles possess buffering capacity through the presence of basic amino acids like lysine, histidine, and arginine, which aid in proton sequestration within the cytoplasm (Baker-Austin and Dopson 2007b). Metal tolerance in acidophiles is achieved through both passive and active mechanisms. Metal efflux proteins transport metals out of the cytoplasm, and active processes convert the metal into less harmful forms. Furthermore, the formation of metal sulfate complexes and the establishment of an internal positive membrane potential create chemiosmotic gradients that counteract the flow of metal cations in passive processes. Active efflux and metal ion trapping by metal chaperones are additional mechanisms employed by acidophiles to resist high levels of metals (Navarro et al. 2013; Dopson and Holmes 2014). These extraordinary adaptations and defense mechanisms enable acidophiles to thrive in extreme acidic environments while maintaining cellular homeostasis.

Alkaliphiles Alkaliphilic microorganisms, also known as "alkaliphiles," exhibit remarkable adaptability to thrive in environments with pH levels higher than 9, typically ranging from 10 to 13 (Preiss et al. 2015). These microorganisms can be categorized into different groups based on their pH and salinity requirements. Obligate alkaliphiles exclusively grow at pH values around 9 and above, while facultative alkaliphiles can thrive in extreme alkaline conditions but also tolerate near-neutral pH levels. Haloalkaliphiles, on the other hand, not only require an alkaline pH but also high salinity levels, often up to 33% NaCl (Preiss et al. 2015). However, it is crucial to recognize the distinct response of alkaliphiles to high pH. Unlike their acidic counterparts, alkaliphiles maintain a cytoplasmic pH close to neutrality (Slonczewski et al. 2009), which presents a unique challenge of countering an imbalance of H+ ions (Mamo 2020). While acidophiles thrive in an abundance of H+ ions, alkaliphiles face a comparatively arid environment in terms of H+ concentration. The ability of alkaliphiles to withstand and adapt to such alkaline conditions highlights their incredible resilience and specialized mechanisms (Pikuta et al. 2007; Slonczewski et al. 2009; Mamo 2020).

Various alkaliphilic microorganisms have been isolated from diverse environments such as soda lakes, hydrothermal vents, and carbonate-rich soil. Some notable examples include *Bacillus pseudofirmus* and *Bacillus halodurans*, *Alkalibacterium* strains, *Alkaliphilus metalliredigens*, *Thioalkalivibrio*, and *Pseudomonas alcaliphila* (Preiss et al. 2015). These microorganisms have evolved specialized mechanisms to adapt and thrive in these extreme conditions, highlighting their remarkable ability to survive and function in environments that are hostile to most other life forms.

Alkaliphiles and halophiles are often mentioned together when discussing adaptations, primarily due to their common occurrence in saline environments (Pikuta et al. 2007). Alkaliphiles possess astounding mechanisms to maintain intracellular pH homeostasis, distinct from their external surroundings. To counter the alkaline conditions, alkaliphiles actively maintain a lower cytoplasmic pH, requiring metabolic energy expenditure (Ali et al. 2023). They achieve this by employing various strategies, including the uptake of solutes from the environment and regulating intracellular pH through a Na^+/H^+ antiporter in conjunction with H^+-coupled respiration (Van De Vossenberg et al. 2000). The presence of the Mrp antiporter, a cation/proton antiporter predominantly found in alkaliphiles, further aids in their pH regulation (Takjl et al. 2010). Alkaliphiles possess cell walls rich in negatively charged residues, particularly glutamic and glucuronic acids, which interact with cations like H^+ and delay their rapid loss from the cell surface, thus contributing significantly to pH homeostasis and bioenergetics (Mamo 2020). In certain alkaliphilic *Bacillus* species, ATP synthesis is driven by Na^+ or K^+ antiporters that facilitate the exchange of outwardly moving ions (Na^+ or K^+) with an increased influx of H+. Alkaliphiles commonly employ antiporters, such as Na^+/H^+ and K^+/H^+, to regulate internal pH and produce acids when metabolic conditions necessitate pH reduction (Krulwich et al. 2011). These organisms also exhibit adaptations in their proteins, with alkaliphilic proteins having a lower isoelectric point and a reduced number of basic amino acids but an increased number of acidic amino acids, such as glutamate and aspartate. Many alkaliphiles are also thermophilic or halophilic, suggesting that their survival strategies extend beyond pH homeostasis (Krulwich 2006). The intricate mechanisms employed by alkaliphiles enable them to thrive in extreme alkaline environments while maintaining essential biological processes (Matsuno and Yumoto 2015).

4.4.2 Temperature

Microorganisms inhabit a wide range of temperatures on Earth, spanning from −98 °C to 495 °C (McDermott et al. 2018; Scambos et al. 2018). Such extreme thermal conditions have led to the classification of microorganisms into three distinct categories based on their temperature preferences and survival abilities. Psychrophiles, thriving optimally at temperatures below 15–20 °C, exhibit exceptional adaptations to survive in cold environments (Turchetti et al. 2020). Mesophiles, on the other hand, are microorganisms that flourish within a temperature range of 20–45 °C, representing the majority of microbial life found in moderate environments. Finally, thermophiles, capable of growing optimally at temperatures above 45 °C, have adapted to thrive in geothermal areas and hydrothermal vents (Kushkevych et al. 2020). These extremophiles, encompassing both psychrophiles and thermophiles, push the boundaries of life's resilience and offer valuable insights into the adaptability and diversity of microbial organisms in extreme habitats.

Alleviation of Low-Temperature Stress

Psychrophiles, organisms that thrive in cold environments, have been identified across various phyla, including *Actinobacteria, Ascomycota, Basidiomycota, Chlamydiae, Chloroflexi, Cyanobacteria, Firmicutes, Euryarchaeota, Gemmatimonadetes, Nitrospirae, Mucoromycota, Planctomycetes, Proteobacteria, Thaumarchaeota, Verrucomicrobia, Flavobacterium, Shewanella, Clostridium*, and others (Sinclair and Stokes 1964; Yadav and Saxena 2018; Chaudhary et al. 2019). The term "psychrophiles" is derived from the Greek words "psukhros" meaning cold and "philein" meaning love (Sinclair and Stokes 1964). These cold-loving organisms have adapted to extreme cold conditions and continue to fascinate scientists studying their unique adaptations and survival strategies in frigid habitats.

Psychrophilic microorganisms have evolved various strategies to adapt to cold temperatures. One of these strategies involves modifying the composition of the cell membrane through the upregulation of genes responsible for long-chain fatty acid production (D'Amico et al. 2001). These fatty acids, characterized by their highly branched and long-chained structure with cis-isomerization, prevent the formation of ice crystals in the cell membrane (He et al. 2015; Goordial et al. 2016). Additionally, the presence of long-chain polyunsaturated fatty acids (LC-PUFAs) such as eicosapentaenoic acid (EPA), docosahexaenoic acid (DHA), and arachidonic acid (ARA) helps maintain membrane fluidity and act as antioxidants against reactive oxygen species (ROS) like H_2O_2 (Feng et al. 2014; Yoshida et al. 2016). Furthermore, membrane pigments such as carotenoids play a crucial role in stabilizing the membrane and protecting the organisms from freeze–thaw cycles (Jagannadham et al. 2000; Dieser et al. 2010). Moreover, psychrophilic organisms employ hydrophobic encrustations, modification of lipopolysaccharide (LPS) structure, and the production of compatible solutes as additional adaptive mechanisms (Mykytczuk et al. 2013; Ghobakhlou et al. 2015; Corsaro et al. 2017). They also produce antifreeze proteins (AFPs) to maintain fluidity within cells by lowering the freezing temperature and inhibiting ice crystal growth (Kim et al. 2017). In contrast, ice nucleating proteins (INPs) are released extracellularly, facilitating ice formation and nutrient acquisition (Tendulkar et al. 2021). The secretion of exopolysaccharides (EPSs) acts as a barrier against ice formation and provides a habitable environment (Garcia-Lopez et al. 2021). Additionally, psychrophilic organisms exhibit enhanced genomic and protein stability through the upregulation of DNA/RNA chaperones and protein chaperones. Lastly, these microorganisms have the ability to accumulate or degrade compounds such as polyhydroxyalkanoate (PHA) and cyanophycin, serving as carbon and nitrogen reservoirs (Tribelli and Lopez 2018). Overall, psychrophilic microorganisms employ a multitude of adaptations to thrive in cold environments. In addition to the strategies, numerous other minor yet crucial adaptations enable microorganisms to thrive in harsh cold environments.

Decreasing temperatures, possibly linked to ongoing climate change, pose a significant peril to worldwide agriculture, leading to diminished crop yields and substantial economic consequences. In the face of these chilling conditions, a study by Fernandez et al. in 2012 demonstrated that *Burkholderia* sp. adapted to low temperatures by altering their carbohydrate metabolism. This adaptation, in turn, enhanced the yield of *Vitis vinifera*. These findings underscore the importance of understanding the mechanisms behind temperature-related stress in plants and the potential benefits of such adaptations (Fernandez et al. 2012). The novel bacterium *Rahnella sikkimica* sp. nov. demonstrated a noteworthy upregulation of cold-stress genes when subjected to frigid environmental conditions (Kumar et al. 2022). A comprehensive genome analysis of this microorganism unveiled the existence of specific plant-growth-promoting factors hinting at its potential significance in enhancing crop yields in the challenging climates of cold hilly regions.

Alleviation of High-Temperature Stress

Thermophiles are organisms that have a strong affinity for heat, as indicated by the Greek roots of the term "thermotita" meaning heat and "philia" meaning love. These extraordinary organisms encompass a diverse range of archaea and bacteria genres, including *Moorella, Gelria, Pseudomonas, Geobacillus, Bacillus, Thermococcus, Thermus, Mycobacterium, Thermotoga, Gallionella, Crenothrix, Sphaerotilus, Leptothrix, Lieskeella, Pyrococcus, Sulfolobus, Metallosphaera,*

Caldicellulosiruptor, and *Thermoanaerobacter*, among others (Zeldes et al. 2015; Elumalai et al. 2019). They thrive in extreme environments characterized by high temperatures, such as hot springs and hydrothermal vents, showcasing their remarkable ability to withstand and even thrive in conditions that are inhospitable to most other life forms.

Thermophilic organisms, similar to psychrophiles, have evolved various strategies to adapt to high temperatures. One crucial adaptation is the modification of their cell membranes to withstand extreme heat. In thermophilic bacteria, the length of lipid acyl chains is increased, along with an altered ratio of iso/anteiso branching and degree of saturation (Deosthali and Sajwani 2022). Thermophilic archaea, on the other hand, utilize lipid monolayers composed of glycerol dialkyl glycerol tetraethers (GDGTs), which lower membrane permeability (Sahonero-Canavesi et al. 2021). Moreover, certain thermophilic strains exhibit increased C40 isoprenoid cyclization, resulting in tightly packed lipids and reduced membrane fluidity (De Rosa and Gambacorta 1988). Additionally, thermostable proteins and enzymes play a crucial role in enabling these organisms to cope with high temperatures. The structure of thermostable proteins is characterized by a decrease in thermolabile amino acids and an increase in charged residues, such as lysine, arginine, and glutamic acid, which contribute to protein stability (Tekaia et al. 2002; Friedman et al. 2004). Furthermore, the presence of intra-helical salt bridges also enhances protein stability at elevated temperatures (Kumar and Nussinov 2001). Chaperones, known as heat shock proteins (HSPs), aid in protein folding, refolding, and preventing misfolding under thermal stress (Boopathy et al. 2021). Various compatible solutes, including cyclic 2,3-diphosphoglycerate (cDPG) and diglycerol phosphate (DGP), are produced by thermophiles to counteract the effects of high temperature (Lamosa et al. 2000; Lentzen and Schwarz 2006). Moreover, extremolytes like ectoin, proline, mannitol, and others have been identified as contributors to thermophilic adaptation (Becker and Wittmann 2020). The presence of high G + C content in structural RNAs and polyadenine tracts in mRNA has been linked to enhanced thermostability (Paz et al. 2004). Furthermore, the discovery of overlapping genes has shed light on a novel strategy employed by thermophiles to overcome thermal stress (Saha et al. 2015). Although the role of G + C content in the thermostability of DNA remains debatable, these various adaptations collectively enable thermophilic organisms to thrive in extreme heat environments.

The increasing frequency of extreme temperatures due to climate change poses a significant threat to global agriculture, leading to potential reductions in crop production and substantial economic consequences. However, there is hope in the form of beneficial microorganism inoculation, which has demonstrated very good efficiency in bolstering plant growth and alleviating the adverse effects of extreme temperature stress. For instance, Ali et al. (2011) found that *Pseudomonas putida* had a positive impact on *Triticum* sp. under high-temperature conditions, while Mukhtar et al. (2020) reported similar benefits of *Bacillus cereus* on *Solanum lycopersicum* under elevated temperatures. This promising research highlights the potential for microorganism-based solutions to mitigate the challenges imposed by climate-induced temperature fluctuations. An ecological niche has been identified for thermophiles, primarily found within a select group of genera, namely, *Bacillus, Geobacillus, Parageobacillus, Ureibacillus,* and *Brevibacillus* (Santana et al. 2013). These thermophilic soil bacteria (STB) play a crucial role in dissimilative organic sulfur oxidation, leading to the mineralization of organically bound sulfur. Through this metabolic process, sulfate and ammonium are generated from the consumption of proteinaceous substances. These resulting compounds can then be utilized by plants or other microorganisms (Santana et al. 2020).

4.4.3 Alleviation of Drought Stress

Life relies on liquid water for vital functions and survival. Desiccation, the removal of water from living organisms, is crucial for adaptability. Complete desiccation is when water content falls below 0.1 grams per gram of dry mass (Leprince and Buitink 2010). Water activity (aw) measures water vapor pressure in a material. These concepts help us understand the requirements and limits of life in various environments.

Desiccation tolerance, the ability to withstand extreme dry conditions, is a phenomenal trait observed across various living organisms, ranging from bacteria and fungi to plants. These organisms that exhibit the ability to survive in desiccated environments are known as xerophiles. Several studies have identified specific species as xerophiles, such as *Bradyrhizobium japonicum, Rhodococcus jostii RHA1, Mycobacterium, Chloroflexus aurantiacus, Staphylococcus aureus, Methanothermobacter thermautotrophicus, Sulfolobus metallicus, Thermoproteus tenax, Hydrogenothermus marinus, Aquifex aeolicus,* and *Archaeoglobus fulgidus* (Cytryn et al. 2007; LeBlanc et al. 2008; Ding et al. 2016). These findings highlight the diverse range of organisms capable of thriving in desiccated conditions and provide valuable insights into the mechanisms underlying desiccation tolerance in nature.

Xerophiles, organisms adapted to survive in arid environments, employ various strategies to withstand desiccation. One such strategy is the formation of biofilms and the increased secretion of extracellular polymeric substances (EPS) and extracellular DNA (eDNA) (Tenore et al. 2023). This phenomenon has been observed in organisms like *Klebsiella* spp., *Salmonella* spp., and certain mucoid strains of *Escherichia coli,* which produce a capsular polysaccharide known as colanic acid (Esbelin et al. 2018). The production of EPSs provides resistance against desiccation, significantly enhancing survival rates. For instance, mucoid variants of *E. coli* showed survival rates ranging from 0.7 to 5%, while EPS-producing strains exhibited survival rates as high as 35%. Another mechanism employed by xenophiles involves the accumulation of compatible solutes or osmolytes, such as trehalose, sucrose, and betaine, which contribute to the vitrification of the cytoplasmic matrix. Cyanobacteria called *Chroococcidiopsis* upregulate genes responsible for the biosynthesis of trehalose and sucrose during desiccation, enhancing their survival (Fagliarone et al. 2020). Additionally, the accumulation of Mn (II) and the formation of HSPs play crucial roles in desiccation resistance. *Deinococcus radiodurans* accumulates Mn(II) to protect proteins from oxidative stress induced by desiccation (Jin et al. 2019), while HSPs compensate for water loss by forming hydrogen bonds with other molecules. The upregulation of HSPs has been observed in cyanobacterium *Anabaena* sp. 7120 and *D. radiodurans*. Furthermore, protein aggregation has been found to confer resistance to desiccation in certain organisms, such as *Acinetobacter baumannii* and yeast, although the exact mechanisms remain unknown (Saad et al. 2017; Wang et al. 2020). These survival strategies employed by xenophiles enable them to thrive in environments with limited water availability, showcasing their exceptional adaptability and resilience.

Precipitation plays a pivotal role as the primary water source for crop cultivation worldwide (Enebe and Babalola 2019). However, fluctuations in water availability, be it due to drought or excessive rainfall, can induce abiotic stress conditions that severely hinder crop yields (Ipek et al. 2017; Danish et al. 2020). One promising approach to enhance plant resilience to drought stress is the utilization of plant-growth-promoting microorganisms (PGPM), as demonstrated by studies by Fleming et al. (2019). For instance, research on *Mentha pulegium* L. reveals the positive impact of *Azotobacter chroococcum* and *Azospirillum brasilense* in enhancing drought tolerance (Asghari et al. 2020). Likewise, in the case of *Cymbopogon citratus* and *Zea mays*, the application of *Pseudomonas* sp. and *Azotobacter* sp. has been found to improve water stress resilience (Danish et al. 2020; Mirzaei et al. 2020).

4.4.4 Alleviation of Salinity Stress

Salinity is a crucial factor in the study of seawater, representing the concentration of salts dissolved in it. It is commonly measured in parts per thousand (ppt) or as the percentage of salt in grams per kilogram of seawater. According to Loganathan et al. (2017), the average salinity of seawater is approximately 35 ppt or 3.5%. Organisms capable of thriving in high salt concentrations are known as halophiles, and they can be categorized based on their tolerance to different salt levels. These classifications include slight, moderate, borderline extreme, and extreme halophiles, as described by Gunjal and Badodekar (2022). Halophiles can be found in brackish or hypersaline environments and encompass a range of genera such as *Bacillus*, *Brevibacterium*, *Dunaliella* (Algae), *Dactylococcopsis* (Cyanobacteria), *Halobacillus*, *Halococcus*, *Haloferax*, *Halogeometricum*, *Halomonas*, *Haloterrigena*, *Marinococcus*, *Natrialba*, *Natrinema*, *Salinibacter*, *Salinicoccus*, *Salinivibrio*, and *Virgibacillus*, among others (Menasria et al. 2018; Villanova et al. 2021; Ruginescu et al. 2020). The study of halophiles and their ability to adapt to extreme salinity levels provides valuable insights into the diverse range of life forms that can survive and reproduce in such challenging environments.

Halophilic organisms, known for their adaptation to high salt concentrations, rely on cellular membranes and specific mechanisms to maintain osmotic balance (Amoozegar et al. 2012). These organisms possess membrane proteins with a higher proportion of acidic and polar amino acids compared to non-halophiles, which helps them cope with osmotic pressure (Gunjal and Badodekar 2022). Unlike non-halophiles, halophiles lack the tetraethers that typically lock the two halves of the lipid bilayer together, allowing for membrane fluidity in high-salinity environments. The lipid composition in halophiles is primarily phosphatidylglycerol phosphate, with some instances of phosphate substitution by sulfate in certain halobacterium species (Kanekar and Kanekar 2022). This adaptation results in a higher concentration of negative charge, which aids in dealing with the extracellular high Na^+ concentrations alongside internal K^+ concentration (Russell 1989). Additionally, studies have shown that cold-adapted proteins in halophilic archaea from Antarctica exhibit amino acid substitutions, particularly an increase in acidic and polar amino acids, and the replacement of larger amino acids with smaller ones. These alterations contribute to protein misfolding, providing the flexibility necessary for functioning under stressful conditions (DasSarma et al. 2013). Halophilic microorganisms employ two main strategies to maintain osmotic pressure within their cytoplasm: the salt-in cytoplasm strategy, where K+ and Cl- ions are accumulated as counter ions and later replaced with Na^+ ions, and the organic-osmolyte strategy, where organic compatible solutes such as sugars, amino acids, and polyols are produced or acquired from the environment. These organic solutes not only assist in maintaining osmotic balance but also contribute to protein and cell stabilization (Kunte 2009). In addition to these strategies, halophiles employ various other mechanisms like upregulation of heat shock proteins, stress proteins, DNA repair proteins, proteasome subunits, and biofilm formation to combat the stress caused by high salinity (Matarredona et al. 2020). Based on research conducted by Mayak et al. in 2004, it was hypothesized that bacteria containing the enzyme ACC deaminase could effectively reduce ethylene levels within plants when associated with their roots (Mayak et al. 2004). Strains of plant-growth-promoting bacteria (PGPB) from soil samples were screened for their ACC deaminase production capabilities, with the bacterium *Achromobacter piechaudii* ARV8 showing promising results. These findings collectively highlight the potential of ACC deaminase–producing bacteria in promoting plant growth and salt tolerance. It was also observed that *Stenotrophomonas* and *Exiguobacterium*, two genera of bacteria, exhibited exceptional halophilic characteristics while simultaneously demonstrating their potential to enhance plant growth under conditions of salt stress (Santos et al. 2023). For instance, these particular bacterial species significantly elevated soybean germination rates, showcasing an impressive increase ranging between 35% and 43% when compared to non-inoculated seeds. Moreover, the inoculation of these bacteria into the soil environment led to a remarkable twofold expansion in both the length and dry biomass of soybean roots, even in the presence of a challenging 250 mM NaCl concentration (Santos et al. 2023).

4.4.5 Alleviation of Radiation Stress

Low earth orbit (LEO) and space beyond LEO are exposed to various types of radiation, such as galactic cosmic rays (GCR), solar energetic particles (SEP), and ultraviolet (UV) radiation. Such radiation has been shown to interact with cells at the DNA level, potentially causing alterations. However, studies have revealed that microorganisms display resistance to these radiations. Microorganisms employ diverse strategies to counteract the harmful effects, including upregulation of DNA repair proteins, enhanced ability to combat oxygen-free radicals, genomic rearrangements, upregulation of specific enzymes, and pigment formation (Senatore et al. 2018). This resistance demonstrated by microorganisms highlights their amazing adaptability and survival mechanisms in extreme environments, offering valuable insights for understanding the potential impact of radiation on living organisms beyond Earth's protective atmosphere.

Radiation resistance mechanisms have been extensively studied in *Deinococcus radiodurans*, a highly resilient bacterium known for its polyextremophilic properties. The unique cell envelope of *D. radiodurans*, comprising six layers, including the plasma layer, peptidoglycan-containing cell wall, fine compartments, outer membrane, and an electron-responsive fifth layer, contributes to its exceptional radiation resistance (Thompson et al. 1982). To mitigate the detrimental effects of radiation, *D. radiodurans* employs cellular cleansing processes to eliminate damaged DNA fragments and prevent the incorporation of disrupted bases during DNA synthesis, reducing mutagenesis (Battista 1997). Enzyme MutT, belonging to the pyrophosphohydrolase family, facilitates the removal of remaining mutagens within the cell (Slade and Radman 2011). Additionally, the bacterium employs superoxide dismutases (SodA) and catalases (KatA) to enzymatically remove reactive oxygen species, while deinoxanthin and pyrroloquinoline-quinone (PQQ) act as antioxidants, further enhancing its gamma radiation resistance (Jung et al. 2017). *D. radiodurans* possesses various DNA repair pathways, including the recognition and excision of pyrimidine cyclobutane dimers by UV endonuclease beta enzyme and the involvement of enzymes such as thymine glycol glycosylase, uracil DNA glycosylase, and deoxyribophosphodiestrase (Makarova et al. 2001). Notably, PprA and DrRecA are crucial for gamma radiation resistance in *D. radiodurans*, with DrRecA facilitating extended synthesis-dependent strand annealing (ESDSA) to restore double-stranded DNA fragments (Rajpurohit et al. 2021). The bacterium also employs nucleotide excision repair pathways to eliminate bipyrimidine photoproducts induced by UV-C radiation (Tanaka et al. 2005). Moreover, the secondary structures of DNA/RNA, known as guanine quadruplexes (G4), contribute to radioresistance in *D. radiodurans*, as observed when exposed to G4-stabilizing drugs (Mishra et al. 2019). In other radiation-resistant organisms, such as *Rubrobacter radiotolerans* and *Rubrobacter xylanophilus*, the presence of Mn^{2+} ions and the intracellular accumulation of compatible solute trehalose aid in scavenging reactive oxygen species generated by radiation (Webb and DiRuggiero 2012). Archaea, like *Thermococcus gammatolerans*, employ base excision repair enzymes, such as endonuclease III and DNA lyase, to repair oxidative damage in DNA caused by gamma radiation (Barbier et al. 2016). Additionally, archaea possess various proteins, including thioredoxin reductase, peroxiredoxin, and superoxide reductase, to counteract oxidative stress (Rodrigues et al. 2006). These diverse strategies and mechanisms contribute to the remarkable radiation resistance observed in these microorganisms.

4.4.6 Alleviation of Heavy Metal Stress

Metal resistance in bacteria and archaea can be achieved through various mechanisms, both passive and active. Passive mechanisms involve sequestering metals either extracellularly or intracellularly, while active mechanisms include intensified efflux of specific metals accompanied by the repression of uptake and enzymatic detoxification through reduction or volatilization of metallic ions (Nies 1999). Additionally, these microorganisms exhibit inducible genomic responses to high levels of heavy metals, leading to systemic transcriptional adjustments aimed at reducing metal intake and minimizing subsequent toxicity. Studies have shown rapid transcriptional responses in certain organisms, such as *Halobacterium* NRC-1, with a significant proportion of genes being translated within a short time after exposure to toxic metal stress (Kaur et al. 2006). However, these responses are usually transient, and once the microorganisms adapt to the metal concentrations in the environment, the transcription levels of the genes encoding the initial response return to normal (Srivastava and Kowshik 2013).

Biopolymers, particularly extracellular polysaccharides (EPSs), play a crucial role in microbial metal tolerance by binding metal cations. Halophilic bacteria and archaea secrete EPS rich in negatively charged residues, enabling them to tolerate high concentrations of metal ions such as Pb^{2+} and Cd^{2+}. For instance, *Halomonas* strains rely on EPS-mediated adsorption for metal tolerance (Amoozegar et al. 2012; Rajesh and Rajesh 2015). Additionally, certain halotolerant cyanobacteria and haloarchaea employ their cell surfaces and EPS to adsorb metals like Mn^{2+} and Zn^{2+} (Naik and Furtado 2014). In halophilic Gram-positive bacteria, heavy metal resistance is attributed to cell wall trapping, while Gram-negative bacteria employ periplasmic proteins for sequestration (Osman et al. 2010). Furthermore, the extracellular matrix of microbial cells may influence metal bioavailability in saline ecosystems (Gutierrez et al. 2012).

Heavy metal uptake in bacteria occurs through transport systems designated for essential ions or organic compounds, followed by intracellular sequestration or active efflux. For instance, in Gram-negative bacteria, arsenate enters cells via transport systems such as Pit and the Pst operon, which are primarily involved in phosphate uptake. This is due to the structural analogy between arsenate and inorganic phosphate, leading to increased tolerance across prokaryotes (Zhang et al. 2013). Metal homeostasis is achieved through cytoplasmic sequestration or metal-specific efflux complexes like RND proteins and ATP-coupled pumps.

Metalloregulators, such as VNG1179C, respond to cytosolic metal levels and regulate the expression of metal transport proteins (Kaur et al. 2006). Metallochaperones, like VNG0702H and VNG2581H, facilitate the transfer of metal ions to specific intracellular sites, fine-tuning the buffering of copper levels (Pang et al. 2013). Additionally, cytoplasmic chelation is achieved through γ-glutamylcysteine and polyphosphate, which act as metal chelators (Orell et al. 2012). These intricate mechanisms ensure the maintenance of metal homeostasis in halophilic organisms, enabling them to adapt to their extreme environments.

Enzyme-mediated reduction of hexavalent chromate (Cr^{6+}) to less toxic Cr^{3+} is a common strategy for mitigating heavy metal toxicity. Haloalkaline conditions have been found to promote efficient chromate reduction, with *Amphibacillus* and *Halomonas* strains demonstrating high reduction rates. Membrane-associated chromate reductases, such as the NADH-dependent reductase of *Halomonas* sp. TA-04 and the copper-dependent reductase of *Amphibacillus* sp. KSUCr3, play crucial roles in enzymatic Cr^{6+} detoxification. However, the purification of cytoplasmic chromate reductases in halophiles remains unexplored (Ibrahim et al. 2011; Focardi et al. 2012; Mabrouk et al. 2014).

Bioremediation, a prominent method in the mitigation of environmental heavy metal pollution, relies heavily on the intricate interplay between microorganisms and plants to alleviate heavy metal stress. Recent research has showcased the remarkable efficacy of an extremophile microorganism, *Deinococcus radiodurans*, particularly its mutant strain Δdr2577, characterized by its deficiency in cell surface-layer structure. This microorganism has exhibited remarkable capabilities in preventing translocation and mitigating damages caused by toxic heavy metals like Cd or Pb in rice (Dai et al. 2021). In another study, *Bacillus subtilis* showcased its remarkable potential in immobilizing cadmium (Cd) metal using cutting edge bioaugmentation technology (Wang et al. 2014). This groundbreaking research involved the introduction of the *Bacillus subtilis* strain into cadmium-contaminated soil, specifically targeting carrot cultivation. The results were nothing short of astounding, as the inoculation of this bacterial strain not only alleviated the stress imposed by Cd contamination but also led to a substantial boost in plant growth. The carrot shoots experienced a remarkable 16% increase, while the root development saw a substantial uptick of 55%. In another noteworthy investigation, Gan et al. (2015) explained how to harness the power of a thermophile consortium that includes *Acidithiobacillus caldus* and *Sulfobacillus thermotolerans*. Their research was focused on the bioleaching of heavy metals, which included a spectrum of toxic elements such as Zn, Cu, Mn, Cd, As, Hg, and Pb. This innovative approach holds the promise of more sustainable and environmentally friendly methods for dealing with heavy metal contamination in various ecosystems.

4.5 Extremophiles Application in Agriculture for Improved Abiotic Stress Tolerant Crop Varieties

Molecular techniques have revolutionized the utilization of beneficial microbes in soil for enhancing plant growth and combating soilborne pathogens. These microbes exhibit diverse nutritional and environmental requirements, with their inoculation proving more effective in nutrient-deficient soils compared to nutrient-rich soils. Understanding the perspectives of microbial diversity in agricultural settings is crucial for identifying indicators of soil quality and plant productivity (Yadav et al. 2015b). Various groups of plant-associated microbes, such as archaea, eubacteria, and fungi, encompassing phyla like *Acidobacteria, Actinobacteria, Ascomycota, Bacteroidetes, Basidiomycota, Deinococcus-Thermus, Euryarchaeota, Firmicutes,* and *Proteobacteria,* have been reported. Among these, *Proteobacteria* and *Actinobacteria* are the most prevalent (Yadav et al. 2017), while *Deinococcus-Thermus* (Sun et al. 2008) and *Acidobacteria* exhibit the lowest representation (Sahay et al. 2017).

Soil saltiness poses a significant challenge for agricultural crops, particularly in arid and semiarid regions worldwide (Yadav et al. 2017; Verma et al. 2019). While various technologies have been developed to enhance salt tolerance, recent studies have highlighted the role of plant-growth-promoting (PGP) microorganisms in inducing plant resistance against salt stress. These microorganisms hold great potential for agriculture as they can enhance plant growth even under restrictive and stressful conditions. Furthermore, microorganisms from extreme environments are of particular ecological importance, given that many terrestrial and aquatic ecosystems are exposed to cold temperatures either permanently or seasonally. Plants have evolved different mechanisms in response to environmental stimuli, including the activation of various metabolic defense molecules (Chagas et al. 2018). Plants produce various metabolic molecules like salicylic acid, ethylene, calcium, and jasmonic acid to boost their defense against alkaline stress. Salicylic acid reduces reactive oxygen species and enhances antioxidant defense in tomato plants. Combining salicylic acid and silicon benefits alkalinity resistance in tomatoes (Khan et al. 2020). Plant breeding is vital for crop productivity in alkaline soil areas, and different plants have diverse tolerance mechanisms. For instance, alkalinity-tolerant lentil cultivars have thicker epidermis, and some plants reduce sodium uptake to handle alkaline stress (Singh et al. 2019). However, more research is needed on related stresses like salinity and drought. Microorganisms that thrive in cold environments have proven to be highly adaptable and are found abundantly in natural habitats, often coexisting with dominant vegetation. These microorganisms play a crucial role as the primary colonizers of our planet. Increasingly extreme temperatures caused by climate change pose a significant threat to global agriculture, potentially reducing crop yields and leading to substantial economic consequences. Researchers have identified promising solutions through the use of microorganisms. Studies by Fernandez et al. in 2012 and Kumar et al. in 2022 demonstrated that certain bacteria can adapt to low temperatures and upregulate cold-stress genes, benefiting plant growth in challenging environments. Additionally, research by Ali et al. in 2011 and Mukhtar et al. in 2020 highlighted the positive effects of microorganism inoculation on plant growth under high-temperature conditions. Notably, thermophilic soil bacteria, such as those from *Bacillus*, *Geobacillus*, *Parageobacillus*, *Ureibacillus*, and *Brevibacillus* genera, play a crucial role in organic sulfur oxidation, providing essential compounds for plants and microorganisms, offering potential solutions for mitigating climate-induced temperature fluctuations in agriculture. Precipitation serves as the primary water source for global crop cultivation, but fluctuations in water availability due to drought or excessive rainfall can cause abiotic stress and reduce crop yields. PGPM have shown promise in enhancing plant resilience to drought stress. Certain bacteria with salt-tolerant properties, such as *Klebsiella* sp., *Kosakonia cowanii*, and *Sinorhizobium meliloti*, have been isolated, exhibiting plant-growth-promoting characteristics and high salt tolerance. Bacteria-producing ACC deaminases, like *Achromobacter piechaudii* ARV8, have the potential to reduce ethylene levels in plants. Additionally, bacteria from the *Stenotrophomonas* and *Exiguobacterium* genera have demonstrated exceptional halophilic traits and the ability to improve plant growth under salt-stress conditions. These findings suggest that PGPM and salt-tolerant bacteria could be valuable tools for enhancing crop yields in challenging environmental conditions. Bioremediation, a key method for mitigating heavy metal pollution in the environment, relies on the synergy of microorganisms and plants to alleviate metal stress. Notably, *Deinococcus radiodurans*, specifically its mutant strain Δdr2577, has shown remarkable efficacy in preventing translocation and reducing damage caused by toxic heavy metals like cadmium (Cd) and lead (Pb) in rice. Another study highlighted the potential of *Bacillus subtilis* in immobilizing Cd metal in contaminated soil, leading to significant improvements in carrot growth. Additionally, a thermophile consortium consisting of *Acidithiobacillus caldus* and *Sulfobacillus thermotolerans* was found to be effective in bioleaching various toxic metals, promising more sustainable approaches for addressing heavy metal contamination in ecosystems.

4.6 Conclusions

The rapid growth of the global population and the increasing demand for food products have placed significant pressure on global food production systems. However, various factors such as climate change, soil degradation, and land loss have contributed to decreased crop yields, highlighting the urgent need for sustainable agricultural practices and innovative solutions to ensure food security for the growing population. Microbial communities, especially extremophilic microbiomes, have garnered attention for their remarkable adaptability and survival mechanisms in extreme and harsh environments. These stress-adaptive microbes produce bioactive compounds and secondary metabolites that aid in their resilience and proliferation. Extremophilic microbiomes, consisting of archaea, bacteria, and eukaryotes, exhibit the ability to thrive in diverse conditions, including high salinity, extreme pH values, and a wide temperature range. The study of extremophilic microbiomes has significant potential applications in various fields, including biotechnology, medicine, and food production. These microbiomes can contribute to the development of novel biotechnological solutions and offer sustainable approaches to enhance agricultural productivity while reducing reliance on chemical fertilizers. By harnessing the homeostasis maintenance potential of extremophilic microbiomes and addressing both abiotic and biotic stresses, researchers and farmers can work toward sustainable agricultural practices that promote increased productivity and environmental resilience. Advancements in technologies like metagenomics and next-generation sequencing have revolutionized our understanding of extremophilic microorganisms, providing valuable insights into their functions and mechanisms of homeostasis. These advancements enable deeper exploration of their remarkable abilities to thrive and adapt in harsh surroundings. Looking ahead, future research should focus on further uncovering the diverse mechanisms and properties of extremophilic microbiomes, as well as their potential applications in different industries. Continued exploration of their interactions with plants and their role in nutrient circulation can lead to the development of effective bioinoculants and sustainable agricultural practices. Moreover, ongoing technological advancements will contribute to a deeper understanding of extremophilic microorganisms and their potential for solving global challenges related to food production, environmental sustainability, and human well-being. In summary, the study of extremophilic microbiomes offers promising prospects for addressing the challenges posed by population growth, food security, and environmental degradation. By leveraging their adaptability and resilience, we can pave the way for sustainable practices that ensure a prosperous future for agriculture and the global population.

References

Ali SZ, Sandhya V, Grover M, Linga VR, Bandi V (2011) Effect of inoculation with a thermotolerant plant growth promoting Pseudomonas putida strain AKMP7 on growth of wheat (Triticum spp.) under heat stress. J Plant Interact 6(4):239–246

Ali N, Nughman M, Shah SM (2014) Extremophiles and limits of life in a cosmic perspective. In: Extreme environments-diversity I, adaptability of bioactive molecules. IntechOpen valuable resources of bioactive Molecules. IntechOpen 2023. Besser Valuable Resources of Bioactive Molecules. IntechOpen 2023. Besser John Valuable Resources of Bioactive Molecules. IntechOpen 2023. Besser John Heather A VR, et al. (eds) Next-generation sequencing technologies and their application to the study and control of bacterial infections. Elsevier, Comprehensive Gut Microbiota, vol. 3, pp 246–256. https://doi.org/10.1016/B978-0-12-819265-8.00001-2. Suneja G, Srivastav R (2021) Impact of microbial genome sequencing advancements in understanding extremophiles. Book -Extreme Environments, Unique, pp 335–341

Ali N, Nughman M, Shah SM (2023) Extremophiles and Limits of Life in a Cosmic Perspective. In Life in Extreme Environments-Diversity, Adaptability and Valuable Resources of Bioactive Molecules. IntechOpen.

Amoozegar MA, Ghazanfari N, Didari M (2012) Lead and cadmium bioremoval by Halomonas sp., an exopolysaccharide-producing halophilic bacterium? Progress Biol Sci 2:1–11

Asghari B, Khademian R, Sedaghati B (2020) Plant growth promoting rhizobacteria (PGPR) confer drought resistance and stimulate biosynthesis of secondary metabolites in pennyroyal (Mentha pulegium L.) under water shortage condition. Sci Hortic (Amsterdam) 263:109132

Baker-Austin C, Dopson M (2007a) Life in acid: pH homeostasis in acidophiles. Trends Microbiol 15:165

Baker-Austin C, Dopson M (2007b) Life in acid: pH homeostasis in acidophiles. Trends Microbiol 15:4

Baltrus DA (2017) Adaptation, specialization, and coevolution within phytobiomes. Curr Opin Plant Biol 38:109–116

Barbier E, Lagorce A, Hachemi A et al (2016) Oxidative DNA damage and repair in the radioresistant archaeon Thermococcus gammatolerans. Chem Res Toxicol 29:1796–1809

Battista JR (1997) Against all odds: the survival strategies of Deinococcus radiodurans. Ann Rev Microbiol 51:203–224

Becker J, Wittmann C (2020) Microbial production of extremolytes High-value active ingredients for nutrition, health care, and well-being. Curr Opin Biotechnol 65:118

Beneduzi A, Ambrosini A, Passaglia LM (2012) Plant growth-promoting rhizobacteria (PGPR): their potential as antagonists and biocontrol agents. Genet Mol Biol 35:1044–1051

Biswas A, Dasgupta S, Das S, Abraham A (2007) A synergy of differential evolution and bacterial foraging optimization for global optimization. Neural Netw World 17:607

Boopathy S, Appavoo MS, Radhakrishnan I (2021) Sunflower seed husk combined with poultry droppings to degrade petroleum hydrocarbons in crude oil-contaminated soil. Environ Eng Res 26:200361

Boyer JS (1982) Plant productivity and environment. Science (80-) 218:443–448

Canganella F, Wiegel J (2011) Extremophiles: from abyssal to terrestrial ecosystems and possibly beyond. Naturwissenschaften 98:253–279

Chagas FO, de Cassia Pessotti RC-R et al (2018) Chemical signaling involved in plant?microbe interactions. Chem Soc Rev 47:1652–1704

Chaudhary DK, Kim DU, Kim D, Kim J (2019) Flavobacterium petrolei sp. nov., a novel psychrophilic, diesel degrading bacterium isolated from oil-contaminated Arctic soil. Sci Rep 9:1–9

Chen X (2021) Thriving at low pH: adaptation mechanisms of acidophiles. In: Acidophiles-fundamentals and applications. IntechOpen

Corsaro MM, Casillo A, Parrilli E, Tutino ML (2017) Molecular structure of lipopolysaccharides of cold-adapted bacteria. In: Psychrophiles: from biodiversity to biotechnology. Springer, Cham, pp 285–303

Cytryn EJ, Sangurdekar DP, Streeter JG et al (2007) Transcriptional and physiological responses Bradyrhizobium japonicum to desiccation-induced stress. J Bacteriol 189:6751–6762

D'Amico S, Claverie P, Collins T, et al (2001) Cold-adapted enzymes: an unachieved symphony. In Cell and Molecular Response to Stress 2:31–42. Elsevier

Dai S, Wang B, Song Y et al (2021) Astaxanthin and its gold nanoparticles mitigate cadmium toxicity in rice by inhibiting cadmium translocation and uptake. Sci Total Environ 786:147496

Danish S, Zafar-Ul-Hye M, Mohsin F, Hussain M (2020) ACC-deaminase producing plant growth promoting rhizobacteria and biochar mitigate adverse effects of drought stress on maize growth. PLoS One 15:e0230615. https://doi.org/10.1371/JOURNAL.PONE.0230615

DasSarma S, Capes MD, Karan R, DasSarma P (2013) Amino acid substitutions in cold-adapted proteins from Halorubrum lacusprofundi, an extremely halophilic microbe from Antarctica. PLoS One 8:3

De Rosa M, Gambacorta A (1988) The lipids of Archaebacteria. Prog Lipid Res 27:153–175

Deosthali C, Sajwani D (2022) Extremophiles: applications and adaptive strategies. Int J Res Trends Innov

Dhakar K, Pandey A (2016) Wide pH range tolerance in extremophiles: towards understanding an important phenomenon for future biotechnology. Appl Microbiol Biotechnol 100:2499–2510

Dieser M, Greenwood M, Foreman CM (2010) Carotenoid pigmentation in Antarctic heterotrophic bacteria as a strategy to withstand environmental stresses. Arctic, Antarct Alp Res 42:396–405

Ding T, Yu YY, Hwang CA et al (2016) Modeling the effect of water activity, pH, and temperature on the probability of enterotoxin a production by Staphylococcus aureus. J Food Prot 79:148–152

Dopson M, Holmes DS (2014) Metal resistance in acidophilic microorganisms and its significance for biotechnologies. Appl Microbiol Biotechnol 98:8133–8144

Elferink M, Schierhorn F (2016) Global demand for food is rising. Can we meet it. Harv Bus Rev 7:2016

Elnahal ASM, El-Saadony MT, Saad AM et al (2022) The use of microbial inoculants for biological control, plant growth promotion, and sustainable agriculture: a review. Eur J Plant Pathol 162:759–792

Elumalai P, Parthipan P, Narenkumar J et al (2019) Role of thermophilic bacteria (Bacillus and Geobacillus) on crude oil degradation and biocorrosion in oil reservoir environment. 3 Biotech 9:1–11

Enebe MC, Babalola OO (2019) The impact of microbes in the orchestration of plants resistance to biotic stress: a disease management approach. Appl Microbiol Biotechnol 103:9–25. https://doi.org/10.1007/s00253-018-9433-3

Esbelin J, Santos T, Hebraud M (2018) Desiccation: an environmental and food industry stress that bacteria commonly face. Food Microbiol 69:82–88

Fagliarone C, Napoli A, Chiavarini S et al (2020) Biomarker preservation and survivability under extreme dryness and Mars-like UV flux of a desert cyanobacterium capable of trehalose and sucrose accumulation. Front Astron Space Sci 7:31

FAO (2008) Coping with Water Scarcity. An Action Framework for Agriculture and Food Security; FAO: Rome, Italy, p 100

Feng S, Powell SM, Wilson R, Bowman JP (2014) Extensive gene acquisition in the extremely psychrophilic bacterial species Psychroflexus torquis and the link to sea-ice ecosystem specialism. Genome Biol Evol 6:133–148

Fernandez O, Theocharis A, Bordiec S et al (2012) A Burkholderia phytofirmans PsJN Acclimates grapevine to cold by Modulating Carbohydrate Metabolism. Mol Plant-Microbe Interact 25:496–504

Ferrante A, Mariani L (2018) Agronomic management for enhancing plant tolerance to abiotic stresses: high and low values of temperature, light intensity, and relative humidity. Horticulturae 4:21

Fleming TR, Fleming CC, Levy CCB et al (2019) Biostimulants enhance growth and drought tolerance in Arabidopsis thaliana and exhibit chemical priming action. Ann Appl Biol 174:153–165

Focardi S, Pepi M, Landi G et al (2012) Hexavalent chromium Reduct by whole cells cell Free Extr moderate halophilic Bact strain Halomonas sp TA-0. Int Biodeterior Biodegradation 66:63–70

Friedman R, Drake JW, Hughes AL (2004) Genome-wide patterns of nucleotide substitution reveal stringent functional constraints on the protein sequences of thermophiles. Genetics 167:1507–1512

Garcia-Lopez E, Alcazar P, Cid C (2021) Identification of Biomolecules Involved in the Adaptation to the Environment of Cold-Loving Microorganisms and Metabolic Pathways for Their Production. Biomol Ther 11:1155

Ghobakhlou AF, Johnston A, Harris L et al (2015) Microarray transcriptional profiling of Arctic Mesorhizobium strain N33 at low temperature provides insights into cold adaption strategies. BMC Genomics 16:1–14

Goordial J, Raymond-Bouchard I, Zolotarov Y et al (2016) Cold adaptive traits revealed by comparative genomic analysis of the eurypsychrophile Rhodococcus sp. JG3 isolated from high elevation McMurdo Dry Valley permafrost, Antarctica. FEMS Microbiol Ecol 92(2):fiv154

Guan N, Liu L (2020) Microbial response to acid stress: mechanisms and applications. Appl Microbiol Biotechnol 104:51–65

Gunjal AB, Badodekar NP (2022) Halophiles. In: Physiolgoy, genomics and biotechnological applications of extremophiles. IGI Global, USA, pp 13–34

Gutierrez T, Biller DV, Shimmield T, Green DH (2012) Metal binding properties of the EPS produced by Halomonas sp. TG39 and its potential in enhancing trace element bioavailability to eukaryotic phytoplankton. Biometals 39:1185–1194

He J, Yang Z, Hu B et al (2015) Correlation of polyunsaturated fatty acids with the cold adaptation of Rhodotorula glutinis. Yeast 32:683–690

Ibrahim AS, El-Tayeb MA, Elbadawi YB, Al-Salamah AA (2011) Isolation and characterization of novel potent Cr (VI) reducing alkaliphilic Amphibacillus sp. KSUCr3 from hypersaline soda lakes. Electron J Biotechnol 14:4

Ipek M, Aras S, Arıkan Ş et al (2017) Root plant growth promoting rhizobacteria inoculations increase ferric chelate reductase (FC-R) activity and Fe nutrition in pear under calcareous soil conditions. Sci Hortic (Amsterdam) 219:144–151

Jagannadham MV, Chattopadhyay MK, Subbalakshmi C et al (2000) Carotenoids an Antarct psychrotolerant bacterium, Sphingobacterium Antarct a mesophilic bacterium, Sphingobacterium multivorum. Arch Microbiol 173:418–424

Jin M, Xiao A, Zhu L et al (2019) The diversity and commonalities of the radiation-resistance mechanisms of Deinococcus and its up-to-date applications. AMB Express 9:1–12

Jung KW, Lim S, Bahn YS (2017) Microbial radiation-resistance mechanisms. J Microbiol 55:499–507

Kanekar PP, Kanekar SP (2022) Piezophilic or Barophilic Microorganisms. In: Diversity and biotechnology of extremophilic microorganisms from India. Nature Singapore, Singapore, pp 269–280

Kaur A, Pan M, Meislin M et al (2006) A systems view of haloarchaeal strategies to withstand stress from transition metals? Genome Res 16:841–854

Khan N, Bano A, Rahman MA et al (2019) Comparative physiological and metabolic analysis reveals a complex mechanism involved in drought tolerance in chickpea (Cicer arietinum L.) induced by PGPR and PGRs. Sci Rep 9:1–19

Khan MA, Asaf S, Khan AL et al (2020) Extending thermotolerance to tomato seedlings by inoculation with SA1 isolate of Bacillus cereus and comparison with exogenous humic acid application. PLoS One 15:e0232228

Kim HJ, Lee JH, Hur YB et al (2017) Marine antifreeze proteins: structure, function, and application to cryopreservation as a potential cryoprotectant. Mar Drugs 15:27

Knack JJ, Wilcox LW, Delaux PM et al (2015) Microbiomes of streptophyte algae and bryophytes suggest that a functional suite of microbiota fostered plant colonization of land. Int J Plant Sci 176:405–420

Kour D, Kaur T, Devi R et al (2020) Biotechnological applications of beneficial microbiomes for evergreen agriculture and human health. In: Rastegari AA, Yadav AN, Yadav N (eds) Trends of microbial biotechnology for sustainable agriculture and biomedicine systems: perspectives for human health. Elsevier, Amsterdam, pp 255–279

Krulwich TA (2006) Alkaliphilic prokaryotes. Prokaryotes 2:283–308

Krulwich TA, Sachs G, Padan E (2011) Molecular aspects of bacterial pH sensing and homeostasis. Nat Rev Microbiol 9:330–343

Kumar S, Nussinov R (2001) Fluctuations in ion pairs and their stabilities in proteins. Proteins Struct Funct Bioinform 43:433–454

Kumar A, Le Flèche-Matéos A, Kumar R et al (2022) Rahnella sikkimica sp. nov., a novel cold-tolerant bacterium isolated from the glacier of Sikkim Himalaya with plant growth-promoting properties. Extremophiles 26:35

Kunte HJ (2009) Osmoregulation in halophilic bacteria. Extremophiles 2:263–277

Kushkevych I, Cejnar J, Vítězová M et al (2020) Occurrence of thermophilic microorganisms in different full scale biogas plants. Int J Mol Sci 21:283

Lamosa P, Burke A, Peist R et al (2000) Thermostabilization proteins by diglycerol phosphate, a new Compat solute from hyperthermophile Archaeoglobus fulgidus. Appl Environ Microbiol 66:1974–1979

LeBlanc JC, Gonçalves ER, Mohn WW (2008) Global response to desiccation stress in the soil actinomycete Rhodococcus jostii RHA1. Appl Environ Microbiol 74:2627–2636

Lentzen G, Schwarz T (2006) Extremolytes: natural compounds from extremophiles for versatile applications. Appl Microbiol Biotechnol 72:623–634

Leprince O, Buitink J (2010) Desiccation tolerance: from genomics to the field. Plant Sci 179:554–564

Loganathan P, Naidu G, Vigneswaran S (2017) Mining valuable minerals from seawater: a critical review. Environ Sci Water Res Technol 3:37–53

Mabrouk ME, Arayes MA, Sabry SA (2014) Hexavalent chromium reduction by chromate-resistant haloalkaliphilic Halomonas sp. M-Cr New Isol from Tannery effluent. Biotechnol Biotechnol Equip 28:659–667

Mahajan S, Tuteja N (2005) Cold, salinity and drought stresses: an overview. Arch Biochem Biophys 444:139–158

Makarova KS, Aravind L, Wolf YI et al (2001) Genome of the extremely radiation-resistant bacterium Deinococcus radiodurans viewed from the perspective of comparative genomics. Microbiol Mol Biol Rev 65:44–79

Mamo G (2020) Challenges and adaptations of life in alkaline habitats. In: Alkaliphiles in biotechnology, pp 85–133

Mariani L, Ferrante A (2017) Agronomic management for enhancing plant tolerance to abiotic stresses?drought, salinity, hypoxia, and lodging. Horticulturae 3:52

Matarredona L, Camacho M, Zafrilla B et al (2020) The role of stress proteins in Haloarchaea and their adaptive response to environmental shifts. Biomol Ther 10:1390

Matsuno T, Yumoto I (2015) Bioenergetics and the role of soluble cytochromes c for alkaline adaptation in Gram-negative alkaliphilic Pseudomonas. Biomed Res Int 2015:847945

Mayak S, Tirosh T, Glick BR (2004) Plant growth-promoting bacteria that confer resistance to water stress in tomatoes and peppers. Plant Sci 166:525–530. https://doi.org/10.1016/J.PLANTSCI.2003.10.025

McDermott JM, Sylva SP, Ono S et al (2018) Geochemistry of fluids from Earth?s deepest ridge-crest hot-springs: Piccard hydrothermal field, Mid-Cayman Rise. Geochim Cosmochim Acta 228:95–118

Meharg AA, Killham K (1990) Carbon distribution within the plant and rhizosphere in laboratory and field-grown Lolium perenne at different stages of development. Soil Biol Biochem 22:471–477

Menasria T, Aguilera M, Hocine H et al (2018) Diversity and bioprospecting of extremely halophilic archaea isolated from Algerian arid and semi-arid wetland ecosystems for halophilic-active hydrolytic enzymes. Microbiol Res 207:289–298

Mendes R, Garbeva P, Raaijmakers JM (2013) The rhizosphere microbiome: significance of plant beneficial, plant pathogenic, and human pathogenic microorganisms. FEMS Microbiol Rev 37:634–663

Mesa-Marin J, Perez-Romero JA, Redondo-Gomez S et al (2020) Impact of plant growth promoting bacteria on Salicornia ramosissima ecophysiology and heavy metal phytoremediation capacity in estuarine soil. Front Microbiol 11:55301

Mirete S, Morgante V, Gonzalez-Pastor JE (2017) Acidophiles: diversity and mechanisms of adaptation to acidic environments. Adaption of microbial life to environmental extremes: novel research results and application, 227–251. https://doi.org/10.1007/-319-48327-6-9

Mirzaei M, Ladan Moghadam A, Hakimi L, Danaee E (2020) Plant growth promoting rhizobacteria (PGPR) improve plant growth, antioxidant capacity, and essential oil properties of lemongrass (Cymbopogon citratus) under water stress. Iran J Plant Physiol 10:3155–3166

Mishra S, Chaudhary R, Singh S et al (2019) Guanine quadruplex DNA regulates gamma radiation response of genome functions in the radioresistant bacterium Deinococcus radiodurans. J Bacteriol 201:119–154

Mukhtar S, Malik KA, Mehnaz S (2019) Microbiome of halophytes: diversity and importance for plant health and productivity. Microbiol Biotechnol Lett 47:1–10

Mukhtar T, Rehman SU, Smith D, Sultan T, Seleiman MF, Alsadon AA, Amna et al. (2020) Mitigation of heat stress in Solanum lycopersicum L. by ACC-deaminase and exopolysaccharide producing Bacillus cereus: effects on biochemical profiling. Sustainability 12(6):2159

Mykytczuk N, Foote SJ, Omelon CR et al (2013) Bacterial growth at? 15 C; molecular insights from the permafrost bacterium Planococcus halocryophilus Or1. ISME J 7:1211–1226

Mylona P, Pawlowski K, Bisseling T (1995) Symbiotic nitrogen fixation. Plant Cell 7:869

Naik S, Furtado I (2014) Equilibrium and kinetics of adsorption of Mn2+ by haloarchaeon Halobacterium sp. GUSF (MTCC3265). Geomicrobiol J 31:708–715

Navarro CA, von Bernath D, Jerez CA (2013) Heavy metal resistance strategies of acidophilic bacteria and their acquisition: importance for biomining and bioremediation. Biol Res 46:363–371

Nies DH (1999) Microbial heavy-metal resistance.? Appl Microbiol Biotechnol 51:730–750

Orell A, Navarro CA, Rivero M et al (2012) Inorganic polyphosphates in extremophiles and their possible functions? Extremophiles 16:573–583

Osman O, Tanguichi H, Ikeda K et al (2010) Copper-resistant halophilic bacterium isolated from the polluted Maruit Lake, Egypt. J Appl Microbiol 108:1459–1470. https://doi.org/10.1111/j.1365-2672.2009.04574.x

Pang WL, Kaur A, Ratushny AV et al (2013) Metallochaperones regulate intracellular copper levels. PLoS Comput Biol 9:1

Paz A, Mester D, Baca I et al (2004) Adaptive role of increased frequency of polypurine tracts in mRNA sequences of thermophilic prokaryotes. Proc Natl Acad Sci 101:2951–2956

Pikuta EV, Hoover RB, Tang J (2007) Microbial extremophiles at the limits of life. Crit Rev Microbiol 33:183–209

Preiss L, Hicks DB, Suzuki S et al (2015) Alkaliphilic bacteria with impact on industrial applications, concepts of early life forms, and bioenergetics of ATP synthesis. Front Bioeng Biotechnol 3:75

Rajesh V, Rajesh N (2015) An indigenous Halomonas BVR1 strain immobilized in crosslinked chitosan for adsorption of lead and cadmium.? Int J Biol Macromol 79:300–308

Rajpurohit YS, Sharma DK, Misra HS (2021) PprA protein inhibits DNA strand exchange and ATP hydrolysis of Deinococcus RecA and regulates the recombination in gamma-irradiated cells. Front Cell Dev Biol 9:636178

Rodrigues JV, Abreu IA, Cabelli D, Teixeira M (2006) Superoxide reduction mechanism of Archaeoglobus fulgidus one-iron superoxide reductase. Biochemistry 45:9266–9278

Ruginescu R, Gomoiu I, Popescu O et al (2020) Bioprospecting for novel halophilic and halotolerant sources of hydrolytic enzymes in brackish, saline and hypersaline lakes of Romania. Microorganisms 8:1903

Russell NJ (1989) Adaptive modifications in membranes of halotolerant and halophilic microorganisms. J Bioenerg Biomembr 21:93–113

Saad S, Cereghetti G, Feng Y et al (2017) Reversible protein aggregation is a protective mechanism to ensure cell cycle restart after stress. Nat Cell Biol 19:1202–1213

Saha D, Panda A, Podder S, Ghosh TC (2015) Overlapping genes: a new strategy of thermophilic stress tolerance in prokaryotes. Extremophiles 19:345–353

Sahay H, Yadav AN, Singh AK et al (2017) Hot springs of {I}ndian Himalayas: potential sources of microbial diversity and thermostable hydrolytic enzymes. 3 Biotech 7:1–11

Sahonero-Canavesi DX, Villanueva L, Bale NJ et al (2021) Changes in the distribution of membrane lipids during growth of Thermotoga maritima at different temperatures: indications for the potential mechanism of biosynthesis of ether-bound diabolic acid (membrane-spanning) lipids. Appl Environ Microbiol 88:e01763

Salwan R, Sharma V (2020) Physiology of extremophiles. In: Physiological and biotechnological aspects of extremophiles. Academic Press, London, pp 13–22

Santana MM, Portillo MC, Gonzalez JM, Clara MIE (2013) Characterization of new soil thermophilic bacteria potentially involved in soil fertilization. J Plant Nutr Soil Sci 176:47–56

Santana MM, Carvalho L, Melo J et al (2020) Unveiling the hidden interaction between thermophiles and plant crops: wheat and soil thermophilic bacteria. J Plant Interact 15:127–138

Santos AP, Belfiore C, Úrbez C et al (2023) Extremophiles as plant probiotics to promote germination and alleviate salt stress in soybean. J Plant Growth Regul 42:946–959

Saxena AK, Yadav AN, Rajawat M et al (2016) Microbial diversity of extreme regions: an unseen heritage and wealth. Indian J Plant Genet Resour 29:246–248

Scambos TA, Campbell GG, Pope A et al (2018) Ultralow surface temperatures in East Antarctica from satellite thermal infrared mapping: the coldest places on Earth. Geophys Res Lett 45:6124–6133

Senatore G, Mastroleo F, Leys N, Mauriello G (2018) Effect of microgravity & space radiation on microbes. Future Microbiol 13:7

Sinclair NA, Stokes JL (1964) Isolation of obligately anaerobic psychrophilic bacteria. J Bacteriol 87:562–565

Singh P, Jain K, Desai C, Tiwari O, Madamwar D (2019) Microbial community dynamics of extremophiles/extreme environment. In Microbial diversity in the genomic era (pp. 323–332). Academic Press

Slade D, Radman M (2011) Oxidative stress resistance in Deinococcus radiodurans. Microbiol Mol Biol Rev 75:133–191

Slonczewski JL, Fujisawa M, Dopson M, Krulwich TA (2009) Cytoplasmic pH measurement and homeostasis in bacteria and archaea. Adv Microb Physiol 55:1–317

Smith DL, Gravel V, Yergeau E (2017) Signaling in the Phytomicrobiome. Front Plant Sci 8:611

Srivastava P, Kowshik M (2013) Mechanisms of metal resistance and homeostasis in Haloarchaea. Archaea 2013:732864. https://doi.org/10.1155/2013/732864

Subramanian S, Souleimanov A, Smith DL (2016) Proteomic studies on the effects of lipo-chitooligosaccharide and thuricin 17 under unstressed and salt stressed conditions in Arabidopsis thaliana. Front Plant Sci 7:1314

Sun L, Qiu F, Zhang X et al (2008) Endophytic bacterial diversity in rice (Oryza sativa L.) roots estimated by 16S rDNA sequence analysis. Microb Ecol 55:415–424

Takjl M, Fujisawa MM, Midb H (2010) Adaptive mechanisms of extreme alkaliphiles. In: Extremophiles handbook, p 119

Tanaka M, Narumi I, Funayama T et al (2005) Characterization of Pathways Dependent on the uvsE, uvrA1, or uvrA2 Gene Product for UV Resistance in Deinococcus radiodurans. J Bacteriol 187:3693–3697

Tekaia F, Yeramian E, Dujon B (2002) Amino acid composition of genomes, lifestyles of organisms, and evolutionary trends: a global picture with correspondence analysis. Gene 297:51–60

Tendulkar S, Hattiholi A, Chavadar M, Dodamani S (2021) Psychrophiles: a journey of hope. J Biosci 46:1–15

Tenore A, Wu Y, Jacob J et al (2023) Water activity in subaerial microbial biofilms on stone monuments. Sci Total Environ 900:165790

Thompson BG, Murray RGE, Boyce JF (1982) The association of the surface array and the outer membrane of Deinococcus radiodurans. Can J Microbiol 28:1081–1088

Toral L et al (2021) Identification of Volatile Organic Compounds in Extremophilic Bacteria and Their Effective Use in Biocontrol of Postharvest Fungal Phytopathogens. Front Microbiol 12:12

Tribelli PM, Lopez NI (2018) Reporting key features in cold-adapted bacteria. Life 8:8

Tripathi S, Singh K, Chandra R (2021) Adaptation of bacterial communities and plant strategies for amelioration and eco-restoration of an organometallic industrial waste polluted site. In: Microbes in land use change management. Elsevier, Amsterdam, pp 45–90

Turchetti B, Marconi G, Sannino C et al (2020) DNA methylation changes induced by cold in psychrophilic and psychrotolerant Naganishia yeast species. Microorganisms 8:296

Van De Vossenberg JLCM, Driessen AJ, Konings WN (2000) Adaptations of the cell membrane for life in extreme environments. In: Environmental stressors and gene responses. Elsevier. In Cell and molecular response to stress 1:71–88. https://doi.org/10.1016/S1568-1254(00)80008-X

Verma P, Yadav AN, Khannam KS et al (2019) Appraisal of diversity and functional attributes of thermotolerant wheat associated bacteria from the peninsular zone of India. Saudi J Biol Sci 26:1882–1895

Villanova V, Galasso C, Fiorini F et al (2021) Biological and chemical Characterization of new isolated halophilic microorganisms from saltern ponds Trapani, Sicily. Algal Res 54:10219

Wang T, Sun H, Jiang C et al (2014) Immobilization of Cd in soil and changes of soil microbial community by bioaugmentation of UV-mutated Bacillus subtilis 38 assisted by biostimulation. Eur J Soil Biol 65:62–69

Wang X, Cole CG, DuPai CD, Davies BW (2020) Protein aggregation is associated with Acinetobacter baumannii desiccation tolerance. Microorganisms 8:343

Webb KM, DiRuggiero J (2012) Role of Mn2+ and compatible solutes in the radiation resistance of thermophilic bacteria and archaea. Archaea 2012:845756

Yadav AN, Saxena AK (2018) Biodiversity and biotechnological applications of halophilic microbes for sustainable agriculture. J Appl Biol 6:1

Yadav AN, Sachan SG, Verma P et al (2015a) Culturable diversity and functional annotation of psychrotrophic bacteria from cold desert of Leh Ladakh (India). World J Microbiol Biotechnol 31:95–108

Yadav AN, Verma P, Kumar M et al (2015b) Diversity and phylogenetic profiling of niche-specific Bacilli from extreme environments of India. Ann Microbiol 65:611–629

Yadav AN, Verma P, Kumar V et al (2017) Extreme cold environments: a suitable niche for selection of novel Psychrotrophic microbes for biotechnological applications. Adv Biotechnol Microbiol 2:1–4

Yadav AN, Kour D, Rana KL et al (2019) Metabolic engineering to synthetic biology of secondary metabolites production. In: Gupta KV, Pandey A (eds) New and future developments in microbial biotechnology and bioengineering. Elsevier, Amsterdam, pp 279–320

Yazdani M, Bahmanyar MA, Pirdashti H, Esmaili MA (2009) Effect of phosphate solubilization microorganisms (PSM) and plant growth promoting rhizobacteria (PGPR) on yield and yield components of corn (Zea mays L.). World Acad Sci Eng Technol 49:90–92

Yoshida K, Hashimoto M, Hori R et al (2016) Bacterial long-chain polyunsaturated fatty acids: their biosynthetic genes, functions, and practical use. Mar Drugs 14:94

Zeldes BM, Keller MW, Loder AJ et al (2015) Extremely thermophilic microorganisms as metabolic engineering platforms for production of fuels and industrial chemicals? Front Microbiol 6:1209

Zhalnina K, Dias R, de Quadros PD et al (2015) Soil pH determines microbial diversity and composition in the park grass experiment. Microb Ecol 69:395–406

Zhang Y, Zhang F, Li X et al (2013) Transcription activator-like effector nucleases enable efficient plant genome engineering. Plant Physiol 161:20–27

Part II
Potential Application of Extremophiles in Agricultural and Improving Soil Health

Chapter 5
Exploration of Extremophiles: Potential Applications in Agriculture and Soil Health Improvement Utilizing Extremophiles

Vinay Mohan Pathak, Nitika Rana, Sumitra Pandey, A. K. Sarkar, Abhishek Chauhan, Tanu Jindal, Mukta Sharma, Seeta Dewali, Satpal Singh Bisht, Divya, Akansha Sharma, Monika Yadav, Balwant Singh Rawat, Somdatt Tyagi, D. P. Uniyal, Pradeep Kumar, Sevaram Singh, Baljinder Kaur, Jose Maria Cunill, and Satish Chandra Garkoti

5.1 Introduction

Extremophiles are a class of organisms that possess the remarkable ability to thrive and proliferate in environments characterized by extreme temperatures, both exceptionally high and low, which are typically inhospitable to conventional organisms. The term extremophiles originates from the Latin words 'extremus', denoting 'extreme', and 'philia', signifying 'love'. The term 'extremophile' was originally

V. M. Pathak (✉) · S. Pandey · S. Tyagi
Research and Development, Pritam International Private Ltd. Bhagwanpur,
Roorkee, Haridwar, India

N. Rana
Department of Environmental Science, Dr Yashwant Singh Parmar University of Horticulture
and Forestry, Solan, India

A. K. Sarkar
Department of Life Science, Shri Rawatpura Sarkar University, Raipur, Chhattisgarh, India

A. Chauhan · T. Jindal
Institute of Environmental Toxicology Safety and Management, Amity University, Noida,
India

M. Sharma
School of Life Science technology IIMT University, Meerut, India

S. Dewali · S. S. Bisht
Department of Zoology, Kumaun University, Nainital, Uttarakhand, India

© The Author(s), under exclusive license to Springer Nature Switzerland AG 2024
A. Ranjan et al. (eds.), *Extremophiles for Sustainable Agriculture and Soil
Health Improvement*, https://doi.org/10.1007/978-3-031-70203-7_5

introduced by MacElroy in the year 1974. Extremophiles can be categorized into two distinct types based on their ability to thrive in extreme environments: extremophilic species and extremotolerant species. This classification is determined by the specific conditions under which these organisms can grow and survive. The 'extremotolerant organisms' pertain to a category of organisms that typically thrive in optimal environmental conditions but also possess the remarkable ability to withstand and adapt to adverse environmental conditions. According to the study conducted by Gupta et al. (2014), it has been observed that extremophilic organisms possess the remarkable ability to flourish in diverse and highly challenging environmental conditions. The evolutionary diversity observed among extremophiles is typically characterized by its vastness and intricacy. This is because certain taxonomic groups exclusively comprise extremophilic organisms, while others encompass a combination of extremophiles and non-extremophiles. Extremophiles, organisms that thrive in extreme environments, exhibit a remarkable distribution across the phylogenetic tree of life due to their evolutionary adaptations enabling survival in these inhospitable conditions. This phenomenon holds for a variety of

Divya
Food Customization Research Lab, Centre for Rural Development and Technology, Indian Institute of Technology Delhi, New Delhi, India

A. Sharma
Allergy and Immunology Section, CSIR-IGIB, New Delhi, India

M. Yadav
Cancer Biology Laboratory, School of Life Sciences, Jawaharlal Nehru University, New Delhi, India

B. S. Rawat
Department of Pharmaceutical Sciences, Gurukul Kangri Deemed to be University, Haridwar, India

D. P. Uniyal
Uttarakhand State Council for Science & Technology (UCOST), Vigyan Dham, Jhajra, Dehradun, Uttarakhand, India

P. Kumar
Agricultural and Biosystems Engineering, South Dakota State University, Brookings, SD, USA

S. Singh
Multidisciplinary Clinical Translational Research, Translational Health Science and Technology Institute, NCR Biotech Science Cluster, Faridabad, Haryana, India

Jawaharlal Nehru University, New Delhi, India

B. Kaur
Indian Institute of Technology Bombay, Mumbai, Maharashtra, India

J. M. Cunill
Biotechnology Engineering, Universidad Politécnica Metropolitana de Puebla, Mexico, Mexico

S. C. Garkoti
School of Environmental Sciences, Jawaharlal Nehru University, New Delhi, India

organisms known as barophiles or psychrophiles, which exhibit a wide distribution across the three domains of life.

5.2 Physiological Adaptations of Extremophiles

Earth exhibits an extraordinary array of life forms, producing a remarkable level of diversity. These organisms have demonstrated their ability to adapt and flourish in environments that were previously considered inhospitable or incapable of supporting life. One of the most intriguing facets of this remarkable adaptability lies in the presence of extremophiles, and these microorganisms possess the remarkable ability to not only endure but also thrive in exceedingly harsh environmental conditions that would prove fatal to the majority of other life forms. The scientific community has been captivated by the physiological adaptations exhibited by extremophiles, which enable them to flourish in exceedingly hostile environments. These adaptations have provided researchers and scientists with invaluable knowledge regarding the boundaries of life and the prospects of habitable conditions in extraterrestrial environments (Fig. 5.1).

Extremophilic microorganisms, which are characterized by their ability to thrive in extreme environmental conditions, have developed adaptive mechanisms to ensure the stability of their cellular metabolites. Extremophiles are a fascinating group of organisms that possess the remarkable ability to thrive in diverse and challenging environments characterized by extreme conditions. These conditions encompass a wide range of factors, such as elevated temperatures, high salt concentrations, alkaline or acidic pH levels, the presence of organic solvents, and the

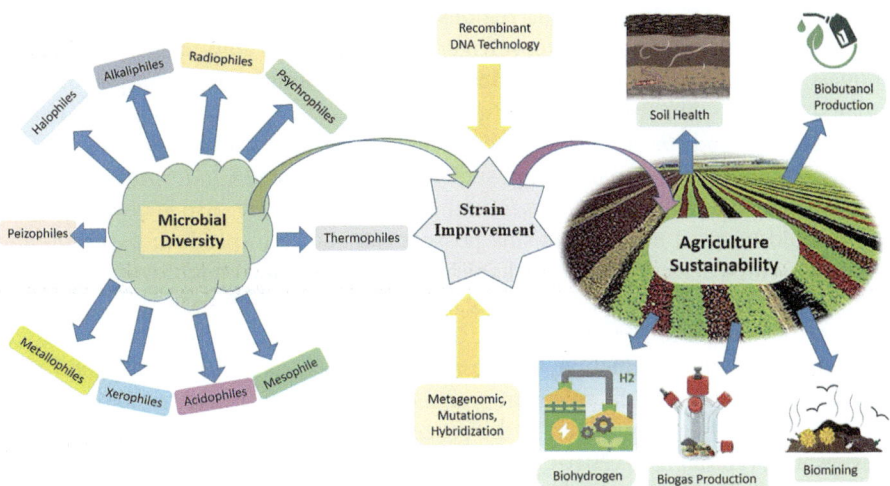

Fig. 5.1 Microbial diversity and approaches used in sustainable agriculture practices (Kaur et al. 2022)

abundance of metal ions, among others. Certain extremophiles have unique adaptation mechanisms that render them highly pertinent for comprehending not only fundamental principles but also their potential biotechnological applications. Notably, these extremophiles exhibit the remarkable ability to endure multiple extreme conditions.

5.2.1 Physiology of Psychrophiles and Impact on Agriculture

Psychrophiles are a class of organisms that exhibit the remarkable ability to thrive and persist within a temperature range of 0–20 °C. Notably, these organisms display optimal growth conditions at a temperature of 5 °C. Psychrophiles are organisms that thrive in cold environments and have developed several unique characteristics that enable them to survive and function effectively at low temperatures. One such adaptation is the presence of unsaturated fatty acids in their cell membranes. These unsaturated fatty acids facilitate the transport of solutes across the membrane, even under cold conditions. Additionally, psychrophiles possess cold shock proteins, which help protect their cellular components from damage caused by sudden temperature changes. Furthermore, these organisms produce cryoprotectants, which are substances that aid in the synthesis of cold-adapted enzymes. These enzymes are crucial for the psychrophiles' metabolic processes in cold environments (Cavicchioli et al. 2013). In the context of low temperatures, various alterations occur within the cellular machinery. These modifications encompass a shift in membrane fluidity, a reduction in the processes of transcription and translation, as well as a modification in the structure of ribosomes. Psychrophiles, also known as cold-adapted organisms, exhibit the remarkable ability to thrive in exceedingly frigid environments. These habitats include the polar regions, glaciers, deep oceans, shallow underground regions, high atmosphere, refrigerators, as well as cold-adapted plants and animals. The microorganisms under investigation predominantly belong to the bacterial family, encompassing *Pseudoalteromonas*, *Pseudomonas*, *Vibrio*, *Arthrobacter*, and *Bacillus*, as well as archaea, specifically *Halorubrum* and *Methanogenium*. Additionally, fungi such as *Penicillium* and *Cladosporium*, as well as yeast species including *Candida* and *Cryptococcus*, are also included in this study. Cold-adapted microorganisms possess the ability to produce psychrophilic enzymes, which demonstrate remarkable catalytic efficiency under low-temperature conditions. These enzymes hold significant potential in various industries, including food processing, leather production, detergent manufacturing, pharmaceutical development, wine production, textile manufacturing, as well as paper and pulp production. Psychrophiles are microorganisms adapted to thrive in cold environments and have been employed in the process of bioremediation to treat soils and wastewaters (Hasan et al. 2010).

Psychrophiles exhibit significant potential in revolutionizing agricultural practices due to their involvement in nutrient cycling, organic matter decomposition, and their ability to be used as biofertilizers and biostimulants. The capacity of these organisms to operate optimally under cold temperatures and their potential to support sustainable agricultural practices are in accordance with the worldwide pursuit of resilient agricultural systems capable of adapting to changing climatic circumstances. The investigation of psychrophiles and their utilization in the field of agriculture represents the convergence of scientific exploration, technological advancement, and pragmatic implementation. Through the elucidation of the enigmatic characteristics possessed by these microorganisms adapted to cold environments, we may discover techniques that not only improve the productivity of crops and the vitality of soil but also foster an agricultural ecosystem that is more ecologically sustainable and capable of withstanding adverse conditions. In the face of contemporary challenges, the agricultural sector is currently exploring novel avenues to enhance scientific knowledge and ensure food security. Among these promising frontiers, the potential contributions of psychrophiles, organisms adapted to thrive in cold environments, have garnered considerable attention.

5.2.2 Thermophiles' Role and Benefits in Agriculture Practices

Thermophiles are microorganisms that exhibit and thrive in high-temperature environments. The term 'thermo' derives from the Greek word for heat, while 'phile' denotes an affinity or attraction towards a particular condition or element. Thermophilic microorganisms exhibit optimal growth conditions at temperatures exceeding 50 °C. According to Gupta et al. (2014), the Earth encompasses diverse ecological zones, such as hot springs, the deep sea, and geothermal soils, which exhibit insulation properties against thermophiles.

Hyperthermophiles, sometimes also referred to as extreme thermophiles, are a distinct category of microorganisms that exhibit optimal growth conditions at temperatures as high as 80 °C. The Archean genera *Pyrodictium*, *Pyrobaculum*, *Melanopyrus*, and *Pyrococcus* are known to exhibit remarkable thermal tolerance, with maximal growth temperatures ranging from 103 to 110 °C. The families Ascomycete and Zygomycete, belonging to the kingdom Fungi, exhibit notable tolerance to high temperatures in their growth environments. Additionally, there are two bacterial species, namely, *Thermotoga maritime* and *Aquifex pyrophilus*, which are known to possess the highest recorded growth temperatures of 90 °C and 95 °C, respectively. Thermophile enzymes have shown considerable potential in various industries, including food processing, starch production, leather manufacturing, textile production, detergent formulation, pulp and paper production, genetic engineering, and pharmaceutical development (Kumar et al. 2011). One of the primary areas of interest lies in the investigation of the pivotal role played by thermophilic microorganisms in the processes of composting and the decomposition of organic matter. The ability of these organisms to withstand high temperatures and exhibit enzymatic activities enables them to effectively degrade organic substances, thereby resulting in the generation of compost that is abundant in nutrients. In addition, thermophilic microorganisms play a significant role in the efficient decomposition and recycling of agricultural waste, thereby helping to alleviate the environmental impact associated with the accumulation of organic residues. Another important aspect that warrants attention is the inherent capacity of thermophiles to synthesize enzymes that possess exceptional thermal stability, thereby rendering them highly suitable for agricultural applications.

The utilization of these enzymes has the potential to augment a wide range of processes, like encompassing the bioconversion of biomass, as well as the synthesis of biofertilizers and biopesticides. The inherent stability exhibited by these substances when exposed to high temperatures assures their ability to maintain optimal functionality in the face of the various environmental conditions encountered within agricultural systems. In addition, thermophilic microorganisms have the potential to make valuable contributions towards enhancing soil health and fertility. The enzymatic activities of these organisms play a crucial role in the process of nutrient cycling and facilitating the release of vital elements that are essential for the uptake by plants. Furthermore, the utilization of biostimulants derived from thermophilic organisms has the potential to augment both the growth and resilience of plants, particularly in environments that are prone to experiencing fluctuations in temperature. The use of thermophiles in agricultural practices provides a comprehensive and multifaceted strategy for tackling current challenges in the field of agriculture. Due to their unique physiological adaptations and enzymatic abilities, thermophiles have the potential to significantly improve soil health, facilitate waste recycling, and support sustainable crop production. In the pursuit of novel approaches within agricultural systems, there is an increasing interest in exploring the unexploited capabilities of thermophilic organisms. The area of study holds great promise, as it sheds light on the intricate interaction between extremophile biology and the progress of both worldwide food security and environmental sustainability.

5.2.3 Alkaliphiles: Physiology and Their Agriculture Benefits

Alkaliphiles are classified as a distinct category of extremophiles that exhibit the remarkable ability to thrive in environments characterized by alkaline conditions, specifically those with a pH level exceeding 8. These environments include soda lakes and soils abundant in carbonate compounds. Alkaliphilic organisms are capable of thriving in environments characterized by high alkalinity or basicity. These organisms have evolved unique adaptations that enable them to withstand and even flourish in such conditions, which are typically hostile to most other life forms.

Alkaliphiles can adapt to alkaline conditions by utilizing electrogenic, secondary cation/proton antiporters. These antiporters play a crucial role in maintaining the homeostasis of cytoplasmic pH and facilitating the absorption of protons (H1) (Mirete et al. 2017). To protect themselves from the adverse effects of environmental alkalinity, these microorganisms use a defensive mechanism by synthesizing an external coating composed of acidic compounds. These compounds include acidic amino acids, teichuronic acid, and teichurono peptide. Alkaliphiles and acidophiles exhibit significant potential for utilization in the field of biotechnology. Alkaliphiles, a group of microorganisms that thrive in alkaline environments, possess a diverse array of pH-stable enzymes. Proteases, amylases, cellulases, lipases, chitinases, pullulanases, pectinases, and xylanases are a few of these enzymes, and these enzymes exhibit remarkable stability under alkaline conditions, enabling alkaliphiles to efficiently carry out various biochemical processes in their extreme habitats. Enzymes find widespread application in various industries, including high pH operations, detergent manufacturing, the pulp and paper sector, hide-dehairing processes, food processing, and starch-hydrolysis (Gomes and Steiner 2004).

The distinctive physiological adaptations of alkaliphiles enable them to thrive in environments that pose challenges for conventional crops, thereby presenting novel opportunities for improving soil fertility, nutrient cycling, and crop productivity. Alkaliphilic microorganisms have developed distinct adaptive mechanisms to effectively endure and respond to the challenging conditions posed by extreme alkaline pH levels and elevated salinity. The remarkable capacity of these organisms to flourish in extremely adverse environments serves to highlight their considerable potential in effectively mitigating soil health issues. The utilization of alkaliphiles' capabilities is being pursued by researchers to formulate sustainable agricultural approaches that effectively address the challenges posed by climate change and conventional farming methods.

The study aims to explore the potential application of alkaliphiles in agricultural practices, with a specific focus on their contribution to enhancing soil health, augmenting nutrient availability, and optimizing crop performance. Through a comprehensive understanding of the distinctive adaptations exhibited by alkaliphiles, scholars and professionals in the field of agriculture can endeavour to cultivate novel biotechnological interventions that effectively enhance the sustainability and resilience of contemporary agricultural practices (Horikoshi 1999; Oren 2013; Kharade et al. 2015; Yadav et al. 2015; Zhang et al. 2017).

5.2.4 Acidophiles' Role in Crop Productivity

Acidophilic organisms are a class of living organisms that exhibit robust growth and survival in environments characterized by elevated levels of acidity, typically with a pH range between 3 and 4. Acidophilic microorganisms, known as acidophiles, have been observed to inhabit diverse environments characterized by extreme acidity.

This microorganism includes environments like volcanic vents, sulfur springs exhibiting high sulfur concentrations, acid rock drainage sites, areas affected by metal mining activities, and regions impacted by acid mine drainage. Microbial organisms have undergone evolutionary adaptations to maintain cellular pH homeostasis at a neutral level and acquire resistance to metals. If environment conditions are characterized by high acidity levels, it is possible to observe the presence of various taxonomic groups, namely, *Acidobacterium*, *Picrophilus*, *Ferroplasma*, and

Leptospirillum, as reported by Gupta et al. (2014). Typically, these microorganisms exhibit a preference for maintaining an intracellular pH that is close to neutral on the pH scale. The regulation of proton influx into the cell, known as membrane impermeability, plays a vital role in determining intracellular pH. The cell membrane of archaea exhibits a notable characteristic of low proton influx permeability, which has led to their evolutionary adaptation in order to withstand acidic conditions. The genomic composition of *Leptospirillum ferriphilum* encompasses a repertoire of genes that are responsible for the synthesis of cellular membrane components and structural constituents of the cell wall, thereby conferring the organism with the ability to withstand and adapt to acidic pH conditions. The process of proton sequestration is facilitated by various adaptations. One such adaptation involves the reduction in the size of membrane pores, which restricts the movement of protons across the membrane. Additionally, proton efflux mechanisms, such as antiporters, symporters, and H1 ATPases, play a crucial role in aiding proton sequestration. These mechanisms actively transport protons out of the cell or into specific compartments. Furthermore, the accumulation of buffering elements, such as arginine, histidine, and lysine, contributes to the sequestration of protons. These buffering elements help maintain the pH balance within the cell by binding to excess protons, thereby preventing their detrimental effects. In addition to the aforementioned adaptation strategies, organisms also employ other mechanisms to cope with low pH conditions. One such strategy involves the breakdown of organic acids, which facilitates proton dissociation. This process helps maintain the appropriate pH balance within the cellular environment. Furthermore, organisms utilize chaperones to protect vital biomolecules, such as deoxyribonucleic acid (DNA) and proteins, from oxidative damage that may occur under low pH conditions. These chaperones act as molecular shields for safeguarding the structural integrity and functionality of these biomolecules. Acidophilic microorganisms have been extensively utilized in various biotechnological applications, such as the bioprocessing of minerals, energy production, and the bioleaching of metals from ores with low-grade mineral content (Mirete et al. 2017).

Acidophiles are microorganisms that exhibit an exceptional capacity to flourish in environments characterized by high acidity levels. The aforementioned habitats characterized by high acidity levels, which were previously deemed unsuitable for the sustenance of a wide array of organisms, encompass various settings such as acid mine drainage, acid soils, and acidic hot springs. The distinctive physiological adaptations exhibited by acidophilic organisms confer upon them the ability not only to endure harsh acidic environments but also to actively participate in a multitude of ecologically significant processes. In light of the various challenges encountered in the field of agriculture, such as those pertaining to soil health, nutrient availability, and the implementation of sustainable practices, there has been a growing interest in exploring the potential of acidophilic microorganisms as a viable means of enhancing agricultural methodologies (Ghorbel et al. 2017; Breccia and Vázquez 2018).

5.2.5 Halophiles in Sustainable Agriculture Production

Halophiles are microorganisms that exhibit optimal growth and survival in environments characterized by high salt concentrations and have gathered significant attention in the scientific community due to their potential applicability in addressing various agricultural challenges. Halophiles exhibit distinctive biochemical and physiological adaptations that render them highly suitable for utilization in agricultural practices, particularly in areas characterized by soils affected by salinity. Through the utilization of halophiles, a group of organisms capable of thriving in high-salinity environments, scientists and agricultural specialists aspire to bring about a paradigm shift in conventional farming methodologies. This endeavour not only seeks to address the pressing issue of global food security but also endeavours to uphold the principles of environmental sustainability. Halophiles, a group of microorganisms that thrive in high-salt environments, present numerous promising prospects for utilization in the field of agriculture. The management of soil salinity involves the utilization of halophiles, which are organisms that have developed adaptive mechanisms to thrive in environments with high salt concentrations. The inclusion of halophiles in soil ecosystems can potentially contribute to the mitigation of soil salinity, thereby reducing the overall salt levels in the soil.

The organisms under consideration exhibit a maximum growth rate of 15% or when subjected to a salt concentration of 2.5 M. To thrive in environments characterized by elevated salt concentrations, microorganisms necessitate specific adaptations in their cellular and enzymatic machinery, which enable them to uphold osmotic equilibrium. Cellular adaptations can be achieved through the regulation of salt concentrations within the cell and its surrounding environment, as well as the synthesis of osmoprotectants using suitable organic solutes. Various strategies have been employed to enhance survival in environments characterized by high salinity. These strategies encompass the reduction of salt intake, utilization of organic solutes, and adoption of high salt tolerance mechanisms. According to Verma et al. (2017), halophiles have developed a greater abundance of negatively charged amino acids. The increased presence of negatively charged amino acids allows halophiles to effectively compete for ions during the process of ion exchange. As a result, proteins are solubilized more efficiently, leading to an enhanced adaptation of enzymes in halophiles.

The microorganisms under consideration possess the ability to enhance soil quality in regions affected by salinity through the promotion of halophytic plant growth and the facilitation of excess salt removal from the soil. The utilization of halophiles, a group of microorganisms adapted to high salt concentrations, holds significant potential in the field of agriculture. These halophiles have been found to produce a diverse array of bioactive compounds, which have been shown to effectively enhance both plant growth and stress tolerance. Consequently, the application of these bioactive compounds used as biofertilizers and biostimulants and has garnered considerable attention in scientific research and agricultural practices. The aforementioned components encompass a variety of substances that play a crucial role in promoting plant growth. These substances consist of plant-growth-promoting hormones, osmoprotectants, and enzymes. The utilization of these inherent characteristics has the potential to facilitate the advancement of biofertilizers and biostimulants derived from halophiles, which can effectively enhance agricultural output, particularly in saline soil conditions. The investigation of halophiles, organisms capable of thriving in high salinity environments, and their genetic adaptations, holds significant potential for the advancement of salt-tolerant crop development through the breeding of resilient plant varieties. The primary aim of researchers is to cultivate crops that are capable of flourishing in regions affected by high salinity through the identification and subsequent transfer of pertinent genes or metabolic pathways.

The enhancement of soil health is a critical aspect of agricultural and environmental management. In this context, halophiles, which are microorganisms adapted to high salinity conditions, have been identified as key contributors to nutrient cycling and organic matter decomposition in saline environments. The introduction of these microorganisms into conventional agricultural soils has the potential to significantly improve soil structure, enhance nutrient availability, and promote overall soil health. The role of biopesticides in the management of plant diseases certain species of halophilic microorganisms can synthesize antimicrobial compounds, which possess the potential to be utilized in the development of biopesticides. This presents a promising avenue for the creation of environmentally sustainable alternatives to conventional chemical pesticides. Furthermore, the facilitation of advantageous microbial assemblages by halophilic organisms has the potential to play a significant role in the suppression of diseases. The incorporation of halophiles into agricultural methodologies presents considerable promise in addressing the issue of soil salinity, enhancing crop productivity, and fostering the development of sustainable farming systems. Researchers seek to derive innovative solutions that align with the objectives of ensuring food security and preserving the environment by drawing inspiration from the natural adaptations exhibited by halophiles.

5.2.6   Piezophiles Effects on Crop Production and Productivity

Piezophiles, which refer to microorganisms possessing the capability to flourish in fragmented or modified environments, have attracted considerable interest due to their remarkable capacity to colonize and restore disturbed soil ecosystems. In the context of contemporary agriculture, which is currently confronted with various issues of soil degradation, nutrient depletion, and the imperative of sustainable land management, it becomes imperative to delve into the significance of piezophiles in fostering soil health and augmenting agricultural productivity. Microorganisms that exhibit a preference for inhabiting environments characterized by elevated pressure levels are commonly denoted as 'piezophiles' or 'barophiles'. These particular species can thrive in environments characterized by elevated salt concentrations, surpassing the typical levels found in their natural habitats. These microorganisms are commonly encountered in the profound depths of the oceanic environment, specifically within the genera *Shewanella*, *Pyrococcus*, *Methanococcus*, and *Moritella*. Certain organisms, referred to as piezophiles, have the remarkable ability to endure and thrive in environments characterized by exceedingly elevated hydrostatic pressure levels. Piezophiles are organisms that exhibit the expression of specific genes that are involved in responding to adapt to high-pressure environments. These genes are activated in response to the pressure exerted on the organisms, allowing them to function effectively under such conditions. In the realm of the food industry, the application of high-pressure techniques for the preparation and sterilization of food products has gained significant attention. This approach involves the utilization of halophiles and microorganisms capable of thriving in high-pressure environments. Additionally, the potential utilization of piezophilic bacteria and their enzymes in the field of biotechnology has emerged as a promising avenue for further exploration. Following the findings of Gupta et al. (2014), it has been determined that these particular organisms possess a growth potential of 15%. Piezophiles, which refer to microorganisms exhibiting the capacity to flourish in fragmented or modified environments, have attracted considerable scientific interest due to their remarkable capability to colonize and remediate disrupted soil ecosystems. In light of the contemporary challenges faced by the agricultural industry, such as soil degradation, nutrient loss, and the need for sustainable land management practices, it is imperative to delve into the significance of piezophiles in fostering soil health and augmenting agricultural productivity. The captivating realm of piezophiles and their prospective contributions to agricultural practices has garnered considerable interest. Through a comprehensive exploration of the underlying mechanisms that facilitate the thriving of piezophiles in environments characterized by disruption, we can unveil potential avenues for leveraging their distinctive attributes in the context of soil restoration, nutrient cycling, and the enhancement of overall ecosystem resilience. The study conducted by Smith et al. (2020) revealed that the researchers examined the influence exerted by piezophilic microorganisms on the soil structure and nutrient accessibility within agriculturally depleted land. Furthermore, the study conducted by Jones and Patel (2018) provides valuable insights into the intricate molecular adaptations that facilitate the successful colonization and proliferation of piezophiles within habitats that have undergone fragmentation.

5.2.7   Xerophiles in Sustainable Agriculture Practices

Xerophiles are a class of extremophiles that have developed adaptations to thrive in arid and water-limited habitats and have emerged as highly valuable assets with the potential to significantly transform agricultural methodologies in regions confronted with the pressing issues of water scarcity and drought. The microorganisms under consideration have undergone notable evolutionary changes in their physiological and biochemical makeup, enabling them to flourish in environments where water is scarce. Xerophiles are a unique group of microorganisms that possess remarkable adaptability and resilience in environments characterized by significant aridity or limited water availability. In the context of this particular ecological environment, there is a limited range of distinct genera comprising bacteria, fungi, algae, yeasts, and lichens that possess an inherent capacity for long-term survival and thriving. Dry commodities, such as grains, spices, nuts, and oilseeds, are susceptible to degradation due to the presence of xerophilic microorganisms. These microorganisms have adapted to thrive in low moisture environments and can cause spoilage and deterioration of the commodities. The microorganisms in question exhibit a predilection for environments characterized by low moisture levels, thereby leading to a decline in the overall quality and safety of the affected commodities. The observation mentioned earlier was given additional emphasis through a comprehensive investigation conducted by Rothschild and Mancinelli in the year 2001.

In light of the escalating pressure on water resources and the imperative for sustainable approaches in agriculture, the utilization of xerophiles presents a potential avenue for augmenting crop productivity, ameliorating soil health, and safeguarding food security. The utilization of xerophiles in the field of agriculture and their capacity to effectively tackle water-related predicaments. Through a comprehensive analysis of the potential contributions of these microorganisms, the study aims to elucidate their role in augmenting water-use efficiency, improving soil structure, enhancing nutrient availability, and bolstering plant resilience. By shedding light on these intricate mechanisms, novel strategies may emerge that have the potential to revolutionize conventional agricultural methodologies. Bacterial communities play a pivotal role in various aspects of plant physiology and soil health. They have been found to significantly contribute to the enhancement of drought tolerance in crops, primarily through their ability to modulate plant responses to water scarcity. Additionally, these bacterial communities have been observed to positively influence soil structure by secreting exopolysaccharides, which aid in improving soil aggregation and stability. Furthermore, they play a crucial role in nutrient cycling, particularly in nutrient-poor soils, by facilitating the availability and uptake of essential nutrients by plants. Lastly, these bacterial communities have the potential to alleviate the detrimental effects of salinity stress on plants, thereby offering a promising avenue for mitigating the negative impacts of high salt concentrations on crop productivity. Xerophilic organisms, which are adapted to thrive in dry environments, offer a strong effective approach to mitigating the urgent issues associated with limited water availability in agricultural practices (Smith and Des Marais 2004).

Through a comprehensive understanding and effective utilization of the various adaptations exhibited by these microorganisms, it is plausible to explore innovative approaches aimed at augmenting both the quantity and quality of crop production, all the while ensuring the sustainable conservation of water resources. In light of the global challenge posed by climate change and the increasing scarcity of water resources, the incorporation of xerophiles into agricultural systems holds great promise as a pivotal measure in advancing the goals of resilient and sustainable food production.

5.2.8 Radiophiles Use in Sensitive Agriculture Area

Radiophiles are a group of microorganisms that demonstrate exceptional resilience when exposed to heightened levels of ionizing and ultraviolet (UV) radiation. The application of microorganisms possesses resistance to hold radiations by considerable promise in the field of radioactive environmental waste management and remediation. The identification of numerous bacteria exhibiting resistance to radiation has been achieved by subjecting their cultures to intense gamma radiation (γ) in controlled laboratory conditions. *Deinococcus radiodurans* is an organism of significant interest due to its extraordinary ability to withstand high levels of radiation and holds a unique position as the only species for which a successful method of genetic transformation and manipulation has been developed (Gupta et al. 2014). Presently, there are ongoing endeavours to manipulate this particular organism to harness its distinctive capacities to efficiently eradicate radioactive waste materials. In recent times, the scientific community has been captivated by radiophiles, a distinct assemblage of extremophiles that have evolved to thrive in environments characterized by elevated levels of ionizing radiation. This intriguing group of organisms has garnered significant interest from researchers and scientists due to their prospective utility across diverse domains. Radiation is commonly acknowledged as detrimental to living organisms owing to its mutagenic properties and ability to inflict damage upon cellular structures. However, certain organisms known as radiophiles have developed distinctive adaptations that enable them to flourish in environments characterized by high levels of radiation. The remarkable adaptability exhibited by radiophiles presents fascinating prospects for leveraging their capabilities in pioneering agricultural methodologies that seek to enhance crop productivity, soil vitality, and ecological sustainability. The capacity of radiophiles to endure and exploit ionizing radiation offers a promising and innovative approach to tackling the various obstacles encountered in contemporary agriculture. The concept of employing radiation as a stress-inducing agent to elicit advantageous reactions in plants and soil ecosystems is under the overall comprehension of hormesis, a phenomenon wherein minimal levels of stress can instigate favourable adaptive alterations. The present study aims to investigate the prospective utilization of radiophiles in the realm of agricultural practices, with a specific focus on comprehending the underlying mechanisms responsible for their resistance to radiation. Furthermore, the research endeavours to elucidate how these mechanisms could be effectively employed to augment both crop productivity and soil health (Singh et al. 2017; Zhang et al. 2019; Wang 2020).

5.2.9 Metallophiles in Agriculture Applications

Metallophiles are a taxonomic group of microorganisms that possess the extraordinary capacity to flourish in habitats that are distinguished by high levels of metallic substances, commonly known as trace elements. In recent years, there has been a growing focus on the extraction and elimination of perilous heavy metals from various environmental matrices such as soils, sediments, and wastewater. The increased attention towards this subject arises from the acknowledgement of the substantial peril by heavy metal pollution, particularly involving chromium, copper, zinc, cadmium, cobalt, lead, and mercury, which presents to human well-being, as well as the well-being of aquatic ecosystems and wildlife. According to the findings of Valls and de Lorenzo (2002), the existence of metallophiles presents a noteworthy opportunity for the retrieval of valuable metals from industrial effluents via the biomining process. To successfully achieve this objective, a wide array of interactions between microorganisms and metals are utilized. These interactions involve various processes, including reduction for anaerobic respiration, reduction for detoxification purposes, bioleaching, biosorption, biomineralization, and bioaccumulation. The inhibition of microbial growth and activity is frequently achieved through the interference of heavy metals with essential functional groups, displacement of important metal ions, or alteration of the active conformation of biological components.

Metallophiles, a distinct group of microorganisms that have evolved specialized adaptations to flourish in environments abundant in metals, present a promising and innovative approach to augmenting soil fertility and facilitating the adoption of sustainable agricultural methodologies. Metallophiles exhibit a remarkable capacity to endure and accumulate substantial quantities of heavy metals, including copper, zinc, and nickel, which are commonly toxic to other living organisms. The captivating phenomenon of this adaptation has sparked considerable interest in exploring the utilization of metallophiles as a means to augment soil health, amplify nutrient accessibility, and alleviate metal contamination in agricultural soils. The utilization of the metabolic capacities exhibited by metallophiles has the potential to enhance the efficiency, environmental sustainability, and resilience of agricultural practices when confronted with challenges associated with metals. This discourse delves into the prospective applications of metallophiles in the realm of agricultural practices, with a particular emphasis on their pivotal role in the domains of nutrient cycling, soil remediation, and biofortification. By comprehensive exploration of the underlying mechanisms that govern the metal tolerance and accumulation characteristics exhibited by metallophiles, valuable insights can be gained to develop novel approaches aimed at addressing significant agricultural challenges while simultaneously mitigating the adverse ecological consequences associated with heavy metal contamination.

5.3 Extremophiles for Sustainable Agriculture

The adoption and integration of sustainable agricultural practices are of predominantly importance in enhancing agricultural productivity as well as minimizing resource inputs and in response to the increasing global demand for food and various agricultural by-products. The significance of the interaction between plants and microorganisms extends beyond the realm of crop yield, making it a matter of paramount importance in the broader context. The utilization of precision agriculture management (PMI) methodologies has been observed to yield substantial improvements in crop productivity, concomitantly safeguarding the holistic welfare and vitality of the cultivated plants. The recognition of extremophilic microorganisms in symbiosis with plants has garnered significant attention due to their remarkable capacity to augment agricultural yield, bolster plant resilience, and enhance overall plant adaptability under arduous and extraordinary physiological circumstances. Therefore, the utilization of extremophilic microorganisms as biofertilizers or biopesticides offers substantial benefits and advantages in diverse applications within the realm of agricultural practices and pest control management. Extremophiles, a taxonomic group of organisms characterized by their ability to thrive in environments that are considered extreme by conventional standards, have developed a diverse range of resilient mechanisms to endure and acclimate to their inhospitable surroundings. These organisms exhibit distinctive molecular mechanisms that facilitate the synthesis of bioactive compounds and extremozymes, thereby presenting considerable prospects for diverse scientific and industrial utilization. Extremophiles, organisms that thrive in extreme environmental conditions, possess distinctive physiological adaptations that have been observed to have a significant impact on promoting effective and environmentally sustainable agricultural methodologies (Yadav and Saxena 2018).

5.3.1   Biomining

The application of microorganisms is used to extract metals from their respective oxides and sulfides are commonly referred to as bioleaching. Within this context, extremophilic enzymes play a crucial role. In a comparative analysis, it can be argued that this method offers a higher degree of safety and security. Two distinct bioleaching methods are commonly employed in various scientific and industrial applications, namely, direct bioleaching and indirect bioleaching. In the process of direct leaching, it has been observed that metal sulfide, a compound whose existence has been subject to some uncertainty, undergoes electron transfer directly to the cells that are interconnected with mineral surfaces. Iron (II) ions can undergo oxidation reactions with planktonic or microorganisms that are attached to mineral surfaces. This process, known as indirect leaching, results in the formation of iron (III) ions. These iron (III) ions then proceed to oxidize metal sulfides. This phenomenon has been observed and documented by Pattanaik et al., in the study conducted in 2020. Li and Wen (2021) have reported that certain acidophilic microorganisms, such as *Acidihalobacter ferrooxidans* and *Acidihalobacter prosperus*, exhibit potential applications in the field of biometallurgy. Archaebacteria, such as *Sulfolobus* and *Metallosphaera*, exhibit remarkable adaptability to extreme thermal environments, rendering them highly suitable candidates for biomining operations. These microorganisms thrive under conditions characterized by exceedingly elevated temperatures. In a study conducted by Deshpande et al. (2018), it was observed that certain microorganisms involved in biomining possess the capacity to sequester carbon dioxide (CO_2) and thrive in an oxygen-rich setting. When examining the efficiency of energy utilization, biomining processes, specifically the roasting and smelting techniques, demonstrate a notable advantage over conventional mining methods. The cost-effective leaching of both low- and high-grade ores presents a favourable approach that avoids the generation of hazardous gases, such as sulfur dioxide (SO_2). One effective strategy for mitigating environmental degradation caused by dumping or leakage is the implementation of measures aimed at reducing acid mine drainage, as highlighted by Jerez (2012).

5.3.2   Biofuel Production

A viable approach to mitigate the dependence on fossil fuels involves the utilization of biomass obtained from diverse sources, including corn, wheat, and sugar cane (Coker 2016; Chakraborty et al. 2020). The production of biofuels can be classified into two discrete generations based on technological advancements and feedstock utilization. The initial iteration of biofuels encompasses the process of transforming readily available agricultural commodities, including starches, oils, and hydrolysed sugars, into usable energy sources. These crops possess the inherent potential to be efficiently converted into biofuels via a multitude of diverse conversion methodologies. Conversely, the subsequent iteration of biofuels is derived from cellulosic materials characterized by a heightened capacity to withstand hydrolysis. The aforementioned materials are employed in the manufacturing process of second-generation biofuels. Microbial-derived biofuels comprise a wide array of energy sources, including biobutanol, bioethanol, hydrogen, and methane. These biofuels are derived from microorganisms and hold significant potential as sustainable alternatives to conventional fossil fuels. The achievement of synthesizing biofuels has required the application of highly acidic or alkaline pH levels and elevated temperatures at different stages of the procedure. In traditional methodologies, mesophilic enzymes are frequently utilized; however, they can be easily replaced by extremophiles, which thrive in extreme environmental conditions. An illustrative instance of an organism that generates methane as an integral component of its physiological processes can be observed in the case of methanogens. In the study conducted by Barnard in 2010, it was observed that the microorganism *Thermoanaerobacterium saccharolyticum* demonstrates a metabolic capability to effectively utilize hemicellulose and pentose sugar as its principal substrates for ethanol production.

5.4 Industrial Applications of Extremophiles

5.4.1 In Agricultural Industry

Extremophiles are a class of microorganisms characterized by their ability to flourish in highly challenging environmental conditions and have attracted considerable interest due to their potential utility in diverse agricultural methodologies and practices. These microorganisms have been extensively investigated for their potential applications as biofertilizers, bioinoculants, and biocontrol agents, with the primary objective of enhancing agricultural productivity and facilitating plant growth in areas that are known for their adverse environmental conditions. The aforementioned factors encompass a diverse array of challenges that span a broad spectrum. These challenges include but are not limited to the presence of exceedingly low temperatures, alterations induced by climate change, issues pertaining to irrigation, the detrimental effects of deforestation, elevated levels of salinity, and the prevalence of drought conditions. Through the utilization of the distinctive attributes possessed by extremophiles, agricultural systems have the potential to effectively alleviate the deleterious consequences associated with unfavourable environmental conditions, thereby augmenting the overall productivity of crops. Extremophiles possess the inherent capacity to function as a feasible alternative to conventional agricultural practices, as they can contribute significantly to diverse facets of plant growth and development. They encompass crucial processes such as nutrient cycling, fixation, mineralization, and solubilization. Furthermore, it is worth noting that these microorganisms possess certain characteristics that render them capable of serving as biofertilizers and bioinoculants, as emphasized by Yadav and Saxena in their comprehensive investigation carried out in the year 2018. Within the field of biological control, these organisms have exhibited their potential as highly efficient agents for the management of pests, thereby highlighting their remarkable capacity to elicit resistance in the target organisms. The organisms under consideration display a remarkable range of genetic diversity, which holds promise as a feasible substitute for products reliant on chemicals. This, in turn, can contribute to the advancement of economically efficient, environmentally friendly, and enduring agricultural practices within the agro-industrial sector (Yadav 2021) (Fig. 5.2).

One of the significant challenges faced in the agricultural sector pertains to soil salinity. This phenomenon is characterized by the accumulation of excessive amounts of soluble sodium salts within the soil, leading to land degradation and detrimental effects on plant growth and development. Extremophiles, such as *Enterobacter* and *Gluconacetobacter*, play a significant role in the process of nitrogen fixation. Phytohormones, which are essential signalling molecules in plants, are known to be synthesized by various bacterial genera, including *Methylobacterium*, *Ochrobactrum*, and *Microbacterium*. In the presence of salinity stress, halophilic extremophiles exhibit enhanced biomass production and yield, alongside the facilitation of seedling germination, promotion of root growth, elongation of shoot length, and augmentation of chlorophyll content. *Haloarcula argentinensis* and

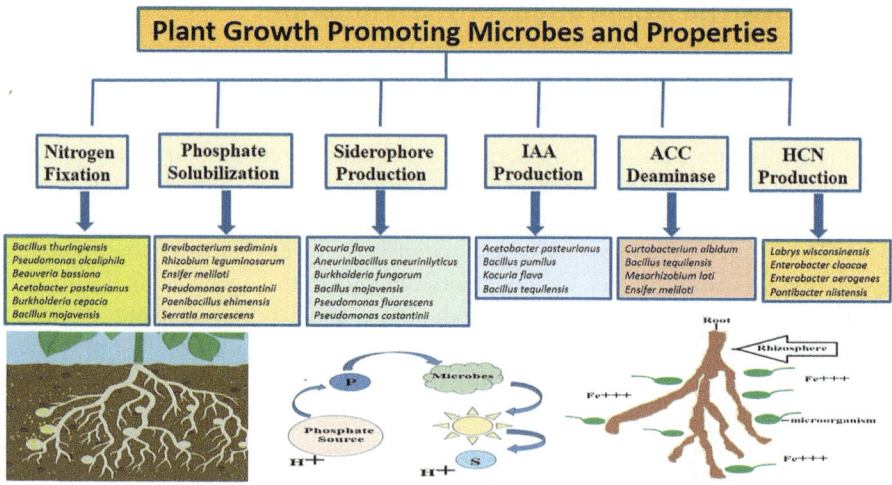

Fig. 5.2 Soil fertility properties and distribution of plant-growth-promoting microbiomes (Kaur et al. 2022)

Haloferax alexandrinus have been observed to exhibit the ability to solubilize phosphorus and accumulate phosphorus in hypersaline soils. In addition, these organisms possess the enzymatic capability of 1-aminocyclopropane-1-carboxylic acid (ACC) deaminase, an enzyme that facilitates the degradation of ACC. This enzymatic activity results in the reduction of ethylene, a plant hormone known to impede plant growth and exacerbate the effects of salinity stress. Psychrophilic extremophiles, due to their ability to solubilize nutrients, fix nitrogen, and produce phytohormones, and siderophores, have been employed as bioinoculants to enhance plant growth in cold environments and enhance resistance against infections. The growth of microorganisms in desert plants is facilitated by a combination of various environmental factors, including pronounced temperature fluctuations, elevated levels of radiation, limited water availability, and high salinity.

The bacterial phyla *Actinobacteria*, *Cyanobacteria*, and *Bacteroidetes* have been observed to exhibit adaptive responses to dry conditions through the upregulation of dormancy and osmoregulatory genes while concurrently downregulating catabolic processes. In the context of soil acidity, acidophilic extremophiles such as *Azotobacter*, *Bacillus*, and *Pseudomonas* have been employed as bioinoculants, bioremediation agents, and biocontrol agents due to their possession of specific characteristics that contribute to the enhancement of plant growth and development. The utilization of extremophiles that are both drought-tolerant and capable of solubilizing phosphorus holds the potential for enhancing agricultural productivity in arid regions, thereby addressing the pressing issue of global food security in the face of human overpopulation. Extremophiles are a class of biocontrolling organisms that possess distinct genetic characteristics, which have been extensively studied for their potential applications in various industrial and biotechnological fields. These organisms have evolved to thrive in extreme environments, exhibiting remarkable adaptability to harsh conditions. *Rhizobacteria* employ various mechanisms to safeguard plants against diseases, including the production of siderophores, chitinases, ammonia, hydrogen cyanide, and other secondary metabolites. Siderophores are compounds that effectively chelate iron, facilitating its acquisition by the rhizobacteria and limiting its availability to potential pathogens. Chitinases, on the other hand, are enzymes that degrade chitin, a major component of fungal cell walls, thereby impeding the growth and proliferation of fungal pathogens. Additionally, rhizobacteria release ammonia, which can alter the pH of the surrounding environment, creating unfavourable conditions for pathogenic microorganisms. Furthermore, the production of hydrogen cyanide by rhizobacteria serves as a potent antimicrobial agent, inhibiting the growth of various pathogens. These secondary metabolites collectively contribute to the protective role of rhizobia. The biocontrol agent exhibits sustained activity across diverse environmental conditions, effectively mitigating the incidence and severity of diseases caused by a wide range of nematodes. Through the manipulation of their reproductive cycle and the engagement in nutritional competition, several bacterial species have been identified as effective agents in the prevention of plant diseases. These include *Bacillus*, *Clavibacter*, *Microbacterium*, and *Pseudomonas* (Yadav 2017). Agricultural biosurfactants refer to a class of naturally derived compounds that are used in agricultural practices to enhance various aspects of crop production and biosurfactants have been employed as a means to enhance the bioremediation process in soil environments, as well as to serve as alternatives to chemical surfactants in the production of insecticides. In addition, these substances demonstrate antibacterial properties and play a role in enhancing the defence mechanisms of plants. Rhamnolipids have been identified as a potential solution for mitigating the impact of *Phytophthora* zoospores. According to Kochhar et al. (2022), the augmentation of hydrophilization and reduction of water infiltration in arid soils play a crucial role in facilitating the advancement of sustainable agricultural practices within arid regions. The utilization of extremophiles that are capable of tolerating drought conditions and solubilizing phosphorus has the potential to make a significant contribution towards addressing the global challenge of food security, particularly in arid regions. This is of utmost importance considering the exponential growth of the world's human population (Verma et al. 2017).

5.4.2 Medicinal Aspects

In addition to mesophilic bacteria, extremophiles are employed in the development of antibiotic, antifungal, and antimutagenic agents. Extremophiles, organisms thriving in extreme environments, have been found to produce diketopiperazine, a compound possessing notable antibacterial, antifungal, and antiviral properties. Additionally, these extremophiles also produce antimicrobial peptides, which have the potential to influence various processes related to blood coagulation. The activation and inhibition of quorum-sensing pathways have been observed in halophiles, which are organisms known for their ability to thrive in high-salt environments. This phenomenon is particularly intriguing as it is also employed by drug-resistant organisms such as *Pseudomonas aeruginosa*. Two examples of such microorganisms are *Haloterrigena hispanica* and *Natronococcus occultus*. Consequently, this particular substance exhibits potential efficacy in the treatment of drug-resistant strains of *Pseudomonas aeruginosa*. Polyhydroxyalkanoates (PHA), a diverse class of polyesters, are utilized in the production of water-resistant and biodegradable bioplastics. These PHAs are synthesized by numerous species of halophilic archaea. In this particular scenario, it has been observed that microbial cells serve as a reservoir for carbon sequestration, as documented by Basak et al. (2020).

The plant-associated extremophiles have been documented from various phyla and groupings, including Ascomycota, Actinobacteria, Bacteroidetes, Basidiomycota, Crenarchaeota, Euryarchaeota, Proteobacteria, and Firmicutes. These phyla and groupings represent a diverse range of organisms spanning all three domains of life: archaea, bacteria, and eukarya. The aforementioned bacteria exhibit functional characteristics and possess the potential to serve as valuable microorganisms in the capacity of biofertilizers, thereby enhancing both crop productivity and soil quality in the context of sustainable agricultural practices. These organisms demonstrate a diverse range of characteristics that contribute to the promotion of plant growth. The utilization of these advantageous and potentially valuable microorganisms can be effectively employed as biofertilizers, thereby enhancing crop productivity and promoting soil health in the context of sustainable agricultural practices. Bacteria exhibit a diverse array of characteristics that promote and enhance plant growth. Microorganisms obtained from diverse plant species thriving in environments characterized by abiotic stressors such as extreme temperatures, high salinity levels, varying pH levels, and limited water availability encompass a range of distinct microbial groups. These groups consist of psychrophiles, which exhibit optimal growth between temperatures of −2 °C and 20 °C; thermophiles, which thrive within the temperature range of 60 °C to 116 °C; halophiles capable of tolerating salinity levels of 2–5 M; acidophiles capable of thriving in environments with a pH as low as 4; alkaliphiles that flourish in alkaline conditions with a pH greater than 9; and xerophiles, which can withstand water potentials as low as 0.75 kPa. Extreme habitats and niches exemplify unique ecosystems that harbour exceptional biodiversity and possess the capacity to adapt to a diverse array of environmental pressures. The organisms referred to as extremophiles have undergone evolutionary processes that have bestowed upon them adaptive characteristics enabling them to exhibit optimal growth in the presence of one or more extreme environmental conditions. Conversely, polyextremophiles have developed the ability to thrive optimally across a diverse range of environments. These organisms, commonly referred to as extremophiles, have undergone evolutionary processes that have endowed them with adaptive characteristics enabling them to flourish in environments characterized by one or more extreme conditions. Conversely, polyextremophiles exhibit optimal growth across a diverse range of environments. There is a prevailing consensus in the scientific community that certain strains of bacteria residing in the rhizosphere, commonly referred to as plant-growth-promoting bacteria (PGPB), play a crucial role in enhancing the overall well-being and development of plants. Our current comprehension of the mechanisms underlying the promotion of plant growth can be attributed, to some extent, to our knowledge of the communication signals exchanged among species inhabiting the rhizosphere. The preservation of agriculture production systems' sustainability is believed to heavily rely on the diversity of microbes associated with crops (Yadav 2017).

Microorganisms play a crucial role in facilitating the development, production, and adaptability of plants. The classification of microbes associated with crops can be organized into three distinct groups, namely, rhizospheric, phyllospheric, and endophytic microbes. These groups encompass a wide range of microorganisms that inhabit different ecological niches within the plant ecosystem. The rhizosphere is a distinct soil region that is subject to the influence of plant roots, as they release various substrates that have a significant impact on the microbial activity within this zone. The plant rhizosphere is known to harbour a diverse array of microbial genera, including but not limited to *Azospirillum, Arthrobacter, Burkholderia, Bacillus, Paenibacillus, Enterobacter, Methylobacterium, Pseudomonas, Rhizobium,* and *Serratia*. Epiphytic microorganisms have been identified and documented within a diverse range of phyllospheric regions of various plant species. The phyllosphere harbours a wide array of potential microorganisms and plants, which can establish synergistic relationships within various ecological niches. Epiphytic bacteria exhibit remarkable adaptability in their natural environment, owing to their ability to withstand high temperatures ranging from 40 °C to 55 °C and their exceptional tolerance towards ultraviolet (UV) radiation. A wide range of crops has been documented to harbour various bacterial species, including *Agrobacterium, Pantoea, Methylobacterium,* and *Pseudomonas*, within their phyllosphere. These bacteria have demonstrated the ability to flourish under diverse environmental conditions, both favourable and unfavourable. The term 'endophytic microbes' pertains to microorganisms that establish colonization within various plant tissues, including the root, stem, or seeds, without exerting any detrimental effects on the host plant. Various plant species have been identified as sources of these microorganisms, including rice, wheat, soybean, pea, common bean, chickpea, and pearl millet (Kochhar et al. 2022).

5.4.3 Food Industry

Extremophilic microorganisms, due to their ability to thrive in extreme environments, have garnered significant attention in the food processing industry. This is primarily attributed to their capacity to produce a diverse array of bioactive substances, such as secondary metabolites, as well as valuable products including flavours, food components, and vitamins (Saini and Keum 2019). These substances can enhance the potential, optimize the health-promoting properties of food products, and mitigate the impact of various chronic diseases. Carotenoids, owing to their diverse and significant properties, find extensive applications in the food industry as additives, colour enhancers, antioxidants, and various other functional agents. In addition to enhancing the nutritional content and oxidative stability of meat and poultry products, these advancements offer a multitude of health advantages to consumers. Pro-vitamin A, also known as beta-carotene, is found in significant quantities in various food sources. This compound exhibits notable antiageing properties, enhances the immune response, and demonstrates resistance against certain types of cancer, including breast, cervical, and prostate cancer. Additionally, pro-vitamin A has been associated with the mitigation of various physiological issues (Nabi et al. 2020). These visually appealing probiotic health products have the potential to be incorporated into a variety of consumables, including food items, fruit-based products, and energy beverages. Canthaxanthin, a compound commonly employed as a food and beverage colouring agent and as a pigment in salmon flesh, has been identified in two specific microbial species, namely, *Bradyrhizobium* sp. and *Haloferax alexandrines*. The utilization of riboflavin in *Ashbya Gossypii* and carotene in *Blakeslea trispora* has been employed for the production of food-grade colours derived from microbial sources. Furthermore, it is worth noting that carotenoids employed as nutritional supplements are derived from the biomass of microalgae.

5.4.4 Biobutanol Production

The majority of bacterial strains exhibit limited tolerance to concentrations exceeding 2% of butanol due to its inhibitory effect on microbial growth and development. According to a recent study conducted by Benninghoff et al. (2021), it has been observed that multiple species belonging to the *Bacillus* genus exhibit resistance to varying concentrations of butanol, specifically ranging from 2.5% to 7%. Extremophiles, which are organisms capable of thriving in extreme environments, have shown potential for genetic modification. One example of such an extremophile is *Pseudomonas putida*, which has been utilized for the production of butanol and enhancement of tolerance. Green Biologics, a business operating in the United Kingdom, specializes in the production of biobutanol through the utilization of waste biomass, specifically maize stock, in conjunction with the thermophilic bacterium *Clostridium* (Karthick and Nanthagopal 2021).

5.4.5 Biohydrogen Production

Currently, the production of hydrogen gas (H_2) is conducted on a somewhat restricted scale, yet it fulfils a multifaceted role as an adaptable energy resource for diverse applications encompassing transportation, electricity generation, and heat generation. Baeyens et al. (2020) assert that anaerobic fermentations have become the favoured methodology for the production of hydrogen. The utilization of thermophilic microorganisms in the production of hydrogen gas exhibits superior performance compared to mesophilic microorganisms. This is primarily attributed to the thermophiles' ability to minimize the production of fermentation byproducts and their wider metabolic capability to utilize diverse organic waste materials. *Thermoanaerobacterium* is a distinct bacterial species that demonstrates remarkable efficiency in the production of hydrogen gas. The taxonomic categorization of microorganisms holds significant importance in comprehending their evolutionary interconnections and ecological implications within scientific research. Within the domain of archaea, a distinct and separate branch of life, numerous genera have been identified and subjected to comprehensive and in-depth scientific investigation. One notable genus that exemplifies thermophilic properties and exhibits remarkable capability in the degradation of complex substances is *Caldicellulosiruptor saccharolyticus*. The study conducted by An et al. (2018) provides evidence that *Pyrococcus* and *Thermococcus* play a crucial role in the facilitation of the conversion of organic molecules into hydrogen.

5.4.6 Biogas Production

Methanogens, known as extremophilic microorganisms, play a crucial role in the generation of methane (CH_4) and carbon dioxide (CO_2), which serve as the primary components of biogas. Biogas, a sustainable and environmentally friendly energy source, is produced through the process of anaerobic digestion. This process involves the decomposition of organic materials, such as animal manure sourced from agricultural operations, domestic garbage, municipal solid waste, and rendered animal fat (Coker 2016). Several distinct species of extremophiles, including *Methanosarcina thermophila*, *Methanobacterium* sp., and *Methanothermococcus okinawensis*, exhibit the remarkable capability of methane production. In addition to the aforementioned, it is worth noting the existence of extremophiles that exhibit psychrotolerance or psychrophilic characteristics. Notable examples include *Methanolobus psychrophilus* and *Methanosarcina lacustris*. The utilization of these distinct microorganisms presents promising prospects for diverse industrial sectors, thereby providing a viable avenue for cost reduction in operational processes. Barnard et al. (2010) have provided empirical evidence indicating that these organisms exhibit accelerated rates of growth and display heightened levels of responsiveness, thereby facilitating a more rapid advancement of the underlying biological process.

5.4.7   Extremophiles' Help in Soil Health

The remarkable progress in the field of biotechnology has made it feasible to harness the potential of advantageous bacterial species within soil ecosystems, thereby facilitating the enhancement of plant growth and the effective mitigation of soilborne diseases through biological means. There exists a multitude of dietary and environmental prerequisites that apply to the microorganisms. In stark contrast to soil that is abundant in nutrients, the process of microbial inoculation plays a significant role in enhancing plant growth and development, particularly in the context of soil that lacks essential nutrients. Numerous taxonomic groups of microorganisms, including Actinobacteria, Acidobacteria, Ascomycota, Bacteroidetes, Basidiomycota, Deinococcus-Thermus, Euryarchaeota, Firmicutes, and Proteobacteria, have been identified as having associations with plants. The phyla Actinobacteria and Proteobacteria exhibited the highest levels of microbial diversity across various distributions. The phyla Deinococcus-Thermus and Acidobacteria exhibit a relatively low abundance of microorganisms within their respective taxonomic groups. Soil salinity poses a substantial constraint on crops, particularly in regions characterized by aridity and low precipitation levels. In recent times, a plethora of innovative technologies have been associated with the augmentation of salt tolerance. The utilization of microbe-associated crops in agriculture exhibits significant potential due to their capacity to enhance plant growth in the presence of environmental stressors or temperature limitations, as demonstrated by Suman et al. (2016).

The microorganisms derived from extreme environments play a crucial role in the overall ecological dynamics of our planet, as a significant proportion of both terrestrial and aquatic ecosystems experience either permanent or seasonal exposure to low temperatures. The prevalence of low-temperature-tolerant microorganisms in natural environments, where they often constitute the predominant flora, is reasonable to conceptualize them as the most triumphant colonizers on our planet. Extensive research has been conducted in recent years to investigate the diverse array of microorganisms that inhabit cold environments. The growth and productivity of crops are significantly hindered by the presence of drought stress, particularly in regions characterized by arid and semi-arid climatic conditions. Acidic and alkaline environments exhibit notable concentrations of microbial diversity, thereby serving as focal points for the proliferation of various microorganisms. These environments possess distinct characteristics that facilitate and promote the growth and development of plants. A multitude of acidophilic bacterial genera have been identified within plant species that thrive in acidic environments. In summary, it can be concluded that plants exert a substantial impact on the types of bacteria that are specifically chosen and enriched by the constituents present in their root exudates. The microbial population undergoes a transformation into either endophytic, epiphytic, or rhizospheric communities, which is contingent upon the characteristics and quantity of organic material present in exudates, as well as the microbes' ability to utilize these compounds as an energy source. Crop-associated bacteria play a vital role in agriculture due to their ability to enhance plant nutrition through biological nitrogen fixation and other intricate processes, thereby facilitating plant development. Crop-associated microorganisms play a pivotal role in the field of agriculture due to their significant contribution to enhancing plant nutrition through the process of biological nitrogen fixation. There is a potential to increase crop yields, the removal of pollutants, the inhibition of diseases, as well as the ability of microbes to generate fixed nitrogen or novel compounds. In the contemporary global context, there exists a pressing need to attain high output yields, enhance crop production, optimize yield potential, and promote soil fertility in order to effectively engage in environmentally sustainable agricultural practices. Extremophiles play a pivotal role in exerting control over plant development and enhancing crop productivity in regions characterized by adverse environmental conditions. These organisms exhibit exceptional qualities that render them highly suitable for various ecological processes such as nutrient cycling, nutrient fixation, mineralization, and solubilization. Consequently, they possess significant potential as bioinoculants and biofertilizers, offering valuable contributions to agricultural and environmental practices. Recent scientific research has focused on investigating the rhizosphere microbiome due to its significant implications for plant development and defence against pathogens, particularly in the presence of stressors. The utilization of the term 'minimal and core rhizosphere stress' is deemed suitable in this context, as suggested by Verma et al. (2017) in their work.

The concept of the minimal and core rhizosphere microbiome is of utmost significance in the field of microbiology and plant science. The dairy biorefinery's synergistic strategy is designed to achieve a state of zero waste, wherein all resources and by-products are efficiently utilized and no waste is generated. It is anticipated that the advancement of the dairy biorefinery's process will attain a state of zero waste at an industrial scale through the utilization of a synergistic approach. Extremophilic microorganisms harbour a diverse array of bioactive substances, secondary metabolites, and a multitude of vitamins within their biological makeup, thereby concealing a plethora of enigmatic secrets. Extremophiles are group of organisms capable of thriving in extreme environments and have garnered significant attention in scientific research due to their potential applications in various fields. One notable area of interest is their utilization in the production of valuable compounds such as halocins, Diketopiperazines, polyhydroxyalkanoates, deoxyribonucleic acid polymerase, and lipases, which hold promise for their incorporation into recombinant vesicles for medical purposes. These extremophile-derived compounds exhibit properties that make them suitable for use in the field of medicine, thereby highlighting the significance of extremophiles in the advancement of recombinant technology. Carotenoids, specifically macula, offer a protective effect against solar radiation. The production of antibiotics is facilitated by carotenoids, as stated by Herreara et al. (2006). Carotenoids play a significant role in the production of antibiotics. Hence, it can be inferred that extremophilic bacteria possess distinctive capabilities that contribute to the enhancement of crop productivity, as well as the production of biofuels and biosurfactants. Microorganisms play a crucial role in various aspects of our society, particularly in the realms of food processing, therapeutics, and nutrient cycling. Hot springs, along with various other extreme environments, are employed as processing sites for food. Psychrophilic environments, such as glaciers, have been found to possess antibiotic-resistance genes, which are genes that enable microorganisms to withstand the effects of antibiotics. However, it is noteworthy that these antibiotic-resistance genes often coexist with genes that confer tolerance to heavy metals. According to Yadav (2017), it has been observed that extremophilic environments, such as hot springs, exhibit a notable absence of antibiotic-resistance genes.

5.5   Conclusion

Extremophiles have the remarkable ability to flourish in settings that are defined by severe conditions. These organisms have specific characteristics and hold potential for use in agriculture and soil improvement. The use of these compounds as biofertilizers and biostimulants has been shown to augment plant growth and facilitate nutrient uptake. Extremophiles have promising possibilities in the realm of sustainable agriculture by effectively mobilizing nutrients. Soils that have been contaminated with pollutants are known to have extremophiles, which possess the capability to either break down or immobilize detrimental chemicals. Extremophilic microorganisms residing in dry or saline environments have the potential to contribute to the advancement of drought and salt-resistant agricultural crops. Soil microorganisms have the capacity to produce antimicrobial enzymes, hence playing a crucial role in safeguarding crops from many illnesses. The study noted the advantages associated with the decrease of soil erosion and the improvement of soil health. Extremophilic microorganisms have a crucial role in promoting nitrogen fixation in agriculturally significant crops under harsh and demanding environmental conditions. The reduction of dependency on nitrogen fertilizer serves to mitigate its environmental effect. Extremophiles play a significant role in the processes of carbon cycling and soil carbon sequestration. Metabolic activities have a significant role in the storage of carbon in soil, hence assisting in the mitigation of climate change in agricultural contexts. Extremophiles have the potential to augment the resistance of agricultural plants to stressors associated with climate change, such as variations in temperature. Microbial inoculants originating from extremophiles have been shown to enhance the composition and functioning of soil microbiomes, hence promoting more efficient nutrient cycling.

Acknowledgements   The authors are thankful to Pritam International Pvt. Ltd. Bhagwanpur, Roorkee, Haridwar-247667(UK) for providing the necessary facilities.

Credit Authorship Contribution Statement   Vinay Mohan Pathak: Writing, review & editing; Nitika Rana: Writing, review & editing, conceptualization; Sumitra Pandey: Supervision; A. K. Sarkar: Supervision; Abhishek Chauhan: Validation; Tanu Jindal: Supervision; Mukta Sharma: Validation & review; Seeta Dewali: Data curation; Satpal Singh Bisht: Supervision, Divya: Editing & review; Akansha Sharma: Formal analysis; Monika Yadav: Review; Balwant

Singh Rawat: Validation; Somdatt Tyagi: Formal analysis; D. P. Uniyal: Supervision; Pradeep Kumar: Review & editing; Sevaram Singh: Editing; Baljinder Kaur: Validation; Jose Maria Cunill: Review; Satish Chandra Garkoti: Supervision.

Declaration of Competing Interest The authors declare that they have no known competing financial interests or personal relationships that could have appeared to influence the work reported in this paper.

Data Availability Statement The data used to support this investigation are available in the article.

References

An Q, Wang JL, Wang YT, Lin ZL, Zhu MJ (2018) Investigation on hydrogen production from paper sludge without inoculation and its enhancement by *Clostridium thermocellum*. Bioresour Technol 263:120–127

Baeyens J, Zhang H, Nie J, Appels L, Dewil R, Ansart R, Deng Y (2020) Reviewing the potential of bio-hydrogen production by fermentation. Renew Sust Energ Rev 131:110023

Barnard D, Casanueva A, Tuffin M, Cowan D (2010) Extremophiles in biofuel synthesis. Environ Technol 31(8–9):871–888

Basak P, Biswas A, Bhattacharyya M (2020) Exploration of extremophiles genomes through gene study for hidden biotechnological and future potential. In: Physiological and biotechnological aspects of extremophiles. Academic Press, pp 315–325. https://doi.org/10.1016/B978-0-12-818322-9.00024-1

Benninghoff JC, Kuschmierz L, Zhou X, Albersmeier A, Pham TK, Busche T, Wright PC, Kalinowski J, Makarova KS, Brasen C, Flemming HC, Wingender J, Siebers B (2021) Exposure to 1-butanol exemplifies the response of the thermoacidophilic archaeon sulfolobus acidocaldarius to solvent stress. Appl Environ Microbiol 87(11):e02988

Breccia G, Vázquez M (2018) Soil acidity management in agriculture through plant growth-promoting bacteria. Adv Agron 150:73–114. https://doi.org/10.1016/bs.agron.2017.07.001

Cavicchioli R, Amlis R, Wagner D, Genity MC (2013) Life and applications of extremophiles. Environ Microbiol 13(8):1903–1907. https://doi.org/10.1111/j.1462-2920.2011.02512.x

Chakraborty D, Efthi JH, Khanom M, Mahbubul IM (2020) Prospective and challenging issues of biofuels. EDU J Comput Electr Eng 1(1):4–10

Coker JA (2016) Extremophiles and biotechnology: current uses and prospects [version 1; peer review: 2 approved]. F1000Research 2016, 5(F1000 Faculty Rev):396. https://doi.org/10.12688/f1000research.7432.1

Deshpande AS, Kumari R, Prem RA (2018) A delve into the exploration of potential bacterial extremophiles used for metal recovery. Glob J Environ Sci Manag 4(3):373–386

Ghorbel S, Kthiri D, Sellami H, Cherif M (2017) Prospects of plant growth-promoting rhizobacteria application in sustainable agriculture: an overview. In: Rhizotrophs: plant growth promotion to bioremediation. Springer, Cham, pp 1–27

Gomes J, Steiner W (2004) The biocatalytic potential of extremophiles and extremozymes. Food Technol Biotechnol 42(4):223–225

Gupta GN, Khare S, Srivastava S, Prakash V (2014) Extremophiles: an overview of microorganism from extreme environment. Int J Agric Environ Biotechnol 7(2):371–380. https://doi.org/10.5958/2230-732X.2014.00258.7

Hasan F, Shah AA, Javed S, Hameed AS (2010) Enzymes used in detergents: lipases. Afr J Biotechnol 9(31):4836–4844

Herreara FC, Santos JA, Otero A, Garcialopez ML (2006) Occurrence of foodborne pathogenic bacteria in retail prepackaged portions of marine fish in Spain. J Appl Microbiol 100:527–536

Horikoshi K (1999) Alkaliphiles: some applications of their products for biotechnology. Microbiol Mol Biol Rev 63(4):735–750. https://doi.org/10.1128/MMBR.63.4.735-750.1999

Jerez CA (2012) The use of extremophilic microorganisms in the industrial recovery of metals. In: Extremophiles: sustainable resources and biotechnological implications, Wiley-Blackwell-John Wiley & Sons Inc.pp 319–334

Jones MB, Patel SS (2018) Molecular adaptations of piezophilic microorganisms. Microb Ecol 76(2):285–296

Karthick C, Nanthagopal K (2021) A comprehensive review on ecological approaches of waste to wealth strategies for production of sustainable biobutanol and its suitability in automotive applications. Energy Convers Manag 239:114219

Kaur T, Kour D, Pericak O, Olson C, Mohan R, Yadav A et al (2022) Structural and functional diversity of plant growth promoting microbiomes for agricultural sustainability. J Appl Biol Biotechnol 10(1):70–89

Kharade SS, Jadhav SD, Pal D (2015) Alkaliphiles: diversity and potential applications in agriculture. In: Maheshwari M, Saraf M (eds) Halophiles: biodiversity and sustainable exploitation. Springer, pp 207–219. https://doi.org/10.1007/978-3-319-14518-8_10

Kochhar N, Shrivastava S, Ghosh A, Rawat VS, Sodhi KK, Kumar M (2022) Perspectives on the microorganism of extreme environments and their applications. Curr Res Microb Sci 3:100134. https://doi.org/10.1016/j.crmicr.2022.100134

Kumar L, Awasthi G, Singh B (2011) Extremophiles: a novel source of industrially important enzymes. Biotechnology 10:121–135

Li M, Wen J (2021) Recent progress in the application of omics technologies in the study of biomining microorganisms from extreme environments. Microb Cell Factories 20(1):1–11

Mirete S, Morgante V, Gonzalez-Pastor JE (2017) Acidophiles: diversity and mechanisms of adaptation to acidic environments. In: StanLotter H, Fendrihan S (eds) Adaption of microbial life to environmental extremes. Springer International Publishing AG. Available from:10.1007/978-3-319-48327-6-9

Nabi F, Arain MA, Rajput N, Alagawany M, Soomro J, Umer M, Soomro F, Wang Z, Ye R, Liu J (2020) Health benefits of carotenoids and potential application in poultry industry: a review. J Anim Physiol Anim Nutr (Berl) 104(6):1809–1818

Oren A (2013) Life at high salt concentrations. In: Extremophiles handbook. Springer, pp 487–520. https://doi.org/10.1007/978-3-642-30141-4_17

Pattanaik A, Samal DK, Sukla LB, Pradhan D (2020) Advancements and use of OMIC technologies in the field of bioleaching: a review. Biointerface Res Appl Chem 11:10185–10204

Rothschild LJ, Manicinelli RL (2001) Life in extreme environments. Nature 409:1092–1101

Saini RK, Keum YS (2019) Microbial platforms to produce commercially vital carotenoids at industrial scale: an updated review of critical issues. J Ind Microbiol Biotechnol 46(5):657–674

Singh V, Kumar P, Jaiswal A, Singh JS (2017) Adaptive strategies of plants for abiotic stress tolerance. In: Plant acclimation to environmental stress. Wiley, pp 19–47

Smith P, Des Marais DJ (2004) Adaptive strategies of xerophytic desert plants: results of a 3-year greenhouse study. Oecologia 141(1):114–126

Smith JD, Johnson AR, Brown S (2020) Harnessing piezophiles for sustainable agriculture. J Soil Sci 47(3):215–224

Suman A, Yadav AN, Verma P (2016) Endophytic microbes in crops: diversity and beneficial impact for sustainable agriculture. In: Microbial inoculants in sustainable agricultural productivity, research perspectives. Springer-Verlag, India, pp 117–143

Valls M, Lorenzo De V (2002) Exploiting the genetic and biochemical capacities of bacteria for the remediation of heavy metal pollution. FEMS Microbiol Rev 26:327–338

Verma P, Yadav AN, Kumar V, Singh DP, Saxena AK (2017) Beneficial plant-microbes interactions: biodiversity of microbes from diverse extreme environments and its impact for crop improvement. In: Plant-microbe interactions in agro- ecological perspectives. Springer Nature, Singapore, pp 543–580

Wang L (2020) Enhancing plant abiotic stress tolerance through genetic manipulation of ABA receptors. Front Plant Sci 11:604

Yadav AN (2017) Beneficial role of extremophilic microbes for plant health and soil fertility. J Agric Sci Bot 1(1):30–33

Yadav AN, Saxena AK (2018) Biodiversity and biotechnological applications of halophilic microbes for sustainable agriculture. J Appl Biol Biotechnol 6(1):48–55

Yadav AN, Sachan SG, Verma P, Saxena AK (2015) Prospecting cold deserts of North Western Himalayas for microbial diversity and plant growth promoting attributes. J Biosci Bioeng 119(6):683–693. https://doi.org/10.1016/j.jbiosc.2014.11.010

Yadav D, Singh A, Mathur N, Agarwal A, Sharma J (2021) Isolation of halophilic bacteria and their screening for extracellular enzyme production. J Sci Ind Res 80:617–622

Zhang C, Cui X, Wang S, Shi L (2017) Alkaline agricultural soil harbors more diverse bacterial communities than neutral soil in different habitats. J Soils Sediments 17(9):2233–2243. https://doi.org/10.1007/s11368-017-1695-y

Zhang D, Luo G, Zhu X, Liu Y, Chen W (2019) Radiotrophic bacteria: application and perspective in bioremediation. In: Radiotrophic bacteria. Springer, pp 117–132

Chapter 6
Bioactive Compounds/Metabolites from Extremophiles for Biocontrol and Plant Disease Management

Monika Shrivastava, Adesh Kumar, and Pallavi Saxena

6.1 Introduction

In recent years, scientific exploration of extreme environments has added several aspects to the diversity of microorganisms. Extremophiles, which are organisms thriving in harsh environmental conditions lethal to most life forms (Singh et al. 2019), inhabit these extreme settings, which harbour a wealth of microbial biodiversity. These extremophiles are increasingly recognized as promising candidates for the management of plant disease and growth promoters along with other biotechnological and industrial applications (Toral et al. 2021). They are commonly referred to as extremophiles and serve as prolific sources of bioactive compounds, including enzymes, antibiotics, and pigments (Sahli et al. 2018). Extremophilic organisms are capable of adapting to a range of varying environmental conditions, allowing them to thrive while maintaining their metabolic processes through various dynamic mechanisms. These microorganisms have evolved defence strategies to endure extreme aridity (xerophilic), low nutrient availability (oligotrophic), pH levels (acidophilic or alkaliphilic), pressures (piezophilic), temperatures (psychrophilic, thermophilic, or hyperthermophilic), and salinity (halophilic) or even within solid rocks or the pores of mineral grains (endolithic). Such adaptation is brought about within the extremophiles by the production of unique and diverse biologically active biomolecules (Alam and Tiwary 2023).

M. Shrivastava · A. Kumar
Jawaharlal Nehru University, Aruna Asaf Ali Marg, New Delhi, India

P. Saxena (✉)
Academy of Biology and Biotechnology, Southern Federal University,
Rostov-on-Don, Russia

Centre of Research Impact and Outcome, Chitkara University Institute of Engineering and Technology, Chitkara University, Rajpura, Punjab, India

© The Author(s), under exclusive license to Springer Nature Switzerland AG 2024
A. Ranjan et al. (eds.), *Extremophiles for Sustainable Agriculture and Soil Health Improvement*, https://doi.org/10.1007/978-3-031-70203-7_6

Piezophilic organisms, also known as barophiles, are a fascinating class of extremophiles that have adapted to thrive in the extreme pressures of the deep ocean. These microorganisms have evolved to withstand the crushing forces of the deep sea, where pressures can exceed 1000 times that of the surface. Moreover, piezophilic organisms have practical applications in biotechnology and the pharmaceutical industry, where their enzymes and biochemical adaptations are valuable for various industrial processes and bioprospecting endeavours (Santos et al. 2021).

Psychrophilic, thermophilic, and hyperthermophilic organisms represent the diverse spectrum of extremophiles adapted to different temperature extremes. Psychrophiles thrive in freezing conditions, such as polar ice caps and deep-sea trenches, where they have evolved unique cellular mechanisms to function optimally at near-freezing temperatures. In contrast, thermophiles flourish in high-temperature environments like geothermal hot springs, withstanding temperatures up to 80 °C or more. Hyperthermophiles push the boundaries even further, inhabiting hydrothermal vents and volcanic areas where temperatures can exceed 100 °C (Yadav and Saxena 2018).

Halophilic are uniquely adapted to thrive in highly saline environments, such as salt flats, saline ponds, and salt mines. These organisms have developed complex strategies to manage the osmotic stress caused by extreme salinity. They often maintain a higher internal salt concentration than their surroundings, preventing water loss and ensuring cell stability. Halophiles encompass various domains of life, including bacteria, archaea, and even some algae. Their study not only provides insights into the limits of life on the earth but also has practical applications in biotechnology and the food industry, where their enzymes and salt-tolerance mechanisms are harnessed for several applications, namely, salt-tolerant enzymes and preservation of foods in saline conditions (Dixit and Kumari 2023).

Xerophilic extremophiles have adaptations to thrive in extreme dryness and low water availability conditions. They are often found in arid deserts, salt flats, and even inside sealed food containers, where water content is minimal. These extremophiles have evolved specialized mechanisms to endure desiccation, such as the production of protective compounds, the ability to form spores or cysts, and a high tolerance for dehydration. Their resilience in the face of extreme dryness provides an in-depth limit of life on the earth and has applications in food preservation, where they can help prevent spoilage in low-moisture environments (Kanekar and Kanekar 2022).

Acidophilic and alkaliphilic extremophiles are microorganisms that thrive in environments with extreme pH levels. Acidophiles are adapted to highly acidic conditions, often found in places like sulfuric acid-rich volcanic pools or acidic mine drainage sites. They possess unique mechanisms to maintain cellular function and integrity in such harsh environments. In contrast, alkaliphiles flourish in highly alkaline conditions, like soda lakes with high pH levels. They too have evolved specialized adaptations to cope with extreme pH, including protective cell membranes and pH-regulating mechanisms. The study of acidophilic and alkaliphilic extremophiles not only enhances our understanding of extremophiles' adaptability but also has been helpful in the development of techniques for industrially useful enzymes and processes suited for extreme pH conditions unlike the conventional methods which may not hold the same potential under extreme environmental conditions (Hui et al. 2021).

Owing to the vastness and diversity of microorganisms inhabiting in extreme terrains, knowledge in this research field is still in its early stages (Gupta et al. 2023). The key factor behind extremophile adaptation is attributed to biochemical and physiological changes. They produce bioactive metabolites, biosurfactants, extremolytes, extremozymes, and so on. These compounds also hold potential applications in sustainable agriculture along with food, cosmetics, and pharmaceutical industries (Santos et al. 2021). These biomolecules are promising for agricultural sectors as plant pathogen biocontrol and plant growth promoters in promoting sustainable agriculture practices. Numerous biocontrol agents of microbial origin have been developed to combat bacterial diseases. *Bacillus* species, for instance, produce a variety of metabolites like amylolysin, bacteriocins, surfactin, and so on with potent antimicrobial properties which are effective in restricting the growth of plant pathogens such as *Erwinia*, *Ralstonia*, and *Penicillium*. Additionally, VOCs produced by *Bacillus amyloliquefaciens* (BA) inhibit the growth of *Penicillium digitatum* and *Ralstonia solanacearum*. *Streptomyces* also function as biological control agents (BCAs) by producing antimicrobials, exoenzymes, and VOCs (Newitt et al. 2019). *Streptomyces* are believed to be significant BCAs, particularly in the management of various wheat diseases, owing to their saprotrophic and spore-forming capabilities, enabling them to thrive in challenging environments (Van der Meij et al. 2017; Coombs 2004). They exhibit the ability to inhibit the growth of *Fusarium*, *Rhizoctonia*, *Gaeumannomyces*, and *Magnaporthe*. An example is *Streptomyces lydicus* WYEC108, which secretes chitinase to degrade the cell walls of fungi like *Pythium*.

Similarly, *Pseudomonas* species employ diverse mechanisms for suppressing plant diseases, with secondary metabolites playing a crucial role in biocontrol. They are reported to inhibit *Clavibacter*, *Hyaloperonospora*, *Pantoea*, and *Thielaviopsis*. The phenazines produced by *Pseudomonas* demonstrate broad-spectrum activity against several bacterial and fungal plant pathogens, especially against *Rhizoctonia solani*, *Streptomyces scabies*, and *Phytophthora infestans* (Morrison et al. 2017). The genus *Serratia* also serves as BCAs in plants, engaging in direct or indirect competition with other pathogens. *Serratia marcescens* produces serrawettin W2, displaying antibacterial activity against *Staphylococcus aureus*. Sodorifen is a VOC produced by *Serratia plymuthica* found to have significant potential against plant pathogens (Domik et al. 2016).

This chapter delivers a comprehensive overview of the extremophiles and their metabolites involved in biocontrol, exploring how extremophilic bacteria can be effectively utilized for managing plant diseases (Kamat et al. 2022). Furthermore, the chapter also highlighted the role of non-ribosomal peptides, polyketides, siderophores, and VOCs in the biocontrol of plant pathogens and supporting plant growth to ensure crop yield. Such extremophiles are significant as they could serve in areas with challenging environmental conditions, such as low temperatures, high salinity, and drought by assisting as biofertilizers, bioinoculants, and biocontrol agents.

6.2 Bioactive Compounds/Metabolites from Extremophiles for Biocontrol

The survival strategies and adaptation among extremophiles are responses to challenging environmental conditions (Fig. 6.1) which make them valuable resources in sustainable agricultural practices and other biotechnological and industrial applications. Some of these strategies are closely associated with the production of secondary metabolites or bioactive compounds which are significantly potential for biological control of plant pathogens. Notably, psychrophiles bacteria have emerged as a novel source of bioactive compounds with antibiotic activity along with its activity and resilience in extreme environmental conditions (Mehetre et al. 2021). Extremophiles employ specific gene expression patterns to thrive in stressful conditions, and these genes are largely studied to explore their potential application in employing them suitable for agricultural applications (Mehetre et al. 2021).

Biological control of plant pathogens plays a critical role in sustainable crop production, with many plant-associated microbes possessing antimicrobial potential. For a microbe to be formulated as a biocontrol agent, it must exhibit activity against a diverse array of bacterial and fungal plant pathogens, along with harmful insects or nematodes (Pikuta et al. 2007). The emphasis lies on beneficial microbes colonizing the rhizosphere, where they promote healthy plant growth through various mechanisms. Beneficial microbes with biocontrol activity have evolved to

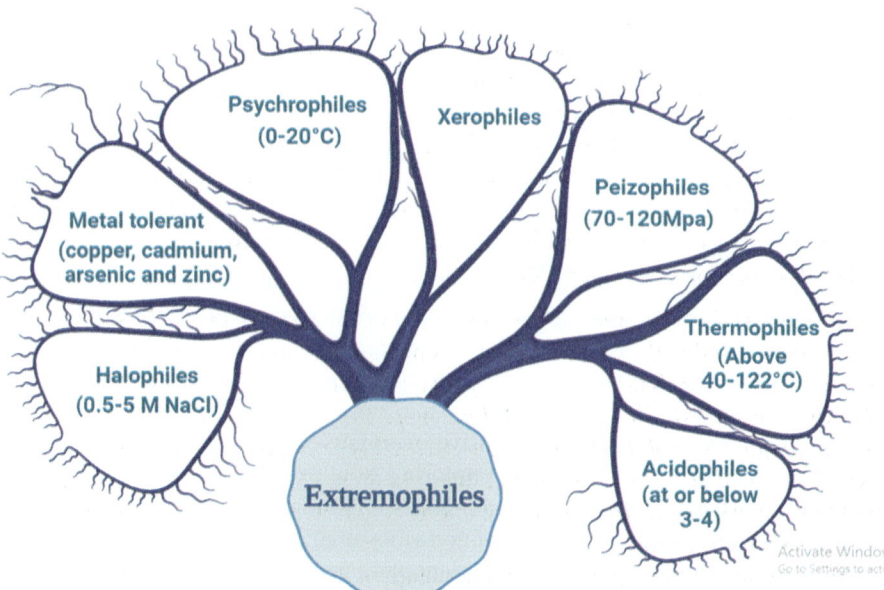

Fig. 6.1 Categories of extremophiles, including psychrophiles, thermophiles, halophiles, acidophiles, piezophiles, and metal tolerance, based on temperature, pH, salt, pressure, and metal

coexist with root tissues without causing disease and prove to be highly effective against plant pathogens, encompassing microorganisms, insects, and nematodes (Gupta et al. 2023).

As an example, *Rhizobacteria* protect plants from pathogens by producing compounds such as non-ribosomal peptides (NRPs), and polyketides along with NH4+, hydrogen cyanide (HCN), siderophores, and many others which also ensure plant's optimal health (Pandey et al. 2021).

Such microbes work over a range of extreme conditions to inhibit the growth of pathogens. Genera such as *Bacillus, Clavibacter, Microbacterium,* and *Pseudomonas* possess several species which have adapted well to extreme conditions and exhibit significant biocontrol potential on plant pathogens (Verma et al. 2017). Bacterial extremophiles are recognized as a rich source of novel bioactive compounds of microbial origin, constituting approximately 45% of all microbial metabolites. Filamentous extremophilic actinobacteria like *Streptomyces* and rare actinomycetes such as *Actinomadura, Streptoverticillium, and Micromonospora* have yielded several thousand bioactive compounds, encompassing a wide range of substances like antibiotics, pesticides, enzyme inhibitors, toxins, and pigments (Dixit and Kumari 2023). The combination of these metabolites acts on pathogens to stop their growth and also contributes to deterrence and toxicity against insects and nematodes (Silva et al. 2022).

The genus *Streptomyces* is actively employed as a biocontrol agent for various commercial crops to combat fungal pathogens. Numerous *Streptomyces* species isolated from grapevines exhibit potent antifungal activity against fungi and yeast from the same habitat. Actinobacteria capable of producing antimicrobial agents and degrading keratin have been utilized to transform poultry farm feather waste through composting into an odourless biofertilizer, effectively eliminating pathogens. Gomes et al. (2000) reported that *Streptomyces* spp. obtained from Brazilian forest soil displayed remarkable activity against three phytopathogenic fungi: *Fusarium solani, Magnaporthe grisea*, and *Aspergillus parasiticus*. Therefore, microorganisms equipped to thrive in adverse conditions hold promise as efficient BCAs.

In the last decade, the world's population has grown enormously, posing a serious threat to environmental degradation and human health, due to the use of chemical fertilizers in agriculture (Goswami and Deka 2020). The application of enormous amounts of agrochemicals is not limited to agricultural production; however, post-harvest usage is also a concern. Post-harvest losses due to fungal pathogens are estimated to contribute to nearly 20–40% of post-harvest decay in fruits and vegetables, and these losses result in substantial quality deterioration, nutrient composition changes, mycotoxin contamination, and reduced market value for fruits (FAO 2019). To control such post-harvest loss, BCAs can also be employed as they offer a natural, safe, and highly efficient alternative for controlling pathogen infections, shedding light on the critical mechanisms required for effective disease management.

6.3 Bioactive Compounds/Metabolites from Extremophiles for Plant Disease Management

The utilization of microbes for biocontrol in the agriculture sector has been proven highly effective, cost-efficient, and environmentally friendly. Consequently, they can serve as substitutes, to curtail the use of harmful chemical fertilizers and pesticides (Pikuta et al. 2007). Secondary metabolites with diffusible properties, such as phenazine (specifically phenazine-1-carboxylic acid), pyrrolnitrin, iturin, surfactin, fengycin, and hydrogen cyanide, exhibit potent antifungal properties. A *Pseudomonas* sp. ST–TJ4 has been reported to demonstrate significant control over 11 fungal plant pathogens; some of them are *Botryosphaeria berengeriana*, *Colletotrichum tropicale*, and *Phytophthora cinnamomic*. Such antagonistic activity was attributed to the production of HCN, pyrrolnitrins, and phenazines (Zhang et al. 2020). Additionally, many *Pseudomonas* spp. are reported to produce compounds such as aerugines, azomycins, butyrolactones, cepaciamides, cepafungins, ecomycins, oomycin A, pyoluteorin, pseudomonic acid, rhamnolipids, and so on (Goswami et al. 2016).

Bacillus spp. are significantly known to produce NRPs such as fengycin, iturin, and surfactin, which exhibit bioactivity against various bacterial and fungal pathogens (Penha et al. 2017). For instance, *B. velezensis* HC6 demonstrated inhibitory activity against *Aspergillus* and *Fusarium* on *Zea mays*, inhibiting their mycelial growth and reducing toxin production through the secretion of lipopeptides (Liu et al. 2020). Such metabolites produced by *Bacillus* are ribosomal or non-ribosomal peptides and polyketide synthetases (PKS), some examples are surfactin, subtilosin, iturin, fengycin, bacillaene, bacilysin, sublancin, chlorotetain, subtilin, and so on (Olanrewaju et al. 2017). Biological control mechanisms utilized by the *Bacillus cereus* group against pathogens include competition for nutrients, antibiosis, space, biofilm formation, and mycoparasitism, yet the genetic aspects of these functions are partially explored (Olanrewaju et al. 2017).

Fungi such as *Trichoderma*, *Trichothecium*, *Talaromyces*, and *Aspergillus* and bacteria such as *Bacillus*, *Agrobacterium*, and *Pseudomonas* are known to produce a variety of compounds with antimicrobial activities. Some of the actinomycetes, namely, *Actinoplanes*, *Streptomyces*, *Spirillospora*, and *Streptoverticillium* have also exhibited antimicrobial properties against plant pathogens (Penha et al. 2017). Metabolites such as non-ribosomal peptide synthetase (NRPS), PKs, and siderophores produced by BCAs are known to mitigate plant biotic stress either directly by destroying pathogens or indirectly by inducing oxidative stress, disrupting the pathogen's Fe source (Ranjan et al. 2023). Endophytes, lichens, and extremophiles are explored using techniques like genome mining and metagenomics which have helped in their screening for novel antimicrobial compounds. However, a few metabolites with low yield and solubility and their potential impacts on human health and the environment should be considered for careful analysis to ensure the successful transition of these biopesticides from the laboratory to practical agricultural use (Liu et al. 2020).

Some of the key groups of metabolites being discussed hold the utmost importance in the biocontrol of pathogens in agriculture.

6.3.1 Non-ribosomal Peptides

NRPs are distinct compounds obtained from several microorganisms, including extremophiles, and are often known for their antimicrobial properties. Extremophiles, found in environments, namely, deep-sea hydrothermal vents or highly saline salt fields, have evolved the machinery to produce NRPs with unique chemical structures and functions. Halophiles also offer significant applications as a valuable source for discovering novel molecules currently used in pharmaceutical and food applications, particularly NRPs known for their antimicrobial properties (Penha et al. 2017). Two salt-tolerant strains, namely, QSLA16 and QSLA17 isolated from saline ponds demonstrated effectiveness in controlling the growth of halo-tolerant pathogens like *Staphylococcus aureus*, *Salmonella typhi*, and *Acinetobacter baumannii* (Kong et al. 2010). The extremophilic bacterium *Bacillus halotolerans* KKD1, isolated from Hoh Xil, displays significant biological efficacy against plant pathogens and shows resilience in environments characterized by high salinity and alkalinity. KKD1 demonstrates encouraging biocontrol capabilities against phytopathogens owing to the potential of surfactin, fengycin, and bacteriocins (Wu et al. 2022). Biosurfactants, particularly surfactin produced by *Pseudomonas* and *Bacillus*, have the potential to serve as environmentally friendly and less toxic green surfactants which have shown biocontrol activity over a range of microbial pathogens and find applications in agriculture and other industries, namely, food, pharmaceuticals, and cosmetics. Several rhizospheres and PGPR can produce surfactin which ensures signalling, motility, and biofilm formation. In agriculture, surfactin-producing microbes can be employed as biocontrol agents against plant pathogens and enhance nutrient bioavailability for beneficial PGPR (Olanrewaju et al. 2017). Among bacterial biosurfactants, rhamnolipids are among the most extensively studied (Tavares et al. 2012). They fall into a category of metabolites with diverse chemical compositions produced by various bacteria and fungi (Hausmann and Syldatk 2014).

NRPs are synthesized by NRPS enzymes, typically organized in operons or gene clusters. They operate in modules, each adding one amino acid, often resembling polyketide and fatty acid biosynthesis. Some NRPS are non-modular, and deviations from standard structures accommodate modifications or unusual amino acids. Each module consists of domains with defined functions, such as adenylation, condensation, and thiolation, separated by short spacer regions. These domains work in concert to activate and transfer amino acids, ultimately facilitating the synthesis of diverse NRPs.

6.3.2 Polyketides

Polyketides are a class of natural products synthesized by polyketide synthases (PKSs) from simple carboxylic acid building blocks. They exhibit structural complexity and diverse biological activities, including antibiotic, antifungal, antiparasitic, and anticancer properties. Extremophilic bacteria, from varying environmental conditions such as high temperatures salinity, or extreme pH levels, have garnered significant attention as potential sources of unique polyketides. Marine extremophiles, with their unique adaptation mechanisms, represent attractive targets for studying novel natural products (Kwak 2009). These extremophilic polyketides have shown remarkable bioactive properties, including antimicrobial, anticancer, and immunosuppressive activities. These complex molecules are synthesized through polyketide synthase (PKS) enzymes, which are modular and enable the production of a wide range of structurally complex polyketides. A polyketide compound avermectins which is produced by *Streptomyces* is a derivative of abamectin, plays a crucial role as a nematicidal metabolite, and serves as an essential component in commercial biological nematicides. Actinomycetes stand out as promising candidates, given their remarkable capacity to produce metabolites with insecticidal properties. Polyketide compounds, predominantly sourced from the actinomycetes group, form a significant family of natural insecticides, encompassing avermectins, spinosyns, polynactins, tetramycins, and analogues. Notably, *Pseudomonas*-derived polyketides, such as the well-studied secondary metabolites phenazines and 2,4-diacetylphloroglucinol (DAPG), have demonstrated biocontrol potential against plant pathogens, contributing to sustainable agriculture practices (Clough 2020).

6.3.3 Siderophores

Siderophores obtained from extremophiles hold significance in agricultural, environmental, and industrial applications. These small molecules (typically <2000 Da), produced under iron-limited conditions, play a pivotal role in facilitating iron uptake, essential for vital metabolic processes. Extremophiles, such as those found in acidic mine drainage or alkaline soda lakes, have evolved to produce unique siderophores tailored to their extreme habitats (Khan and Patra 2018). Siderophores and their derivatives have significant applications in agriculture, both for enhancing soil fertility and biocontrol of fungal pathogens. As part of the efforts to reduce the dependency on fungicides, one biocontrol strategy being explored implies the use of siderophores-producing BCAs. These are capable of chelating iron and are synthesized by various organisms, including bacteria, fungi, and plants particularly under low-iron conditions (Hider and Kong 2010). For example, *Pseudomonas* sp. NCIMB 10586, known for producing mupirocin, exhibits a broad-spectrum activity against Gram-positive and, to a lesser extent, Gram-negative bacteria also (Sutherland et al. 1985) and is recognized for its production of the siderophore pyoverdine PYO13525

(Matthijs et al. 2014). In addition to its known siderophore production, *Pseudomonas* sp. NCIMB 10586 has been found to produce two newly identified siderophores: mupirochelin, a relatively weak siderophore that exhibits strong antagonistic activity against *Globisporangium ultimum*, and the potent siderophore triabactin (Grosse et al. 2023). Polyketide compounds such as equisetin, fumonisins, and nigerloxin are utilized as phytopathogen control agents. Equisetin, from *Fusarium equiseti*, demonstrates antifungal properties (O'Donnel et al. 2022), while fumonisins, produced by *Fusarium* species, inhibit ceramide synthase in plants (Wangia-Dixon et al. 2020). Additionally, nigerloxin, derived from *Aspergillus niger*, exhibits antifungal activity against plant pathogens (Rao et al. 2002), offering potential alternatives to chemical pesticides.

6.3.4 Volatile Organic Compounds

Volatile organic compounds originating from extremophilic microorganisms, thriving in extreme environments like hydrothermal vents or arid deserts, are getting interest due to their unique properties and applications in the agriculture sector. Microbes considered as PGPR produce a diverse range of VOCs, including acetaldehyde, butanoic acid, camphene, and more. These VOCs serve various functions, from inhibiting the growth of phytopathogenic fungi to triggering systemic resistance in plants (Navarro et al. 2023). VOCs are special particularly because they suppress nearby pathogens in the rhizosphere and communicate with plants. This provides them with promising environmentally friendly alternatives to traditional agrochemicals. The study of VOCs from extremophiles not only enhances our understanding of their adaptability but also provides sustainable solutions for agriculture and pest management.

Microbial volatile organic compounds (mVOCs) are synthesized through various metabolic pathways by microorganisms. The mVOCs have gained attention due to their promising biocontrol potential, which is capable of exerting control over pathogens, both in proximity and at a distance (Toral et al. 2021). The VOCs do not require physical contact with pathogens or plant parts to exhibit their effects on both unlike other methods for phytopathogens control and the use of plant growth promoters. General bacteria such as *Arthrobacter*, *Bacillus*, *Pseudomonas*, *Serratia*, and *Stenotrophomonas* are well-known for producing VOCs for biocontrol of phytopathogens, apart from helping in plant growth promotion and induced systemic resistance. *Bacillus subtilis* FB17 has been identified as a primary producer of VOC acetoin (3-hydroxy-2-butanone), which plays a role in triggering beneficial effects (Grahovac et al. 2023). The VOC produced by *B. amyloliquefaciens* SQR-9 has been reported to have antibacterial activity against *Ralstonia solanacearum* ZJ3721 which causes bacterial wilt in tomatoes. Application of such VOCs for biocontrol agents resulted in a 70% suppression of bacterial pathogen growth efficiency (Raza et al. 2016). VOCs released by PGPR possess the capability to disperse over considerable distances, enabling indirect interactions between organisms, even at low concentrations. VOCs include molecules such as propanoic acid, camphor, 5-hydroxy-methyl-furfural, acetaldehyde, camphene, methanol, geosmin, butanoic acid, β-caryophyllene, and so on as outlined by Kanchiswamy et al., in 2015. Several studies have illustrated the suppressive effects of bacterial VOCs on phytopathogenic fungi. For example, *B. velezensis* strain ZSY-1 emits VOCs like 2,4-bis(1,1-dimethyl ethyl)-phenol, 2,5-dimethylpyrazine, 4-chloro-3-methyl-phenol, and benzothiazole, displaying significant biocontrol activity against *Botrytis cinerea* and *Alternaria solani*. This effectively controls fungal diseases in tomatoes (Gao et al. 2017). Pyrrolnitrin, a VOC prominently produced by *Burkholderia* spp., exhibits activity against crucial phytopathogens, including *R. solani*, *Bacillus cinerea*, *Verticillium dahliae*, and *Sclerotinia sclerotiorum* (Kanchiswamy et al. 2015a, b). VOCs released by *Pseudomonas* spp. ST–TJ4, particularly 1-undecene, have been identified to inhibit phytopathogens such as *Colletotrichum tropicale*, *Fusarium oxysporum*, *Phytophthora cinnamomi*, and *R. solani* (Zhang et al. 2020).

Furthermore, *Staphylococcus sciuri* MarR44 has demonstrated a 72.17% reduction in anthracnose disease in post-harvest strawberries caused by *Colletotrichum nymphaeae*. This reduction is attributed to the presence of 24 identified VOCs, with mesityl oxide (81.4%) being the most predominant (Alijani et al. 2019).

6.3.5 Others (Alkaloids, Phenolics, Flavonoids and Terpenes)

Medicinal plants associated with rhizobacteria have received relatively scant attention concerning their potential in combating plant pathogens despite the potential advantages they may offer. *Glycyrrhiza uralensis* is a well-studied plant acknowledged in Chinese traditional medicine for its diverse array of compounds, including flavonoids, phenolics, saponins, and coumarins (Zhang and Ye 2009). Licorice demonstrates significant pharmacological properties, namely, antimicrobial, anti-inflammatory, antioxidant, and antitumor activities (Liao et al. 2012). The economic importance and wide-ranging pharmacological benefits associated with the traditional use of *Glycyrrhiza uralensis* have been explored well; however, research on its endophytic bacteria and their potential as biocontrol agents has been limited (Asl and Hosseinzadeh 2008). Alkaloids, phenolics, flavonoids, and terpenes obtained from extremophiles, resilient microorganisms thriving in extreme environments, represent a source of unique bioactive compounds with potential applications across various industries. Extremophiles, whether residing in acidic hot springs or salt-saturated salt flats, have evolved to produce these secondary metabolites as part of their adaptation strategies. These compounds have drawn significant interest due to their diverse biological activities, ranging from antimicrobial and antioxidant properties to potential pharmaceutical and therapeutic applications. The study of extremophilic alkaloids, phenolics, flavonoids, and terpenes not only expands our knowledge of life's adaptability but also holds the promise of discovering novel bioactive molecules that can address challenges in healthcare, agriculture, and biotechnology, providing innovative solutions for both human needs and environmental concerns. It is noteworthy that this report represents the first instance of *Bacillus atrophaeus* being identified as a producer of bioactive compounds with antimicrobial activity. The shikimate pathway in bacteria not only supplies precursors like chorismate for the synthesis of aromatic amino acids but also serves as a source for diverse carbocyclic aromatic secondary metabolites, such as phenazines. These phenazine derivatives exhibit antibacterial and antifungal properties, making them noteworthy for pharmaceutical and biocontrol applications (Pierson and Pierson 2010). For instance, Glandorf et al. (2001) modified a *Pseudomonas putida* strain for producing phenazine-1-carboxylic acid, a yellow pigment modulating the expression of the phzABCDEFG operon from *Pseudomonas fluorescens* 2–79. Such modification enhanced its ability to stop fungal growth significantly and made the modified *P. putida* a suitable BCA (Glandorf et al. 2001). Another strategy which *Streptomyces* involves for the biocontrol of fungus are secretion of lytic enzymes such as chitinases. Bacteriocins also exhibit biocontrol activity against *Penicillium citrinum*, *A. niger*, and *Aspergillus flavus* (Adebayo and Aderiye 2010). Archaea synthesises bacteriocin-like antimicrobial peptides called archaeocins which have been identified and characterized within the *Haloarchaea* and *Sulfolobus* classes of archaea (Shand and Leyva 2008).

6.4 Mode of Action

The effective control of plant diseases has consistently posed a significant challenge in the pursuit of enhancing plant productivity and yields. Microbes-based products are sustainable and commercially viable alternatives to harmful agrochemicals (Rao et al. 2022). Extremophilic bacteria of genera *Bacillus*, *Pseudomonas*, and actinomycetes are resilient to environmental conditions and more suitable to undergo industrial application and processing (Fig. 6.2). Actinomycetes, including the genus *Streptomyces*, are often abundant producers of secondary metabolites, including antibiotics and antifungal compounds. They are emerging as promising candidates for crop protection, and their agroactive antibiotics are gaining commercial traction (Wang et al. 2023). Metabolites from these have been reported in several species of extremophilic microbes which are evident to alleviate plant biotic stress in controlled environments. They are affected through direct and indirect inhibition of pathogens. In direct inhibition, compounds with antimicrobial properties act on the cell wall, or cellular machinery to destroy the microbial pathogen however in case of indirect inhibition, it is brought about by induction of oxidative stress, competing for space and nutrients, and so on. The search for and discovery of novel microbes producing novel antimicrobial compounds is crucial for antibiotic-resistant pathogens. The extremolytes are molecules synthesized by extremophiles, aiding in their survival by stabilizing proteins, membranes, and cellular structures. Compounds, like trehalose, glycine betaine, ectoine, and hydroxyectoine, have garnered interest for their potential applications in biotechnology, medicine, and industry, offering opportunities to stabilize biomolecules, enhance pharmaceutical stability, and

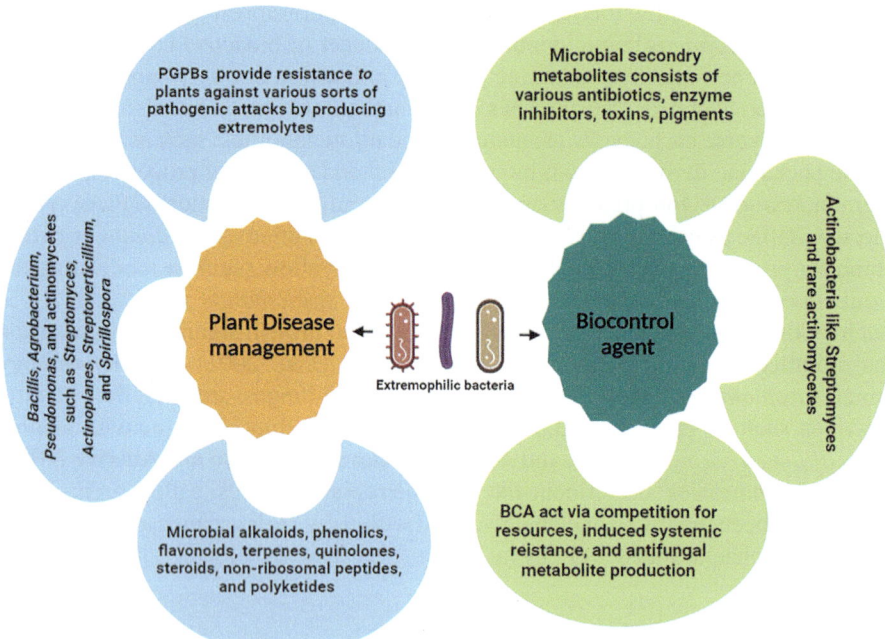

Fig. 6.2 Extremophiles and their use for plant diseases biocontrol and management

bolster crop resilience to stressors. Understanding extremolytes illuminates adaptation mechanisms to extreme conditions and holds promise for innovative biotechnological and medical advancements that provide resistance to various pathogenic attacks, promote plant growth, and protect plants against pathogens. They have been highly effective in controlling diseases like tomato spotted wilt viruses and cucumber mosaic viruses in crops such as tomato and pepper (Kumar et al. 2022).

Extremophiles primarily thrive in the root region, seeking optimal conditions for survival. In some cases, these bacteria work in concert with mycorrhizae to provide protection and resistance against various fungal pathogens (Rizvi et al. 2021). Additionally, they inhibit the growth of various root pathogens. Consequently, there has been significant attention on isolating, characterizing, and exploring these bacterial extremophiles to tap into their unexplored microbiome and extract various novel compounds. A soil-borne pathogen, *Macrophomina phaseolina* (Tassi) which is a pathogen to over 500 plant species globally, is inhibited by *Exiguobacterium* sp. S58 and *Stenotrophomonas* sp. AG3. In a study, they exhibited in vitro antifungal effects against *M. phaseolina* strains Mp02 (40.7% inhibition) and Mp06 (52.2% inhibition) on soybean seedlings. The study also reported altered fungal hyphae morphology via the secretion of lytic enzymes and polyamines. *Exiguobacterium* sp. S56a reduces the growth of *M. phaseolina* strains through putrescine secretion suggesting potential biocontrol in soybean seedlings. Polyamines play a significant role in regulating cellular metabolism and aiding plant acclimation to extreme environmental conditions. However, the polyamines produced by PGPBs and their antagonistic activity have received limited attention (Tuesta-Popolizio et al. 2021). The use of extremophile as PGPB is pivotal in the biological management of plant diseases, providing a sustainable alternative for fungicides.

To conclude, biological control offers an effective and sustainable alternative to conventional pesticides for managing bacterial plant diseases. Extremophilic bacterial-based biocontrol agents hold great promise in this regard. However, there remains a significant gap in our understanding of their role in controlling plant-borne diseases and their application in commercial agriculture. Various factors, both biotic and abiotic, influence the effectiveness of extremophilic bacteria as biocontrol agents, including the complex interplay between plants, pathogens, and associated bacteria.

6.5 Conclusions

Sustainable agriculture has emerged as an alternative approach to address the issues arising from the excessive and indiscriminate use of PGPBs by replacing chemical fertilizers and pesticides to boost crop yields. Consequently, there is a growing need for the identification of more potential alternatives to these chemicals. The PGPB-based biocontrol agents, with their diverse and unique direct and indirect traits, offer an appealing and practical alternative. The research in the field of extremophile microbial diversity and the discovery of novel bioactive molecules for biocontrol and plant disease management signifies a key chance for finding potential and resilient strains of PGPBS. The pursuit of extremophilic bacteria, which can thrive under extreme environmental stress conditions, has recognised several promising strains that can produce several active bacterial metabolite molecules such as NRPs, PKs, and VOCs for pre- and post-harvest applications. This chapter highlights the diverse potential of such microbial metabolites for biocontrol in agriculture. Ongoing research in this area holds the promise of delivering sustainable and environmentally friendly solutions to tackle agricultural challenges and enhance crop yields due to the growing agroclimatic crisis.

References

Adebayo AO, Aderiye BI (2010) Bacteriocin activity against some selected pathogenic Candida species. Afr J Microbiol Res 4(4):272–278

Alam M, Tiwary BK (2023) Extremophiles: diversity, adaptation and applications. Bentham Science Publishers

Alijani Z, Amini J, Ashengroph M, Bahramnejad B (2019) Antifungal activity of volatile compounds produced by Staphylococcus sciuri strain MarR44 and its potential for the biocontrol of Colletotrichum nymphaeae, causal agent strawberry anthracnose. Int J Food Microbiol 307:108276

Asl MN, Hosseinzadeh H (2008) Review of pharmacological effects of Glycyrrhiza sp. and its bioactive compounds. Phytotherapy Research: An International Journal Devoted to Pharmacological and Toxicological Evaluation of Natural Product Derivatives, 22(6):709–724.

Clough S (2020) The potential of Pseudomonas bacteria as biocontrol agents against multiple plant pathogens (Doctoral dissertation, University of York)

Coombs WT (2004) Impact of past crises on current crisis communication: insights from situational crisis communication theory. J Bus Commun (1973) 41(3):265–289

Dixit R, Kumari M (2023) Microbial metabolites in plant disease management. In: Microbial Biomolecules. Academic Press, pp 159–179

Domik D, Thürmer A, Weise T, Brandt W, Daniel R, Piechulla B (2016) A terpene synthase is involved in the synthesis of the volatile organic compound sodorifen of Serratia plymuthica 4Rx13. Front Microbiol 7:737

FAO (2019) The State of Food and Agriculture 2019. Moving forward on food loss and waste reduction. Rome. https://doi.org/10.4060/CA6030EN

Gao Z, Zhang B, Liu H, Han J, Zhang Y (2017) Identification of endophytic Bacillus velezensis ZSY-1 strain and antifungal activity of its volatile compounds against Alternaria solani and Botrytis cinerea. Biol Control 105:27–39. https://doi.org/10.1016/j.biocontrol.2016.11.007

Glandorf DCM, Verheggen P, Janssen T, Joritsma J, Thomashow LS, Leeflang P, Smit E, Wernars K, Lauereijs E, Thomas-Oates JE, Bakker PAHM, van Loon LC (2001) Effect of *Pseudomonas putida* WCS358r, genetically modified with the *phz* biosynthetic locus, on the fungal rhizosphere of microflora of field-grown wheat as determined by 18S rDNA analysis. Appl Environ Microbiol 67:3371–3378

Gomes J, Gomes I, Steiner W (2000) Thermolabile xylanase of the Antarctic yeast Cryptococcus adeliae: production and properties. Extremophiles 4:227–235

Goswami M, Suresh DEKA (2020) Plant growth-promoting rhizobacteria—alleviators of abiotic stresses in soil: a review. Pedosphere, 30(1), 40–61.

Grahovac J, Pajčin I, Vlajkov V (2023) *Bacillus* VOCs in the context of biological control. Antibiotics 12(3):581. https://doi.org/10.3390/antibiotics12030581

Grosse C, Brandt N, Van Antwerpen P, Wintjens R, Matthijs S (2023) Two new siderophores produced by Pseudomonas sp. NCIMB 10586: the anti-oomycete non-ribosomal peptide synthetase-dependent mupirochelin and the NRPS-independent triabactin. Front Microbiol 14:1143861

Gupta RK, Fuke P, Khardenavis AA, Purohit HJ (2023) In silico genomic characterization of Bacillus velezensis strain AAK_S6 for secondary metabolite and biocontrol potential. Curr Microbiol 80(2):81

Hider RC, Kong X (2010) Chemistry and biology of siderophores. Nat Prod Rep 27(5):637–657

Hui MLY, Tan LTH, Letchumanan V, He YW, Fang CM, Chan KG, Lee LH (2021) The extremophilic actinobacteria: from microbes to medicine. Antibiotics 10(6):682

Hausmann R, Syldatk C (2014) Types and classification of microbial surfactants. Biosurfactants: production and utilization—processes, technologies, and economics, 159, 1.

Kamat S, Dixit R, Kumari M (2022) Endophytic microbiome in bioactive compound production and plant disease management. In: Microbial biocontrol: food security and post harvest management: volume 2. Springer International Publishing, Cham, pp 79–128

Kanchiswamy CN, Malnoy M, Maffei ME (2015a) Chemical diversity of microbial volatiles and their potential for plant growth and productivity. Front Plant Sci 6:151

Kanchiswamy CN, Malnoy M, Maffei ME (2015b) Bioprospecting bacterial and fungal volatiles for sustainable agriculture. Trends Plant Sci 20(4):206–211. https://doi.org/10.1016/j.tplants.2015.01.004

Kanekar PP, Kanekar SP (2022) Xerophilic and xerotolerant microorganisms. In: Diversity and biotechnology of extremophilic microorganisms from India, pp 281–288. https://doi.org/10.1007/978-981-19-1573-4_10

Khan MF, Patra S (2018) Deciphering the rationale behind specific codon usage pattern in extremophiles. Sci Rep 8(1):15548

Kong DX, Jiang YY, Zhang HY (2010) Marine natural products as sources of novel scaffolds: achievement and concern. Drug Discov Today 15(21–22):884–886

Kumar R, Merugu R, Mohapatra S, Sharma S (2022) Extremophiles life of microorganisms in extreme environments. In: Extremophiles. CRC Press, pp 43–66

Kwak YS (2009) Impact of 2, 4-diacetylphloroglucinol on Gaeumannomyces graminis var. tritici and wheat in the take-all pathosystem. Washington State University

Liao YH, Lin CC, Li TC, Lin JG (2012) Utilization pattern of traditional Chinese medicine for liver cancer patients in Taiwan. BMC Complement Altern Med, 12, 1–9.

Liu Y, Teng K, Wang T, Dong E, Zhang M, Tao Y, Zhong J (2020) Antimicrobial Bacillus velezensis HC6: production of three kinds of lipopeptides and biocontrol potential in maize. J Appl Microbiol 128(1):242–254

Matthijs S, Vander Wauven C, Cornu B, Ye L, Cornelis P, Thomas CM, Ongena M (2014) Antimicrobial properties of Pseudomonas strains producing the antibiotic mupirocin. Res Microbiol 165(8):695–704

Mehetre G, Leo VV, Singh G, Dhawre P, Maksimov I, Yadav M, Singh BP (2021) Biocontrol potential and applications of extremophiles for sustainable agriculture. In: Microbiomes of extreme environments. CRC Press, pp 230–242

Morrison CK, Arseneault T, Novinscak A, Filion M (2017) Phenazine-1-Carboxylic acid production by Pseudomonas fluorescens LBUM636 alters Phytophthora infestans growth and late blight development. Phytopathology 107:273–279. https://doi.org/10.1094/phyto-06-16-0247-r

Nayarro LT, Rodríguez M, Martínez-Checa F, Montaño A, Cortés-Delgado A, Smolinska A, Sampedro I (2023) Corrigendum: identification of volatile organic compounds in extremophilic bacteria and their effective use in biocontrol of postharvest fungal phytopathogens. Front Microbiol 14:1267324

Newitt JT, Prudence SM, Hutchings MI, Worsley SF (2019) Biocontrol of cereal crop diseases using streptomycetes. Pathogens 8(2):78

O'Donnell K, Gräfenhan T, Laraba I, Busman M, Proctor RH, Kim H, Wiederhold NP, Geiser DM, Seifert KA (2022). Fusarium abutilonis and F. guadeloupense, two novel species in the Fusarium buharicum clade supported by multilocus molecular phylogenetic analyses. Mycologia 114(4):682–696.

Olanrewaju OS, Glick BR, Babalola OO (2017) Mechanisms of action of plant growth promoting bacteria. World J Microbiol Biotechnol 33:1–16

Pandey KD, Patel AK, Singh M, Kumari A (2021) Secondary metabolites from bacteria and viruses. In: Sinha RP, Häder HCD-P (eds) Natural bioactive compounds. Academic Press. https://doi.org/10.1016/B978-0-12-820655-3.00002-1

Penha CB, Bonin E, da Silva AF, Hioka N, Zanqueta ÉB, Nakamura TU, Mikcha JMG (2017) Photodynamic inactivation of foodborne and food spoilage bacteria by curcumin. LWT-Food Sci Technol 76:198–202

Pierson LS, Pierson EA (2010) Metabolism and function of phenazines in bacteria: impacts on the behavior of bacteria in the environment and biotechnological processes. Appl Microbiol Biotechnol 86:1659–1670

Pikuta EV, Hoover RB, Tang J (2007) Microbial extremophiles at the limits of life. Crit Rev Microbiol 33(3):183–209

Ranjan A, Rajput VD, Prazdnova EV, Gurnani M, Bhardwaj P, Sharma S et al (2023) Nature's antimicrobial arsenal: non-ribosomal peptides from PGPB for plant pathogen biocontrol. Fermentation 9(7):597. https://doi.org/10.3390/fermentation9070597

Rao KCS, Divakar S, Babu KN, Rao AGA, Karanth NG, Sattur AP (2002) Nigerloxin, a Novel Inhibitor of Aldose Reductase and Lipoxygenase with Free Radical Scavenging Activity from Aspergillus niger CFR-W-105. J. Antibiot. 55(9), 789–793. https://doi.org/10.7164/antibiotics.55.789

Rao AS, Nair A, More VS, Anantharaju KS, More SS (2022) Extremophiles for sustainable agriculture. In: New and future developments in microbial biotechnology and bioengineering. Elsevier, pp 243–264

Raza W, Ling N, Yang L, Huang Q, Shen Q (2016) Response of tomato wilt pathogen Ralstonia solanacearum to the volatile organic compounds produced by a biocontrol strain Bacillus amyloliquefaciens SQR-9. Sci Rep 6(1):24856. https://doi.org/10.1038/srep24856

Rizvi A, Ahmed B, Khan MS, Umar S, Lee J (2021) Psychrophilic bacterial phosphate-biofertilizers: a novel extremophile for sustainable crop production under cold environment. Microorganisms 9(12):2451

Sahli R, Rivière C, Siah A, Smaoui A, Samaillie J, Hennebelle T, Sahpaz S (2018) Biocontrol activity of effusol from the extremophile plant, Juncus maritimus, against the wheat pathogen Zymoseptoria tritici. Environ Sci Pollut Res 25:29775–29783

Santos AP, Muratore LN, Solé-Gil A, Farías ME, Ferrando A, Blázquez MA, Belfiore C (2021) Extremophilic bacteria restrict the growth of Macrophomina phaseolina by combined secretion of polyamines and lytic enzymes. Biotechnology Reports 32:e00674

Shand RF, Leyva KJ (2008) Archaeocins. FEMS Microbiol Rev 32(2):103–112

Silva GDC, Kitano IT, Ribeiro IADF, Lacava PT (2022) The potential use of actinomycetes as microbial inoculants and biopesticides in agriculture. Frontiers in Soil Science 2:833181

Singh P, Jain K, Desai C, Tiwari O, Madamwar D (2019) Microbial community dynamics of extremophiles/extreme environment. In: Microbial diversity in the genomic era. Academic Press, pp 323–332

Sutherland R, Boon RJ, Griffin KE, Masters PJ, Slocombe B, White AR (1985) Antibacterial activity of mupirocin (pseudomonic acid), a new antibiotic for topical use. Antimicrob Agents Chemother 27(4):495–498. https://doi.org/10.1128/aac.27.4.495

Tavares LFD, Silva PM, Junqueira M, Mariano DCO, Nogueira FCS, Domont GB, Freire DMG, Neves BC (2012) Characterization of rhamnolipids produced by wild-type and engineered Burkholderia kururiensis. Appl Microbiol Biotechnol 97:1909–1921. https://doi.org/10.1007/s00253-012-4454-9

Toral L, Rodríguez M, Martínez-Checa F, Montaño A, Cortés-Delgado A, Smolinska A et al (2021) Identification of volatile organic compounds in extremophilic bacteria and their effective use in biocontrol of postharvest fungal phytopathogens. Front Microbiol 12. https://doi.org/10.3389/fmicb.2021.773092

Tuesta-Popolizio DA, Velázquez-Fernández JB, Rodriguez-Campos J, Contreras-Ramos SM (2021) Isolation and identification of extremophilic bacteria with potential as plant growth promoters (PGPB) of a geothermal site: a case study. Geomicrobiol J 38(5):436–450

Van der Meij A, Worsley SF, Hutchings MI, van Wezel GP (2017) Chemical ecology of antibiotic production by actinomycetes. FEMS Microbiol Rev 41(3):392–416

Verma P, Yadav AN, Kumar V, Singh DP, Saxena AK (2017) Beneficial plant-microbes interactions: biodiversity of microbes from diverse extreme environments and its impact for crop improvement. In: Plant-microbe interactions in agro-ecological perspectives, pp 543–580. https://doi.org/10.1007/978-981-10-6593-4_22

Wang D, Luo WZ, Zhang DD, Li R, Kong ZQ, Song J, Chen JY (2023) Insights into the biocontrol function of a Burkholderia gladioli Strain against Botrytis cinerea. Microbiol Spectr 11(2):e04805–e04822

Wangia-Dixon RN, Nishimwe K (2020) Molecular toxicology and carcinogenesis of fumonisins: A review. Journal of Environmental Science and Health, Part C, 39(1):44–67.

Wu X, Fan Y, Wang R, Zhao Q, Ali Q, Wu H et al (2022) Bacillus halotolerans KKD1 induces physiological, metabolic and molecular reprogramming in wheat under saline condition. Front Plant Sci 13. https://doi.org/10.3389/fpls.2022.978066

Yadav AN, Saxena AK (2018) Biodiversity and biotechnological applications of halophilic microbes for sustainable agriculture. J Appl Biol Biotechnol. https://doi.org/10.7324/jabb.2018.60109

Zhang Z, Song W, Su J, Tian H (2020) Vibration-induced emission (VIE) of N, N'-disubstituted-dihydribenzo [a, c] phenazines: fundamental understanding and emerging applications. Adv Funct Mater 30(2):1902803.

Zhang Q, Ye M (2009) Chemical analysis of the Chinese herbal medicine Gan-Cao (licorice). J Chromatogr A, 1216(11):1954–1969.

Zhang Y, Kong WL, Wu XQ, Li PS (2022) Inhibitory effects of Phenazine compounds and volatile organic compounds produced by Pseudomonas aurantiaca ST-TJ4 against Phytophthora cinnamomi. Phytopathology 112(9):1867–1876. https://doi.org/10.1094/phyto-10-21-0442-r

Chapter 7
Bioactive Molecules Derived from Extremophilic Fungi and Their Agro-Biotechnological Application

Namita Ashish Singh, Avinash Marwal, Juhi Goyal, and Nitish Rai

7.1 Introduction

The demand for novel compounds with clinical and industrial importance has always been ever-increasing and voluminous studies reporting the isolation of such compounds continue to rise. This demand is partly due to the known limitations of drugs and existing molecules. Novel Bioactive molecules from extremophiles provide an attractive alternative (Giordano 2021). As defined by Bérdy (2005), bioactive secondary metabolites are naturally occurring small chemical compounds isolated from different living organisms. Bioactive molecules are produced by microorganisms as secondary metabolites that are non-essential for their growth and development. These molecules have been produced due to the evolution of microbes over billions of years and therefore unique in terms of structure and function (Giordano 2021).

The bioactive molecules from the microbes found in common habitats have been explored quite well. Now, there is a need to explore the molecules from less common and extreme habitats. These natural habitats include hot springs, thermal vents, freezing temperatures of Arctic and Antarctica regions, deep seas and oceans, regions around nuclear and thermal power plants, extremely hot and cold deserts, glaciers and several other habitats. The habitats provide an excellent source of

N. A. Singh
Department of Microbiology, Vigyan Bhawan – Block B, New Campus, Mohanlal Sukhadia University, Udaipur, Rajasthan, India

A. Marwal · J. Goyal
Department of Biotechnology, Vigyan Bhawan – Block B, New Campus, Mohanlal Sukhadia University, Udaipur, Rajasthan, India

N. Rai (✉)
Department of Zoology, Lucknow University, Lucknow, Uttar Pradesh, India
e-mail: nitish.rai@mlsu.ac.in

© The Author(s), under exclusive license to Springer Nature Switzerland AG 2024
A. Ranjan et al. (eds.), *Extremophiles for Sustainable Agriculture and Soil Health Improvement*, https://doi.org/10.1007/978-3-031-70203-7_7

molecules with novel structure and function due to their extreme environment and microbial flora. This is due to the unique metabolome of organisms to thrive under such hostile environments and the expression of novel molecules at specific time points. To survive in these conditions, extremophiles have evolved mechanisms to maintain intracellular temperature, pH, metabolites, biochemical reactions and genomic and proteomic integrity (Deshmukh et al. 2017b).

Earth harbours some of the most varied forms of fungi in virtually almost all ecosystems. There is a special interest in extremophilic fungi thriving in deep-sea environments, deserts, frozen soil and so on to produce novel bioactive compounds to cope with the harsh conditions of their surroundings (Coleine et al. 2022). There is a growing interest among researchers to isolate the novel and unique natural products from marine as well as terrestrial extreme fungi, with wide-ranging applications from agriculture to industries (Chen et al. 2014a; Agrawal et al. 2016; Deshmukh et al. 2017a; Ibrar et al. 2020). Since the discovery of penicillin, fungi have been considered an excellent source of bioactive compounds and rightly so. However, the majority of fungal species remain undiscovered to this date, and only a small fraction has been described so far with a smaller proportion explored for potential application.

In this chapter, an overview is presented of novel bioactive metabolites isolated from marine, terrestrial and permafrost soil fungi. The key molecules isolated from fungi thriving in the extreme environment with their potential application in agriculture have been discussed in detail.

7.2   Overview of Extremophilic Fungi and Bioactive Molecules

Organisms flourishing/growing/thriving in extreme/harsh environments are known for their extremophilic nature. Extremophilic fungi are those that thrive in environments, including places with low and high temperatures (psychrophiles and thermophiles), low and high pH (acidophiles and alkaliphiles), radiation (radioresistant), high pressure (barophilic), high salinity (halophilic) and high metallic content (metallotolerants) (Table 7.1).

7.3   Bioactive Molecules Isolated from Marine Extremophilic Fungi

7.3.1   Polyketide Compounds

Several marine extremophilic fungi from the deep sea have been used to isolate many polyketide compounds with vital biological activities. The compounds penilactones and methyl isoverrucosidinol were isolated from the deep-sea fungus

Table 7.1 Fungi thriving under extreme environment and their limits

S. no.	Factor	Fungi	Environment/source	Limits	Examples	References
1.	High temperature	Thermophilic	Submarine hydrothermal vents, oceanic crust	110 °C to 121 °C	*Mucor pusillus, Thermomyces lanuginosus, Thermoidium sulfureum, Thermoascus aurantiacus*	Luo et al. (2018)
2.	Low temperature	Psychrophilic	Ice	−20 °C to −25 °C	*Pseudogymnoascus destructans, Ophiocordyceps sinensis*	Toledo et al. (2014)
3.	Alkaline systems	Alkaliphilic	Soda lakes	pH > 11	*Sodiomyces* sp. (*Plectosphaerellaceae*), *Acrostalagmus luteoalbus* (*Plectosphaerellaceae*), *Emericellopsis alkaline* (*Hypocreales*), *Thielavia* sp. (*Chaetomiaceae and Pleosporaceae*)	Grum-Grzhimaylo et al. (2015)
4.	Acidic systems	Acidophilic	Volcanic springs, acid mine drainage	pH −0.06 to 1.0	*Leucosporidium, Mortierella* and *Penicillium* taxa	Gross and Robbins (2000)
5.	Ionizing radiation	Radioresistant	Cosmic rays, X-rays, radioactive decay	1500 to 6000 Gy	*Epicoccum nigrum, Rhizopus nigricans, Cladosporium herbarum*	Rateb et al. (2011)
6.	UV radiation	Radioresistant	Sunlight	5000 J/m²	*Epicoccum nigrum, Rhizopus nigricans, Cladosporium herbarum*	Rateb et al. (2011)
7.	High pressure	Barophilic	Mariana Trench	1100 bar	*Phaeotheca* sp.	Uchida et al. (2016)
8.	Salinity	Halophilic	High salt concentration	a_w ~ 0.6	*Hortaea werneckii, Phaeotheca triangularis, Trimmatostroma salinum,* halotolerant *Aureobasidium pullulans*	Kogej et al. (2005)
9.	Deep crust	Metallotolerants	Some gold mines		*Alternaria, Aspergillus, Cladosporium, Phoma, Penicillium, Trichoderma*	Urík et al. (2014)
10.	Sub-zero temperature for at least 2 years	Permafrost fungi	Frozen ground or underwater sediment		*Penicillium citrinum, Aspergillus ochraceopetaliformis, Oidiodendron truncatum, P. melanoconidium*	Lu et al. (2014)

Penicillium crustosum PRB-2 and hydrothermal vent fungus *Penicillium* sp. Y-50-10, respectively (Wu et al. 2012; Pan et al. 2016). The penilactones possess a new carbon skeleton formed from two 3,5-dimethyl-2,4-diol-acetophenone units and a g-butyrolactone moiety, while methyl isoverrucosidinol possesses a novel conformational isomer of verrucosidin backbone. The marine *Aspergillus* sp. is reported to possess several molecules having diverse applications (Li et al. 2023). Novel compounds Asperalin A–F structurally possess benzoic acid derivatives coupled with dihydroquinolone alkaloid. Notably, Asperalin F possesses a novel bactericidal activity against the fish pathogenic bacterium, *Edwardsiella ictaluri* and *Streptococcus parauberis* (Hu et al. 2023).

Reticulol, a polyketide, was isolated from *Graphostroma* sp., a deep-sea fungus. Reticulol (6,8-dihydroxy-7-methoxy-3-methylisocoumarin) is an isocoumarin derivative and exhibited antifungal activity against *Trichophyton mentagrophytes* (Niu et al. 2018). A novel macrocyclic lactone polyketide, curvularin, was isolated from *Penicillium* sp. SF-5859, thriving under the Ross Sea. The compound is known to promote growth and increase the efficiency of cattle feed (Ha et al. 2017).

7.3.2 Nitrogenous Compounds

There are novel alkaloid bioactive compounds like cyclopiamide, penipanoid, 2-(4-hydroxybenzyl)quinazolin-4(3H)-one, oximoaspergillimide, and so on isolated from deep-sea *Penicillium* spp. Penipanoid A is isolated from *Penicillium paneum* and belongs to the class of benzoic acids. The compound possesses benzoic acid substituted by a 1H-1,2,4-triazol-1-yl group at a second position which in turn is substituted by a 4-hydroxybenzyl group at position 5. 2-(4-Hydroxybenzyl) quinazolin-4(3H)-one is a novel alkaloid isolated from *Penicillium paneum* (Ghosh and Nagarajan 2016). A 4-hydroxybenzyl group is replaced at position 2 of a quinazoline alkaloid to form 2-(4-hydroxybenzyl)quinazolin-4(3H)-one. A novel alkaloid derivative called oximoaspergillimide was discovered in a cultured marine-derived *Aspergillus* fungus (Cardoso-Martínez et al. 2015). The first natural compound to be reported to have an oxime-imide functionality is oximoaspergillimide.

The compounds like brevianamide, neoaspergillic and neohydroxyaspergillic were isolated from *Penicillium* spp. and inhibited the growth of soil-inhabiting bacteria having agricultural relevance like *Pseudomonas aeruginosa, Bacillus subtilis, Escherichia coli, Bacillus cereus*, and *Candida albicans* (Zhang et al. 2018; Li et al. 2011a; Xu et al. 2015). Varioxepine A, isolated from marine endophytic fungus *Paecilomyces variotii*, inhibited the growth of plant-growth-related bacteria like *Micrococcus luteus* and fish pathogens (Zhang et al. 2014) like *Aeromonas hydrophila, Vibrio anguillarum* and *Vibrio harveyi*. Neochinuline A is an indole alkaloid that is derived from the marine fungus *Microsporum* sp. It has three structural moieties, including diketopiperazine, indole and an isoprenyl moiety (Wijesekara et al. 2013).

7.3.3 Polypeptides

Polypeptide canescenin A and B, isolated from *Penicillium canescens*, inhibited the growth of plant pathogen *Pseudomonas aeruginosa* and *Bacillus amyloliquefaciens* at 100 μM (Dasanayaka et al. 2020). Several compounds isolated from marine *Aspergillus* spp. like clavatustides A and B are novel cyclodepsipeptides that have d-phenyllactic acid residues and a unique anthranilic acid dimer (Jiang et al. 2013). Butyrolactone I is a butan-4-olide that is tetrahydrofuran substituted by an oxo group at position 2 (Xiao-Wei et al. 2019).

The compound simplicilliumtides A–I, isolated from *Simplicillium obclavatum*, is a novel linear peptide and showed antibacterial activity against *Escherichia coli* and *Bacillus amyloliquefaciens* (Liang et al. 2016).

7.3.4 Ester and Phenolic Derivatives

The marine fungus *Aspergillus unguis* is shown to be an excellent source of novel ester derivatives. The ester compounds isolated from *Aspergillus unguis* include folipastatin, chlorofolipastatin, unguinol and nornidulin. These compounds belong to the chemical group of chlorinated depsidone (Uchida et al. 2016). The phenolic derivatives are found in marine *Aspergillus*, and *Penicillium* species like *Aspergillus versicolor* are reported to possess phenolics like Coccoquinone and Aspergilol G, H, and I (Huang et al. 2017). The pestalotionol, isolated from *Penicillium* sp. is a phenol antibiotic and showed potent activity against biocontrol bacteria *Bacillus subtilis* with minimum inhibitory concentration (MIC) values of 8 μg/mL (Pan et al. 2017).

7.3.5 Piperazine Derivatives

Several novel piperazine derivatives are reported from marine fungi in recent times. The piperazine compounds from marine fungi *Aspergillus versicolor* include versicolorin, brevianamide R, 7,8-epoxy-brevianamide Q, 8-epoxy-brevianamide R and 8-hydoxy-brevianamide R (Hu et al. 2019). The compounds pretrichodermamide C and N-methylpretrichodermamide B isolated from *Penicillium* sp. thriving in hypersaline lake belong to a group of epidithiodiketopiperazines which are a diverse group of bioactive compounds possessing disulfide linkage in the dioxopiperazine ring (Orfali et al. 2015). A deep-sea-derived fungus *Dichotomomyces cejpii* is reported to possess a piperazine compound dichotocejpin which showed an inhibited α-glucosidase enzyme with an IC50 value of 138 μM. α-Glucosidase is an endocellular enzyme, present in non-germinated and germinated cereals (Fan et al. 2016).

7.3.6   Terpenoid Compounds

The sediment-derived *Penicillium* sp. thriving at a depth of approximately 5115 m possessed brevione F–I terpenoid which structurally belongs to breviane spiroditerpenoids class of metabolites (Li et al. 2012b). Compounds spirograterpene A and conidiogenone C and I were isolated from the deep-sea-derived fungus *Penicillium granulatum*. Spirograterpene A is a spiro-tetracyclic diterpene while conidiogenone C and I are cyclopianes which are tetracyclic diterpenes featuring a highly fused and strained 6/5/5/5 ring system (Niu et al. 2017). The terpenoid compounds (7S)-(+)-7-O-methylsydonol, (S)-(+)-sydonol and 7-deoxy-7,14-didehydrosydonol were obtained from *Aspergillus sydowii* thriving in the hydrothermal vent (Chung et al. 2013). Liu et al. (2020) reported the novel diterpenoids, longidiacids A and B, polyketides, as well as the cytochalasin analogues, longichalasins A and B from the fungus *Diaporthe longicolla* FS429 collected from deep-sea sediment at a depth of 3000 m.

7.3.7   Sorbicillionids

Meng et al. (2018) reported saturnispols A–H (Sorbicillionids) isolated from sponge-derived fungus *Trichoderma saturnisporum* DI-IA owning antibacterial activity against gram-negative bacteria. Six new sorbicillinoids, trichoreeseione A & B, trichodermolide B, 13-hydroxy-trichodermolide, 24-hydroxy-trichodimerol, and 15-hydroxy-bisvertinol, along with three known analogues, trichodimerol, 24-hydroxy-bisvertinol, and bisvertinol, were isolated from the sponge-derived fungus *Trichoderma reesei* (HN-2016-018) (Rehman et al. 2020). *Phialocephala* sp. FL30r (piezophilic fungi) has been reported to produce various sorbicillin-type compounds, two novel being dihydrooxosorbiquinol (bisorbicillinoids) and oxosorbiquinol (Li et al. 2010).

7.4   Bioactive Molecules Isolated from Terrestrial Extremophilic Fungi

7.4.1   Peptaibiotics

Peptaibiotics include two subgroups, namely, peptaibols and lipopeptaibols. Peptaibols are a huge family of bioactive peptides that contain 7–20 amino acid residues. Peptaibols comprise a high proportion of aminoisobutyric acid and an acetate group in the N-terminal residue and a C-terminal amino alcohol (Marik et al. 2017; Das et al. 2018). Lipopeptaibols are a novel group of naturally occurring, short peptides which is distinguished by a lipophilic acyl chain at the N-terminus, an elevated content of α-aminoisobutyric acid along with a 1,2-amino alcohol at C-terminus. Thirty-nine lipopeptaibols have been reported from different *Trichoderma* species, including *T. longibrachiatum*, *T. phellinicola*, *T. Strigosum*, *T. Pubescens*, and *T. stromaticum* (Degenkolb et al. 2006; Neumann et al. 2015).

7.4.2 Anthraquinone

Anthraquinones are the largest group of innate pigments of quinoid nature produced by fungi. Anthraquinone are derived from anthracenes and contains two keto groups, mostly in positions 9 and 10. Three new anthraquinone derivatives (1–3) were isolated from the extremophilic fungus *Penicillium* sp. OUCMDZ-4736 was isolated from the sediment around the roots of the mangrove at pH 2.5. Anthraquinones, emodin is a bioactive molecule produced from the thermophilic fungus *Penicillium* sp., which was isolated from the Ghamiqa hot spring in Saudi Arabia (Orfali and Perveen 2019).

7.4.3 Phomopsolides

Phomopsolides and phomopsolidones are two classes of natural products that contain oxidized decanoic acid lactones and share an ester at the C-4/C-5 position. The phomopsolides comprised a group of five dihydropyranone structures with a C-4 tiglate ester (Stierle and Stierle 2014). Huang et al. (2008) isolated three metabolites named phomopsin A, B, and C, together with two known compounds cytosporone B and C, from the mangrove endophytic fungus, *Phomopsis* sp. *ZSU-H76* is derived from the South China Sea. Cytosporone is a fungal metabolite that is closely related to phomposin C. Phomopsin A, B and C showed insignificant antibiotic activities, but cytosporone B and C showed inhibition against *Candida albicans* and *Fusarium oxysporum* with MIC ranging from 32 to 64 µg/ml. Stierle and Stierle (2014) reported phomopsolide A and C (57–58) compounds that were isolated by the acidophilic fungi (*Penicillium clavigerum*) collected from acid mine waste.

7.4.4 Meroterpenes

Meroterpenoids are a combination of polyketide and terpenoid structures. Ten meroterpenoids, namely, asperterpenes D–M, were isolated from the soil fungi *Aspergillus terreus* (Qi et al. 2018). The chloroform extract of *Penicillium rubrum* produced two novel meroterpenes: berkeleydione and berkeleytrione (Stierle et al. 2011). Stierle et al. (2011) reported berkeleyones A–C from *Penicillium rubrum* that showed inhibitory activity against signalling enzyme caspase-1 and interleukin 1-β.

Three new sesquiterpenes, namely, xylarenones C–E, have been isolated from Camarops-like endophytic fungus isolated from *Alibertia macrophylla*. Four new diterpenes: scoparasin B, libertellenone H and eutypenoids A and C, have been identified from the psychrophilic fungal *Eutypella strain* isolated from Arctic soil on London Island (Liu et al. 2014). The meroterpenoids chrodrimanins I and J together with five chrodrimanins were found in the Antarctic moss-derived fungus *Penicillium funiculosum* (Zhou et al. 2015). Chen et al. (2019) isolated three novel diterpenes koninginols A–C and two new sesquiterpenoids: 11-hydroxy-15-drimeneoic acid and koninginol D, from the endophytic fungus *Trichoderma koningiopsis* A729, which was derived from *Morinda officinalis*. Out of three, two new diterpenes, koninginols A and B, showed significant antibacterial activity against *B. subtilis* at minimum inhibitory concentrations of 10 and 2 µg mL^{-1}, respectively. Desert fungi have also been reported as a source of unique terpenes. Wallemino and walleminone are two novel terpenes produced by *Wallemia sebi*, a xerophile isolated from the Atacama Desert (Jančič et al. 2016).

7.4.5 Polyketide

Rateb et al. (2011) isolated *Streptomyces leeuwenhoekii* C34 from Atacama desert soil which generates chaxalactins A–D, which is a 22-membered macrolactone polyketides. Chaxalactins D exhibited antibacterial activity against *Staphylococcus aureus* ATCC 25923. *Streptomyces leeuwenhoekii* strain C38, isolated from a hyperarid soil collected from the Atacama Desert produces 22-membered macrolactone antibiotics, the atacamycins A–C.

Chen et al. (2020) isolated the six polyketides from the terrestrial endophytic fungus *Trichoderma spirale* A725 of the medicinal plant *Morinda officinalis* via silica gel, reversed-phase silica column chromatography, and HPLC. Out of six, two compounds were new compounds, and one compound was a new natural product.

7.4.6 Other Compounds

Zhang et al. (2009) reported the cytotoxicity of the compound secalonic acid D that was separated from the secondary metabolites of the mangrove endophytic fungus No. ZSU44 against HL60 and K562 cells with IC(50) of values 0.38 and 0.43 mumol/L, respectively. Secalonic acid D induces cell cycle arrest of G(1) phase related to the downregulation of c-Myc by activation of GSK-3beta. Secalonic acid D reduced the production of multidrug-resistant (MDR) as well as their parental cancer cells and induced apoptosis by the activation of JNK/c-Jun signalling pathway followed by inhibition of Src/STAT3 signalling (Zhang et al. 2019).

7.5 Bioactive Molecules Isolated from Permafrost Soil Fungi

7.5.1 Polyketides

Wang et al. (2016b) isolated asteltoxin B, ochraceopones A–E, and isoasteltoxin from *Aspergillus ochraceopetaliformis* SCSIO 05702, an Antarctic soil-derived fungus. Isoasteltoxin was found to be an isomer of asteltoxin, with one new double bond, and ochraceopones A-D were among the initial examples of α-pyrone merosesquiterpenoids with a previously unreported linear tetracyclic carbon structure (Wang et al. 2016b). An Antarctic marine sponge-derived fungus *Pseudogymnoascus* spp. produced 3-nirtoasterric acid and pseudogymnoascin A–C that were found to be nitro derivatives of asterric acid, a recognized fungal metabolite (Figueroa et al. 2015). *Streptomyces leeuwenhoekii* strain C34T produces the aminoglycoside antibiotics hygromycin as well as the chaxamycins A–D and an ansamycin-type macrocyclic polyketide (Rateb et al. 2011).

7.5.2 Nitrogenous Compounds

The *Eutypella* sp. D1, found in the Arctic soil, was reported to produce three novel tyrosine-derived Cytochalasins Z_{24-26} as well as one previously identified substance, scoparasin B (Liu et al. 2014). Li et al. (2012a) reported two unique epipolythiodioxopiperazines (chetracin B and chetracin C) and five new diketopiperazines (chetracin D and oidioperazines A–D) from an Antarctica-derived psychrophilic fungus, *Oidiodendron truncatum*. Nitrogen-containing diterpenes, Eutypenoid B and Libertellenone G, were isolated from *Eutypella* spp. (Lu et al. 2014; Zhang et al. 2016a, b).

7.5.3 Quinolines and Quinazolines

Permafrost-derived fungal strain *Penicillium citrinum* was reported to synthesize two compounds of quinoline nature, quinocitrinin A and quinocitrinin B (Kozlovsky et al. 2011). Later, *P. waksmani* VKM FW2875, which was isolated from frozen Arctic soil, was shown to produce a similar range of metabolites (Antipova et al. 2011). Fumiquinazolines F and G have been discovered from two *P. thymicola* strains (Kozlovskii et al. 2012), and Meleagrin, a benzylisoquinoline alkaloid, was identified from *P. chrysogenum*, inhabiting permafrost region (Kozlovskii et al. 2002).

7.5.4 Terpenes

Zhou et al. (2015) reported two novel meroterpenoids, chrodrimanins I and J, from *Penicillium funiculosum* GWT2-24, a moss-derived fungus from Antarctica. *Eutypella* strain, isolated from arctic soil, was found to produce four unique diterpenes, named eutypenoids A and C, libertellenone H and scoparasin B (Lu et al. 2014).

7.5.5 Other Compounds

The genus *Penicillium* could be regarded as a source of bioactive molecules. *Penicillium citrinum*, isolated from permafrost regions, was reported to produce metabolites, epoxyagro-clavinet-I and agroclavine-I ergot alkaloids (E) (Antipova et al. 2011). Roquefortine, a derivative of diketopiperazine and dihydroroquefortine, has been obtained from *P. melanoconidium* and *P. chysogenum*. Findings reported the synthesis of cyclopenine and cyclopeptin from *P. solitum* (Kozlovskii et al. 2013). Ochratoxin A and B are secondary metabolites containing a dihydro-isocoumarin moiety associated with phenylalanine and are produced by *P. verrucos* (Zhelifonova et al. 2009).

7.6   Applications of Bioactive Molecule from Extremophilic Fungi

A variety of bioactive molecules have been isolated from marine and terrestrial extremophilic fungi which have agricultural and industrial applications (Figs. 7.1 and 7.2) (Table 7.2). Some of the important applications are discussed here.

7.6.1 Agricultural Applications

Bioactive molecules from extremophilic fungi have several applications in the agriculture sector (Fig. 7.1 and Table 7.2).

Plant Growth Enhancement

Cotylenins are the famous bioactive compounds secreted by species of *Cladosporium* that support better plant growth (Honma 2002). Psychrotolerant strain *Aureobasidium pullullans* synthesizes siderophores and auxin-like compounds that promote the

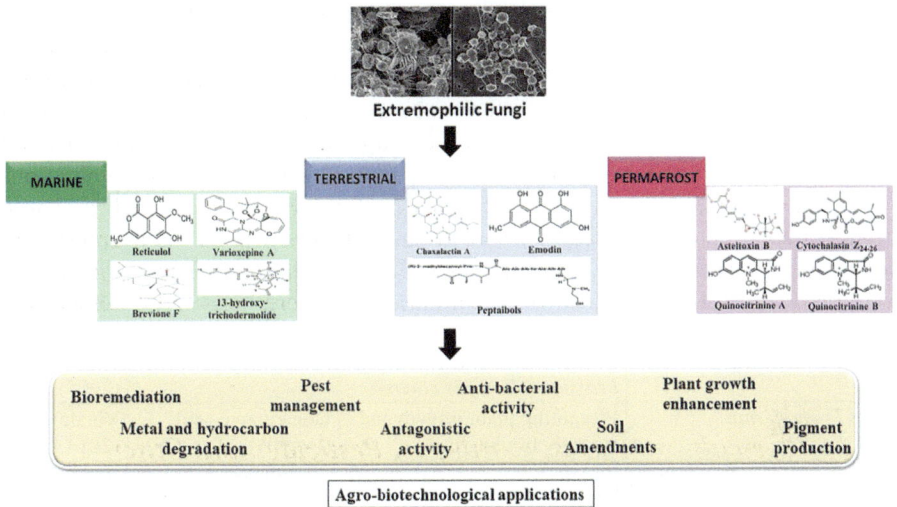

Fig. 7.1 Agro-biotechnological application of bioactive molecules from extremophilic fungi

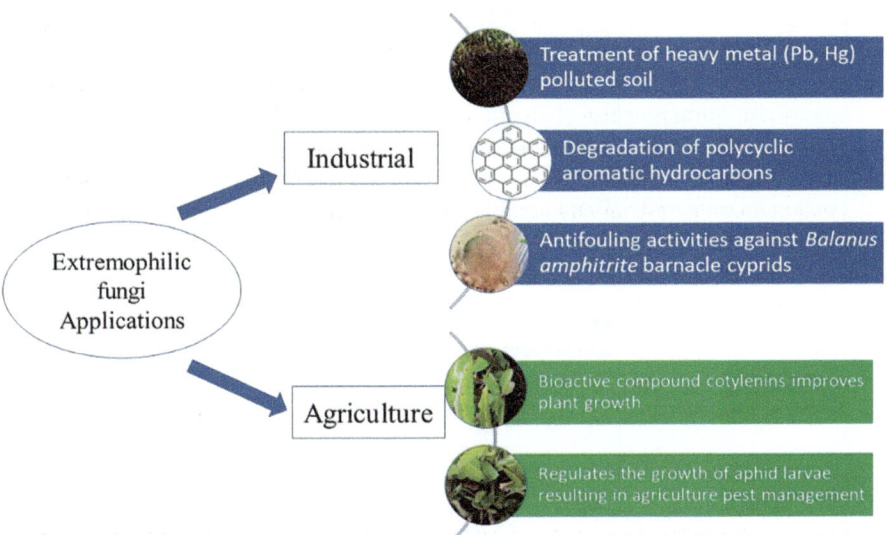

Fig. 7.2 Extremophilic fungi applications in industries and agriculture

growth of the plant. Further, *Candida maritima* and *Holtermaniella takashimae* display phosphate solubilizing ability, hence enhancing plant growth. Such strains could be further applied to increase plant production in cold regions (Mestre et al. 2016). It is recognized that a variety of xylanolytic and cellulolytic fungi, improve root systems, raise crop yields, and increase seed germination, hence promoting

Table 7.2 Fungus-derived bioactive molecules applications in the agriculture sector

S. no.	Fungal species	Application/activity	Compound	References
1	*Cryptosporiopsis quercina*	Biopesticide for rice blast (*Pyricularia oryzae*)	Cryptocin	Strobel et al. (1999)
2	*Phomopsis* sp.	Biopesticide for *Sclerotinia sclerotiorum*, *Fusarium oxysporum*, and *Botrytis cinerea*	Cytochalsins	Fu et al. (2011)
3	*Colletotrichum gloeosporioides*	Biopesticide for brown spot of rice (*Helminthosporium sativum*)	Colletotric acid	Zou et al. (2000)
4	*Lactarius rufus*	Biopesticide for *Alternaria brassicae*, *Botrytis cinerea*	Rufuslactone	Luo et al. (2005)
5	*Trichoderma harzianum*	Promoting plant growth and strongly binding iron	Isoharzianic acid (iso-HA)	Vinale et al. (2014)
6	*Strobilurus tenacellus*	Fungicides for a broad range of plant pathogenic fungi	Strobilurins A–D and other strobilurins and oudemansins	Ypema and Gold (1999)
7	*Fusarium fujikuroi*	GA signalling promoting plant growth by overcoming DELLA-mediated growth restraint	Gibberellic acid (GA)	Grennan (2006)
8	*Glomus mosseae*	Disease resistance in plants	Rosmarinic and caffeic acids	Cooper and Grandisons (1986)
9	*G. intraradices*	Disease resistance in plants	Phosphate related compounds	Heald et al. (1989)
10	*Gigaspora margarita* and *G. fasciculatum*	Disease resistance in plants	Phyroalexins and phenolic compounds	Matsubara et al. (2001)
11	*G. etunicatum*	Disease resistance in plants	Hydroxyproline-rich glycoprotein	Hao et al. (2005)
12	*G. monosporum* and *Phytophthora parasitica*	Disease resistance in plants and helping in drought tolerance	β-1,3-glycanase isoform	Utkhede (2006)
13	*Trichoderma koningii*	Disease resistance in plants	Prone isoforms	Oyetunji and Salami (2011)
14	*G. aggregatum*	Disease resistance in plants	Methylated compounds	Ararsa (2012)
15	*Funneliformis mosseae*	Enhanced disease resistance in plants and drought tolerance capacity, reduced heavy metal toxicity, and helpful in nutrient absorption	Sulfur-containing compounds	Song et al. (2015)
16	*G. versiforme*	Phenolic compounds in trifoliate orange	Homobrassinolide related compounds	Li et al. (2015)
17	*Rhizophagus irregularis*	Total sugar, proline, and chlorophyll concentrations in Apple	Fatty acids	Huang et al. (2020)

(continued)

Table 7.2 (continued)

S. no.	Fungal species	Application/activity	Compound	References
18	*Rhizophagus intraradices*	Anthocyanins and organic acids in sweet basil	Fatty acids	Scagel and Lee (2012)
19	*Aspergillus* sp., *Botryotinia* sp., *Colletotrichum* sp.	Bioactive compounds production	Pyranones, polyketides, and pyranones	Zheng et al. (2017)
20	*Cladosporium cladosporioides*	Antimicrobial metabolites, viz., 3-phenylpropionic acid, 1-acetyl-17-methoxyaspidospermidin-20-ol, Isocladosporin, Cladosporin, etc.	3-phenylpropionic acid	Yehia et al. (2020)
21	*Cryptosporiopsis* sp., *Phialocephala sphaeroides*	New antimicrobial metabolites being secreted by both endophytes		Terhonen et al. (2016)
22	*Trichoderma asperellum*	Antagonistic activity with mycoparasitism	Prone isoforms	Baiyee et al. (2019)
23	*Trichoderma viride*	Antagonistic activity with mycoparasitism	Prone isoforms	Talapatra et al. (2017)
24	*Penicillium simplicissimum*, *Leptosphaeria sp*	Antagonistic activity with mycoparasitism	Andrastin isomers and related compounds	Yuan et al. (2017)
25	*Diaporthe* sp., *Leptosphaeria* spp., *Nigrospora oryzae*	Antagonistic activity with mycoparasitism	Sesquiterpenoids, esters, and alcohols	Landum et al. (2016)
26	*Fomitopsis* sp., *Fusarium solani*, *Nigrospora sphaerica*, *Purpureocillium lilacinum*	Antagonistic activity with mycoparasitism		Yao et al. (2017)
27	*Trichoderma citrinoviride*	Antagonistic activity with mycoparasitism	Prone isoforms	Park et al. (2019)
28	*Paenibacillus polymyxa*	Antagonistic activity with mycoparasitism	Paenilan, Fusaricidans and isomers	Liu et al. (2018)
29	*Rhizopycnis vagnum*	Antagonistic activity with mycoparasitism		Anisha et al. (2018)
30	*Paraconiothyrium variabile*	The biocontrol process include the production of metabolites, viz., 13-oxo-9,11-octadecadienoic acid, beauvericin	Terpenoids, furanones, and polyketides	Combès et al. (2012)
31	*Induratia cofeana*, *I. yucatanensis*	Antagonistic activity with mycoparasitism		Mota et al. (2021)

(continued)

Table 7.2 (continued)

S. no.	Fungal species	Application/activity	Compound	References
32	*Hypoxylon anthochroum, Nodulisporium* spp.	Volatile organic compounds synthesized include phenylethyl alcohol, 2-methyl-1-butanol, ocimene, terpinolene, etc. that showed antifungal activity	1,8-cineole and terpenes	Medina-Romero et al. (2017)
33	*Chaetomium* sp., *Paecilomyces* sp.	Antagonistic activity with mycoparasitism	Xanthones and some cyclic compounds	Kusari et al. (2013)
34	*Bipolaris* sp., *Phoma* sp.	Antibiosis, parasitism, and production of lytic enzymes	Phomasetin and Terpenes	Felber et al. (2016)
35	*Muscodor yucatanensis*	Antagonistic activity by direct contact is suspected to have bioactive compounds	Propionic acid and isomers	Katoch and Pull (2017)
36	*Xylaria sp*	Endophytes inhibited the growth of pathogens (especially *Oidium* sp.)	Sesquiterpenoids, esters and alcohols	Yu et al. (2018)
37	*Diaporthe citri*	Amylase, pectinase, and cellulase produced along with Phytoprotective properties were observed against pathogens	Pyrones, Polyketides, etc.	Dos Santos et al. (2019)
38	*Ophiognomonia sp*	Antagonistic effect against phytopathogens	CAZymes, Cytochrome P450, etc.	Bongiorno et al. (2016)
39	*Muscodor coffeanum, M. vitigenus, M. yucatanensis, Simplicillium* sp.	VOC produced by endophytes, which helps inhibit the growth of pathogens	Propionic acid and isomers	Monteiro et al. (2017)

plant development. *Aspergillus fumigatus*, *Penicillium oxalicum*, *Penicillium citrinum*, and *Humicola insolens*Y1 (Du et al. 2009) are examples of xylanase-producing extremophilic fungi. They exhibit optimal activity at pH 8–9 and 45–55 °C, with *H. insolens*Y1 optimum temperature range being 70–80 °C, indicating thermophilic nature. When screened from the corn silage, *Rhizomucor pusillus* and *Aspergillus fumigatus* demonstrated optimal xylanase activity at pH 6 and 75 °C (Robledo et al. 2016).

Pest Management

Brevicoryne brassicae is a well-known aphid serving as a host and carrier of plant diseases. *Trichoderma harzianum*, *Aspergillus versicolor*, *Penicillium dipodomyicola*, and *Aspergillus sydowii* have been isolated from the marine environment for regulating the growth of aphid larvae, assisting in agriculture pest management.

From the above-said extremophilic fungus, *Aspergillus versicolor* showed the best results (85% efficiency) against *Brevicoryne brassicae* (Pacheco et al. 2017). The halophilic fungi *Thielavia* sp. UST030930–004 and *P. brevicompactum* DFFSCS025 were isolated from Hong Kong and South China Sea, respectively. Several bioactive compounds were identified from these two species that showed strong antifouling activities against *Balanus amphitrite* barnacle cyprids and larval settlement of *Bugula neritina*, respectively (Ha et al. 2017; Xu et al. 2017). As biopesticides, anthraquinone-based repellents have been used for defending rice seeds and sunflowers from blackbirds (Icteridae), whole kernel, as well as ripening corn from sandhill cranes (Grus canadensis) and blackbirds (DeLiberto and Werner 2016).

Soil Amendments

Soil amendments is a general phrase that encompasses additives used to a soil to modify its structure, pH, water level, nutrient and/or organic matter content, stability, and so on (Purkis et al. 2022). Out of these many concerns, the occurrence of saline or sodic soil is very common. Extremophilic fungi of various types can be added to alter the status of such soils. By releasing organic acids, eliminating salt ions, and contributing biomolecules like cellulases, for instance, the haloalkaliphilic fungi can fix such soils and even improve their physical characteristics and fertility (Wei and Zhang 2019). Alkalitolerant fungi belonging to several families, such as *Emericellopsis alkaline* (Hypocreales), *Alteraria* sp., *Thielavia* spp. (Chaetomiaceae), *Sodiomyces* species (Plectosphaerellaceae), and *Acrostalagmus luteoalbus* (Plectosphaerellaceae) could potentially enhance soil health (Grum-Grzhimaylo et al. 2015). The species of *Fusarium*, *Cladosporium*, and an *Acremonium*-like fungus from the Bionectriaceae family as well as members of Scopulariopsis (*Microascales*) were found to be moderately alkaliphilic (Sahay 2022).

7.6.2   Bioremediation

Coniochaeta fodinicola, *Teratosphaeria acidotherma*, *Hortaea acidophilia*, and *Acidomyces acidophilus* are a few extremophile fungi that synthesize novel enzymes and metabolites that enable their survival in extreme environments. These can be employed in bioremediation, as they act mostly by eliminating harmful metals and metalloids from soil (Selbmann et al. 2008). Bonatti et al. (2023) reported $22 \pm 11\%$ biodegradation of anthracene (polycyclic aromatic hydrocarbon) by *Exophiala oligosperma* LaBioMMi 1217, suggesting its potential use in the bioremediation of soil. Numerous metal transporters with specialized functions for iron, copper, zinc, magnesium, calcium, and nickel were discovered by proteome and transcriptome investigations of *Acidomyces richmondensis*. In addition to these transporters, it also had genes for numerous transcripts of ferric-chelate reductase, a siderophore-dependent iron transporter called sideroflexin, and a ferrochelatase (Mosier et al.

2016). According to Chan et al. (2018), an investigation of the biosorption of arsenic and antimony by *A. acidophilus* WKC-1 revealed that As^{5+} and Sb^{5+} bind to functional groups such as -OH, -CH, -NH, PO_4, and $-SO_3$. The isolated *A. acidophilus* WKC-1 strain removed about 170 mg As^{5+} per gram of dry biomass and can tolerate both low pH and high As concentration making it a viable candidate for application in bioremediation. Halophilic fungi, *Sterigmatomyces halophilus* and *Aspergillus flavus*, demonstrated 85% biosorption ability for zinc and iron (Bano et al. 2018).

Some fungal strains isolated from Antarctica regions have been found to produce enzymes such as esterases and lipases with a potential role in bioremediation. When the sites near fuel storage tanks were screened for the lipase/esterase-positive isolates, various fungi species were reported such as *Guehomyces pullulans*, *Phenoliferia glacialis*, *Cryptococcus Adeliensis*, and *Pichia caribbica*. Out of all the strains *P. caribbica* displayed the highest enzymatic activities and can be employed for the degradation of hydrocarbons (Yang et al. 2009a).

Metal and Hydrocarbon Degradation

Metallotolerant fungi (*Aspergillus niger* M1, *Aspergillus fumigatus* M3, *Aspergillus terreus* M6, and *Aspergillus flavus* M7) are employed in the treatment of heavy metal-polluted soil (industrially contaminated) (Fig. 7.1). The isolates M3 and M7 were good enough in clearing the lead (Pb) content, whereas M1 and M6 were found to be more efficient in the removal of mercury (Hg) from the test sample (Khan et al. 2019). Similarly, the halophilic fungus *Aspergillus sydowii* was helpful in the degradation of Polycyclic aromatic hydrocarbons (PAHs). Peidro-Guzmán et al. (2021) successfully identified the chloroperoxidase gene responsible for PAHs detoxification.

7.6.3   Antimicrobial Activity

Peptaibols are biologically active peptides produced from *Trichoderma asperellum* isolated from soil having antimicrobial activity. The peptaibols extract inhibited plant pathogens and exhibited inhibition rates similar to commercially available fungicides in the market. The growth inhibition rates against *Colletotrichum gloeosporioides*, *Botrytis cinerea*, *Alternaria alternata*, and *Fusarium oxysporum* were 92.2%, 74.2%, 58.4%, and 36.2%, respectively. Subsequently, the antifungal activity was evaluated in tomatoes inoculated with *A. alternata*; the incidence of the disease in tomatoes treated with the extract was nil, while the untreated tomato fruit showed a 92.5% incidence (Alfaro-Vargas et al. 2022). Gehan and Samir (2017) evaluated six monoterpenes for their antifungal activity against eight plant pathogenic fungi. Furthermore, (−) menthone exhibited strong antifungal activity against *Alternaria solani* (EC_{50}–9.31 mg/l), *Penicillium digitatum* (EC_{50}–16.14 mg/l), and *Rhizoctonia solani* (EC_{50}–24.69 mg/l).

Bacillaene, an antibacterial polyketide synthesized by *B. velezensis* FZB42, showed bacteriostatic activity against *Erwinia amylovora*, a causal agent of fire blight disease, while bacillaene-A synthesized by *Bacillus* spp. exhibited antifungal activity against antagonist fungi *Termitomyces* (Chen et al. 2009; Um et al. 2013). Difficidin and bacilysin, polyketide produced from *Bacillus amyloliquefaciens*, displayed antibacterial activity against two rice pathogens, i.e., *Xanthomonas oryzae* pv. oryzae and *Xanthomonas oryzae* pv. oryzicola, which causes bacterial blight and bacterial leaf streak disease, respectively, in rice (Wu et al. 2015).

7.6.4 Industrial Applications

Extremophilic fungi possess several industrially relevant bioactive compounds and their applications are highlighted in this subsection (Fig. 7.2).

Pigment Production

Rhodotorula sp. RY1801 isolated from the marine environment have been found to produce carotenoid, a natural pigment, with a crucial role in active oxygen scavenging, acting as a precursor of vitamin A, and colouring agents in food industries and cosmetic industries. At a pH of 5.0 and temperature of 28 °C, a total of 987 µg/L of carotenoids were obtained suggesting the utilization of *Rhodotorula* sp. RY1801 for the synthesis of carotenoids for commercial purposes (Zhao et al. 2019). *Dioszegia patagonica*, isolated from the Antarctic region, has also been shown to synthesize carotenoid (Trochine et al. 2017). Torulene and torularhodin belong to the carotenoids group and have also been recognized as the novel fungal carotenoids produced by *Sporobolomyces salmonicolor* and *Rhodotorula mucilaginosa* (Dimitrova et al. 2013). Recently, other fungal species such as *Cryptococcus albidus*, *C. laurentii*, *Sporobolomyces metaroseus*, and *S. salmonicolor* also derived from the Antarctic have been found to produce β-carotene (Dimitrova et al. 2010). Lycopene, a type of carotene, is produced by *Rhodotorula larynges* and *R. mucilaginosa* (Cavalcante et al. 2023). Mycosporine, a pink or cream pigment, has been identified from *Cryptococcus* and *Torrubiella* sp. and is known as the UV absorbing pigment (Barahona et al. 2016).

7.6.5 Food and Beverage

Recent studies focus on the use of enzymes in place of chemicals for processing food materials as it results in products with higher yields and improved nutritional value and aesthetic appeal of the items. Psychrophiles produce cold-active enzymes that enhance catalytic activity at low temperatures (Hamid et al. 2022). These enzymes are employed as wine and juice clarifying agents, as well as to soften frozen meat products, and hasten ripening of cheese. An Antarctic yeast strain called *Rhodotorula mucilaginosa* L7 generates acid protease with an activity range of pH 5 (Lario et al. 2015). Another class of enzymes that is widely employed in the food sector is amylases. It has been found that several thermophilic fungus species can secrete amylases. Thermophilic fungi that generate the enzyme amylase are *Thermomyces ibadanensis, Thermomyces lanuginosus, Myriococcum thermophilum, and Rhizomucor pusillus*. Extremophilic fungal strains are also a great source of pectinases and xylanases which aid in juice and wine clarification, pulping, oil extraction, and so on (Sahay 2022).

7.6.6 Others

Another species of the same genus, named *R. glutinis*, isolated from sour milk has also been reported to contain various fatty acids and could be used as a promising source of biodiesel. *R. glutinis* also produces the enzyme phenylalanine ammonia-lyase (PAL), which plays a significant role in the synthesis of aspartame, a sweetener (Kot et al. 2016).

7.7 Conclusion

Extreme environmental fungi are one of the rich and unexploited sources of novel lead compounds having widespread applications. The extreme environments fungi like *Penicillium* spp. and *Aspergillus* spp. have the potential to synthesize novel bioactive molecules. Despite such significant potential, studies concerning in extraction of metabolites from extremophilic fungi are lacking. This is partly due to the difficulty in sample collection and the lack of advanced fungal culturing techniques. However, recent programmes like deep-sea drilling and genome technology have helped in exploring new compounds with significant biological activities and are expected to be much more fruitful in the near future. The discussed compounds from extremophilic fungi exhibited strong bioactivities and possess great potential as drugs, plant growth activators, industrial proteases, and antiviral compounds. Various enzymes secreted by the fungi could act efficiently with nanoparticle synthesis through the biological approach for various applications. The bioconjugation could also aid in enhancing the potency and ability of the nanoparticles for enhanced functionalities. Hence, it can be concluded that extremophilic fungi could serve as a potent and efficient source for the generation of bioactive compounds with several advantages over the conventional chemical methods.

Acknowledgement The authors are grateful to the authorities of Mohanlal Sukhadia University, Udaipur, and MHRD RUSA 2.0, Government of India for supporting this work.

Conflict of Interest The authors declare no conflict of interest.

References

Agrawal S, Adholeya A, Deshmukh SK (2016) The pharmacological potential of non-ribosomal peptides from marine sponge and tunicates. Front Pharmacol 7:333. https://doi.org/10.3389/fphar.2016.00333

Alfaro-Vargas P, Bastos-Salas A, Muñoz-Arrieta R, Pereira-Reyes R, Redondo-Solano M, Fernández J, Mora-Villalobos A, López-Gómez JP (2022) Peptaibol production and characterization from Trichoderma asperellum and their action as biofungicide. J Fungi 8:1037. https://doi.org/10.3390/jof8101037

Anisha C, Jishma P, Bilzamol VS, Radhakrishnan EK (2018) Effect of ginger endophyte Rhizopycnis vagum on rhizome bud formation and protection from phytopathogens. Biocatal Agric Biotechnol 14:116–119. https://doi.org/10.1016/j.bcab.2018.02.015

Antipova TV, Zhelifonova VP, Baskunov BP, Ozerskaya SM, Ivanushkina NE, Kozlovsky AG (2011) New producers of biologically active compounds—fungal strains of the genus Penicillium isolated from permafrost. Appl Biochem Microbiol 47(3):288–292. https://doi.org/10.1134/S0003683811030033

Ararsa L (2012) Evaluation of Arbuscular Mycorrhizal Fungi and Trichoderma species for the control of onion white rot (Sclerotium cepivorum Berk). J Plant Pathol Microbiol 04:159. https://doi.org/10.4172/2157-7471.1000159

Baiyee B, Ito S, Sunpapao A (2019) Trichoderma asperellum T1 mediated antifungal activity and induced defense response against leaf spot fungi in lettuce (Lactuca sativa L.). Physiol Mol Plant Pathol 106:96–101. https://doi.org/10.1016/j.pmpp.2018.12.009

Bano A, Hussain J, Akbar A et al (2018) Biosorption of heavy metals by obligate halophilic fungi. Chemosphere 199:218–222. https://doi.org/10.1016/j.chemosphere.2018.02.043

Barahona S, Yuivar Y, Socias G, Alcaíno J, Cifuentes V, Baeza M (2016) Identification and characterization of yeasts isolated from sedimentary rocks of Union Glacier at the Antarctica. Extremophiles 20(4):479–491. https://doi.org/10.1007/s00792-016-0838-6

Bérdy J (2005) Bioactive Microbial Metabolites. J Antibiot 58:1–26. https://doi.org/10.1038/ja.2005.1

Bonatti E, Dos Santos A, Birolli WG, Rodrigues-Filho E (2023) Endophytic, extremophilic and entomophilic fungi strains biodegrade anthracene showing potential for bioremediation. World J Microbiol Biotechnol 39:152. https://doi.org/10.1007/s11274-023-03590-8

Bongiorno VA, Rhoden SA, Garcia A et al (2016) Genetic diversity of endophytic fungi from Coffea arabica cv. IAPAR-59 in organic crops. Ann Microbiol 66:855–865. https://doi.org/10.1007/s13213-015-1168-0

Cardoso-Martínez F, de la Rosa JM, Díaz-Marrero AR, Darias J, D'Croz L, Cerella C, Diederich M, Cueto M (2015) Oximoaspergillimide, a fungal derivative from a marine isolate of Aspergillus sp. Eur J Org Chem 10:2256–2261. https://doi.org/10.1002/ejoc.201403668

Cavalcante SB, dos Santos BC, Kreusch MG, da Silva AF, Duarte RTD, Robl D (2023) The hidden rainbow: the extensive biotechnological potential of Antarctic fungi pigments. Braz J Microbiol 54(3):1675–1687. https://doi.org/10.1007/s42770-023-01011-4

Chan WK, Wildeboer D, Garelick H, Purchase D (2018) Competition of As and other Group 15 elements for surface binding sites of an extremophilic Acidomyces acidophilus isolated from a historical tin mining site. Extremophiles 22:795–809. https://doi.org/10.1007/s00792-018-1039-2

Chen XH, Scholz R, Borriss M et al (2009) Dicidin and bacilysin produced by plant-associated Bacillus amyloliquefaciens are efficient in controlling fire blight disease. J Biotechnol 140:38–44

Chen G, Wang H-F, Pei Y-H (2014a) Secondary metabolites from marine-derived microorganisms. J Asian Nat Prod Res 16:105–122. https://doi.org/10.1080/10286020.2013.855202

Chen S, Li H, Chen Y et al (2019) Three new diterpenes and two new sesquiterpenoids from the endophytic fungus Trichoderma koningiopsis A729. Bioorg Chem 86:368–374. https://doi.org/10.1016/j.bioorg.2019.02.005

Chen S, Liu H, Liu Z et al (2020) Two new polyketide compounds from the endophytic fungus *Trichoderma spirale* A725 of *Morinda officinalis*. Chin J Org Chem 40:209. https://doi.org/10.6023/cjoc201907041

Chung Y-M, Wei C-K, Chuang D-W et al (2013) An epigenetic modifier enhances the production of anti-diabetic and anti-inflammatory sesquiterpenoids from Aspergillus sydowii. Bioorg Med Chem 21:3866–3872. https://doi.org/10.1016/j.bmc.2013.04.004

Coleine C, Stajich JE, Selbmann L (2022) Fungi are key players in extreme ecosystems. Trends Ecol Evol 37(6):517–528. https://doi.org/10.1016/j.tree.2022.02.002

Combès A, Ndoye I, Bance C et al (2012) Chemical communication between the endophytic fungus Paraconiothyrium variabile and the phytopathogen Fusarium oxysporum. PLoS One 7:e47313. https://doi.org/10.1371/journal.pone.0047313

Cooper KM, Grandisons GS (1986) Interaction of vesicular-arbuscular mycorrhizal fungi and root knot nematode on cultivars of tomato and white clover susceptible to Meloidogyne hapla. Ann Appl Biol 108:555–565. https://doi.org/10.1111/j.1744-7348.1986.tb01994.x

Das S, Ben Haj Salah K, Djibo M, Inguimbert N (2018) Peptaibols as a model for the insertions of chemical modifications. Arch Biochem Biophys 658:16–30

Dasanayaka SAHK, Nong X-H, Liang X et al (2020) New dibenzodioxocinone and pyran-3,5-dione derivatives from the deep-sea-derived fungus Penicillium canescens SCSIO z053. J Asian Nat Prod Res 22:338–345. https://doi.org/10.1080/10286020.2019.1575819

Degenkolb T, Gräfenhan T, Berg A et al (2006) Peptaibiomics: screening for polypeptide antibiotics (peptaibiotics) from plant-protective Trichoderma species. Chem Biodivers 3:593–610. https://doi.org/10.1002/cbdv.200690063

DeLiberto ST, Werner SJ (2016) Review of anthraquinone applications for pest management and agricultural crop protection. Pest Manag Sci 72(10):1813–1825. https://doi.org/10.1002/ps.4330

Deshmukh S, Prakash V, Ranjan N (2017a) Recent advances in the discovery of bioactive metabolites from Pestalotiopsis. Phytochem Rev 16:883–920. https://doi.org/10.1007/s11101-017-9495-3

Deshmukh SK, Prakash V, Ranjan N (2017b) Marine fungi: a source of potential anticancer compounds. Front Microbiol 8:2536. https://doi.org/10.3389/fmicb.2017.02536

Dimitrova S, Pavlova K, Lukanov L, Zagorchev P (2010) Synthesis of coenzyme Q10 and beta-carotene by yeasts isolated from antarctic soil and lichen in response to ultraviolet and visible radiations. Appl Biochem Biotechnol 162(3):795–804. https://doi.org/10.1007/s12010-009-8845-z

Dimitrova S, Pavlova K, Lukanov L, Korotkova E, Petrova E, Zagorchev P, Kuncheva M (2013) Production of metabolites with antioxidant and emulsifying properties by Antarctic strain Sporobolomyces salmonicolor AL1. Appl Biochem Biotechnol 169(1):301–311. https://doi.org/10.1007/s12010-012-9983-2

Dos Santos CM, Ribeiro ADS, Garcia A, Polli AD, Polonio JC, Azevedo JL, Pamphile JA (2019) Actividad enzimatica y antagonista de los hongos endofiticos de *Sapindus saponaria* L. Revista Acta Biolo Colomb 24(2):322–331

Du L, Yang X, Zhu T et al (2009) Diketopiperazine alkaloids from a deep ocean sediment derived fungus Penicillium sp. Chem Pharm Bull (Tokyo) 57:873–876. https://doi.org/10.1248/cpb.57.873

Fan Z, Sun Z-H, Liu Z et al (2016) Dichotocejpins A–C: New Diketopiperazines from a Deep-Sea-Derived Fungus Dichotomomyces cejpii FS110. Mar Drugs 14:164. https://doi.org/10.3390/md14090164

Felber AC, Orlandelli RC, Rhoden SA et al (2016) Bioprospecting foliar endophytic fungi of Vitis labrusca Linnaeus, Bordô and Concord cv. Ann Microbiol 66:765–775. https://doi.org/10.1007/s13213-015-1162-6

Figueroa L, Jiménez C, Rodríguez J et al (2015) 3-Nitroasterric acid derivatives from an Antarctic Sponge-Derived Pseudogymnoascus sp. Fungus. J Nat Prod 78:919–923. https://doi.org/10.1021/np500906k

Fu J, Zhou Y, Li H-F et al (2011) Antifungal metabolites from Phomopsis sp. By254, an endophytic fungus in Gossypium hirsutum. Afr J Microbiol Res 5:1231–1236. https://doi.org/10.5897/AJMR11.272

Gehan IKM, Samir AMA (2017) Antifungal potential and biochemical effects of monoterpenes and phenylpropenes on plant. Plant Prot Sci 54:9–16

Ghosh SK, Nagarajan R (2016) Total synthesis of penipanoid C, 2-(4-hydroxybenzyl) quinazolin-4(3H)-one and NU1025. Tetrahedron Lett 57(38):4277–4279. https://doi.org/10.1016/j.tetlet.2016.08.018

Giordano D (2021) Bioactive molecules from extreme environments II. Mar Drugs 19(11):642. https://doi.org/10.3390/md19110642

Grennan AK (2006) Gibberellin metabolism enzymes in rice. Plant Physiol 141:524–526. https://doi.org/10.1104/pp.104.900192

Gross S, Robbins EI (2000) Acidophilic and acid-tolerant fungi and yeasts. Hydrobiologia 433:91–109. https://doi.org/10.1023/A:1004014603333

Grum-Grzhimaylo AA, Georgieva ML, Bondarenko SA, Debets AJM, Bilanenko EN (2015) On the diversity of fungi from soda soils. Fungal Divers 76(1):27

Ha TM, Ko W, Lee SJ et al (2017) Anti-inflammatory effects of curvularin-type metabolites from a marine-derived fungal strain Penicillium sp. SF-5859 in lipopolysaccharide-induced RAW264.7 macrophages. Mar Drugs 15:282. https://doi.org/10.3390/md15090282

Hamid B, Bashir Z, Yatoo AM, Mohiddin F, Majeed N, Bansal M, Poczai P, Almalki WH, Sayyed RZ, Shati AA, Alfaifi MY (2022) Cold-active enzymes and their potential industrial applications—a review. Molecules 27(18):5885. https://doi.org/10.3390/molecules27185885

Hao Z, Christie P, Qin L et al (2005) Control of fusarium wilt of cucumber seedlings by inoculation with an arbuscular mycorrhizal fungus. J Plant Nutr 28:1961–1974. https://doi.org/10.1080/01904160500310997

Heald CM, Bruton BD, Davis RM (1989) Influence of glomus intraradices and soil phosphorus on Meloidogyne incognita infecting Cucumis melo. J Nematol 21:69–73

Honma Y (2002) Cotylenin A--a plant growth regulator as a differentiation-inducing agent against myeloid leukemia. Leuk Lymphoma 43:1169–1178. https://doi.org/10.1080/10428190290026222

Hu J, Li Z, Gao J et al (2019) New Diketopiperazines from a marine-derived fungus strain Aspergillus versicolor MF180151. Mar Drugs 17:262. https://doi.org/10.3390/md17050262

Hu Z, Zhu Y, Chen J, Chen J, Li C, Gao Z, Li J, Liu L (2023) Discovery of novel bactericides from Aspergillus alabamensis and their antibacterial activity against fish pathogens. J Agric Food Chem 71(10):4298–4305. https://doi.org/10.1021/acs.jafc.2c09141

Huang Z, Cai X, Shao C et al (2008) Chemistry and weak antimicrobial activities of phomopsins produced by mangrove endophytic fungus Phomopsis sp. ZSU-H76. Phytochemistry 69:1604–1608. https://doi.org/10.1016/j.phytochem.2008.02.002

Huang Z, Nong X, Ren Z et al (2017) Anti-HSV-1, antioxidant and antifouling phenolic compounds from the deep-sea-derived fungus Aspergillus versicolor SCSIO 41502. Bioorg Med Chem Lett 27:787–791. https://doi.org/10.1016/j.bmcl.2017.01.032

Huang R, Li Z, Mao C et al (2020) Natural variation at OsCERK1 regulates arbuscular mycorrhizal symbiosis in rice. New Phytol 225:1762–1776. https://doi.org/10.1111/nph.16158

Ibrar M, Ullah MW, Manan S et al (2020) Fungi from the extremes of life: an untapped treasure for bioactive compounds. Appl Microbiol Biotechnol 104:2777–2801. https://doi.org/10.1007/s00253-020-10399-0

Jančič S, Frisvad JC, Kocev D et al (2016) Production of secondary metabolites in extreme environments: food- and airborne Wallemia spp. produce toxic metabolites at hypersaline conditions. PLoS One 11:e0169116. https://doi.org/10.1371/journal.pone.0169116

Jiang W, Ye P, Chen C-TA et al (2013) Two novel hepatocellular carcinoma cycle inhibitory cyclodepsipeptides from a hydrothermal vent crab-associated fungus Aspergillus clavatus C2WU. Mar Drugs 11:4761–4772. https://doi.org/10.3390/md11124761

Katoch M, Pull S (2017) Endophytic fungi associated with Monarda citriodora, an aromatic and medicinal plant and their biocontrol potential. Pharm Biol 55:1528–1535. https://doi.org/1 0.1080/13880209.2017.1309054

Khan I, Ali M, Aftab M et al (2019) Mycoremediation: a treatment for heavy metal-polluted soil using indigenous metallotolerant fungi. Environ Monit Assess 191:622. https://doi.org/10.1007/s10661-019-7781-9

Kogej T, Ramos J, Plemenitas A, Gunde-Cimerman N (2005) The halophilic fungus Hortaea werneckii and the halotolerant fungus Aureobasidium pullulans maintain low intracellular cation concentrations in hypersaline environments. Appl Environ Microbiol 71:6600–6605. https://doi.org/10.1128/AEM.71.11.6600-6605.2005

Kot AM, Błażejak S, Kurcz A et al (2016) Rhodotorula glutinis—potential source of lipids, carotenoids, and enzymes for use in industries. Appl Microbiol Biotechnol 100:6103–6117. https://doi.org/10.1007/s00253-016-7611-8

Kozlovskii AG, Zhelifonova VP, Adanin VM, Antipova TV, Shnyreva AV, Viktorov AN (2002) The biosynthesis of low-molecular-weight nitrogen-containing secondary metabolites—alkaloids—by the resident strains of Penicillium chrysogenum and Penicillium expansum isolated on board the Mir Space Station. Microbiology 71(6):666–669. https://doi.org/10.102 3/A:1021475722091

Kozlovskii AG, Zhelifonova VP, Antipova TV, Baskunov BP, Kochkina GA, Ozerskaya SM (2012) Secondary metabolite profiles of the Penicillium fungi isolated from the arctic and antarctic permafrost as elements of polyphase taxonomy. Microbiology 81(3):306–311. https://doi.org/10.1134/S0026261712030071

Kozlovskii AG, Zhelifonova VP, Antipova TV (2013) Fungi of the genus Penicillium as producers of physiologically active compounds (review). Appl Biochem Microbiol 49(1):1–10. https://doi.org/10.1134/S0003683813010092

Kozlovsky AG, Zhelifonova VP, Antipova TV, Zelenkova NF (2011) Physiological and biochemical characteristics of the genus Penicillium fungi as producers of ergot alkaloids and quinocitrinins. Appl Biochem Microbiol 47(4):426–430. https://doi.org/10.1134/S0003683811040065

Kusari P, Kusari S, Spiteller M, Kayser O (2013) Endophytic fungi harbored in Cannabis sativa L.: diversity and potential as biocontrol agents against host plant-specific phytopathogens. Fungal Divers 60:137–151. https://doi.org/10.1007/s13225-012-0216-3

Landum MC, do Rosário Félix M, Alho J et al (2016) Antagonistic activity of fungi of Olea europaea L. against Colletotrichum acutatum. Microbiol Res 183:100–108. https://doi.org/10.1016/j.micres.2015.12.001

Lario LD, Chaud L, das Graças Almeida M, Converti A, Durães Sette L, Pessoa A (2015) Production, purification, and characterization of an extracellular acid protease from the marine Antarctic yeast Rhodotorula mucilaginosa L7. Fungal Biol 119(11):1129–1136. https://doi.org/10.1016/j.funbio.2015.08.012

Li D, Cai S, Zhu T et al (2010) Three new sorbicillin trimers, trisorbicillinones B, C, and D, from a deep ocean sediment derived fungus, Phialocephala sp. FL30r. Tetrahedron 66:5101–5106. https://doi.org/10.1016/j.tet.2010.04.111

Li C-S, An C-Y, Li X-M et al (2011a) Triazole and Dihydroimidazole alkaloids from the marine sediment-derived fungus Penicillium paneum SD-44. J Nat Prod 74:1331–1334. https://doi.org/10.1021/np200037z

Li L, Li D, Luan Y et al (2012a) Cytotoxic metabolites from the Antarctic psychrophilic fungus Oidiodendron truncatum. J Nat Prod 75:920–927. https://doi.org/10.1021/np3000443

Li Y, Ye D, Shao Z et al (2012b) A sterol and spiroditerpenoids from a Penicillium sp. isolated from a deep sea sediment sample. Mar Drugs 10:497–508. https://doi.org/10.3390/md10020497

Li J-F, He X-H, Li H et al (2015) Arbuscular mycorrhizal fungi increase growth and phenolics synthesis in Poncirus trifoliata under iron deficiency. Sci Hortic 183:87–92. https://doi.org/10.1016/j.scienta.2014.12.015

Li H, Fu Y, Song F (2023) Marine Aspergillus: a treasure trove of antimicrobial compounds. Mar Drugs 21(5):277. https://doi.org/10.3390/md21050277

Liang X, Zhang X-Y, Nong X-H et al (2016) Eight linear peptides from the deep-sea-derived fungus Simplicillium obclavatum EIODSF 020. Tetrahedron 72:3092. https://doi.org/10.1016/j.tet.2016.04.032

Liu J-T, Hu B, Gao Y et al (2014) Bioactive tyrosine-derived cytochalasins from fungus Eutypella sp. D-1. Chem Biodivers 11:800–806. https://doi.org/10.1002/cbdv.201300218

Liu Y, Bai F, Li T, Yan H (2018) An endophytic strain of genus Paenibacillus isolated from the fruits of Noni (Morinda citrifolia L.) has antagonistic activity against a Noni's pathogenic strain of genus Aspergillus. Microb Pathog 125:158–163. https://doi.org/10.1016/j.micpath.2018.09.018

Liu Z, Chen Y, Li S et al (2020) Bioactive metabolites from the Deep-Sea-Derived Fungus Diaporthe longicolla FS429. Mar Drugs 18:381. https://doi.org/10.3390/md18080381

Lu X-L, Liu J-T, Liu X-Y et al (2014) Pimarane diterpenes from the Arctic fungus Eutypella sp. D-1. J Antibiot (Tokyo) 67:171–174. https://doi.org/10.1038/ja.2013.104

Luo D-Q, Wang F, Bian X-Y, Liu J-K (2005) Rufuslactone, a new antifungal sesquiterpene from the fruiting bodies of the basidiomycete Lactarius rufus. J Antibiot (Tokyo) 58:456–459. https://doi.org/10.1038/ja.2005.60

Luo X, Yang J, Chen F et al (2018) Structurally diverse polyketides from the mangrove-derived fungus Diaporthe sp. SCSIO 41011 with their anti-influenza a virus activities. Front Chem 6:282. https://doi.org/10.3389/fchem.2018.00282

Marik T, Urban P, Tyagi C et al (2017) Diversity profile and dynamics of peptaibols produced by green mould Trichoderma species in interactions with their hosts Agaricus bisporus and Pleurotus ostreatus. Chem Biodivers 14:e1700033

Matsubara Y, Ohba N, Fukui H (2001) Effect of arbuscular mycorrhizal fungus infection on the incidence of Fusarium root rot in asparagus seedlings. J Jpn Soc Hortic Sci 70:202–206. https://doi.org/10.2503/jjshs.70.202

Medina-Romero, Y. M., Roque-Flores, G., & Macías-Rubalcava, M. L. (2017). Volatile organic compounds from endophytic fungi as innovative postharvest control of Fusarium oxysporum in cherry tomato fruits. Appl Microbiol Biotechnol 101:8209–8222

Meng J, Cheng W, Heydari H et al (2018) Sorbicillinoid-based metabolites from a sponge-derived fungus Trichoderma saturnisporum. Mar Drugs 16:226. https://doi.org/10.3390/md16070226

Mestre MC, Fontenla S, Bruzone MC et al (2016) Detection of plant growth enhancing features in psychrotolerant yeasts from Patagonia (Argentina). J Basic Microbiol 56:1098–1106. https://doi.org/10.1002/jobm.201500728

Monteiro MCP, Alves NM, Queiroz MVD, Pinho DB, Pereira OL, Souza SMCD, Cardoso PG (2017) Antimicrobial activity of endophytic fungi from coffee plants. Biosci J (Online):381–389

Mosier AC, Miller CS, Frischkorn KR et al (2016) Fungi contribute critical but spatially varying roles in nitrogen and carbon cycling in acid mine drainage. Front Microbiol 7:238

Mota SF, Pádua PF, Ferreira AN et al (2021) Biological control of common bean diseases using endophytic Induratia spp. Biol Control 159:104629. https://doi.org/10.1016/j.biocontrol.2021.104629

Neumann NKN, Stoppacher N, Zeilinger S et al (2015) The peptaibiotics database--a comprehensive online resource. Chem Biodivers 12:743–751. https://doi.org/10.1002/cbdv.201400393

Niu S, Fan Z-W, Xie C-L et al (2017) Spirograterpene A, a Tetracyclic Spiro-Diterpene with a fused 5/5/5/5 ring system from the Deep-Sea-Derived Fungus Penicillium granulatum MCCC 3A00475. J Nat Prod 80:2174–2177. https://doi.org/10.1021/acs.jnatprod.7b00475

Niu S, Liu Q, Xia J-M et al (2018) Polyketides from the Deep-Sea-Derived Fungus Graphostroma sp. MCCC 3A00421 showed potent antifood allergic activities. J Agric Food Chem 66:1369–1376. https://doi.org/10.1021/acs.jafc.7b04383

Orfali R, Perveen S (2019) New bioactive metabolites from the thermophilic fungus Penicillium sp. isolated from Ghamiqa Hot Spring in Saudi Arabia. Journal of Chemistry 2019:e7162948. https://doi.org/10.1155/2019/7162948

Orfali RS, Aly AH, Ebrahim W et al (2015) Pretrichodermamide C and N-methylpretrichodermamide B, two new cytotoxic epidithiodiketopiperazines from hyper saline lake derived Penicillium sp. Phytochem Lett 11:168–172. https://doi.org/10.1016/j.phytol.2014.12.010

Oyetunji OJ, Salami AO (2011) Study on the control of *Fusarium* wilt in the stems of mycorrhizal and trichodermal inoculated pepper (*Capsicum annum* L.). J Appl Biosci 45:3071–3080

Pacheco JC, Poltronieri AS, Porsani MV et al (2017) Entomopathogenic potential of fungi isolated from intertidal environments against the cabbage aphid Brevicoryne brassicae (Hemiptera: aphididae). Biocontrol Sci Tech 27:496–509. https://doi.org/10.1080/09583157.2017.1315053

Pan C, Shi Y, Auckloo BN et al (2016) An unusual conformational isomer of Verrucosidin backbone from a hydrothermal vent fungus, Penicillium sp. Y-50-10. Mar Drugs 14:156. https://doi.org/10.3390/md14080156

Pan C, Shi Y, Auckloo BN et al (2017) Isolation and antibiotic screening of fungi from a hydrothermal vent site and characterization of secondary metabolites from a Penicillium isolate. Mar Biotechnol (NY) 19:469–479. https://doi.org/10.1007/s10126-017-9765-5

Park Y-H, Chandra Mishra R, Yoon S et al (2019) Endophytic Trichoderma citrinoviride isolated from mountain-cultivated ginseng (Panax ginseng) has great potential as a biocontrol agent against ginseng pathogens. J Ginseng Res 43:408–420. https://doi.org/10.1016/j.jgr.2018.03.002

Peidro-Guzmán H, Pérez-Llano Y, González-Abradelo D et al (2021) Transcriptomic analysis of polyaromatic hydrocarbon degradation by the halophilic fungus Aspergillus sydowii at hypersaline conditions. Environ Microbiol 23:3435–3459. https://doi.org/10.1111/1462-2920.15166

Purkis JM, Bardos RP, Graham J, Cundy AB (2022) Developing field-scale, gentle remediation options for nuclear sites contaminated with 137Cs and 90Sr: the role of nature-based solutions. J Environ Manag 308:114620. https://doi.org/10.1016/j.jenvman.2022.114620

Qi C, Liu M, Zhou Q et al (2018) BACE1 inhibitory meroterpenoids from Aspergillus terreus. J Nat Prod 81(9):1937–1945

Rateb ME, Houssen WE, Arnold M et al (2011) Chaxamycins A-D, bioactive ansamycins from a hyper-arid desert Streptomyces sp. J Nat Prod 74:1491–1499. https://doi.org/10.1021/np200320u

Rehman SU, Yang L-J, Zhang Y-H et al (2020) Sorbicillinoid derivatives from sponge-derived fungus Trichoderma reesei (HN-2016-018). Front Microbiol 11:1334. https://doi.org/10.3389/fmicb.2020.01334

Robledo A, Aguilar CN, Belmares-Cerda RE, Flores-Gallegos AC, Contreras-Esquivel JC, Montañez JC, Mussatto SI (2016) Production of thermostable xylanase by thermophilic fungal strains isolated from maize silage. CyTA-J Food 14(2):302–308. https://doi.org/10.1080/19476337.2015.1105298

Sahay S (ed) (2022) Extremophilic Fungi: ecology, physiology and applications. Springer Nature, Singapore

Scagel CF, Lee J (2012) Phenolic composition of basil plants is differentially altered by plant nutrient status and inoculation with mycorrhizal fungi. HortScience 47(5):660–671

Selbmann L, de Hoog GS, Zucconi L et al (2008) Drought meets acid: three new genera in a dothidealean clade of extremotolerant fungi. Stud Mycol 61:1–20. https://doi.org/10.3114/sim.2008.61.01

Song Y, Chen D, Lu K et al (2015) Enhanced tomato disease resistance primed by arbuscular mycorrhizal fungus. Front Plant Sci 6:786. https://doi.org/10.3389/fpls.2015.00786

Stierle AA, Stierle DB (2014) Bioactive secondary metabolites from acid mine waste extremophiles. Nat Prod Commun 9:1037–1044

Stierle DB, Stierle AA, Patacini B et al (2011) Berkeleyones and related meroterpenes from a deep water acid mine waste fungus that inhibit the production of interleukin 1-β from induced inflammasomes. J Nat Prod 74:2273–2277. https://doi.org/10.1021/np2003066

Strobel GA, Miller RV, Martinez-Miller C et al (1999) Cryptocandin, a potent antimycotic from the endophytic fungus Cryptosporiopsis cf. quercina. Microbiology (Reading) 145(Pt 8):1919–1926. https://doi.org/10.1099/13500872-145-8-1919

Talapatra K, Das AR, Saha AK, Das P (2017) In vitro antagonistic activity of a root endophytic fungus towards plant pathogenic fungi. J Appl Biol Biotech 5:068–071. https://doi.org/10.7324/JABB.2017.50210

Terhonen E, Sipari N, Asiegbu FO (2016) Inhibition of phytopathogens by fungal root endophytes of Norway spruce. Biol Control 99:53–63. https://doi.org/10.1016/j.biocontrol.2016.04.006

Toledo TR, Dejani NN, Monnazzi LGS et al (2014) Potent anti-inflammatory activity of Pyrenocine A isolated from the marine-derived fungus Penicillium paxilli Ma(G)K. Mediat Inflamm 2014:767061. https://doi.org/10.1155/2014/767061

Trochine A, Turchetti B, Vaz ABM, Brandao L, Rosa LH, Buzzini P, Rosa C, Libkind D (2017) Description of Dioszegia patagonica sp. nov., a novel carotenogenic yeast isolated from cold environments. Int J Syst Evol Microbiol 67(11):4332–4339. https://doi.org/10.1099/ijsem.0.002211

Uchida R, Nakajyo K, Kobayashi K et al (2016) 7-Chlorofolipastatin, an inhibitor of sterol O-acyltransferase, produced by marine-derived Aspergillus ungui NKH-007. J Antibiot (Tokyo) 69:647–651. https://doi.org/10.1038/ja.2016.27

Um S, Fraimout A, Sapountzis P (2013) The fungus-growing termite Macrotermes natalensis harbors bacillaene-producing Bacillus sp. that inhibit potentially antagonistic fungi. Sci Rep 3:3250

Urík M, Hlodák M, Mikušová P, Matúš P (2014) Potential of microscopic fungi isolated from mercury contaminated soils to accumulate and volatilize mercury (II). Water Air Soil Pollut 225:1–11. https://doi.org/10.1007/s11270-014-2219-z

Utkhede R (2006) Increased growth and yield of hydroponically grown greenhouse tomato plants inoculated with Arbuscular Mycorrhizal Fungi and Fusarium oxysporum f. sp. radicis-lycopersici. BioControl 51:393–400. https://doi.org/10.1007/s10526-005-4243-9

Vinale F, Sivasithamparam K, Ghisalberti EL et al (2014) Trichoderma secondary metabolites active on plants and fungal pathogens. Open Mycol J 8:127

Wang J, Wei X, Qin X et al (2016b) Antiviral Merosesquiterpenoids produced by the Antarctic Fungus Aspergillus ochraceopetaliformis SCSIO 05702. J Nat Prod 79:59–65. https://doi.org/10.1021/acs.jnatprod.5b00650

Wei Y, Zhang S-H (2019) Haloalkaliphilic fungi and their roles in the treatment of saline-alkali soil. In: Tiquia-Arashiro SM, Grube M (eds) Fungi in extreme environments: ecological role and biotechnological significance. Springer International Publishing, Cham, pp 535–557

Wijesekara I, Li Y-X, Vo T-S et al (2013) Induction of apoptosis in human cervical carcinoma HeLa cells by neoechinulin A from marine-derived fungus Microsporum sp. Process Biochem 48:68–72. https://doi.org/10.1016/j.procbio.2012.11.012

Wu G, Ma H, Zhu T et al (2012) Penilactones A and B, two novel polyketides from Antarctic deep-sea derived fungus Penicillium crustosum PRB-2. Tetrahedron 68:9745–9749. https://doi.org/10.1016/j.tet.2012.09.038

Wu L, Wu H, Chen L et al (2015) Dicidin and bacilysin from Bacillus amyloliquefaciens FZB42 have antibacterial activity against Xanthomonas oryzae rice pathogens. Sci Rep 5:12975

Xiao-Wei LUO, Yun LIN, Yong-Jun LU, Xue-Feng ZHOU, Yong-Hong LIU (2019) Peptides and polyketides isolated from the marine sponge-derived fungus Aspergillus terreus SCSIO 41008. Chin J Nat Med 17(2):149–154. https://doi.org/10.1016/S1875-5364(19)30017-2

Xu X, Zhang X, Nong X et al (2015) Oxindole alkaloids from the fungus Penicillium commune DFFSCS026 isolated from deep-sea-derived sediments. Tetrahedron 71:610–615. https://doi.org/10.1016/j.tet.2014.12.031

Xu X, Zhang X, Nong X et al (2017) Brevianamides and mycophenolic acid derivatives from the Deep-Sea-derived fungus Penicillium brevicompactum DFFSCS025. Mar Drugs 15:43. https://doi.org/10.3390/md15020043

Yang S-Z, Jin H-J, Wei Z et al (2009a) Bioremediation of oil spills in cold environments: a review. Pedosphere 19:371–381. https://doi.org/10.1016/S1002-0160(09)60128-4

Yao YQ, Lan F, Qiao YM et al (2017) Endophytic fungi harbored in the root of Sophora tonkinensis Gapnep: diversity and biocontrol potential against phytopathogens. Microbiology 6:e00437. https://doi.org/10.1002/mbo3.437

Yehia RS, Osman GH, Assaggaf H et al (2020) Isolation of potential antimicrobial metabolites from endophytic fungus Cladosporium cladosporioides from endemic plant Zygophyllum mandavillei. S Afr J Bot 134:296–302. https://doi.org/10.1016/j.sajb.2020.02.033

Ypema HL, Gold RE (1999) Kresoxim - methyl: modification of a naturally occurring compound to produce a new fungicide. Plant Dis 83:4–19. https://doi.org/10.1094/PDIS.1999.83.1.4

Yu J, Wu Y, He Z et al (2018) Diversity and antifungal activity of Endophytic fungi associated with Camellia oleifera. Mycobiology 46:85–91. https://doi.org/10.1080/12298093.2018.1454008

Yuan Y, Feng H, Wang L et al (2017) Potential of Endophytic fungi isolated from cotton roots for biological control against Verticillium wilt disease. PLoS One 12:e0170557. https://doi.org/10.1371/journal.pone.0170557

Zhang J, Tao L, Liang Y et al (2009) Secalonic acid D induced leukemia cell apoptosis and cell cycle arrest of G(1) with involvement of GSK-3beta/beta-catenin/c-Myc pathway. Cell Cycle 8:2444–2450. https://doi.org/10.4161/cc.8.15.9170

Zhang P, Mándi A, Li X-M, Du F-Y, Wang J-N, Li X, Kurtán T, Wang B-G (2014) Varioxepine A, a 3H-oxepine-containing alkaloid with a new oxa-cage from the marine algal-derived endophytic fungus Paecilomyces variotii. Org Lett 16(18):4834–4837. https://doi.org/10.1021/ol502329k

Zhang L-Q, Chen X-C, Chen Z-Q et al (2016a) Eutypenoids A-C: novel Pimarane Diterpenoids from the Arctic Fungus Eutypella sp. D-1. Mar Drugs 14:44. https://doi.org/10.3390/md14030044

Zhang T, Wang N-F, Liu H-Y et al (2016b) Soil pH is a key determinant of soil fungal community composition in the Ny-Ålesund Region, Svalbard (High Arctic). Front Microbiol 7:227. https://doi.org/10.3389/fmicb.2016.00227

Zhang X, Li S-J, Li J-J et al (2018) Novel natural products from Extremophilic fungi. Mar Drugs 16:194. https://doi.org/10.3390/md16060194

Zhang P, Deng Y, Lin X et al (2019) Anti-inflammatory mono- and dimeric Sorbicillinoids from the marine-derived fungus Trichoderma reesei 4670. J Nat Prod 82:947–957. https://doi.org/10.1021/acs.jnatprod.8b01029

Zhao Y, Guo L, Xia Y et al (2019) Isolation, identification of carotenoid-producing Rhodotorula sp. from marine environment and optimization for carotenoid production. Mar Drugs 17:161. https://doi.org/10.3390/md17030161

Zhelifonova VP, Antipova TV, Ozerskaya SM, Kochkina GA, Kozlovsky AG (2009) Secondary metabolites of Penicillium fungi isolated from permafrost deposits as chemotaxonomic markers. Microbiology 78(3):350–354. https://doi.org/10.1134/S0026261709030138

Zheng Y-K, Miao C-P, Chen H-H et al (2017) Endophytic fungi harbored in Panax notoginseng: diversity and potential as biological control agents against host plant pathogens of root-rot disease. J Ginseng Res 41:353–360. https://doi.org/10.1016/j.jgr.2016.07.005

Zhou H, Li L, Wang W et al (2015) Chrodrimanins I and J from the Antarctic Moss-Derived Fungus Penicillium funiculosum GWT2-24. J Nat Prod 78:1442–1445. https://doi.org/10.1021/acs.jnatprod.5b00103

Zhou Y, Zhang Y-X, Zhang J-P et al (2017) A new sesquiterpene lactone from fungus Eutypella sp. D-1. Nat Prod Res 31:1676–1681. https://doi.org/10.1080/14786419.2017.1286486

Zou WX, Meng JC, Lu H et al (2000) Metabolites of Colletotrichum gloeosporioides, an endophytic fungus in Artemisia mongolica. J Nat Prod 63:1529–1530. https://doi.org/10.1021/np000204t

Chapter 8
Metallotolerant Microbes for Improving the Health of Heavily Polluted Soil

Sarieh Tarigholizadeh, Roghayeh Heydari, Svetlana Sushkova, Saglara Mandzhieva, Sudhir Shende, Vishnu D. Rajput, and Tatiana Minkina

8.1 Introduction

Anthropological activities, such as urbanization, industrial growth, rapid escalation in population, and unsafe agriculture techniques, have all accelerated the build-up of potentially hazardous residues such as HMs (Nivetha et al. 2022). HMs refer to elements with a density exceeding 7 g cm³, including arsenic (As), cadmium (Cd), copper (Cu), lead (Pb), chromium (Cr), mercury (Hg), zinc (Zn), silver (Ag), and others. These persistent and toxic chemicals pose substantial health and environmental consequences, as they contaminate water bodies, air, and fertile soil, posing an immediate threat to every living organism on Earth (Sardar et al. 2013; Mathivanan et al. 2021). Unlike organic pollutants, HMs are toxic and non-biodegradable, making it difficult for biological or physicochemical means to break them down into non-toxic metabolites. Consequently, their concentrations in the environment can increase thousands of times and persist for extended periods (Saharan et al. 2023).

The bioaccumulation and long persistence of HMs can lead to cytotoxic, mutagenic, and carcinogenic effects, causing severe harm to living organisms. Such

S. Tarigholizadeh (✉)
Academy of Biology and Biotechnology, Southern Federal University,
Rostov-on-Don, Russia

Department of Plant, Cell and Molecular Biology, Faculty of Natural Sciences,
University of Tabriz, Tabriz, Iran
e-mail: tarigolizade@sfedu.ru

R. Heydari
Department of Plant, Cell and Molecular Biology, Faculty of Natural Sciences,
University of Tabriz, Tabriz, Iran

S. Sushkova · S. Mandzhieva · S. Shende · V. D. Rajput · T. Minkina
Academy of Biology and Biotechnology, Southern Federal University,
Rostov-on-Don, Russia

© The Author(s), under exclusive license to Springer Nature Switzerland AG 2024
A. Ranjan et al. (eds.), *Extremophiles for Sustainable Agriculture and Soil Health Improvement*, https://doi.org/10.1007/978-3-031-70203-7_8

accumulation of HMs found to be associated with liver damage, cancer, neurological and cardiovascular disorders, central nervous system disorders, and sensory disturbances among humans (Verma et al. 2023). In plants, HMs have been reported to result in deleterious effects such as decreased growth and germination and chlorosis due to reduced mineral nutrition, photosynthesis rate, and essential enzyme activities (Verma et al. 2021; Sharma et al. 2022). This issue has gained worldwide attention as HMs accumulations in the tissues of organisms end up biomagnification in food chains causing a negative influence on the environment and public health (Nnaji et al. 2023). Production of reactive oxygen species (ROS), namely, oxygen radicals and non-radical products of O_2, is another way that HMs exert their toxicity. This results in a decrease in the levels of antioxidant molecules within living cells, resulting in changes in the conformation of DNA and proteins. Moreover, HMs toxicity poses a serious threat to indigenous microorganisms, disrupting cell membranes and cellular functions, denaturing proteins, and nucleic acids, causing chromosomal aberrations, mutations, inhibition of enzyme activities, oxidative stress, and more (Ojuederie and Babalola 2017).

Given the overall risk associated with HMs pollution to ecosystems and public health, the strategies for remediation of HMs contaminated soil are of utmost importance. Traditional soil remediation techniques, like ion exchange, chemical precipitation, and electrochemical removal, are known to be costly and complex and often lead to secondary pollution. These approaches also have negative impacts on biological activity, soil fertility, and soil structure, among other limitations and deficiencies (Ahmed et al. 2023). In contrast, bioremediation offers a low-cost, simple, and efficient solution for cleaning up HM-contaminated soils while ensuring maximum benefits and environmental safety (Yadav et al. 2023).

Bioremediation relies on the utilization of indigenous and exogenous living organisms, particularly microorganisms present at contaminated places, which facilitate the elimination or transformation of HMs into less hazardous forms (Verma et al. 2021). Microorganisms, including algae, bacteria, and fungi, have demonstrated remarkable capabilities in detoxifying HMs due to their rapid mutation and evolutionary adaptation, allowing them to withstand extremophilic environments (Kaushik et al. 2021). However, how precisely these microbes manage to thrive in these conditions and efficiently break down contaminants is yet to be understood well. Additional investigation is required to elucidate the complex mechanisms that govern the survival tactics of microorganisms in HM-contaminated settings and their capacity to remediate pollutants. Advancing our understanding of these mechanisms will not only enhance the efficacy of bioremediation approaches but also aid in the development of novel and sustainable solutions for soil remediation in the future. It is imperative to continue investigating the interactions between microorganisms and HMs to unlock the full potential of bioremediation as an effective strategy for addressing HM pollution (Sharma et al. 2021; Sarker et al. 2023).

Metallotolerant microorganisms exhibit impressive capabilities to flourish in settings containing elevated levels of HMs. This is attributed to their capacity for rapid mutation and evolution, enabling them to utilize pollutants as a resource (Barman et al. 2020). These microorganisms have evolved diverse strategies to deal with the associated toxicity. One such mechanism involves the production of an extracellular barrier in soils with elevated HM levels, which helps mitigate toxicity and promotes microbial growth in metal-contaminated environments. Furthermore, they exhibit efficient efflux systems that actively or passively remove toxic metal ions from their cells, allowing for the bioaccumulation of metal ions. By expressing different metal-resistant genes, they can also genetically adapt to detoxify HMs found in the contaminated soil (Das et al. 2016b; Yin et al. 2019).

These unique capabilities make microorganisms ideal candidates for low-cost, highly efficient, and environmentally friendly bioremediation processes. Metallotolerant microorganisms have been successfully employed in bioremediation practices to eliminate pollutants from the environment, extract valuable metals from industrial waste, and remediate abandoned mines and smelting sites (Maqsood et al. 2023; Yadav et al. 2023). To advance bioremediation strategies and effectively protect and restore the environment in the face of increasing HM pollution, it is crucial to understand the intricate relationships between HM resistance and environmental health (Sharma et al. 2021; Sarker et al. 2023). This chapter aims to address HMs remediation and enhance our understanding of the underlying mechanisms involved in HM resistance, shedding light on their interrelationship. By doing so, more efficient bioremediation strategies can be developed. This review serves as a valuable reference for comprehending the impacts of HM contamination on the environment and the diversity of metallotolerant microbes.

8.2 Heavy Metals and Their Potential Risks on Microbial Community

Heavy metals are stable metals or metallic elements with high density, an atomic weight of 63.5–200.6 u, and a specific gravity higher than 4 ± 1 g/cm^3 that are very prone to pose threat to public health after their bioaccumulation (Verma et al. 2017). They are classified according to their biological functions, chemical characteristics, and density. Trace amounts of some metals, namely, iron (Fe), cobalt (Co), and Zn, are essential as micronutrients to organisms since many biological processes, especially enzymes-mediated catalytic reactions, require these elements (Čvančarová et al. 2020). However, HMs are considered highly toxic pollutants of significant environmental concern, even in small quantities, due to their high density and toxicity (Table 8.1). Metals such as Hg, Pb, Cr, Cd, and As are among the HMs identified as the most hazardous elements by the Environmental Protection Agency (EPA) (Rahman and Singh 2020; Shuaib et al. 2021). Anthropogenic pollution of the environment with HMs is the most deleterious and widely distributed probably due to their instability and solubility as well as bioavailability. Various human activities, including industrial operations, transportation, agriculture, residual organic matter, fuel consumption, and sewage water, contribute to the HMs pollution in the environment. Industrial processes like mining, smelting, and manufacturing release HMs

Table 8.1 The impact of different heavy metals on microorganisms and their toxicity

Heavy metal	Toxic forms	Sources	Impacts on microorganisms	References
Arsenic (As)	$NaAsO_2$, AsH_3, $CH_3AsO(OH)_2$	Contaminated soils, sediments, groundwater, chemical weapons, forest fires, herbicides, land erosion, leaching of weathering rocks, LED, semiconductors, volcanic eruption, wood preservative chemicals	Enzyme deactivation	Bissen and Frimmel (2003), Wuana and Okieimen (2011), Abbas et al. (2018), Kuivenhoven and Mason (2019), Sher and Rehman (2019)
Cadmium (Cd)	CdO, $CdCl_2$, $CdSO_2$	Anthropogenic inputs such as detergent, phosphate fertilizers, industrial refining of Zn and Pb, refined petroleum products, sedimentary rocks	Disruption in cell transcription and division, nucleic acid damage, protein denaturation	Das et al. (1997), Sankarammal et al. (2014), Kuivenhoven and Mason (2019)
Chromium (Cr)	CrO_4^{2-}, $Cr_2O_7^{2-}$, CrO_3	Cement manufacture, chrome plating, coal combustion, pigment manufacture, smoke from wildfires, volcanic eruptions, weathering of rocks	Growth inhibition, hampering oxygen absorption	Cervantes et al. (2001), Shanker et al. (2005), Mishra and Doble (2008), Panda and Sarkar (2012), Singh et al. (2013)
Cobalt (Co)	Inorganic forms of cobalt	Coal combustion, fertilizers, mining of Pb, Fe and Ag, special steels	Inactivating iron-sulfur enzymes	Nies (2000), Aronson (2006), Ranquet et al. (2007), Tripathi and Srivastava (2007)
Copper (Cu)	Toxic in high concentration (Cu^{2+})	Rocks and soils	Cellular metabolic disruption, inhibition of enzyme activities	Fashola et al. (2016), Igiri et al. (2018)
Lead (Pb)	$C_6H_{16}Pb$, C_3H_9ClPb, $C_8H_{20}Pb$, $Pb(OH)_2.2PbCO_3$	Coal burning, burning fuels mixed with Pb, emission from industry and transportation, battery manufacturing, paint containing Pb, mining and smelting operations, use of leaded gasoline, gaseous,	Inhibition of enzyme action and transcription, protein/nucleic acid destruction	Jarosławiecka and Piotrowska-Seget (2014), Fashola et al. (2016), Wani et al. (2016), De and Ghosh (2018), Kumar and Prasad (2018)

(continued)

Table 8.1 (continued)

Heavy metal	Toxic forms	Sources	Impacts on microorganisms	References
Mercury (Hg)	CH_3Hg, $C_2H_5Hg^+$	Dental amalgam, fossil fuels, the pulp and paper industries, mining, forest fires, volcanic eruptions, waste incinerators	Disruption of plasma membrane, inhibition of enzyme activity, protein denaturation	Revis et al. (1990), Irawati et al. (2012), Fashola et al. (2016)
Nickel (Ni)	$NiCl_2$, $Ni(CO)_4$, $Ni(NO_2)_2$, $NiSO_4$	Combustion of fossil fuels, electroplating, industry, metal plating, Ni mining, power plants, sulphide ores, waste incinerators	Cell membrane disruption, hindering enzyme functions, oxidative stress induction	Stoppel et al. (1995), Malik (2004), Osmani et al. (2015), Kuivenhoven and Mason (2019)
Selenium (Se)	SeO_4 2A	Minerals	Hindering growth rate	Dixit et al. (2015), Igiri et al. (2018)
Silver (Ag)	Both organic and inorganic forms are hazardous in high concentrations	Mining, smelting ores battery manufacture, photographic processing,	Cell lysis, inhibition of cell transduction and growth	Percival et al. (2005), Prabhu and Poulose (2012), Qian et al. (2013)
Uranium (U)	$UO_2(NO_3)_2$, UF_6, UF_4, UCl_4	Erosion of tailing from mines and mills, operation of coal-burning power plant, processing of U ores, production of phosphate fertilizers, volcanic eruptions	Cell death, DNA damages	Keith et al. (2013), Suzuki and Banfield (2004), Martinez et al. (2007)
Zinc (Zn)	Toxic in high concentrations	Coal and waste, combustion, mining, soil and sediments, steel processing	Death and decrease in biomass, growth inhibition	Malik (2004), Chaudhary et al. (2017), Vicentin et al. (2018), Nriagu (2019)

into the environment, leading to the accumulation of these metals in soils. Improper waste management practices in unregulated landfills and disposal sites also contribute to soil metal accumulation, contaminating both soil and water sources. Moreover, HMs can be deposited into the soil through air pollution and acid rain, leading to the accumulation of Pb, Hg, and Cd in the soil. Moreover, these metals are also found in natural constituents of the earth's crust and are found in volcanogenic particles, windblown dust, vegetation, forest wildfires, and sea salt (Alloway 2013; Abdu et al. 2017). The presence of HMs induces changes in the physicochemical properties of soil, including alterations in organic matter, pH, inorganic cations and anions, and clay contents (Cui et al. 2023). Understanding the sources and potential risks associated with HMs is crucial for safeguarding public health.

Soil pollution of HMs significantly affects the soil microbiome in terms of microbial activity, diversity, community structure, and population size (Table 8.1). Over time, the accumulation of HMs in the soil can be toxic to microbes, leading to adverse effects on soil enzymatic activities, nutrient cycling, carbon sequestration, soil structure, and eventually hindering plant growth (Shuaib et al. 2021; Wang et al. 2021). Soil microorganisms can use certain HMs as electron acceptors or donors during their biological metabolism, employing specialized enzymes to remove or transform these compounds into part of their metabolic procedures. However, microbes have still been found to be the most sensitive of all soil-living organisms to HMs stress (Abdu et al. 2017; Shuaib et al. 2021). High concentrations of metal pollution have been found to constrain the activity and dynamics of soil microorganisms, resulting in the accumulation of organic matter at the soil surface layer (Enya et al. 2020). In heavily contaminated soils, HMs can lead to higher microbial metabolic entropy and reduced conversion of organic carbon to biocarbon (Inobeme 2021). Numerous studies have reported the interference of HMs with the biochemistry of soil and various microorganisms isolated from their native habitats. Microbes significantly affect soil quality, involve in biochemical reactions in soil, and contribute to the generation of soil organic matter, the breakdown of toxic chemicals, the development of soil structure, and the biochemical cycles within the soil. The interaction between HMs and enzymes, as HM concentrations increase, directly causes a significant decrease in microbial enzyme activity, quality, and quantity in soil. Consequently, it becomes evident that metal-contaminated soil has a detrimental impact on microbial diversity and population (Wyszkowska et al. 2013; Chu 2018; Shuaib et al. 2021).

In regions characterized by elevated levels of metal contamination, plants display biochemical and physiological abnormalities, which can have significant implications for food security (Rehman et al. 2021; Bharti and Sharma 2022). Innovative strategies can be implemented to improve the damaging impacts of metal toxicity in plants (Narayanan and Ma 2023). The accumulation of HMs in soil and water, resulting from industrial activities and human practices, often hampers the physiological processes of plants, namely, photosynthesis, nutrient assimilation, and physiology. Elevated levels of HMs, including Pb, Cd, Cu, and As, induce oxidative stress, damage cell membranes, disrupt hormonal balance, impede root development, and inhibit overall plant growth (Rehman et al. 2021; Bharti and Sharma 2022). For example, Pb in soil negatively impacts germination percentage, growth, biomass, and protein content of *Zea mays*. It also hinders ribulose-bisphosphate carboxylase oxygenase activity, negatively affecting CO_2 fixation (Rizvi and Khan 2018). Similarly, Arsenic results in stunted growth, wilting, and chlorosis in *Brassica napus* while inhibiting the transpiration rate of *Avena sativa* seedlings. In *Oryza sativa*, the soil causes decreases in seed germination, leaf area, seedling height and dry matter production (Chibuike and Obiora 2014). Plants may eventually perish as a result of these impacts, which are linked to diminished metabolic activities, growth, and water-nutrient uptake (Singh and Kalamdhad 2011). Moreover, toxic elements may also enter the food chain through plants contaminated with HMs. This underscores the urgent need for strict environmental regulations and efficient remediation techniques to prevent harmful effects and protect ecosystems and human populations. The uptake of HMs by plant roots represents a key pathway for their integration into the food chain (Sarker et al. 2022). Soil pollution typically involves a combination of HMs rather than a single metal, which results in greater harm to plants (Chibuike and Obiora 2014). A study conducted on *Lythrum salicaria* has demonstrated that combined Pb and Cu, at both low and high concentrations, lead to rapid the demise of leaves and stems (Nicholls and Mal 2003). The uptake and accumulation of HMs in plant tissues are affected by various factors, namely, temperature, pH, moisture, organic matter, and nutrient availability (Kwiatkowska-Malina 2018). For instance, during the summer, spinach is more likely to uptake and accumulate Cd, Zn, Cr, and Mn compared to winter, whereas Cu, Ni, and Pb exhibit higher accumulation during winter. Possibly, the increased intake of HMs during the summer months is linked to the accelerated breakdown of organic matter, influenced by higher transpiration rates compared to the winter season. The low humidity and high ambient temperature can also be attributed to this issue (Sharma et al. 2007).

8.3 Bioremediation and Metal–Microbe Interactions

Bioremediation refers to the biological breakdown or transformation of organic or inorganic waste into less toxic compounds, accomplished through the activities of microorganisms. It is a cumulative phenomenon to detoxify a polluted environment using biological systems (Yin et al. 2019). Microbial bioremediation is utilizing microorganisms to control or destroy hazardous environmental contaminants. The principle behind microbial bioremediation is the enzymatic reactions of the microbes as integral components of their regular metabolic processes. The advantage of this type of bioremediation is that it is a natural and low-cost technology, which can result in the destruction of the target pollutants. It can be improved by introducing an electron acceptor, nutrients, or other factors positively influencing the process (Choudhury and Chatterjee 2022; Jeyakumar et al. 2023).

Microbial remediation mechanisms play a crucial role in addressing the challenges posed by constant exposure to HMs. Over time, microbes become accustomed to and develop resistance against these metals, necessitating a comprehensive understanding of the nature of metal–microbe interactions (Sharma et al. 2023). These interactions encompass various types, including biosorption, bioaccumulation, bioreduction, bioprecipitation, biovolatilization, bioleaching, composting, land farming, bioreactors, biopiles, and biosparging. These mechanisms, influenced by the organism's characteristics and proficiencies, allow for the bioremediation of HMs. Microbes are crucial in altering HMs into different ionic states, thereby influencing their bioavailability, solubility, and mobility in aquatic and soil environments (Jeyakumar et al. 2023). Bioremediation of HMs with the aid of microbes involves processes such as mobilization, immobilization, oxidation-reduction, chelation, alteration of metallic complexes, and biomethylation. The interaction between microbes and HMs depends not only on degradation and removal processes but also on biological factors, namely, cell wall composition of the microbes and physiology, as well as environmental conditions pH, temperature, contact time, initial metal ion concentration, and their ionic strength (Verma et al. 2021). Ultimately, the initial objective of bioremediation is to lower the toxicity of HM ions. Functional groups present in or on microorganisms can capture HM ions, while other ions transform a toxic form to a less toxic product through microorganism-mediated redox state changes. This efficient decrease in the toxicity of HM ions highlights the significant role played by microbial remediation mechanisms (Verma and Kuila 2019; Liu et al. 2021).

8.3.1 Biosorption

Biosorption is a complex process involving the natural uptake of HMs through physicochemical pathways. It relies on the surface adsorption and binding of HMs to charge functional groups, such as hydroxyl, carbonyl, and sulfate groups, on the cell surface of microorganisms. The biosorption capabilities of microbes are influenced by various factors, including temperature, experimental conditions, ionic strength, contact time, and cell wall composition (Javanbakht et al. 2014; Shamim 2018). Fungi and bacteria both play a significant role in biosorption due to their ability to resist drastic environments and grow under controlled settings, enabling them to degrade the pollutants (Ayangbenro and Babalola 2017). Biosorption is an effective and cost-efficient way to remove multiple types of HM contaminants from the environment and perform across varying environmental conditions (Rahman and Singh 2020; Jeyakumar et al. 2023). The process involves the interaction between the sorbate and the biosorbent in which sorbate molecules accumulate at the sorbate–sorbent interface (Dave et al. 2020). Functional groups, namely, thiol, hydroxyl, carbonyl, and phosphate, can effectively bind metal binding groups and participate in biosorption, and the dead biomass plays a key role in enhancing the efficiency of HMs sorption supported by these functional groups. The binding of HMs to the cell wall of the organisms is brought about by adsorption, coordination, complexation, chelation, ion exchange, electrostatic interaction, and microprecipitation with the ability to exchange cations such as K^+, Ca^{2+}, and Mg^{2+} for HM ions. Adsorption and absorption are the main mechanisms by which microbes accumulate HMs, with adsorption being a rapid and energy-independent process, while absorption is slower and dependent on energy metabolism (Rahman and Singh 2020; Jeyakumar et al. 2023).

8.3.2 Bioaccumulation

A metabolically active biological process that implies the uptake and storage of HM ions by microbial cells. The import storage system is crucial in this method, as it facilitates the HM ions into the cytoplasm by crossing the plasma membrane and supported by transporter proteins. Within the cell, metal-binding entities such as proteins and peptide ligands sequester the HM ions, which can occur in particulate and insoluble forms, as well as their by-products. In bacterial membranes, various mechanisms contribute to HMs bioaccumulation, including ion channels, carrier-mediated transport, lipid permeation, endocytosis, and complex permeation. Studies have shown the intracellular and periplasmic accumulation of metals Viz., Pb, Ni, Cd, and Cr in different microbial species (Verma and Sharma 2017; Pande et al. 2022; Sharma et al. 2022). Bioaccumulation takes place in two phases: the first phase is metabolism-independent (much like biosorption) and occurs rapidly; however, the second phase is metabolism-dependent and allows the dissemination of metal ions across the lipid bilayer through ATP-dependent active transport, followed by their binding to intracellular compounds. Contact time, biomass dosage, agitation, pH, temperature, and initial metal ion concentration are key factors that influence the bioaccumulation process (Hansda et al. 2016; Segretin et al. 2018). Metallothionein, a metal-binding protein, is helpful in reducing the free metal ion concentration in the cell, which is significant for intracellular metal bioaccumulation. In bioaccumulation, metal uptake mechanisms in bacterial cells include passive diffusion, facilitated diffusion, and active transport (Jayaram et al. 2022). *Pseudomonas stutzeri*, isolated from the foundry soil, exhibited tolerance to Cr pollution and was capable of reducing Cr(VI) anaerobically (Badar et al. 2000). The extracellular and intracellular accumulation of uranium (U) have been observed in *Saccharomyces cerevisiae* and *P. aeruginosa*, respectively (Shukla et al. 2019). Microorganisms such as *Cupriavidus metallidurans* CH34 have shown the ability to bioaccumulate Au (gold) and Se and volatilize Hg, making them effective in remediating HMs-polluted environments (Sarret et al. 2005; Reith et al. 2006). This process is advantageous as it requires minimal growth medium and has minimal toxic effects on cells (Verma and Kuila 2019; Choudhury and Chatterjee 2022).

8.3.3 Bioleaching

Bioleaching involves the solubilization of metal sulfides and oxides derived from ore deposits and secondary wastes by bacteria and fungi. Bioleaching is an economical and environmentally friendly process that has been employed for centuries in the extraction of metals from low-grade ores as globally profitable methods for Cu, Au, U, Co, Ni, and Zn (Jin et al. 2018; Choudhury and Chatterjee 2022). In this process, two methods are employed, namely, contact and non-contact methods. In the contact mechanism, bacterial cells and the mineral surface interact together leading to oxidation and electron transfer process. In the non-contact mechanism, bacteria produce lixiviant, typically ferric iron, which chemically oxidizes the sulfide mineral (Tayang and Songachan 2021). Acidophiles, such as chemolithotrophs that survive in low pH conditions, play a crucial role in bioleaching by oxidizing ferrous iron to Fe^{3+} and reducing sulfur to sulfuric acid. The resulting Fe^{3+} and H^+ are helpful in solubilizing the metal oxides and sulfides from the ores, facilitating metal extraction (Srichandan et al. 2014). Bioleaching is a simple and efficient technology that transforms sparsely soluble metal compounds into readily soluble forms, enabling their extraction (Jin et al. 2018; Choudhury and Chatterjee 2022). Bacterial species like *Leptospirillum ferrooxidans* and *Thiobacillus ferrooxidans* and fungi such as *Rhizopus* sp., *Aspergillus* sp., *Cladosporium* sp., and *Penicillium* sp. are known for their ability to bioleaching metals due to their tolerance to varying environmental conditions owing to their biochemical adaptation (Abraham et al. 2020; Chatterjee et al. 2020). The efficiency of bioleaching depends on the ore's chemical and mineralogical composition, and optimization of conditions leads to maximum metal extraction yields. Advanced bioleaching techniques, including those applied to electronic waste, show promising results in reducing HM concentrations through microbial interactions (Jin et al. 2018; Choudhury and Chatterjee 2022).

8.3.4 Biomineralization

Biomineralization is natural chemical alterations in which living organisms facilitate the formation of minerals such as silicates, carbonates, oxides, phosphates, and sulfates. The cell wall and its EPS and S layer are significant players in improving the reactivity owing to its varying hydration, structure, and composition, which plays a key role in mineral formation. Organic ligands such as hydroxyl, carboxyl, phosphoryl, amine, and sulfur on the microbial surface contribute to the deprotonation and the imparting of a net negative charge, particularly with increasing pH. Toxic metal ions with positive charges precipitate inconsistently yielding compact and stable mineral products (Tayang and Songachan 2021; Zheng et al. 2021). Metal immobilization or complexation can occur through processes like phosphate, carbonate, and oxalate precipitations (Shan et al. 2021). Studies have demonstrated the prospective of biomineralization for remediating toxic HMs. For instance, *Bacillus* sp. has been found to release free inorganic phosphate, which forms an insoluble metal phosphate layer, thereby trapping HM ions (Zhang et al. 2019). Bacterial and algal species are known to exhibit both intracellular and extracellular mineral formation, leading to the development of crystals that resemble those formed through inorganic precipitation. Mineral formation in bacteria occurs through the reaction of biogenic gases and HM ions; however, in algae, the reduction of HMs using CO_2 leads to the formation of minerals (Dave et al. 2020). Studies have examined the bioremediation potential of microbially induced calcite precipitation using *Sporosarcina ginsengisoli* for As removal. Calcite mineralized is efficient in absorbing and incorporating metalloids into its structure, making it effective across various environments. The presence of NH^{4+} and additional CO_2 released increases the pH and accelerates the rate of precipitation of urease-induced calcite (Achal et al. 2012). Different microbial species such as *Citrobacter freundii*, *Vibrio harveyi*, and *Staphylococcus aureus* have been found to precipitate Pb (II) either as extracellular or intracellular phosphates, while *Providencia alcalifaciens* 2EA precipitates Pb(II) as lead phosphates through phosphatase activity (Naik et al. 2013; Dave et al. 2020). These findings highlight the diverse processes opted by microorganisms in biomineralization for the remediation of HMs.

8.3.5 Biotransformation

Biotransformation is a microbial remediation process wherein a chemical compound is modified to create a polar molecule. This process involves the transformation of both metals and organic compounds from harmful forms to comparatively less toxic forms, allowing microorganisms to adjust to alterations in their environment (Smitha et al. 2017). Microbes are ideal for biotransformation owing to their extended surface-to-volume ratio, metabolic activity, and rapid growth rate, as well as their ability to maintain sterile conditions. Various reactions such as condensation, hydrolysis, oxidation, reduction, alkylation, dealkylation, isomerization, introduction of functional groups, carbon bond formation, and methylation can drive biotransformation. This process can result in the volatilization of HMs, thereby reducing their toxicity (Smitha et al. 2017; Tayang and Songachan 2021).

Biotransformation has been widely used for the remediation of pollutants and microbial cells, vegetative cells, spores, enzymes, and so on are commonly employed in microbial transformation processes (Choudhury and Chatterjee 2022; Smitha et al. 2017). For example, *Bacillus* sp. SFC 500-1E, resistant to Cr (VI), utilizes NADH-dependent reductase to reduce toxic Cr (VI) to less toxic Cr (III) (Thatoi et al. 2014). *Micrococcus* sp. and *Acinetobacter* sp. have been shown to oxidize toxic As (III) into less soluble and harmless forms, reducing their toxicity (Nagvenkar and Ramaiah 2010). Biotransformation process often involves the surface of the microbial cells, in the extracellular matrices, or depending on the specific metal being targeted for recovery or elimination. Recent progress in biological research has further propelled the use of biotransformation-based remediation approaches for HMs in the environment (Dave et al. 2020; Rahman and Singh 2020).

8.3.6 Bioaugmentation

Bioaugmentation is a process of natural attenuation that involves the specific microbial strains or consortia of microbes to enhance the disintegration and elimination of HM ions (Adams et al. 2015). It involves the indigenous or exogenous insertion of microorganisms (bioaugmenting agents) into the affected sites. These agents have higher genetic stability and survivability in comparison with their native counterparts due to the former's ability to degrade the toxins by being resilient to the prevailing environmental conditions. In this method, the utilization of pure cultures has shown greater adaptability to HMs and enhanced degradation capabilities when compared to mixed cultures (Verma et al. 2021; Choudhury and Chatterjee 2022).

8.3.7 Biostimulation

Biostimulation is another bioremediation approach that depends on the inherent capabilities of indigenous microbes. It involves the enhancement of HMs biodegradation by introducing biostimulants such as electron acceptor species, rate-limiting nutrient material, and so on to achieve a nutrient balance for accelerating the growth of microorganisms (Verma et al. 2021). Biostimulation is more effective when employed in specific physical conditions (temperature, humidity, etc.) with selected degrading microorganisms along with their respective organic and chemical (nutrients) sources. Recent research has shown that nanocomposites improve the biostimulation process by enhancing microbial absorption, adsorption, and the occurrence of chemical reactions involved in bioremediation (Tayang and Songachan 2021; Verma et al. 2021).

8.3.8 Bioweathering

Bioweathering, also known as microbial weathering, is a fundamental process that involves the chemical and mechanical disintegration of minerals by a diverse range of microorganisms. Such microbes use the minerals present in rocks as nutrients for normal metabolic function. Bioweathering is usually triggered by acid rain and air pollution. The toxins are remediated through diverse mechanisms, including ion exchange, precipitation, and redox reaction. This process also enhances the bioaccumulation of metal by plants (Choudhury and Chatterjee 2022).

8.4 Microbial Survival Strategies

Since HMs cannot be degraded like toxic organic compounds, in HMs stress conditions, microbes exhibit various resistance mechanisms, such as efflux of HM ions, extracellular barriers, intracellular sequestration, extracellular sequestration, and their reduction (Fig. 8.1) (Tayang and Songachan 2021).

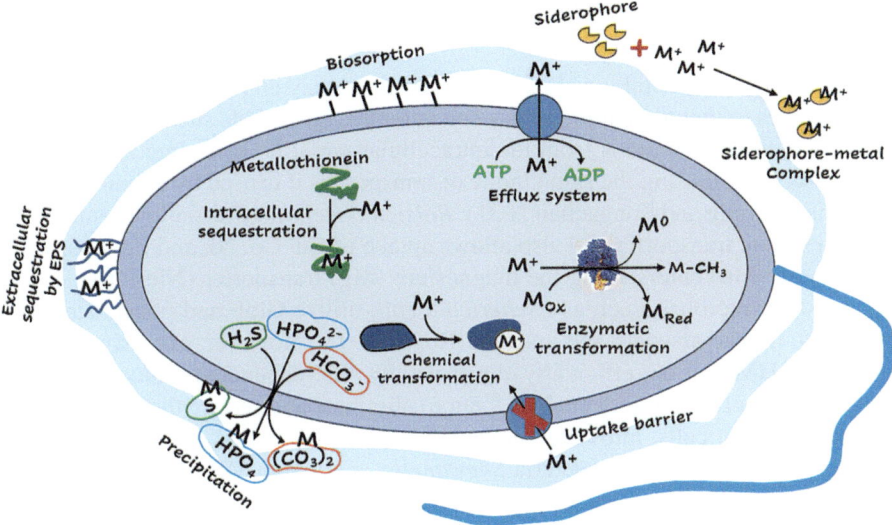

Fig. 8.1 Illustration of biological processes involved in heavy metal detoxification in microbes

8.4.1 Extracellular Barriers

Extracellular barriers such as the cell wall, plasma membrane, extracellular polymeric substances (EPS), and biofilms prevent the entry of HMs into the intracellular environment of microbial cells. These structures act as physical barriers by adsorbing metal ions onto their surfaces. Research studies have demonstrated the biosorbent properties of fungal and bacterial isolates, such as their ability to adsorb HMs. Gram-positive bacteria, such as *Cellulosimicrobium* sp., have shown resistance to multiple HMs through chemisorption sites (Bhati et al. 2019). Biofilms, consisting of extracellular polymers, act as protective barriers by accumulating metal ions and safeguarding the cells within. For instance, biofilm produced by *P. aeruginosa* exhibits tolerance for Zn^{2+}, Cu^{2+}, and Pb^{2+}. Similarly, biofilms produced by *Rhodotorula mucilaginosa* have been found to enhance metal removal efficiency (Grujić et al. 2017). EPS also contributes to metal resistance by adsorbing Pb ions in bacteria like *P. aeruginosa*, *Azotobacter chroococcum* XU1, and *Acinetobacter junii* L. Pb1. Moreover, altering the plasma membrane permeability can resist the uptake of HM ions into the intracellular region (Kazy et al. 2002; Teitzel and Parsek 2003; Rasulov et al. 2013; Kushwaha et al. 2017).

8.4.2   Active Transport of Metal Ions

Active transport of metal ions is a vital survival strategy employed by metallotolerant microbes to maintain homeostasis and defend against HMs stress. Efflux pumps actively remove metal ions from the intracellular parts; however, metals also enter the microbial cell using the same mode of transport as it is used by essential micronutrients (Tayang and Songachan 2021). *Ralstonia metallidurans* allows entry of Cr using a sulfate transporter and also allows uptake of Cd, Co, Ni, and Zn ions, while manganese (Mn) enters using the magnesium (Mg) transporter (Nies et al. 1989). As-resistant microbes, such as *Escherichia coli*, utilize GlpF and phosphate transporters to uptake arsenate and arsenite (Rosen 2002). However, excessive concentrations of HM ions are efficiently eliminated by ATP-dependent efflux. In the host cells, such an efflux system enables them to eliminate and reduce the concentration of Cr within the cells, allowing damaged cells to recover and detoxify over time. This is exemplified by the bacteria *Shewanella oneidensis* MR1, which efficiently ejects Cr using its efflux system (Baaziz et al. 2017). Also, *E. coli* controls Cu efflux through the RND CusCBA multiprotein complex (Franke et al. 2003). *Bacillus* sp. SFC-1E employs a chromate ion transporter protein called ChrA for the efflux of toxic chromate ions $(Cr_2O_4)^{2-}$ (Ontañon et al. 2018). The expression of these transporters depends on the bacterial species and specific HMs and is controlled by resistance genes present on either chromosome or plasmid. Such genes are responsible for encoding membrane transporter proteins that regulate the uptake as well as elimination of HMs. Some examples of metal transporter proteins expressed on the plasma membrane are ABC transporters, P-type efflux ATPase, cation diffusion facilitators, and proton–cation antiporters (Yin et al. 2019). For example, P-type efflux ATPase is commonly found in gram-positive bacteria responsible for exporting Cd(II), Cu(II), and Zn(II) (Maynaud et al. 2014). Arsenite efflux is also ATPase dependent, which is controlled by exporting proteins on the cell membrane. Some bacteria employ efflux systems in conjunction with other HMs resistance mechanisms to eliminate metals. ATPase pump ArsA/ArsB and arsenic reductase ArsC are coded by the *ars* operon structure in both gram-positive and negative bacteria. In the cytoplasm, ArsC converts arsenate to arsenite. The efflux mechanism then exports arsenite across the plasma membrane to the outside of the cell (Mukhopadhyay et al. 2002).

8.4.3   Extracellular Sequestration

In this process, cellular structures on the outer membrane or periplasm form complexes with HM ions, leading to the formation of insoluble compounds through complexation or precipitation. Extracellular sequestration involves various biological structures that can accumulate HMs and decrease their toxicity (Prabhakaran et al. 2016). Proteins nucleic acids, complex carbohydrates, and lipids are found in the EPS which widely possess functional groups such as amino, phosphate, hydroxyl, carboxylic, and sulfhydryl, which enable them to be efficient biosorbents for HMs (Yin et al. 2019). Siderophores produced by bacteria and fungi can chelate Fe^{3+} aiding microbial survival under a Fe-deficient environment. Siderophores can chelate other metals such as Cu, Ni, and Zn, protecting microorganisms from their toxicity (Saha et al. 2016; Sharma et al. 2018). Glutathione, secreted by microorganisms, can bind HMs and prevent their entry into living cells. Furthermore, the cationic and anionic functional groups found in microorganisms' cell walls—such as phosphate, amine, carboxyl, and hydroxyl groups—act as a barrier against the toxicity of HMs by obstructing the entry of HM ions into the intracellular environment through extracellular metal sequestration (Kieu et al. 2011; Joo et al. 2015). Some microbes also produce hydrogen sulfide in the extracellular environment, which offers protection against HMs toxicity. For example, *Klebsiella planticola* produces hydrogen sulfide and precipitates Cd ions as insoluble sulfides under aerobic conditions (Sharma et al. 2000). Cu-resistant strains such as *P. pickettii* US321 and *P. syringae* synthesize Cu-inducible proteins that bind and accumulate Cu ions, resulting in blue bacterial colonies (Cha and Cooksey 1991; Gilotra and Srivastava 1997). Similarly, in a carbon deficit environment, the multi-resistant *P. putida* S4 strain forms an insoluble precipitate of Cu^{2+}, hydroxyl, and phosphate groups (Saxena and Srivastava 1998). *Desulfovibrio desulfuricans* produce hydrogen sulfide in the extracellular surroundings to protect them from HMs toxicity by causing precipitation of HM ions (Yin et al. 2019).

8.4.4 Intracellular Sequestration

In intracellular sequestration, HM ions and metal-binding peptides create complexes in the cytoplasm, keeping hazardous levels of HMs below threshold and safeguarding sensitive cellular components (Tayang and Songachan 2021). Metal-binding peptides can take the form of metallothioneins or phytochelatins, both of which are rich in cysteine and bind HM ions through thiol groups. Metallothioneins possess higher metal-binding affinity and hence act as sinks by utilizing cysteine residues to sequester excess metal ions (Yin et al. 2019; Tayang and Songachan 2021). This mechanism can be observed in bacteria and certain microalgal species such as *Chlorella*, *Scenedesmus*, and *Chlamydomonas* (Balzano et al. 2020).

Phytochelatins are low molecular weight short peptides found in plants and fungi that are glutathione-derived 5–11 amino acids peptides; they possess the ability to sequester specific metal ions like Cd and Cr (Clemens and Simm 2003). *Rhizobium leguminosarum* cells were observed to sequester Cd ions intracellularly using glutathione, and *Aspergillus* sp. converts Cr (IV) ions into a less toxic form through glutathione sequestration, ultimately removing them from the system via the efflux system (Ianeva 2009; Talukdar et al. 2020).

8.4.5   Reduction of Heavy Metal Ions

Reduction is brought about by metallotolerant bacteria to change HM and metalloids from their higher oxidation states into more stable and less hazardous forms. Furthermore, change in the redox state of HM ions is regulated by detoxification enzymes through reduction or oxidation events, which are regulated by certain resistance genes. Oxidized forms of Hms can occasionally be used as energy-producing electron donors or acceptors. This strategy has been observed in bacteria that reduce chromate, molybdate, and vanadate (Barman et al. 2020; Barman and Jha 2021). *Geobacter sulfurreducens*, for example, produces magnetic iron (II)-bearing nanoparticles (NPs) using iron (III) as a source, which reduces Cr(VI) to Cr(III) making it more soluble and less hazardous (Joshi et al. 2018). *Bacillus* sp. exhibits tolerance to Hg facilitated by mercuric ion reductase, which converts mercuric ions to metallic mercury and releases it into the surrounding environment (Noroozi et al. 2017). Microorganisms like *Micrococcus* sp. and *Acinetobacter* sp. oxidize As(III) into As (V), effectively reducing their hazard (Nagvenkar and Ramaiah 2010).

8.5   Diversity of Metallotolerant Microbes

The diversity of metallotolerant microbes includes bacteria, fungi, and algae, and within these broad taxonomic groups, specific genera and species are studied for exhibiting metallotolerance that can withstand and adapt to metal-contaminated environments (Table 8.2) (Pikuta et al. 2007; Barman and Jha 2021).

8.5.1   Bacteria

In HMs bioremediation, bacteria are one of the extensively explored groups of metallotolerant microbes. Within the bacterial domain, numerous phyla and genera, including *Proteobacteria*, *Firmicutes*, *Actinobacteria*, and *Acidobacteria*, are identified to tolerate high metal concentrations (Hemmat-Jou et al. 2018; Yan et al. 2020). They not only can transform and detoxify metals using their enzymatic activities but also support plant health and growth in contaminated soils (Mishra et al. 2017). In recent years, there has been a significant body of research focused on the utilization of bacteria in bioremediation applications. Alboghobeish et al. (2014) conducted a study where they isolated Ni-tolerant bacteria from wastewater samples of the industries, specifically *Methylobacterium* sp. ATHA7, *Cupriavidus* sp. ATHA3, and *Klebsiella oxytoca* ATHA6. These bacterial isolates were found to be effective in remediating Ni-polluted wastewater and sewage. Moreover, the survival and performance of bacteria are not limited to individual cultures; mixed bacterial

Table 8.2 Bioremediation of heavy metals using metallotolerant bacteria, fungi, and algae

Microbe type	Species	Studied heavy metals	Microbial mechanism/survival strategy	References
Bacteria	*Acinetobacter* sp.	As, Cr	Biosorption, enzymatic detoxification	Nagvenkar and Ramaiah (2010), Bhattacharya et al. (2014)
	Acinetobacter sp. B9	Ni	Biosorption	Bhattacharya and Gupta (2013)
	Alcaligenes sp. MMA	Cd, Zn	Bioaccumulation	Sodhi et al. (2020)
	Alteromonas macleodii	Cd, Pd, Zn	Extracellular sequestration	Loaëc et al. (1997)
	Anacystis nidulans	Cd	Intracellular sequestration	Keyhani et al. (1996)
	Arthrobacter viscosus	Cr	Biosorption	Hlihor et al. (2017)
	Bacillus cereus	Cd, Cr, Pb	Biosorption, bioaccumulation, enzyme-mediated	Murthy et al. (2011),Dong et al. (2013), Jan et al. (2015), Nayak et al. (2018)
	Bacillus cereus RPb5-3	Pb	Bioaccumulation	Utami et al. (2020)
	Bacillus firmus	Hg	Enzymatic detoxification	Noroozi et al. (2017)
	Bacillus licheniformis	Fe, Cr, Cu	Biosorption	Samarth et al. (2012)
	Bacillus megaterium	B, Cd, Pb	Biosorption, bioaccumulation	Naik et al. (2012), Esringü et al. (2014)
	Bacillus safensis	Cr	Biodegradation	Kalaimurugan et al. (2020)
	Bacillus sp.	Cr	Reduction	Ontañon et al. (2018)
	Bacillus sp. (Q3 and Q5 Strains)	Pb	Biosorption	Heidari and Panico (2020)
	Bacillus sp. MH778713	Cr	Modulation of phytoremediation efficiency	Ramírez et al. (2019)
	Bacillus sp. KL1	As, Ni	Biosorption	Taran et al. (2019)
	Bacillus sp. SFC	Cr	Biosorption	Ontañon et al. (2018)
	Bacillus subtilis	Au	Biosorption	Niu and Volesky (2000)
	Bacillus subtilis X3	Pb	Bioimmobilization	Qiao et al. (2019)
	Bacillus subtilis BM2	Ni, Pb	Metal detoxification	Igiri et al. (2018)

(continued)

Table 8.2 (continued)

Microbe type	Species	Studied heavy metals	Microbial mechanism/survival strategy	References
	Bacillus thuringiensis	Cd, Hg	Biosorption	Jiang et al. (2015), Saranya et al. 2019
	Burkholderia dabaoshanensis	Cd	Biosorption	Zhu et al. (2012)
	Burkholderia fungorum FM-2	Cd, Pb	Metal accumulation	Liu et al. (2019)
	Cellulosimicrobium sp. (KX710177)	Pb	Biosorption	Bharagava and Mishra (2018)
	Cupriavidus metallidurans	Au, Cr	Active export, EPS accumulation	Zammit et al. (2016), Alviz-Gazitua et al. (2019)
	Cupriavidus sp. strain Cd^{+2}	Cd	Bioprecipitation	Li et al. (2019)
	Desulfovibrio desulfuricans	Cd, Cr, Cu, Ni, Zn	Biosorption, Extracellular sequestration	Joo et al. (2015), Yue et al. (2015), Zhang et al. (2020)
	Enterobacter cloacae	Hg; Pb	Bioreduction, precipitation	Park et al. (2011), Chen et al. (2018)
	Ensifer adhaerens	Pd	Extracellular sequestration	Oves et al. (2017)
	Enterobacter cloacae	Cd	Bioaccumulation	Ghosh et al. (2022)
	Enterobacter sp.	Cd	Bioaccumulation	Mitra et al. (2018)
	Enterobacter asburiae KUNi5	Ni	Metal resistance	Paul and Mukherjee (2016)
	Escherichia coli	Cd, Hg, Zn	Active export	Lerebours et al. (2016)
	Gemella sp.	Cd, Cr, Pb	Plasmid-mediated	Marzan et al. (2017)
	Gloeocapsa gelatinosa	Pb	Bioaccumulation	Mosa et al. (2016)
	Hafnia sp.	Cd	Plasmid-mediated	Wei et al. (2016)
	Klebsiella sp. J1	Pb	Adsorption by EPS	Wei et al. (2016)
	Klebsiella sp. 3S1	Pb	Biosorption	Muñoz et al. (2015)
	Klebsiella sp. USL2D	Pb	Biosorption	Orji et al. (2021)
	Kocuria flava	Cu	Precipitation	Achal et al. (2011)
	Lactobacillus plantarum	Cr, Ni	Biosorption	Ameen et al. (2020)
	Micrococcus luteus	Pb	Biosorption	Burakov et al. (2018)

Microbe	Metals	Mechanism	References
Micrococcus sp.	Cd, Cr, Pb	Plasmid-mediated	Marzan et al. (2017)
Oceanobacillus profundus	Pb, Zn	Biosorption	Abd El-Motaleb et al. (2020), Mwandira et al. (2020)
Ochrobactrum MT180101	Cu	Biosorption	Torres (2020)
Providencia alcalifaciens 2EA	Pb	Precipitation	Naik et al. (2013)
Providencia vermicola strain SJ2A	Pb	Bioaccumulation	Saleem et al. (2017)
Pseudomonas aeruginosa z	Cd, Cu, Hg, Pd, Zn	Biosorption, extracellular sequestration	Teitzel and Parsek (2003), Bhojiya and Joshi (2016), Chellaiah (2018), Hwang and Jho (2018), Kumari and Das (2019), Mat Arisah et al. (2021)
Pseudomonas aeruginosa ASU6a	Pb	Biosorption	Gabr et al. (2008)
Pseudomonas sp.	As, Cd, Co, Cr, Cu, Hg, Ni, Pb	Biosorption, enzymatic detoxification	Kumaran et al. (2011), Nanda et al. (2011), Azzam and Tawfik (2015), Giovanella et al. (2016), Naz et al. (2016)
Pseudomonas azotoformans JAW1	Cd, Pb	Biosorption	Choińska-Pulit et al. (2018), Orji et al. (2021)
Pseudomonas fluorescens	Cr	Biodegradation	Kalaimurugan et al. (2020)
Pseudomonas pseudoalcaligenes	Pb	Biosorption	Burakov et al. (2018)
Pseudomonas veronii	Cd, Cu, Zn	Biosorption	Vullo et al. (2008)
Rhodobacter capsulatus	Zn	Biosorption	Magnin et al. (2014)
Shewanella putrefaciens	Cd	Biosorption	Wang et al. (2019)
Sinorhizobium meliloti	Zn	Active export	Lu et al. (2016)
Sporosarcina saromensis (M52)	Cr	Biosorption	Ran et al. (2016)
Streptomyces rochei ANH	Cr	Biosorption	Hamdan et al. (2021)
Sulfolobus solfataricus	Cu	Active export, intracellular sequestration	Remonsellez et al. (2006), Soto et al. (2019)

(continued)

Table 8.2 (continued)

Microbe type	Species	Studied heavy metals	Microbial mechanism/survival strategy	References
	Synechococcus sp	Cd, Zn	Intracellular sequestration	Blindauer et al. (2008)
	Turbinaria vulgaris	Cr	Biosorption	Boddu et al. (2022)
	Variovorax boronicumulans	Cd, Pb, Zn		Jalilvand et al. (2020)
	Vibrio fluvialis	Hg	Biosorption	Saranya et al. (2017)
	Vibrio harveyi	Pb	Precipitation	Chen et al. (2016)
Fungi	Acremonium persicinum	Cu	Biosorption	Mohammadian et al. (2017)
	Aspergillus flavus	Cu	Bioaccumulation	Kanamarlapudi et al. (2018)
	Aspergillus fumigates	Cd, Cr, Pb	Biosorption	Kumar Ramasamy et al. (2011), Dhal and Pandey (2018), Talukdar et al. (2020), Priyadarshini et al. (2021)
	Aspergillus niger	Cd, Co, Cr, Cu, Hg, Pb	Biosorption	Tsekova et al. (2010), Iram et al. (2015), Acosta-Rodríguez et al. (2018), Cárdenas González et al. (2019), Priyadarshini et al. (2021)
	Aspergillus versicolor	Cr, Cu, Ni	Bioaccumulation	Taştan et al. (2010)
	Paecilomyces sp.	Co	Biosorption	Cárdenas González et al. (2019)
	Penicillium canescens	As, Cd, Hg, Pb	Biosorption	Say et al. (2003)
	Penicillium chrysogenum	Au	Biosorption	Niu and Volesky (2000)
	Penicillium chrysogenum CS1	Pb	Precipitation	Qian et al. (2017)
	Penicillium simplicissimum	Cu	Biosorption	Mohammadian et al. (2017)
	Penicillium sp.	Co	Biosorption	Cárdenas González et al. (2019)
	Saccharomyces cerevisiae	Cr	Bioaccumulation	Kanamarlapudi et al. (2018)
	Sterigmatomyces halophilus	Pb, Zn	Biosorption	Bano et al. (2018)
	Trichoderma brevicompactum QYCD-6	Cd, Cr, Cu, Pb	Bioaccumulation	Bano et al. (2018), Zhang et al. (2020)

Algae			
	Aphanothece sp.	Pb	Biosorption	Keryanti and Mulyono (2021)
	Arthrospira maxima	Fe	Biosorption	Blanco-Vieites et al. (2022)
	Chaetoceros sp.	Pb	Biosorption	Molazadeh et al. (2015)
	Chlamydomonas reinhardtii	Cd	Biosorption	Piña-Olavide et al. (2020)
	Chlorella kessleri	Pb	Biosorption	Sultana et al. (2020)
	Chlorella miniata	Cr	Biosorption	Han et al. (2006)
	Chlorella sorokiniana	Cd	Bioaccumulation	León-Vaz et al. (2021)
	Chlorella sp.	Cd, Pb	Biosorption	Molazadeh et al. (2015), Shen et al. (2018)
	Chlorella vulgaris	Hg, Pb	Biosorption	Solisio et al. (2019), Leong and Chang (2020)
	Mougeotia genuflexa	As	Biosorption	Leong and Chang (2020)
	Neochloris oleoabundans	Pb	Biosorption	Gu et al. (2021)
	Oscillatoria angustissima	Co, Cu, Zn	Biosorption	Mohapatra and Gupta (2005)
	Phormidium sp.	Pb	Biosorption	Das et al. (2016a)
	Pleurococcus sp.	Hg	Biosorption	Vela-García et al. (2019)
	Sargassum fluitans	Au	Biosorption	Niu and Volesky (2000)
	Scenedesmus obtusus XJ-15	Hg	Biosorption	Huang et al. (2019)
	Scenedesmus quadricauda	Cr	Biosorption	Daneshvar et al. (2019)
	Ulothrix cylindricum	As	Biosorption	Tuzen et al. (2009), Leong and Chang (2020)

consortia have demonstrated successful outcomes in metal biosorption and are recommended more significantly for in situ applications (Igiri et al. 2018).

8.5.2   Fungi

Fungi have also evolved unique physiological and molecular mechanisms to remediate HMs. Besides the production of metallothioneins and phytochelatins, metallotolerant fungi can also form symbiotic associations to enhance their metal tolerance in plants. These symbiotic interactions involve the exchange of nutrients and the production of metal chelators or enzymes that aid in metal detoxification. The presence of mycorrhizal associations can contribute to their survival in HM-polluted sites (Fagorzi et al. 2018; Riaz et al. 2021). The diversity of metallotolerant fungi is vast, including various taxonomic groups such as *Ascomycetes*, *Basidiomycetes*, and *Zygomycetes* (Coleine et al. 2022; Huang et al. 2022). Furthermore, some fungi possess the ability to transform metal ions into NPs with unique properties, a process known as biomineralization. These biologically synthesized NPs have applications in diverse fields such as medicine, electronics, and catalysis (Li et al. 2011; Srivastava et al. 2021). Fungi are also used as biosorbents for the removal of HMs. The adhesion of inorganic chemicals is influenced by both active and non-viable fungal cells, demonstrating their significant roles in this process (Kumar et al. 2015). Ayangbenro and Babalola (2017) reported that *Saccharomyces cerevisiae*, *Cephalosporium aphidicola*, and *Aspergillus parasitica* possess the ability to detoxify Pb, Cd, and Zn ions. Marine fungi have demonstrated the capacity to resist higher HM concentrations such as Cu and Pb through enzymatic mechanisms (Deshmukh et al. 2016). White-rot fungi, including *Bjerkandera adusta*, *Phanerochaete chrysosporium*, *Pleurotus* sp., and *Trametes versicolor*, employ various ligninolytic enzymes to transform diverse organic pollutants. Moreover, non-living fungal biomass, such as that of *Saccharomyces cerevisiae* and *Rhizopus oryzae*, can utilize adsorption mechanisms for converting Cr(VI) to Cr(III) which is less toxic by binding the $(Cr_2O_4)^{2-}$ to the $[NR_4]^+$ on the cell wall. Furthermore, the redox reaction that converts Cr (VI) to Cr (III) is also brought about by the dead biomass of *Aspergillus niger* (Park et al. 2005).

8.5.3   Algae

The use of algae for the bioremediation of HMs-polluted effluents is studied, with living algae exhibiting more complexity compared to dead algae. During the growth phase, living algae undergo intracellular absorption of HMs although sorption processes show significant variations depending on the specific growth phase. Furthermore, the environmental factors that affect the growth of algae significantly affect biosorption. On the contrary, dead algal biomass engages in extracellular absorption of HM ions on the cell membrane surface (Zeraatkar et al. 2016; Danouche et al. 2021). For instance, a study explored the potential of *Ulothrix cylindricum* for the detoxification of As(III) ions, as well as *Ulva lactuca* for the removal of Pb(II) and Cd(II) ions (Sarı and Tuzen 2008; Tuzen et al. 2009). Some metallotolerant algae are capable of synthesizing lipids and other metabolites under metal stress conditions. These metabolites can be harnessed for biofuel research, nutraceuticals, and pharmaceuticals, offering sustainable and renewable alternatives to conventional chemical synthesis (Singh et al. 2023).

8.6 Conclusion

Metallotolerant microbes have proven to be invaluable in addressing the challenges posed by heavily polluted soil. These microbes have developed diverse methods to adapt to the toxicity of HMs, enabling them to survive in heavily polluted sites such as mine tailings, polluted soils, and industrial wastewater. Their adaptability and versatility are evident in their colonization of extreme habitats, including hot springs, acidic mine drainage, and deep-sea hydrothermal vents. The implications of studying metallotolerant microbes extend to various fields, with bioremediation being a particularly promising application. These microbes can be exploited to effectively remediate and recover metals from polluted sites, offering an eco-friendly and cost-effective option to traditional remediation approaches. Moreover, their unique abilities to bind and transform metals hold promise for the synthesis or production of novel compounds.

While HM contamination poses risks not only to human health but also to microbial communities, recent research has shed light on the defense mechanisms employed by certain microbes to adapt to such adverse conditions. Understanding the genetic and physiological changes that occur in response to HM stress is significant in developing methods for metal detoxification and restoring contaminated sites. Efflux transport systems, aggregations, biofilm formation, and genetic manipulation are among the mechanisms being explored to enhance microbial resistance and remediation capabilities. To further advance the field, it is essential to continue investigating the molecular basis of the microbial response to HM stress. Research efforts should focus on understanding newer strategies and technologies, including genetically engineered microorganisms (GEMs) while addressing ethical, legal, and biosafety concerns. In addition, integrating biotechnological approaches and considering regulatory risk assessment can lead to the development of designed microbes with improved bioremediation potential and productivity under stress conditions.

The environmental microbiome plays a significant role in controlling biogeochemical cycles that impact soil structure, climate, and fertility. Therefore, microbe-assisted bioremediation should be given high priority due to the inherent functions and mechanisms possessed by microorganisms. Their use for the bioremediation of polluted sites, waste management, and sustainable agriculture offers a substantially effective, cheaper, and realistic solution for HM bioremediation. In the future, there is extensive potential for genetic engineering to enhance the competence of organisms involved in bioremediation. By combining physical, chemical, and biological techniques, scientists can explore a wide range of strategies to address HMs pollution comprehensively. Continued research and collaboration will pave the way for a future where the remediation of HM contamination is more effective and sustainable.

Acknowledgement The research was financially supported by the Strategic Academic Leadership Program of the Southern Federal University (Priority 2030).

References

Abbas G, Murtaza B, Bibi I et al (2018) Arsenic uptake, toxicity, detoxification, and speciation in plants: physiological, biochemical, and molecular aspects. Int J Environ Res Public Health 15:59

Abd El-Motaleb M, El-Sabbagh S, Mohamed W, Wafy K (2020) Biosorption of Cu2+, Pb2+ and Cd2+ from wastewater by dead biomass of *Streptomyces cyaneus* Kw42. Int J Curr Microbiol Appl Sci 9:422–435

Abdu N, Abdullahi AA, Abdulkadir A (2017) Heavy metals and soil microbes. Environ Chem Lett 15:65–84

Abraham J, Chatterjee A, Sharma J (2020) Isolation and characterization of *Bacillus licheniformis* Strain for bioleaching of heavy metals. J Appl Biotechnol Rep 7:139–144

Achal V, Pan X, Zhang D (2011) Remediation of copper-contaminated soil by *Kocuria flava* CR1, based on microbially induced calcite precipitation. Ecol Eng 37:1601–1605

Achal V, Pan X, Fu Q, Zhang D (2012) Biomineralization based remediation of As (III) contaminated soil by *Sporosarcina ginsengisoli*. J Hazard Mater 201:178–184

Acosta-Rodríguez I, Cárdenas-González JF, Rodríguez Pérez AS et al (2018) Bioremoval of different heavy metals by the resistant fungal strain *Aspergillus niger*. Bioinorg Chem Appl 2018:3457196

Adams GO, Fufeyin PT, Okoro SE, Ehinomen I (2015) Bioremediation, biostimulation and bio-augmentation: a review. Int J Environ Bioremediat Biodegrad 3:28–39

Ahmed SN, Baitharu I (2023) Potential of microbes for the remediation of heavy metals–contaminated soil. In: Integrative strategies for bioremediation of environmental contaminants. Academic Press, pp 31–47

Alboghobeish H, Tahmourespour A, Doudi M (2014) The study of Nickel Resistant Bacteria (NiRB) isolated from wastewaters polluted with different industrial sources. J Environ Health Sci Eng 12:1–7

Alloway BJ (2013) Sources of heavy metals and metalloids in soils. In: Heavy metals in soils: trace metals and metalloids in soils and their bioavailability. Springer Dordrecht, pp 11–50

Alviz-Gazitua P, Fuentes-Alburquenque S, Rojas LA et al (2019) The response of *Cupriavidus metallidurans* CH34 to cadmium involves inhibition of the initiation of biofilm formation, decrease in intracellular c-di-GMP levels, and a novel metal regulated phosphodiesterase. Front Microbiol 10:1499

Ameen FA, Hamdan AM, El-Naggar MY (2020) Assessment of the heavy metal bioremediation efficiency of the novel marine lactic acid bacterium, *Lactobacillus plantarum* MF042018. Sci Rep 10:314

Aronson JK (2006) Meyler's side effects of drugs: the international encyclopedia of adverse drug reactions and interactions. Elsevier

Ayangbenro AS, Babalola OO (2017) A new strategy for heavy metal polluted environments: a review of microbial biosorbents. Int J Environ Res Public Health 14:94

Azzam AM, Tawfik A (2015) Removal of heavy metals using bacterial bio-flocculants of *Bacillus* sp. and *Pseudomonas* sp. J Environ Eng Landsc Manag 23:288–294

Badar U, Ahmed N, Beswick A J et al (2000) Reduction of chromate by microorganisms isolated from metal contaminated sites of Karachi, Pakistan. Biotechnol Lett 22:829–836

Baaziz H, Gambari C, Boyeldieu A et al (2017) ChrASO, the chromate efflux pump of *Shewanella oneidensis*, improves chromate survival and reduction. PLoS One 12:e0188516

Balzano S, Sardo A, Blasio M et al (2020) Microalgal metallothioneins and phytochelatins and their potential use in bioremediation. Front Microbiol 11:517

Bano A, Hussain J, Akbar A et al (2018) Biosorption of heavy metals by obligate halophilic fungi. Chemosphere 199:218–222

Barman D, Jha DK (2021) Metallotolerant microorganisms and microbe-assisted phytoremediation for a sustainable clean environment. In: Microbes in microbial communities: ecological and applied perspectives. Springer, Singapore, pp 307–336

Barman D, Jha DK, Bhattacharjee K (2020) Metallotolerant bacteria: insights into bacteria thriving in metal-contaminated areas. In: Microbial versatility in varied environments: microbes in sensitive environments. Springer, Singapore, pp 135–164

Bharagava RN, Mishra S (2018) Hexavalent chromium reduction potential of *Cellulosimicrobium* sp. isolated from common effluent treatment plant of tannery industries. Ecotoxicol Environ Saf 147:102–109

Bharti R, Sharma R (2022) Effect of heavy metals: an overview. Mater Today Proc 51:880–885

Bhati T, Gupta R, Yadav N et al (2019) Assessment of bioremediation potential of *Cellulosimicrobium* sp. for treatment of multiple heavy metals. Microbiol Biotechnol Lett 47:269–277

Bhattacharya A, Gupta A (2013) Evaluation of *Acinetobacter* sp. B9 for Cr (VI) resistance and detoxification with potential application in bioremediation of heavy-metals-rich industrial wastewater. Environ Sci Pollut Res 20:6628–6637

Bhattacharya A, Gupta A, Kaur A, Malik D (2014) Efficacy of *Acinetobacter* sp. B9 for simultaneous removal of phenol and hexavalent chromium from co-contaminated system. Appl Microbiol Biotechnol 98:9829–9841

Bhojiya AA, Joshi H (2016) Study of potential plant growth-promoting activities and heavy metal tolerance of *Pseudomonas aeruginosa* HMR16 isolated from Zawar, Udaipur, India. Curr Trends Biotechnol Pharm 10:161–168

Bissen M, Frimmel FH (2003) Arsenic—a review. Part I: occurrence, toxicity, speciation, mobility. Acta Hydrochim Hydrobiol 31:9–18

Blanco-Vieites M, Suárez-Montes D, Delgado F et al (2022) Removal of heavy metals and hydrocarbons by microalgae from wastewater in the steel industry. Algal Res 64:102700

Blindauer C, Harrison M, Parkinson J et al (2008) Isostructural replacement of zinc by cadmium in bacterial metallothionein. Metal Ions Biol Med 10:167–173

Boddu S, Alugunulla VN, Dulla JB et al (2022) Estimation of biosorption characteristics of chromium (VI) from aqueous and real tannery effluents by treated T. vulgaris: experimental assessment and statistical modelling. Int J Environ Anal Chem 102:4842–4861

Burakov AE, Galunin EV, Burakova IV et al (2018) Adsorption of heavy metals on conventional and nanostructured materials for wastewater treatment purposes: a review. Ecotoxicol Environ Saf 148:702–712

Cárdenas González JF, Rodríguez Pérez AS, Vargas Morales JM et al (2019) Bioremoval of cobalt (II) from aqueous solution by three different and resistant fungal biomasses. Bioinorg Chem Appl 2019:1

Cervantes C, Campos-García J, Devars S et al (2001) Interactions of chromium with microorganisms and plants. FEMS Microbiol Rev 25:335–347

Cha J-S, Cooksey DA (1991) Copper resistance in *Pseudomonas syringae* mediated by periplasmic and outer membrane proteins. Proc Natl Acad Sci 88:8915–8919

Chatterjee A, Das R, Abraham J (2020) Bioleaching of heavy metals from spent batteries using *Aspergillus nomius* JAMK1. Int J Environ Sci Technol 17:49–66

Chaudhary A, Shirodkar S, Sharma A (2017) Characterization of nickel tolerant bacteria isolated from heavy metal polluted glass industry for its potential role in bioremediation. Soil Sediment Contam Int J 26:184–194

Chellaiah ER (2018) Cadmium (heavy metals) bioremediation by *Pseudomonas aeruginosa*: a minireview. Appl Water Sci 8:154

Chen Z, Pan X, Chen H et al (2016) Biomineralization of Pb (II) into Pb-hydroxyapatite induced by *Bacillus cereus* 12-2 isolated from Lead–Zinc mine tailings. J Hazard Mater 301:531–537

Chen SC, Lin WH, Chien CC et al (2018) Development of a two-stage biotransformation system for mercury-contaminated soil remediation. Chemosphere 200:266–273

Chibuike GU, Obiora SC (2014) Heavy metal polluted soils: effect on plants and bioremediation methods. Appl Environ Soil Sci 2014:2014

Choińska-Pulit A, Sobolczyk-Bednarek J, Łaba W (2018) Optimization of copper, lead and cadmium biosorption onto newly isolated bacterium using a Box-Behnken design. Ecotoxicol Environ Saf 149:275–283

Choudhury S, Chatterjee A (2022) Microbial application in remediation of heavy metals: an overview. Arch Microbiol 204:268

Chu D (2018) Effects of heavy metals on soil microbial community. In: IOP conference series: earth and environmental science. IOP Publishing, p 012009

Clemens S, Simm C (2003) *Schizosaccharomyces pombe* as a model for metal homeostasis in plant cells: the phytochelatin-dependent pathway is the main cadmium detoxification mechanism. New Phytol 159:323–330

Coleine C, Stajich JE, Selbmann L (2022) Fungi are key players in extreme ecosystems. Trends Ecol Evol 37:517

Cui W, Li X, Duan W et al (2023) Heavy metal stabilization remediation in polluted soils with stabilizing materials: a review. Environ Geochem Health 45:1–37

Čvančarová M, Shahgaldian P, Corvini PF-X (2020) Enzyme-based nanomaterials in bioremediation. In: Advanced nano-bio technologies for water and soil treatment. Springer, Cham, pp 345–372

Daneshvar E, Zarrinmehr MJ, Kousha M et al (2019) Hexavalent chromium removal from water by microalgal-based materials: adsorption, desorption and recovery studies. Bioresour Technol 293:122064

Danouche M, El Ghachtouli N, El Aroussi H (2021) Phycoremediation mechanisms of heavy metals using living green microalgae: physicochemical and molecular approaches for enhancing selectivity and removal capacity. Heliyon 7:e07609

Das P, Samantaray S, Rout GR (1997) Studies on cadmium toxicity in plants: a review. Environ Pollut 98:29–36

Das D, Chakraborty S, Bhattacharjee C, Chowdhury R (2016a) Biosorption of lead ions (Pb2+) from simulated wastewater using residual biomass of microalgae. Desalination Water Treat 57:4576–4586

Das S, Dash HR, Chakraborty J (2016b) Genetic basis and importance of metal resistant genes in bacteria for bioremediation of contaminated environments with toxic metal pollutants. Appl Microbiol Biotechnol 100:2967–2984

Dave D, Sarma S, Parmar P et al (2020) Microbes as a boon for the bane of heavy metals. Environ Sustain 3:233–255

De S, Ghosh S (2018) Potential risks associated with bacterial strains isolated from heavy metal rich soil of a landfill area. Int Res J Eng Technol (IRJET) 5:355–357

Deshmukh R, Khardenavis AA, Purohit HJ (2016) Diverse metabolic capacities of fungi for bioremediation. Indian J Microbiol 56:247–264

Dhal B, Pandey BD (2018) Mechanism elucidation and adsorbent characterization for removal of Cr (VI) by native fungal adsorbent. Sustain Environ Res 28:289–297

Dixit R, Wasiullah X, Malaviya D et al (2015) Bioremediation of heavy metals from soil and aquatic environment: an overview of principles and criteria of fundamental processes. Sustainability 7:2189–2212

Dong G, Wang Y, Gong L et al (2013) Formation of soluble Cr (III) end-products and nanoparticles during Cr (VI) reduction by *Bacillus cereus* strain XMCr-6. Biochem Eng J 70:166–172

Enya O, Heaney N, Iniama G, Lin C (2020) Effects of heavy metals on organic matter decomposition in inundated soils: microcosm experiment and field examination. Sci Total Environ 724:138223

Esringü A, Turan M, Güneş A, Karaman MR (2014) Roles of *Bacillus megaterium* in remediation of boron, lead, and cadmium from contaminated soil. Commun Soil Sci Plant Anal 45:1741–1759

Fagorzi C, Checcucci A, DiCenzo GC et al (2018) Harnessing rhizobia to improve heavy-metal phytoremediation by legumes. Genes (Basel) 9:542

Fashola MO, Ngole-Jeme VM, Babalola OO (2016) Heavy metal pollution from gold mines: environmental effects and bacterial strategies for resistance. Int J Environ Res Public Health 13:1047

Franke S, Grass G, Rensing C, Nies DH (2003) Molecular analysis of the copper-transporting efflux system CusCFBA of *Escherichia coli*. J Bacteriol 185:3804–3812

Gabr RM, Hassan SHA, Shoreit AAM (2008) Biosorption of lead and nickel by living and non-living cells of *Pseudomonas aeruginosa* ASU 6a. Int Biodeterior Biodegradation 62:195–203

Ghosh A, Pramanik K, Bhattacharya S et al (2022) A potent cadmium bioaccumulating *Enterobacter cloacae* strain displays phytobeneficial property in Cd-exposed rice seedlings. Curr Res Microb Sci 3:100101

Gilotra U, Srivastava S (1997) Plasmid-encoded sequestration of copper by *Pseudomonas pickettii* strain US321. Curr Microbiol 34:378–381

Giovanella P, Cabral L, Bento FM et al (2016) Mercury (II) removal by resistant bacterial isolates and mercuric (II) reductase activity in a new strain of *Pseudomonas* sp. B50A. New Biotechnol 33:216–223

Grujić S, Vasić S, Radojević I et al (2017) Comparison of the *Rhodotorula mucilaginosa* biofilm and planktonic culture on heavy metal susceptibility and removal potential. Water Air Soil Pollut 228:1–8

Gu S, Boase EM, Lan CQ (2021) Enhanced Pb (II) removal by green alga *Neochloris oleoabundans* cultivated in high dissolved inorganic carbon cultures. Chem Eng J 416:128983

Hamdan AM, Abd-El-Mageed H, Ghanem N (2021) Biological treatment of hazardous heavy metals by *Streptomyces rochei* ANH for sustainable water management in agriculture. Sci Rep 11:9314

Han X, Wong YS, Tam NFY (2006) Surface complexation mechanism and modeling in Cr (III) biosorption by a microalgal isolate, *Chlorella miniata*. J Colloid Interface Sci 303:365–371

Hansda A, Kumar V, Anshumali (2016) A comparative review towards potential of microbial cells for heavy metal removal with emphasis on biosorption and bioaccumulation. World J Microbiol Biotechnol 32:1–14

Heidari P, Panico A (2020) Sorption mechanism and optimization study for the bioremediation of Pb (II) and Cd (II) contamination by two novel isolated strains Q3 and Q5 of *Bacillus* sp. Int J Environ Res Public Health 17:4059

Hemmat-Jou MH, Safari-Sinegani AA, Mirzaie-Asl A, Tahmourespour A (2018) Analysis of microbial communities in heavy metals-contaminated soils using the metagenomic approach. Ecotoxicology 27:1281–1291

Hlihor RM, Figueiredo H, Tavares T, Gavrilescu M (2017) Biosorption potential of dead and living *Arthrobacter viscosus* biomass in the removal of Cr (VI): batch and column studies. Process Saf Environ Prot 108:44–56

Huang R, Huo G, Song S et al (2019) Immobilization of mercury using high-phosphate culture-modified microalgae. Environ Pollut 254:112966

Huang W-L, Wu P-C, Chiang T-Y (2022) Metagenomics: potential for bioremediation of soil contaminated with heavy metals. Ecol Genet Genom 22:100111

Hwang SK, Jho EH (2018) Heavy metal and sulfate removal from sulfate-rich synthetic mine drainages using sulfate reducing bacteria. Sci Total Environ 635:1308–1316

Ianeva OD (2009) Mechanisms of bacteria resistance to heavy metals. Mikrobiol Z 71:54–65

Igiri BE, Okoduwa SIR, Idoko GO et al (2018) Toxicity and bioremediation of heavy metals contaminated ecosystem from tannery wastewater: a review. J Toxicol 2018:2568038

Inobeme A (2021) Effect of heavy metals on activities of soil microorganism. In: Microbial rejuvenation of polluted environment: volume 3. Springer, Singapore, pp 115–142

Iram S, Shabbir R, Zafar H, Javaid M (2015) Biosorption and bioaccumulation of copper and lead by heavy metal-resistant fungal isolates. Arab J Sci Eng 40:1867–1873

Irawati W, Soraya Y, Baskoro AH (2012) A study on mercury-resistant bacteria isolated from a gold mine in Pongkor Village, Bogor, Indonesia. Hayati 19:197–200

Jalilvand N, Akhgar A, Alikhani HA et al (2020) Removal of heavy metals zinc, lead, and cadmium by biomineralization of urease-producing bacteria isolated from Iranian mine calcareous soils. J Soil Sci Plant Nutr 20:206–219

Jan AT, Azam M, Siddiqui K et al (2015) Heavy metals and human health: mechanistic insight into toxicity and counter defense system of antioxidants. Int J Mol Sci 16:29592–29630

Jarosławiecka A, Piotrowska-Seget Z (2014) Lead resistance in micro-organisms. Microbiology (N Y) 160:12–25

Javanbakht V, Alavi SA, Zilouei H (2014) Mechanisms of heavy metal removal using microorganisms as biosorbent. Water Sci Technol 69:1775–1787

Jayaram S, Ayyasamy PM, Aiswarya KP et al (2022) Mechanism of microbial detoxification of heavy metals: a review. J Pure Appl Microbiol 16:1562

Jeyakumar P, Debnath C, Vijayaraghavan R, Muthuraj M (2023) Trends in bioremediation of heavy metal contaminations. Environ Eng Res 28:220631

Jiang J, Liu H, Li Q et al (2015) Combined remediation of Cd–phenanthrene co-contaminated soil by *Pleurotus cornucopiae* and *Bacillus thuringiensis* FQ1 and the antioxidant responses in *Pleurotus cornucopiae*. Ecotoxicol Environ Saf 120:386–393

Jin Y, Luan Y, Ning Y, Wang L (2018) Effects and mechanisms of microbial remediation of heavy metals in soil: a critical review. Appl Sci 8:1336

Joo JO, Choi J-H, Kim IH et al (2015) Effective bioremediation of Cadmium (II), nickel (II), and chromium (VI) in a marine environment by using *Desulfovibrio desulfuricans*. Biotechnol Bioprocess Eng 20:937–941

Joshi N, Filip J, Coker VS et al (2018) Microbial reduction of natural Fe (III) minerals; toward the sustainable production of functional magnetic nanoparticles. Front Environ Sci 6:127

Kalaimurugan D, Balamuralikrishnan B, Durairaj K et al (2020) Isolation and characterization of heavy-metal-resistant bacteria and their applications in environmental bioremediation. Int J Environ Sci Technol 17:1455–1462

Kanamarlapudi S, Chintalpudi VK, Muddada S (2018) Application of biosorption for removal of heavy metals from wastewater. Biosorption 18:70–116

Kaushik S, Alatawi A, Djiwanti SR et al (2021) Potential of extremophiles for bioremediation. In: Microbial rejuvenation of polluted environment: volume 1. Springer, Singapore, pp 293–328

Kazy SK, Sar P, Singh SP et al (2002) Extracellular polysaccharides of a copper-sensitive and a copper-resistant Pseudomonas aeruginosa strain: synthesis, chemical nature and copper binding. World J Microbiol Biotechnol 18:583–588

Keith S, Faroon O, Roney N (2013) Toxicological Profile for Uranium. Atlanta (GA): Agency for Toxic Substances and Disease Registry (US); Health Effects

Keryanti K, Mulyono EWS (2021) Determination of optimum condition of Lead (Pb) biosorption using dried biomass microalgae *Aphanothece* sp. Period Polytech Chem Eng 65:116–123

Keyhani S, Lopez JL, Clark DS, Keasling JD (1996) Intracellular polyphosphate content and cadmium tolerance in *Anacystis nidulans* R2. Microbios 88:105–114

Kieu HTQ, Müller E, Horn H (2011) Heavy metal removal in anaerobic semi-continuous stirred tank reactors by a consortium of sulfate-reducing bacteria. Water Res 45:3863–3870

Kuivenhoven M, Mason K (2019) Arsenic (arsine) toxicity. StatPearls [Internet] StatPearls Publishing

Kumar A, Prasad MNV (2018) Plant-lead interactions: transport, toxicity, tolerance, and detoxification mechanisms. Ecotoxicol Environ Saf 166:401–418

Kumar Ramasamy R, Congeevaram S, Thamaraiselvi K (2011) Evaluation of isolated fungal strain from e-waste recycling facility for effective sorption of toxic heavy metal Pb (II) ions and fungal protein molecular characterization—a mycoremediation approach. Asian J Exp Biol Sci 2:342–347

Kumar KS, Dahms H-U, Won E-J et al (2015) Microalgae–a promising tool for heavy metal remediation. Ecotoxicol Environ Saf 113:329–352

Kumaran NS, Sundaramanicam A, Bragadeeswaran S (2011) Adsorption studies on heavy metals by isolated cyanobacterial strain (*Nostoc* sp.) from Uppanar estuarine water, southeast coast of India. J Appl Sci Res 7:1609–1615

Kumari S, Das S (2019) Expression of metallothionein encoding gene bmtA in biofilm-forming marine bacterium *Pseudomonas aeruginosa* N6P6 and understanding its involvement in Pb (II) resistance and bioremediation. Environ Sci Pollut Res 26:28763–28774

Kushwaha A, Rani R, Kumar S et al (2017) A new insight to adsorption and accumulation of high lead concentration by exopolymer and whole cells of lead-resistant bacterium *Acinetobacter junii* L. Pb1 isolated from coal mine dump. Environ Sci Pollut Res 24:10652–10661

Kwiatkowska-Malina J (2018) Functions of organic matter in polluted soils: the effect of organic amendments on phytoavailability of heavy metals. Appl Soil Ecol 123:542–545

Leong YK, Chang J-S (2020) Bioremediation of heavy metals using microalgae: recent advances and mechanisms. Bioresour Technol 303:122886

León-Vaz A, León R, Giráldez I et al (2021) Impact of heavy metals in the microalga *Chlorella sorokiniana* and assessment of its potential use in cadmium bioremediation. Aquat Toxicol 239:105941

Lerebours A, To VV, Bourdineaud J (2016) Danio rerio ABC transporter genes abcb3 and abcb7 play a protecting role against metal contamination. J Appl Toxicol 36:1551–1557

Li X, Xu H, Chen Z-S, Chen G (2011) Biosynthesis of nanoparticles by microorganisms and their applications. J Nanomater 2011:1–16

Li F, Zheng Y, Tian J et al (2019) *Cupriavidus* sp. strain Cd02-mediated pH increase favoring bioprecipitation of Cd2+ in medium and reduction of cadmium bioavailability in paddy soil. Ecotoxicol Environ Saf 184:109655

Liu X, Hu X, Cao Y et al (2019) Biodegradation of phenanthrene and heavy metal removal by acid-tolerant *Burkholderia fungorum* FM-2. Front Microbiol 10:408

Liu P, Zhang Y, Tang Q, Shi S (2021) Bioremediation of metal-contaminated soils by microbially-induced carbonate precipitation and its effects on ecotoxicity and long-term stability. Biochem Eng J 166:107856

Loaëc M, Olier R, Guezennec J (1997) Uptake of lead, cadmium and zinc by a novel bacterial exopolysaccharide. Water Res 31:1171–1179

Lu M, Li Z, Liang J et al (2016) Zinc resistance mechanisms of P1B-type ATPases in *Sinorhizobium meliloti* CCNWSX0020. Sci Rep 6:29355

Magnin J-P, Gondrexon N, Willison JC (2014) Zinc biosorption by the purple non-sulfur bacterium *Rhodobacter capsulatus*. Can J Microbiol 60:829–837

Malik A (2004) Metal bioremediation through growing cells. Environ Int 30:261–278

Maqsood Q, Sumrin A, Waseem R et al (2023) Bioengineered microbial strains for detoxification of toxic environmental pollutants. Environ Res 227:115665

Martinez RJ, Beazley MJ, Taillefert M et al (2007) Aerobic uranium (VI) bioprecipitation by metal-resistant bacteria isolated from radionuclide-and metal-contaminated subsurface soils. Environ Microbiol 9:3122–3133

Marzan LW, Hossain M, Mina SA et al (2017) Isolation and biochemical characterization of heavy-metal resistant bacteria from tannery effluent in Chittagong city, Bangladesh: bioremediation viewpoint. Egypt J Aquat Res 43:65–74

Mat Arisah F, Amir AF, Ramli N et al (2021) Bacterial resistance against heavy metals in *Pseudomonas aeruginosa* RW9 involving hexavalent chromium removal. Sustainability 13:9797

Mathivanan K, Chandirika JU, Mathimani T et al (2021) Production and functionality of exopolysaccharides in bacteria exposed to a toxic metal environment. Ecotoxicol Environ Saf 208:111567

Maynaud G, Brunel B, Yashiro E et al (2014) CadA of *Mesorhizobium metallidurans* isolated from a zinc-rich mining soil is a PIB-2-type ATPase involved in cadmium and zinc resistance. Res Microbiol 165:175–189

Mishra S, Doble M (2008) Novel chromium tolerant microorganisms: isolation, characterization and their biosorption capacity. Ecotoxicol Environ Saf 71:874–879

Mishra J, Singh R, Arora NK (2017) Alleviation of heavy metal stress in plants and remediation of soil by rhizosphere microorganisms. Front Microbiol 8:1706

Mitra S, Pramanik K, Sarkar A et al (2018) Bioaccumulation of cadmium by Enterobacter sp. and enhancement of rice seedling growth under cadmium stress. Ecotoxicol Environ Saf 156:183–196

Mohammadian E, Ahari AB, Arzanlou M et al (2017) Tolerance to heavy metals in filamentous fungi isolated from contaminated mining soils in the Zanjan Province, Iran. Chemosphere 185:290–296

Mohapatra H, Gupta R (2005) Concurrent sorption of Zn (II), Cu (II) and Co (II) by *Oscillatoria angustissima* as a function of pH in binary and ternary metal solutions. Bioresour Technol 96:1387–1398

Molazadeh P, Khanjani N, Rahimi MR, Nasiri A (2015) Adsorption of lead by microalgae *Chaetoceros* sp. and *Chlorella* sp. from aqueous solution. J Community Health Res 4:114–127

Mosa KA, Saadoun I, Kumar K et al (2016) Potential biotechnological strategies for the cleanup of heavy metals and metalloids. Front Plant Sci 7:303

Mukhopadhyay R, Rosen BP, Phung LT, Silver S (2002) Microbial arsenic: from geocycles to genes and enzymes. FEMS Microbiol Rev 26:311–325

Muñoz AJ, Espínola F, Moya M, Ruiz E (2015) Biosorption of Pb (II) ions by *Klebsiella* sp. 3S1 isolated from a wastewater treatment plant: kinetics and mechanisms studies. Biomed Res Int 2015:719060

Murthy S, Bali G, Sarangi SK (2011) Effect of lead on metallothionein concentration in lead-resistant bacteria *Bacillus cereus* isolated from industrial effluent. Afr J Biotechnol 10:15966–15972

Mwandira W, Nakashima K, Kawasaki S et al (2020) Biosorption of Pb (II) and Zn (II) from aqueous solution by *Oceanobacillus profundus* isolated from an abandoned mine. Sci Rep 10:21189

Nagvenkar GS, Ramaiah N (2010) Arsenite tolerance and biotransformation potential in estuarine bacteria. Ecotoxicology 19:604–613

Naik MM, Pandey A, Dubey SK (2012) Biological characterization of lead-enhanced exopolysaccharide produced by a lead resistant *Enterobacter cloacae* strain P2B. Biodegradation 23:775–783

Naik MM, Khanolkar D, Dubey SK (2013) Lead-resistant *Providencia alcalifaciens* strain 2EA bioprecipitates Pb+ 2 as lead phosphate. Lett Appl Microbiol 56:99–104

Nanda M, Sharma D, Kumar A (2011) Isolation and characterization of bacteria resistant to heavy metals cadmium (Cd), arsenic (As), mercury (Hg) from industrial effluent. Global J Appl Environ Sci 2:127–132

Narayanan M, Ma Y (2023) Metal tolerance mechanisms in plants and microbe-mediated bioremediation. Environ Res 222:115413

Nayak AK, Panda SS, Basu A, Dhal NK (2018) Enhancement of toxic Cr (VI), Fe, and other heavy metals phytoremediation by the synergistic combination of native *Bacillus cereus* strain and Vetiveria zizanioides L. Int J Phytoremediation 20:682–691

Naz T, Khan MD, Ahmed I et al (2016) Biosorption of heavy metals by *Pseudomonas* species isolated from sugar industry. Toxicol Ind Health 32:1619–1627

Nicholls AM, Mal TK (2003) Effects of lead and copper exposure on growth of an invasive weed, *Lythrum salicaria* L.(Purple Loosestrife). Ohio J Sci 103(5):129–133

Nies DH (2000) Microbial heavy-metal resistance. Appl Microbiol Biotechnol 51:451–460

Nies DH, Nies A, Chu L, Silver S (1989) Expression and nucleotide sequence of a plasmid-determined divalent cation efflux system from *Alcaligenes eutrophus*. Proc Natl Acad Sci 86:7351–7355

Niu H, Volesky B (2000) Gold-cyanide biosorption with L-cysteine. J Chem Technol Biotechnol 75:436–442

Nivetha N, Srivarshine B, Sowmya B et al (2022) A comprehensive review on bio-stimulation and bio-enhancement towards remediation of heavy metals degeneration. Chemosphere 312:137099

Nnaji ND, Onyeaka H, Miri T, Ugwa C (2023) Bioaccumulation for heavy metal removal: a review. SN Appl Sci 5:125

Noroozi M, Amoozegar MA, Pourbabaei AA et al (2017) Isolation and characterization of mercuric reductase by newly isolated halophilic bacterium, *Bacillus firmus* MN8. Glob J Environ Sci Manag 3:427–436

Nriagu JO (2019) Encyclopedia of environmental health. Elsevier

Ojuederie OB, Babalola OO (2017) Microbial and plant-assisted bioremediation of heavy metal polluted environments: a review. Int J Environ Res Public Health 14:1504

Ontañon OM, Fernandez M, Agostini E, González PS (2018) Identification of the main mechanisms involved in the tolerance and bioremediation of Cr (VI) by *Bacillus* sp. SFC 500-1E. Environ Sci Pollut Res 25:16111–16120

Orji OU, Awoke JN, Aja PM et al (2021) Halotolerant and metalotolerant bacteria strains with heavy metals biorestoration possibilities isolated from Uburu Salt Lake, Southeastern, Nigeria. Heliyon 7(7)

Osmani M, Bani A, Hoxha B (2015) Heavy metals and Ni phytoextractionin in the metallurgical area soils in Elbasan. Albanian J Agric Sci 14:414

Oves M, Khan MS, Qari HA (2017) *Ensifer adhaerens* for heavy metal bioaccumulation, biosorption, and phosphate solubilization under metal stress condition. J Taiwan Inst Chem Eng 80:540–552

Panda J, Sarkar P (2012) Isolation and identification of chromium-resistant bacteria: test application for prevention of chromium toxicity in plant. J Environ Sci Health A 47:237–244

Pande V, Pandey SC, Sati D et al (2022) Microbial interventions in bioremediation of heavy metal contaminants in agroecosystem. Front Microbiol 13:824084

Park D, Yun Y-S, Jo JH, Park JM (2005) Mechanism of hexavalent chromium removal by dead fungal biomass of *Aspergillus niger*. Water Res 39:533–540

Park JH, Bolan N, Megharaj M, Naidu R (2011) Concomitant rock phosphate dissolution and lead immobilization by phosphate solubilizing bacteria (*Enterobacter* sp.). J Environ Manag 92:1115–1120

Paul A, Mukherjee SK (2016) Enterobacter asburiae KUNi5, a nickel resistant bacterium for possible bioremediation of nickel contaminated sites. Pol. J Microbiol 65:115

Percival SL, Bowler PG, Russell D (2005) Bacterial resistance to silver in wound care. J Hosp Infect 60:1–7

Pikuta EV, Hoover RB, Tang J (2007) Microbial extremophiles at the limits of life. Crit Rev Microbiol 33:183–209

Piña-Olavide R, Paz-Maldonado LMT, Alfaro-De La Torre MC et al (2020) Increased removal of cadmium by *Chlamydomonas reinhardtii* modified with a synthetic gene for γ-glutamylcysteine synthetase. Int J Phytoremediation 22:1269–1277

Prabhakaran P, Ashraf MA, Aqma WS (2016) Microbial stress response to heavy metals in the environment. RSC Adv 6:109862–109877

Prabhu S, Poulose EK (2012) Silver nanoparticles: mechanism of antimicrobial action, synthesis, medical applications, and toxicity effects. Int Nano Lett 2:1–10

Priyadarshini E, Priyadarshini SS, Cousins BG, Pradhan N (2021) Metal-fungus interaction: review on cellular processes underlying heavy metal detoxification and synthesis of metal nanoparticles. Chemosphere 274:129976

Qian H, Peng X, Han X et al (2013) Comparison of the toxicity of silver nanoparticles and silver ions on the growth of terrestrial plant model *Arabidopsis thaliana*. J Environ Sci 25:1947–1956

Qian X, Fang C, Huang M, Achal V (2017) Characterization of fungal-mediated carbonate precipitation in the biomineralization of chromate and lead from an aqueous solution and soil. J Clean Prod 164:198–208

Qiao W, Zhang Y, Xia H et al (2019) Bioimmobilization of lead by *Bacillus subtilis* X3 biomass isolated from lead mine soil under promotion of multiple adsorption mechanisms. R Soc Open Sci 6:181701

Rahman Z, Singh VP (2020) Bioremediation of toxic heavy metals (THMs) contaminated sites: concepts, applications and challenges. Environ Sci Pollut Res 27:27563–27581

Ramírez V, Baez A, López P et al (2019) Chromium hyper-tolerant Bacillus sp. MH778713 assists phytoremediation of heavy metals by mesquite trees (*Prosopis laevigata*). Front Microbiol 10:1833

Ran Z, Bi W, Cai QT et al (2016) Bioremediation of hexavalent chromium pollution by *Sporosarcina saromensis* M52 isolated from offshore sediments in Xiamen, China. Biomed Environ Sci 29:127–136

Ranquet C, Ollagnier-de-Choudens S, Loiseau L et al (2007) Cobalt stress in *Escherichia coli*: the effect on the iron-sulfur proteins. J Biol Chem 282:30442–30451

Rasulov BA, Yili A, Aisa HA (2013) Biosorption of metal ions by exopolysaccharide produced by *Azotobacter chroococcum* XU1 J Environ Prot :989–993

Rehman AU, Nazir S, Irshad R et al (2021) Toxicity of heavy metals in plants and animals and their uptake by magnetic iron oxide nanoparticles. J Mol Liq 321:114455

Reith F, Rogers SL, McPhail DC, Webb D (2006) Biomineralization of gold: biofilms on bacterioform gold. Science (1979) 313:233–236

Remonsellez F, Orell A, Jerez CA (2006) Copper tolerance of the thermoacidophilic archaeon *Sulfolobus metallicus*: possible role of polyphosphate metabolism. Microbiology (N Y) 152:59–66

Revis NW, Osborne TR, Holdsworth G, Hadden C (1990) Mercury in soil: a method for assessing acceptable limits. Arch Environ Contam Toxicol 19:221–226

Riaz M, Kamran M, Fang Y et al (2021) Arbuscular mycorrhizal fungi-induced mitigation of heavy metal phytotoxicity in metal contaminated soils: a critical review. J Hazard Mater 402:123919

Rizvi A, Khan MS (2018) Heavy metal induced oxidative damage and root morphology alterations of maize (*Zea mays* L.) plants and stress mitigation by metal tolerant nitrogen fixing *Azotobacter chroococcum*. Ecotoxicol Environ Saf 157:9–20

Rosen BP (2002) Biochemistry of arsenic detoxification. FEBS Lett 529:86–92

Saha M, Sarkar S, Sarkar B et al (2016) Microbial siderophores and their potential applications: a review. Environ Sci Pollut Res 23:3984–3999

Saharan BS, Chaudhary T, Mandal BS et al (2023) Microbe-plant interactions targeting metal stress: new dimensions for bioremediation applications. J Xenobiot 13:252–269

Saleem M, Asghar HN, Ahmad W et al (2017) Prospects of bacterial-assisted remediation of metal-contaminated soils. In: Agro-environmental sustainability: volume 2: managing environmental pollution. Springer, Cham, pp 41–58

Samarth DP, Chandekar CJ, Bhadekar RK (2012) Biosorption of heavy metals from aqueous solution using *Bacillus licheniformis*. Int J Pure Appl Sci Technol 10:12

Sankarammal M, Thatheyus A, Ramya D (2014) Bioremoval of cadmium using *Pseudomonas fluorescens*. Open J Water Pollut Treat 1:92–100

Saranya K, Sundaramanickam A, Shekhar S et al (2017) Bioremediation of mercury by *Vibrio fluvialis* screened from industrial effluents. Biomed Res Int 2017:6509648

Saranya K, Sundaramanickam A, Shekhar S, Swaminathan S (2019) Biosorption of mercury by *Bacillus thuringiensis* (CASKS3) isolated from mangrove sediments of southeast coast India. Indian J Geo Mar Sci 48:143–150

Sardar K, Ali S, Hameed S et al (2013) Heavy metals contamination and what are the impacts on living organisms. Greener J Environ Manage Public Safety 2:172–179

Sarı A, Tuzen M (2008) Biosorption of cadmium (II) from aqueous solution by red algae (*Ceramium virgatum*): equilibrium, kinetic and thermodynamic studies. J Hazard Mater 157:448–454

Sarker A, Kim J-E, Islam ARMT et al (2022) Heavy metals contamination and associated health risks in food webs—a review focuses on food safety and environmental sustainability in Bangladesh. Environ Sci Pollut Res 29:3230–3245

Sarker A, Al Masud MA, Deepo DM et al (2023) Biological and green remediation of heavy metal contaminated water and soils: a state-of-the-art review. Chemosphere 138861:138861

Sarret G, Avoscan L, Carrière M et al (2005) Chemical forms of selenium in the metal-resistant bacterium *Ralstonia metallidurans* CH34 exposed to selenite and selenate. Appl Environ Microbiol 71:2331–2337

Saxena D, Srivastava S (1998) Carbon source-starvation-induced precipitation of copper by *Pseudomonas putida* strain S4. World J Microbiol Biotechnol 14:921–923

Say R, Yilmaz N, Denizli A (2003) Removal of heavy metal ions using the fungus *Penicillium canescens*. Adsorp Sci Technol 21:643–650

Segretin AB, Cazón JP, Donati ER (2018) Bioaccumulation and biosorption of heavy metals. In: Heavy metals in the environment. CRC Press, pp 93–113

Shamim S (2018) Biosorption of heavy metals. Biosorption 2:21–49

Shan B, Hao R, Xu H et al (2021) A review on mechanism of biomineralization using microbial-induced precipitation for immobilizing lead ions. Environ Sci Pollut Res 28:30486–30498

Shanker AK, Cervantes C, Loza-Tavera H, Avudainayagam S (2005) Chromium toxicity in plants. Environ Int 31:739–753

Sharma PK, Balkwill DL, Frenkel A, Vairavamurthy MA (2000) A new *Klebsiella planticola* strain (Cd-1) grows anaerobically at high cadmium concentrations and precipitates cadmium sulfide. Appl Environ Microbiol 66:3083–3087

Sharma RK, Agrawal M, Marshall F (2007) Heavy metal contamination of soil and vegetables in suburban areas of Varanasi, India. Ecotoxicol Environ Saf 66:258–266

Sharma R, Bhardwaj R, Gautam V et al (2018) Microbial siderophores in metal detoxification and therapeutics: recent prospective and applications. In: Plant microbiome: stress response. Springer, Singapore, pp 337–350

Sharma P, Pandey AK, Kim S-H et al (2021) Critical review on microbial community during in-situ bioremediation of heavy metals from industrial wastewater. Environ Technol Innov 24:101826

Sharma A, Kapoor D, Gautam S et al (2022) Heavy metal induced regulation of plant biology: recent insights. Physiol Plant 174:e13688

Sharma A, Kohli SK, Khanna K et al (2023) Salicylic acid: a phenolic molecule with multiple roles in salt-stressed plants. J Plant Growth Regul 42:1–25

Shen Y, Zhu W, Li H et al (2018) Enhancing cadmium bioremediation by a complex of water-hyacinth derived pellets immobilized with *Chlorella* sp. Bioresour Technol 257:157–163

Sher S, Rehman A (2019) Use of heavy metals resistant bacteria—a strategy for arsenic bioremediation. Appl Microbiol Biotechnol 103:6007–6021

Shuaib M, Azam N, Bahadur S et al (2021) Variation and succession of microbial communities under the conditions of persistent heavy metal and their survival mechanism. Microb Pathog 150:104713

Shukla A, Mehta K, Parmar J et al (2019) Depicting the exemplary knowledge of microbial exopolysaccharides in a nutshell. Eur Polym J 119:298–310

Singh J, Kalamdhad AS (2011) Effects of heavy metals on soil, plants, human health and aquatic life. Int J Res Chem Environ 1:15–21

Singh HP, Mahajan P, Kaur S et al (2013) Chromium toxicity and tolerance in plants. Environ Chem Lett 11:229–254

Singh M, Trivedi N, Mal N, Mehariya S (2023) Extremophilic microalgae as a potential source of high-value bioproducts. In: Extremophiles. CRC Press, pp 167–186

Smitha MS, Singh S, Singh R (2017) Microbial biotransformation: a process for chemical alterations. J Bacteriol Mycol Open Access 4:85

Sodhi KK, Kumar M, Singh DK (2020) Multi-metal resistance and potential of *Alcaligenes* sp. MMA for the removal of heavy metals. SN Appl Sci 2:1–13

Solisio C, Al Arni S, Converti A (2019) Adsorption of inorganic mercury from aqueous solutions onto dry biomass of *Chlorella vulgaris*: kinetic and isotherm study. Environ Technol 40:664–672

Soto DF, Recalde A, Orell A et al (2019) Global effect of the lack of inorganic polyphosphate in the extremophilic archaeon *Sulfolobus solfataricus*: a proteomic approach. J Proteome 191:143–152

Srichandan H, Pathak A, Singh S et al (2014) Sequential leaching of metals from spent refinery catalyst in bioleaching–bioleaching and bioleaching–chemical leaching reactor: comparative study. Hydrometallurgy 150:130–143

Srivastava S, Usmani Z, Atanasov AG et al (2021) Biological nanofactories: using living forms for metal nanoparticle synthesis. Mini Rev Med Chem 21:245–265

Stoppel R-D, Meyer M, Schlegel HG (1995) The nickel resistance determinant cloned from the enterobacterium *Klebsiella oxytoca*: conjugational transfer, expression, regulation and DNA homologies to various nickel-resistant bacteria. Biometals 8:70–79

Sultana N, Hossain SMZ, Mohammed ME et al (2020) Experimental study and parameters optimization of microalgae based heavy metals removal process using a hybrid response surface methodology-crow search algorithm. Sci Rep 10:15068

Suzuki Y, Banfield JF (2004) Resistance to, and accumulation of, uranium by bacteria from a uranium-contaminated site. Geomicrobiol J 21:113–121

Talukdar D, Jasrotia T, Sharma R et al (2020) Evaluation of novel indigenous fungal consortium for enhanced bioremediation of heavy metals from contaminated sites. Environ Technol Innov 20:101050

Taran M, Fateh R, Rezaei S, Gholi MK (2019) Isolation of arsenic accumulating bacteria from garbage leachates for possible application in bioremediation. Iran J Microbiol 11:60

Taştan BE, Ertuğrul S, Dönmez G (2010) Effective bioremoval of reactive dye and heavy metals by *Aspergillus versicolor*. Bioresour Technol 101:870–876

Tayang A, Songachan LS (2021) Microbial bioremediation of heavy metals. Curr Sci 120:00113891

Teitzel GM, Parsek MR (2003) Heavy metal resistance of biofilm and planktonic *Pseudomonas aeruginosa*. Appl Environ Microbiol 69:2313–2320

Thatoi H, Das S, Mishra J et al (2014) Bacterial chromate reductase, a potential enzyme for bioremediation of hexavalent chromium: a review. J Environ Manag 146:383–399

Torres E (2020) Biosorption: a review of the latest advances. PRO 8:1584

Tripathi P, Srivastava S (2007) Mechanism to combat cobalt toxicity in cobalt resistant mutants of *Aspergillus nidulans*. Indian J Microbiol 47:336–344

Tsekova K, Todorova D, Dencheva V, Ganeva S (2010) Biosorption of copper (II) and cadmium (II) from aqueous solutions by free and immobilized biomass of *Aspergillus niger*. Bioresour Technol 101:1727–1731

Tuzen M, Sarı A, Mendil D et al (2009) Characterization of biosorption process of As (III) on green algae *Ulothrix cylindricum*. J Hazard Mater 165:566–572

Utami U, Harianie L, Dunyana NR, Romaidi (2020) Lead-resistant bacteria isolated from oil wastewater sample for bioremediation of lead. Water Sci Technol 81:2244–2249

Vela-García N, Guamán-Burneo MC, González-Romero NP (2019) Efficient bioremediation of metallurgical effluents through the use of microalgae isolated from the Amazonic and highlands of Ecuador. Rev Int de Contam Ambient 35:917–929

Verma S, Kuila A (2019) Bioremediation of heavy metals by microbial process. Environ Technol Innov 14:100369

Verma N, Sharma R (2017) Bioremediation of toxic heavy metals: a patent review. Recent Pat Biotechnol 11:171–187

Verma A, Rejendra K, Singh K, Shukla S (2017) Use of low cost adsorbents for the remediation of heavy metals from waste water. Int J Eng Manage Econ 6(7):13–20

Verma S, Bhatt P, Verma A et al (2021) Microbial technologies for heavy metal remediation: effect of process conditions and current practices. Clean Techn Environ Policy:1–23

Verma N, Rachamalla M, Kumar PS, Dua K (2023) Assessment and impact of metal toxicity on wildlife and human health. In: Metals in water. Elsevier, pp 93–110

Vicentin RP, dos Santos JV, Labory CRG et al (2018) Tolerance to and accumulation of cadmium, copper, and zinc by *Cupriavidus necator*. Rev Bras Cienc Solo 42:e0170080

Vullo DL, Ceretti HM, Daniel MA et al (2008) Cadmium, zinc and copper biosorption mediated by *Pseudomonas veronii* 2E. Bioresour Technol 99:5574–5581

Wang N, Qiu Y, Xiao T et al (2019) Comparative studies on Pb (II) biosorption with three spongy microbe-based biosorbents: high performance, selectivity and application. J Hazard Mater 373:39–49

Wang F, Dong W, Zhao Z et al (2021) Heavy metal pollution in urban river sediment of different urban functional areas and its influence on microbial community structure. Sci Total Environ 778:146383

Wani S, Barnes J, Singleton I (2016) Investigation of potential reasons for bacterial survival on 'ready-to-eat' leafy produce during exposure to gaseous ozone. Postharvest Biol Technol 111:185–190

Wei W, Wang Q, Li A et al (2016) Biosorption of Pb (II) from aqueous solution by extracellular polymeric substances extracted from *Klebsiella* sp. J1: adsorption behavior and mechanism assessment. Sci Rep 6:31575

Wuana RA, Okieimen FE (2011) Heavy metals in contaminated soils: a review of sources, chemistry, risks and best available strategies for remediation. Int Sch Res Notices 2011:402647

Wyszkowska J, Borowik A, Kucharski M, Kucharski J (2013) Effect of cadmium, copper and zinc on plants, soil microorganisms and soil enzymes. J Elem 18

Yadav N, Jyoti, Thakur IS, Srivastava S (2023) Omics approach in bioremediation of heavy metals (HMs) in industrial wastewater. In: Genomics approach to bioremediation: principles, tools, and emerging technologies, pp 343–361

Yan C, Wang F, Geng H et al (2020) Integrating high-throughput sequencing and metagenome analysis to reveal the characteristic and resistance mechanism of microbial community in metal contaminated sediments. Sci Total Environ 707:136116

Yin K, Wang Q, Lv M, Chen L (2019) Microorganism remediation strategies towards heavy metals. Chem Eng J 360:1553–1563

Yue Z-B, Li Q, Li C et al (2015) Component analysis and heavy metal adsorption ability of extracellular polymeric substances (EPS) from sulfate reducing bacteria. Bioresour Technol 194:399–402

Zammit CM, Weiland F, Brugger J et al (2016) Proteomic responses to gold (iii)-toxicity in the bacterium *Cupriavidus metallidurans* CH34. Metallomics 8:1204–1216

Zeraatkar AK, Ahmadzadeh H, Talebi AF et al (2016) Potential use of algae for heavy metal bioremediation, a critical review. J Environ Manag 181:817–831

Zhang K, Xue Y, Xu H, Yao Y (2019) Lead removal by phosphate solubilizing bacteria isolated from soil through biomineralization. Chemosphere 224:272–279

Zhang D, Yin C, Abbas N et al (2020) Multiple heavy metal tolerance and removal by an earthworm gut fungus *Trichoderma brevicompactum* QYCD-6. Sci Rep 10:6940

Zheng Y, Xiao C, Chi R (2021) Remediation of soil cadmium pollution by biomineralization using microbial-induced precipitation: a review. World J Microbiol Biotechnol 37:1–15

Zhu H, Guo J, Chen M et al (2012) *Burkholderia dabaoshanensis* sp. nov., a heavy-metal-tolerant bacteria isolated from Dabaoshan mining area soil in China. PLoS One 7:e50225

Chapter 9
Remediation of Soil Organic Pollutants by Microbes from Extreme Environment

Dinoo Gunasekera and Disna Ratnasekera

9.1 Introduction

The major sources of environmental pollution take place due to growth of population, industrialization, emission of greenhouse gases, mining and so on. Environmental pollution can originate from organic or inorganic sources creating negative impacts on soil and water. The organic contaminants include hydrocarbons, pesticides containing organo groups, plastics, pharmaceuticals and other personal care stuffs, while inorganic sources are mainly due to excess application of chemical fertilizers and heavy metals from various sources. Some of such pollutants are extremely persistent, concentrating in soil environments and augmenting through food webs. The long-term accumulation of such pollutants makes the soil environment unfavourable for soil inhabitants, human health and all macro-/microflora and macro-/microfauna. Further, soil and water contamination are continuous and unavoidable processes because of the large-scale industrial wastes and anthropogenic activities with increasing population. Hence, maintaining soil pollutants below the threshold level in a cost-effective manner is compulsory. Bioremediation using biological means such as plants (phytoremediation) or microorganisms has been identified as a cost-effective, simple and environmentally friendly technique in a wide array of applications in cleaning contaminated soils (Cui et al. 2022). However, the efficacy of restoring contaminated soils by biological agents depends upon many factors such as nutrient status, environmental adaptability of the introduced microorganism, toxic resistance and poor pollutant bioavailability and

D. Gunasekera
Department of Information and Communication Technology, Faculty of Technology, University of Ruhuna, Matara, Sri Lanka

D. Ratnasekera (✉)
Department of Agricultural Biology, Faculty of Agriculture, University of Ruhuna, Matara, Sri Lanka

consumes comparatively lengthy time for activating the remedial process (Liu et al. 2023). Thus, site-specific studies are crucial to check the feasibility and to explore the successful cleaning process (Wang et al. 2020).

The plants or microbes that have resistant or excess accumulation properties such as for degradation, transformation, translocation, extraction, sequestration or detoxification of pollutants in soil and water were ideal in bioremediation (Samardjieva et al. 2014). Detoxification through bioremediation has been successfully adopted in restoring pesticide-contaminated fields (Karimi et al. 2021), polycyclic aromatic hydrocarbon polluted sites and polychlorinated biphenyl polluted sites (Zhou et al. 2023) and accepted as an environmentally friendly approach. The slow activity rate of the degradation process is one of the key challenges of bioremedial approaches. The amendment of certain chemicals such as ethylenediaminetetraacetic acid (EDTA) and Tween 80 has proven to escalate contaminant solubility and enhance microbial efficacy (Jelusic and Lestan 2014). The use of such synthetic substances is, however, costly and imposes negative impacts on the environment causing persistent toxicants in the soils. To overcome such drawbacks, alternative eco-friendly approaches, including biofiltration, bioventing, bioreactors, biostimulation, bioaugmentation, bioleaching and biocomposting, have been broadly encountered.

The success of the bioremediation approach depends on the identification of microbes surviving in the particular contaminated site and their capability of detoxifying pollutants (Table 9.1). Such microbe species could be isolated, mass cultured and introduced as an ecological and economical alternative to physiochemical approaches in decontamination.

9.2 Sources of Soil Organic Pollutants

Soil is a non-renewable resource consistently a sink to all types of pollutants and toxic compounds generated by industries, agriculture inputs, waste disposal, mining and other means decreasing biodiversity, soil health and fertility. Soil is frequently subjected to contamination by various kinds of organic pollutants (OPs), which could be generated via large-scale production of synthetic compounds such as perfluoroalkyl and polyfluoroalkyl (PFASs) substances, fossil fuel-derived hydrocarbons, agriculture-associated organo-chemicals, cosmetic, pharmaceutical and personal care products and plastics.

Organic pollutants widely spread throughout the environment via various natural and anthropogenic means causing health hazards to all living organisms. The contaminants are mainly uptake by plant roots on a large scale and by foliar from the atmosphere on a minor scale. The plant uptake efficiency of such contaminants depends upon the physicochemical properties of the soil.

Table 9.1 Bioremediation methods adopted in treating contaminated soils/water

Method	Description
Bioaccumulation	Absorption of certain substances and deposited in an organism
Biostimulation	Stimulating itself or by adding surfactants, sub-substrates, inducers
Biofiltration	Filtration by porous materials colonized by dense microbial colonies
Bioleaching	Mobilizing, separating, extracting or removing toxic substances using microbes in wastewater or sludge
Biotransformation	Conversion of endogenic substances to water-soluble molecules
Biocomposting	Microbial degradation of organic materials under aerobic conditions
Bioaugmentation	Adding pre-grown microbes to remove contaminants
Bioventing	Adding ventilation/oxygen to stimulate microbial degradation, especially in soil hydrocarbon
Bioreactors	Large vessels containing microbes that facilitate biological reactions or processes to happen inside the vessel under a controlled environment
Biosorption	Adsorption of pollutants using biological agents
Biouptake	Transfer of materials from the environment to living organisms
Biomonitoring	Methodical evaluation of toxic/pollutants using microbes/living organisms
Phytodegradation	Degradation of contaminants into simple compounds sequestered by plants
Phytoextraction	Absorbing and storing pollutants within plants
Phytostabilization	Minimizing mobility of pollutants by plants and root systems
Phytovolatilization	Removal of pollutants through stomata after converting to volatile forms
Phytoexcretion	Living organisms that absorb wastes and excrete and in the process purify absorbed pollutants through osmotic balance

9.2.1 Perfluoroalkyl and Polyfluoroalkyl

Perfluoroalkyl and polyfluoroalkyl (PFASs) are man-made synthetic substances with high chemical and thermal stability (Bolan et al. 2021). PFASs have been widely used on a large scale in over 200 groups of products such as surfactants, coating formulations, waterproof textiles and fabrics, non-stick cooking ware, food packing materials, detergents and surface protectants in many industries (Bolan et al. 2021; Zhang et al. 2022). PFASs belong to the synthetic organic group holding fluorinated carbon chains with diverse active groups. PFASs possess higher stability and surface activity and are thus used in many industries such as textile, kitchenware, metal plating, lubrication items, foam manufacturing, food package materials, electronic items and so on. There were nearly 5000 PFAS compounds that have been reported so far that were utilized in various industries, and some of them were serious persistent organic pollutants (OECD 2018). The persistence of PFASs is mainly due to the hydrophilic functional groups and long carbon chains and are likely to be adsorbed easily to soils, sediment materials and particles. The PFASs with short-chain carbon are volatile contaminating the atmosphere and cycling through geohydrological processes. PFASs have been traced in crop plants and shown to accumulate in food webs (Ghisi et al. 2019) causing serious human health issues such as infertility, disorders in the brain and neuro system and weak immunity (Liew et al. 2018). The degradation of PFASs in the soil has been shown via the hydrolysis process or mediated by microbes (Zhao et al. 2019), and the rate of degradation process depends upon PFASs chain size, degree of solubility and soil characteristics (Wang et al. 2012).

The major PFASs soil water contaminant sources were surface water from areas where the application of fluorochemical industry impacted biosolids, seepage and sludge in wastewater treatment plants, landfill leachate, biosolids amended soils, soil reformer by industrial waste, waste biological sludge and treated biosolids (Bolan et al. 2021).

9.2.2 Fossil Fuel-Derived Hydrocarbons

The alkanes and alkane derivatives, aromatic hydrocarbons, polychlorinated biphenyls and polychlorinated dibenzo-p-dioxins are the main fossil fuel-derived hydrocarbon organic pollutants that frequently admix with soil and atmosphere. Hydrocarbons are hydrophobic substances that persist in soil sediments and water for a long period. Pollution due to fossil fuels has become a great concern with the development of maritime transport and oil-related industries. The global daily oil consumption for transportation, other industries and households is gradually increasing which is leading to the release of large quantities of hydrocarbons and crude oils contaminating soil and water (Ma et al. 2021). The oil contaminations further contributed to form toxic substances affecting ecological niches. These chemical compounds are gradually spread-out soils, water bodies atmosphere and so on through biohydrogeochemical cycles (Khan et al. 2018). These compounds are slowly breaking down and mixing with the atmosphere, soils and water causing long-term health hazards (Perelo 2010).

9.2.3 Agriculture Wastes, Fertilizers, Pesticides and Other Chemical Inputs

The agricultural inputs, including pesticides, herbicides that contain organo group and chemical fertilizers with such contaminants are the main sources of soil organic pollutants in large-scale agricultural lands. The application of such agricultural inputs has increased with the growing population resulting in enhanced economic benefits and subsequently causing serious environmental impacts on non-target organisms and human health (Briceno et al. 2018). Nearly 50% of global pesticide production is organo-phosphorus pesticides imposing significant human poisoning and causing death (Ranjan and Jindal 2021a, b). Organo-phosphate chemical residue easily enters terrestrial and aquatic plants via soil solution and certain enzymes are involved in the degradation of such compounds by dichlorination and methanogenesis (Ranjan and Jindal 2021a, b; Chen et al. 2022). Fungicides, nematicides, insecticides, rodenticides, herbicides and antibiotics are the widely used plant protection chemicals in agriculture. Pesticides, in general, are highly toxic for both target and non-target organisms as well as on soil, water and atmosphere causing deadly effects on human health. The contribution of pesticides to soil contamination is one of the huge parts because of their consistent application in agriculture.

9.2.4 Plastics and Organic Dyes

Macro- and microplastics are artificially produced organic polymers with very slow degradability. The use of plastics is increasing day by day in all levels of the community due to convenience, lightweight and durability. Organic dyes are placed into three main groups: anthraquinone, azo and phthalocyanine according to their chemical nature (Singh et al. 2007) and are key water pollutants with low degradation and difficult to remove. The majority of organic dyes are azo dyes and anthraquinone dyes which are chemical compounds with aromatic or aliphatic active groups largely used in the chemical dye industry as synthetic colouring agents. The main sources of dyes are the textile industry, plastics and medicine with various degrees of colour intensities. The dyes pollute aquatic ecosystems, which reduce photosynthesis of aquatic autotrophs causing negative effects on food webs. Further, large accumulations of dyes in aquatic ecosystems cause eutrophication and hypoxia or anoxia which will result in the death of aerobics and favouring anaerobes. The release of toxic amines during the slow decomposition process of azo dyes will lead to water poisoning and the death of aquatic organisms (Ali and Muhammad 2008). Thus, continued large-scale production and utilization of both plastic and dye products has become a serious environmental issue.

9.2.5 Cosmetics, Pharmaceuticals and Personal Care Products

Cosmetics, pharmaceuticals and personal care products (CPPCPs) are emerging serious issues as the usage of such products is increasing tremendously, leading to a huge accumulation of contaminants. Pharmaceuticals are a category of medicament for humans and vet drugs for animal diseases, while personal care products enhance the life quality. Personal care products commonly include shampoo and soap, toothpaste, cosmetics, lotions, skin and UV-protectant creams, disinfectants, preservatives and repellent products that are used for external applications. Thus, there is a high possibility of entering such products as it is to the environment through regular usage. Some personal care products are highly persistent and bioactive; therefore, there is a significant possibility of bioaccumulation. Ketoprofen, caffeine, estrone, estradiol and oestrogen are some common personal care organic compounds detected in soils that cause negative impacts on humans and the environment

(Nguyen et al. 2020). The contaminants related to CPPCPs are released into the environment mainly through industrial and domestic wastewater, hospital discharge and inappropriate disposal of vet drugs and wastes (Liu and Wong 2013) which consistently increase accumulation in soil.

9.3 Microbes for Bioremediation

The inhabitant microorganisms in extreme environments and contaminated soils are the best sources to select and isolate suitable microorganisms for bioremediation. The capability of microbes to produce enzymes for degrading contaminants effectively is the main feature to be considered in the selection and isolation process (Shen et al. 2020). The rate of biodetoxification of contaminants varies with substrate molecular size, microbial colony size, metabolic rate of microbes, growth stage, concentration of pollutant and environmental factors. Certain environmental parameters, including soil temperature, pH, electrical conductivity, moisture level, other contaminants, nutrients, organic materials, soil type and so on could affect the bioremediation process (Topuz et al. 2022). Table 9.2 describes some of the most significant microbes and microbial consortium for degradation of organic pollutants.

9.3.1 Microbial Application in Plastic and Organic Dye Degradation

Microalgae are effectively degraded azo dyes in various contaminated sites. In general, autotrophs and algae show an ability to survive in a vast range of environments contributing to nutrient cycling and oxygen generation. The mechanism of microalgae species such as *Chlorella*, *Scenedesmus* and *Aphanocapsa* have been extensively studied in organic dye degradation (Bai et al. 2022). The subtraction of industrial colours and pigments using microalgae is advocated by biosorption, bioconversion or biodegradation, thereby questing simple organic molecules and nitrogen sources for their growth and development (Li et al. 2022a, b). The efficacy of *Chlorella* species in breaking down the majority of various azo dyes into smaller organic molecules and CO_2 has been found under many circumstances (Touliabah et al. 2022). The key roles of microalgal biomass in the biodegradation approach are the breakdown of dyes for searching nitrogen and carbon sources for their growth, hypersecretion of enzymes that can digest toxic compounds, nutrient recycling, photosynthesis and oxygen production so that keeping the environment favourable to other aerobic organisms. The algal enzymatic secretions can transform toxic large compounds into simple and safe molecules removing harmful contaminants.

Table 9.2 Some of the successful microbial groups and microbe species with potential degradation of environmental organic pollutants

Microbial group	Degradable substrates/ contaminants	References
Bacteria		
Corynebacterium		
Rhodococcus	Hydrocarbons, chlorophenols, polychlorinated biphenyls and sulfonated azo dyes	Alvarez (2010), Nazari et al. (2022)
Nocardia	Persistent organic pollutants (POP), polycyclic aromatic hydrocarbons, polychlorinated biphenyls, chlorophenols, sulfonated azo dyes and alkanes	Azadi and Shojaei (2020)
Gordonae	Alkanes	Liu et al. (2021)
Mycobacteria	Poly-chlorophenols, heavy metals and diverse polycyclic aromatic hydrocarbons	
Pseudomonas aeruginosa, Streptomyces badius, Bacillus subtilis, Dehalococcoides, Comamonas, Burkholderia, Alcaligenes, Deinococcus radiodurans, Acidithiobacillus ferrooxidans, Mesorhizobium huakuii	Plastics, organic pollutes, oil, hydrocarbons, organic solvents	Paço et al. (2019), Patel et al. (2022)
Fungi		
Aspergillus niger, Aspergillus flavus,	Plastics, synthetic dyes, pesticides	Paço et al. (2019)
Microsporidia, Blastocladiomycota	Polycyclic aromatic hydrocarbons	Pinedo-Rivilla et al. (2009)
Mucoromycotina	Biphenyls, polyaromatic hydrocarbons, synthetic dyes, pesticides	Jasu et al. (2021)
Ascomycota	Alkanes, biphenyl, alkylbenzene, chlorophenols, crude oil, diesel, fragrances, pesticides, synthetic dyes and toluene	Patel et al. (2022)

(continued)

Table 9.2 (continued)

Microbial group	Degradable substrates/ contaminants	References
Microalgae and cyanobacteria		
Chlorella spp., *Volvox aureus Nostoc linckia HA 46, N. muscorum, N. linckia, Oscillatoria rubescens, Chroococcus minutus, Gloeocapsa pleurocapsoides*	Organic dyes, azo dyes	Mona et al. (2011), Ishchi and Sibi (2019), Touliabah et al. (2022)
Nannochloris oculate, Chlamydomonas reinhardtii, Prototheca zopfii, Oscillatoria sp., *Anabaena variabilis*	Hydrocarbon (aromatic compounds, saturated aliphatic hydrocarbons), phenol, pesticides and herbicides	Aldaby and Mawad (2018), Verasoundarapandian, et al. (2022), Touliabah et al. (2022)

Being synthetic organic polymers plastics are long persistent in the environment. The biodegradation of such low degradable contaminants using bacteria (*Pseudomonas aeruginosa* and *Streptomyces badius*) and fungi (*Aspergillus niger* and *Aspergillus flavus*) was reported (Paço et al. 2019). The successful bioremediation in heavy plastic contaminated sites with other co-contaminants such as bisphenol-S using *Phingobium fuliginis* OMI was reported (Fang et al. 2019). S*phingobium fuliginis* OMI is an aerobic bacterium capable of degrading recalcitrant alkylphenols and bisphenols.

9.3.2 Biodegradation of Fossil Fuel and Aromatic Compounds

Contaminants of fossil fuel residues and aromatic compounds are global issues that every government pays more attention to. Exploring approaches that are eco-friendly, economical, safe and effective with minimal environmental impacts are the major concerns. The application of extremophiles has been investigated extensively in restoring crude oil-polluted areas. Screening, identification and isolation of microbes dwelling in fossil fuel-contaminated soils is one of the effective approaches (Lahiri et al. 2021). The efficient degradation of polycyclic aromatic hydrocarbons using microalgae was reported with the mechanism of adsorption of contaminants to microalgal cell walls and then bioaccumulation within algal cells resulting in cleaning up the polluted site (Touliabah et al. 2022). The syntrophic anaerobic bacteria and archaea with different metabolic rates have been reported to be highly effective in the biodegradation of hydrocarbons.

The use of genetically engineered bacteria strains in the remediation of offshore fuel spills with combinations of multiple hydrocarbons has shown to be promising in decontamination. For example, *Acinetobacter baumannii* S30 pJES strain has shown higher detoxification efficacy on petroleum hydrocarbons (Zakaria et al. 2021), overexpressed alkane monooxygenase in *Streptomyces coelicolor* M145 showed hyper degrading ability of alkane (Rafeeq et al. 2023), xylE gene cloned in

Acinetobacter sp. BS3 exhibited a broad range of degrading ability including n-alkanes and aromatic hydrocarbons (Ma et al. 2021).

9.3.3 Biodegradation of Perfluoroalkyl and Polyfluoroalkyl

Perfluoroalkyl and polyfluoroalkyl substances (PFASs) are highly persistent in the environment polluting water and soil. Owing to the widespread usage in industries and households, PFASs have been recorded in water, soil, soil and water inhabitants, higher plants, animals and the human body. Because of the high chemical, thermal and biological stability of PFASs with soil and sediment matrices, the remediation methods that can adopt successful removal of PFASs are extremely limited and challenging. Some chemical and physical approaches such as thermal destruction, oxidation, sorption and stabilization have been shown effective in the degradation of PFASs (Kupryianchyk et al. 2016; Bulusu et al. 2020) though such methods are costly. Alternatively, eco-friendly, economical biological techniques have been tested. The microbial performances under aerobic environments in degrading perfluorinated compounds have shown to be less effective compared to anaerobic environments (Fahid et al. 2020). Further, the efficacy of the biodegradation process depends upon the selection of an appropriate microbe or microbe combination. For example, PFASs were resistant to biodegrading by *Escherichia coli*; in contrast, certain *Pseudomonas* spp. and *Acidimicrobium* spp. have efficiently reduced PFAS in contaminated sites (Bolan et al. 2021). The extremophiles such as chemo-organoheterotrophic bacteria and certain yeast strains were reported to reduce PFAS in sludges and industrial sites where high concentrations of PFAS exist (Beskoski et al. 2018).

9.3.4 Biodegradation of Cosmetic, Pharmaceuticals and Personal Care Products

The global daily consumption of cosmetics, pharmaceuticals and personal care products is increasing exponentially. The water body contamination by cosmetics, pharmaceuticals and personal care products is significantly higher than that of soil contamination. The decontamination of such polluted environments using fungal species such as *Trametes pubescens* and *Phanerochaete chrysosporium* has demonstrated effective results (Zdarta et al. 2021). The fungal enzymes laccase and lignin peroxidase from *T. pubescens* and *P. chrysosporium*, respectively, showed the ability to digest cosmetics, pharmaceuticals and personal care products into non-toxic simple molecules such as dimers (Zdarta et al. 2021). The active microbes and membrane reactors with bioactive molecules have been efficiently adopted in cleaning up pharmaceutical industrial wastewater (Leal et al. 2019). The biofiltration using hybrid approaches by a combination of autotrophic/heterotrophic bacteria has efficiently degraded organic carbons by heterotrophic bacteria and toxic compounds (ketoprofen and caffeine) by autotrophic bacteria (Wang et al. 2021). Further, microalgae have demonstrated tremendous efficacy in the decontamination of personal care products because of their characteristic features such as large biomass, enormous surface area, solid binding ability, autotrophy and wide environmental adaptability (Usmani et al. 2020). The microalgal remedial mechanism proceeded by adsorption to the surface followed by cellular bioaccumulation and intra- or extra-cellular degradation (Nguyen et al. 2020).

9.4   Strategies to Enhance the Bioremediation Process

Bioremediation is solely mediated by livening organisms, and the effectiveness of the remedial process depends upon the extent of compatibility of microbes with the external environment. In recent years, many studies have been focused on exploring strategies and mechanisms for enhancing bioremediation processes (Usmani et al. 2020; Wang et al. 2020; Li et al. 2022a, b; Miri et al. 2018). The immobilization of microorganisms, genetically engineered strains, formulation of microbial consortia, and use of combined agents, including nanoparticles, chemical agents, industrial wastes, biosurfactants, farm manure and so on are some of the strategies tested to support microbial activity accelerating bioremediation process (Usmani et al. 2020; Wang et al. 2020; Li et al. 2022a, b).

9.4.1   Microbial Consortium

A combination of group of microorganisms effectively clean up contaminated sites compared to axenic cultures due to the broad spectrum of activity performed by multiple populations of microbes (Corrado and Meyer 2018). The role of microbial consortia is owing to the combination of multiple enzymes that work together in biodegradation. Microbial consortia-like bacteria-algae combinations were well adapted to broader environments and environmental fluctuations so that more competent in decontamination. The mutualistic relationships have been detected in microbial consortia. For example, carbon excreted by algae was an energy source for the bacteria and the mineralized ions by bacteria activate algal development in bacteria–algae consortia (Nguyen et al. 2020). The enzyme production of consortia can happen in different growth phases in different concentrations that depend on the intact types of hydrocarbons to be subjected to breakdown. For example, some hydrocarbon catalytic enzymes such as alkane hydroxylase, catechol dioxygenase, and alcohol dehydrogenase activate synergistically in microbial consortia, thereby degrading hydrocarbons into simple, non-toxic/less-toxic compounds at the site. The synergistic enzyme activity has been observed in microbial consortia, including bacterial strains of *Alcanivorax borkumensis*, *Bacillus licheniformis* MN6, *Geobacillus thermoparaffinivorans* IR2, *Geobacillus stearothermophilus* IR4, *Pseudomonas aeruginosa* PSA5 and *P. putida* producing alcohol dehydrogenase, lipase and alkane hydroxylase and *Pseudomonas* BP10 and *O. intermedium* P2 producing catechol 1,2 dioxygenase (C12D) and catechol 2,3 dioxygenase (C23D) at the degradation process (Elumalai et al. 2021). The consortia enhanced bioconversion and biosorption, improved membrane stability and greater amalgamation. Thus, microbial consortia can play collective functions that are impossible for a single species.

9.4.2 Biosurfactants

Biosurfactants are surface-active biomolecules synthesized by microbes facilitating detachment or attachment to other surfaces. The amphipathic nature of having hydrophilic and hydrophobic regions in biosurfactants allows them to interact with different phases. The biosurfactants are capable of expanding the surface for organic contaminants and desorbing toxic substances. The hydrophilic regions of surfactants showed increased surface solubility and enhanced adsorbing pollutants resulting in higher bioavailability (Li et al. 2022a, b). The rapid metabolic rates, free abundance, higher proliferation ability and high surface-to-volume ratio are excellent characteristics of efficient biosurfactants that can facilitate successive bioremediation (Anoop et al. 2021). The microbial colonies inhabit interfaces and create amphipathic biosurfactants that can alter the chemical nature of the substrate (Markande et al. 2013). This phenomenon is adopted in bioremediation in toxic soils forming microbe–toxic substance interfaces and converting toxic compounds to non-toxic substances. The significantly high number of microbe species, their roles and applications have been reported to produce biosurfactants (Sajid et al. 2020).

9.4.3 Arresting/Immobilizing Microbial Cells

The microbial biomass or microalgal populations arrested in a contaminated substrate or inside the containers thereby enhanced direct interaction facilitating better performances due to restricted space . The immobilized microbes are eco-friendly and cost-effective as they are adapted to a wider range of pH, temperature and higher toxic levels and can be re-employed many times compared to free microbes (Usmani et al. 2020). The immobilized bacteria *Zoogloea* sp. encapsulated by polyvinyl alcohol in-alginate beads has exhibited enhanced biodegradation of aromatic hydrocarbon phenanthrene (Partovinia and Rasekh 2018). Though immobilization approaches are being applied in various detoxification processes, low cell longevity during encapsulation, high cost and complexity of construction are certain weaknesses that are yet to be improved.

9.4.4   Genetic and Metabolic Modifications

The application of synthetic biology in addressing environmental issues has drawn great attention. Enzymes are biological catalysts for stimulating chemical reactions. The use of genetically altered microbes has proven to be remediate various environmental hazards. The creation of synthetic microbes with specialized functional expression has been carried out by combining desirable species. As a result, the bioremediation process would be improved through metabolic engineering that modifies the existing biochemical pathway. The various biochemical metabolic pathways related to enzyme production have been modified for enhanced production of microbial enzymes esterases, oxidases, phenoloxidases and oxygenases which are part of many biochemical breakdown processes (Mujawar et al. 2018) through microbial metabolic engineering. The industrial application of enzymes has shown to be in a wider array of areas, including drug manufacturing, detoxification, food industry and others. The use of genetically engineered microbes capable of hyperexcretion of enzymes is of interest in cleaning contaminants. The enzymes of genetically modified microbes can sustainably degrade contaminants. Genomic studies, protoplast fusion and recombinant DNA technology have been adopted in industrial enzyme production using microbes aiming to uplift the quality and quantity of enzymes (Kazemi Shariat Panahi et al. 2022). The elucidation of multiple genes responsible for biodegradation would lead to the construction of new microbial strains that can enhance detoxification processes. The discovery of new genes that can break down new contaminants is an endless process with high priority under sophisticated advancement of technology and industrialization in a global context. However, special attention is necessary in constructing engineered microbes in field applications, including environmental compatibility, self-viability under extreme environments, risk of horizontal gene transfer with naturally occurring microbes and related issues (Thomas et al. 2021). Thus, field application of genetically altered microbes is still limited due to potential risks to the environment and human beings.

9.4.5   Biosorption

Biosorption is a series of physicochemical activities comprising absorption, adsorption, molecule exchange and precipitation followed by desorption of adsorbed pollutants from the biosorbent agent (Yaashikaa et al. 2021). Biosorption is an eco-friendly, economical, simple and efficient approach using microbes such as bacteria, fungi, microalgae and yeast in de-polluting various contaminants in a wider array of chemical groups which includes dyes, phenolic compounds and pesticides. The biosorption process can happen extra- or intra-cellularly under anaerobic or aerobic conditions. The efficacy of biosorption relies on the interaction between the bioagent and the contaminant that includes ion exchange, coordination, electrostatic interactions, microprecipitation and so on while environmental factors such as temperature, pH of the sludge, ratio of biosorbent/pollutant, shaking speed can influence this process (Enrique 2020). In addition, competition with other pollutants (Debs et al. 2019), type and the number of functional groups in the biosorbent (Wernke et al. 2018; Zhang et al. 2020a, b, c) also affect the biodegradation process. Biosorption is an efficient alternative for eliminating hazardous and toxic contaminants as it adsorbs and degrades toxicants into non-toxic molecules compared to chemical and other approaches. Microalgae and cyanobacteria have shown the decomposition of toxicants in dyes and pesticides through absorption followed by excreting enzymes, such as azo-dye reductase that transform complex dye molecules into simple and harmless molecules like NH_2 and CO_2 (Touliabah et al. 2022).

9.4.6 Amendment-Based Modifications

The efficacy of the biological remediation process can be improved by various amended substances which include chemical substances, industrial and farm wastes, nanomaterials, biochar, a variety of microbial species, compost, sludge and so on (Liu et al. 2023). The amended substances are supported by bioremedial process providing better nutrient accessibility, facilitating redox potential, reducing toxicity, improving favourable environment to microbial growth and modifying contaminants by bioactivity. Reports highlighted that micronutrients such as Cu and Fe and primary nutrients such as N and P can enhance pollutant removal rate significantly (Chen et al. 2022; Miri et al. 2018). Cu can increase the growth of bacteria and expression of methane monooxygenase enzyme activity (Chen et al. 2022), while Fe can generate a favourable redox environment facilitating better decontamination rates (Miri et al. 2018). The application of N- and P-rich composites showed better detoxification of petroleum hydrocarbons owing to enhanced nutrient availability for microbial growth (Miri et al. 2018). Further to the reduction of toxicity, the improved nutrient status in the contaminated sites by adding supplementary amendments promotes plant growth. Moreover, fungal inoculations have proven to efficiently remove contaminants and enhance nutrient availability by hormone and enzyme secretion facilitating biodegradation.

9.4.7 Biostimulation

Biostimulation refers to the modification of the environment to enhance microbial activity using various methods. The contaminated sites are amended by supplements such as nutrients, moisture, aeration, pH, optimum temperature and electron acceptors/donors providing favourable conditions for microbial decomposition (Ghosh et al. 2021). For example, water-soluble nutrients including $NaNO_3$, KNO_3, K_2HPO_4, $MgNH_4 PO_4$ and nutrient slow-releasing materials have shown accelerated microbial growth and activity (Sharma et al. 2022). However, a higher amount of chemical nutrient application in polluted water could lead to eutrophication, increasing algae growth and reducing oxygen levels in the aquatic system. The combinations of amendments should be precisely designed to maximize microbial efficacy by considering pollutant type, site-specific pollutant combinations and other environmental factors. The encapsulation of microbial cells using agar, gelatin and polyurethane-like substances provided additional protection under harsh, toxic environments minimizing cell damage.

9.4.8   Bioventing

Bioventing is an in situ bioremediation approach that provides oxygen or air to soil or water bodies stimulating aerobic biodegradation in unsaturated layers. Bioventing enhances natural microbial growth and activity and is highly suitable for oil spills, persistent organic pollutants and hydrocarbon derivatives in soil or aquatic systems. It is an ideal approach for the removal of pesticide residuals in deeper layers of soils. In more adverse conditions, nutrients and moisture are also infiltrated along with the air to enhance microbial transformation of toxic substances (Duodu et al. 2022). The biodegradation rate and efficacy depend on air quality and the air flow rate and the mechanism of the air infiltration. The uniform air dispersion facilitates escalating toxicant dispersion and increasing affinity for microbial biodegradation (Yadav et al. 2021).

9.4.9   Bioaugmentation

Bioaugmentation is an eco-friendly cost-effective approach by application of exogenic microbes having precise catabolic roles on specific toxic substrates that are highly applicable to toxic and persistent contaminants. Bacteria, fungi, algae and microbes derive enzymes that are frequently used in bioaugmentation of organic polluted environments. However, the field application of exogenous microbes is limited owing to the incompatibility of soil environment, competition between natural–exotic microbes, degradation process, substrate specificity and longevity of enzymes. The fine adjustment of environmental factors, application of broad-spectrum microbes, microbe supportive amendments and application of contaminant-specific microbes are some approaches to overcome negative effects (Gao et al. 2022). Aerobic bacteria, including *Bacillus*, *Arthrobacter*, *Rhodococcus*, *Pseudomonas* and *Mycobacterium*, have played a significant role in the decontamination of organic pollutants in upper layers of the soil with sufficient air circulation

(Gao et al. 2022). The decontamination of deep layers of soil and groundwater relies on the application of anaerobic bacteria such as *Pseudomonas stutzeri*. The degradation of phenanthrene in deep soil was reported by co-culture PheN9 with *Pseudomonas stutzeri by* methyltransferase and carboxylase enzymes (Zhang et al. 2020a, b, c).

9.5 Conclusion

Organic pollutants emission due to anthropogenic or natural activity causes serious risks to soil health and fertility. These pollutants negatively impact soil organisms, plants, animals and humans. Bioremediation approaches, particularly those using extremophiles, are promising for reclaiming contaminated soils. Extremophiles can withstand extreme environmental conditions, such as high salinity, alkalinity and temperatures; hence, they can be most appropriate for harnessing remediation potential. They could be reliable candidates for bioremediation due to their stability and unique enzymes for degrading pollutants.

References

Aldaby ESE, Mawad AMM (2018) Pyrene biodegradation capability of two different microalgal strains. Global NEST J. https://doi.org/10.30955/gnj.002767

Ali H, Muhammad SK (2008) Biosorption of crystal violet from water on leaf biomass of Calotropis procera. J Environ Sci Technol 1(3):143–150. https://doi.org/10.3923/jest.2008.143.150

Alvarez HM (ed) (2010) Biology of Rhodococcus. In: Microbiology Monographs. https://doi.org/10.1007/978-3-642-12937-7

Anoop R. Markande, Divya Patel, Sunita Varjani (2021) A review on biosurfactants: properties, applications and current developments, Bioresource Technology, 330, 124963, ISSN 0960–8524, https://doi.org/10.1016/j.biortech.2021.124963

Azadi D, Shojaei H (2020) Biodegradation of polycyclic aromatic hydrocarbons, phenol and sodium sulfate by Nocardia species isolated and characterized from Iranian ecosystems. Sci Rep 10(1):21860. https://doi.org/10.1038/s41598-020-78821-1

Bai X, Liang W, Sun J, Zhao C, Wang P, Zhang Y (2022) Enhanced production of microalgae-originated photosensitizer by integrating photosynthetic electrons extraction and antibiotic induction towards photocatalytic degradation of antibiotic: a novel complementary treatment process for antibiotic removal from effluent of conventional biological wastewater treatment. J Environ Manag 308:114527. https://doi.org/10.1016/j.jenvman.2022.114527

Beškoski VP, Yamamoto A, Nakano T, Yamamoto K, Matsumura C, Motegi M, Beškoski LS, Inui H (2018) Defluorination of perfluoroalkyl acids is followed by production of monofluorinated fatty acids. Sci Total Environ 636:355–359. https://doi.org/10.1016/j.scitotenv.2018.04.243

Bolan N, Sarkar B, Yan Y, Li Q, Wijesekara H, Kannan K, Tsang DCW, Schauerte M, Bosch J, Noll H, Ok YS, Scheckel K, Kumpiene J, Gobindlal K, Kah M, Sperry J, Kirkham MB, Wang H, Tsang YF, Hou D, Rinklebe J (2021) Remediation of poly- and perfluoroalkyl substances (PFAS) contaminated soils—to mobilize or to immobilize or to degrade? J Hazard Mater 401:123892. https://doi.org/10.1016/j.jhazmat.2020.123892

Briceño G, Fuentes MS, Saez JM, Diez MC, Benimeli CS (2018) Streptomyces genus as biotechnological tool for pesticide degradation in polluted systems. Crit Rev Environ Sci Technol 48(10–12):773–805. https://doi.org/10.1080/10643389.2018.1476958

Bulusu RKM, Wandell RJ, Zhang Z, Farahani M, Tang Y, Locke BR (2020) Degradation of PFOA with a nanosecond-pulsed plasma gas–liquid flowing film reactor. Plasma Process Polym 17(8). https://doi.org/10.1002/ppap.202000074

Corrado N, Vera Meyer (2018) From Axenic to Mixed Cultures: Technological Advances Accelerating a Paradigm Shift in Microbiology. Trends Microbiol. 26(6):538–554, ISSN 0966-842X. https://doi.org/10.1016/j.tim.2017.11.004.

Chen Y, Jiang G, Sivakumar M, Wu J (2022) Enhancing integrated denitrifying anaerobic methane oxidation and Anammox processes for nitrogen and methane removal: a review. Crit Rev Environ Sci Technol 53(3):390–415. https://doi.org/10.1080/10643389.2022.2056391

Cui Q, Zhang Z, Beiyuan J, Cui Y, Chen L, Chen H, Fang L (2022) A critical review of uranium in the soil-plant system: distribution, bioavailability, toxicity, and bioremediation strategies. Crit Rev Environ Sci Technol 53(3):340–365. https://doi.org/10.1080/10643389.2022.2054246

Debs KB, da Silva HDT, de Lourdes Leite de Moraes M, Carrilho ENVM, Lemos SG, Labuto G (2019) Biosorption of 17α-ethinylestradiol by yeast biomass from ethanol industry in the presence of estrone. Environ Sci Pollut Res 26(28):28419–28428. https://doi.org/10.1007/s11356-019-05202-1

Duodu MG, Singh B, Christina E (2022) Waste management through bioremediation technology: an eco-friendly and sustainable solution. In: Relationship between microbes and the environment for sustainable ecosystem services, Volume 2, pp 205–234. https://doi.org/10.1016/b978-0-323-89937-6.00007-3

Elumalai P, Parthipan P, Huang M, Muthukumar B, Cheng L, Govarthanan M, Rajasekar A (2021) Enhanced biodegradation of hydrophobic organic pollutants by the bacterial consortium: impact of enzymes and biosurfactants. Environ Pollut 289:117956. https://doi.org/10.1016/j.envpol.2021.117956

Fahid M, Arslan M, Shabir G, Younus S, Yasmeen T, Rizwan M, Siddique K, Ahmad SR, Tahseen R, Iqbal S, Ali S, Afzal M (2020) Phragmites australis in combination with hydrocarbons degrading bacteria is a suitable option for remediation of diesel-contaminated water in floating wetlands. Chemosphere 240:124890. https://doi.org/10.1016/j.chemosphere.2019.124890

Fang Z, Gao Y, Wu X, Xu X, Sarmah AK, Bolan N, Gao B, Shaheen SM, Rinklebe J, Ok YS, Xu S, Wang H (2019) A critical review on remediation of bisphenol S (BPS) contaminated water: efficacy and mechanisms. Crit Rev Environ Sci Technol 50(5):476–522. https://doi.org/10.1080/10643389.2019.1629802

Gao D, Zhao H, Wang L, Li Y, Tang T, Bai Y, Liang H (2022) Current and emerging trends in bioaugmentation of organic contaminated soils: a review. J Environ Manag 320:115799. https://doi.org/10.1016/j.jenvman.2022.115799

Ghisi R, Vamerali T, Manzetti S (2019) Accumulation of perfluorinated alkyl substances (PFAS) in agricultural plants: a review. Environ Res 169:326–341. https://doi.org/10.1016/j.envres.2018.10.023

Ghosh S, Sharma I, Nath S, Webster TJ (2021) Bioremediation—the natural solution. In: Microbial ecology of wastewater treatment plants, pp 11–40. https://doi.org/10.1016/b978-0-12-822503-5.00018-7

Ishchi T, Sibi G (2019) Azo dye degradation by Chlorella vulgaris: optimization and kinetics. Int J Biol Chem 14(1):1–7. https://doi.org/10.3923/ijbc.2020.1.7

Jasu A, Lahiri D, Nag M, Ray RR (2021) Fungi in bioremediation of soil organic pollutants. In: Fungi bio-prospects in sustainable agriculture, environment and nano-technology, pp 381–405. https://doi.org/10.1016/b978-0-12-821925-6.00017-4

Jelusic M, Lestan D (2014) Effect of EDTA washing of metal polluted garden soils. Part I: toxicity hazards and impact on soil properties. Sci Total Environ 475:132–141. https://doi.org/10.1016/j.scitotenv.2013.11.049

Karimi H, Mahdavi S, Asgari Lajayer B, Moghiseh E, Rajput VD, Minkina T, Astatkie T (2021) Insights on the bioremediation technologies for pesticide-contaminated soils. Environ Geochem Health 44(4):1329–1354. https://doi.org/10.1007/s10653-021-01081-z

Kazemi Shariat Panahi H, Dehhaghi M, Dehhaghi S, Guillemin GJ, Lam SS, Aghbashlo M, Tabatabaei M (2022) Engineered bacteria for valorizing lignocellulosic biomass into bioethanol. Bioresour Technol 344:126212. https://doi.org/10.1016/j.biortech.2021.126212

Khan AHA, Ayaz M, Arshad M, Yousaf S, Khan MA, Anees M, Sultan A, Nawaz I, Iqbal M (2018) Biogeochemical cycle, occurrence and biological treatments of polycyclic aromatic hydrocarbons (PAHs). Iran J Sci Technol Trans A: Sci 43(3):1393–1410. https://doi.org/10.1007/s40995-017-0393-8

Kupryianchyk D, Hale SE, Breedveld GD, Cornelissen G (2016) Treatment of sites contaminated with perfluorinated compounds using biochar amendment. Chemosphere 142:35–40. https://doi.org/10.1016/j.chemosphere.2015.04.085

Lahiri D, Nag M, Dey A, Sarkar T, Joshi S, Pandit S, Das AP, Pati S, Pattanaik S, Tilak VK, Ray RR (2021) Biofilm mediated degradation of petroleum products. Geomicrobiol J 39(3–5):389–398. https://doi.org/10.1080/01490451.2021.1968979

Leal CS, Mesquita DP, Amaral AL, Amaral AM, Ferreira EC (2019) Environmental impact and biological removal processes of pharmaceutically active compounds: the particular case of sulfonamides, anticonvulsants and steroid estrogens. Crit Rev Environ Sci Technol 50(7):698–742. https://doi.org/10.1080/10643389.2019.1642831

Li S, Li F, Zhu X, Liao Q, Chang J-S, Ho S-H (2022a) Biohydrogen production from microalgae for environmental sustainability. Chemosphere 291:132717. https://doi.org/10.1016/j.chemosphere.2021.132717

Li S-N, Zhang C, Li F, Ren N-Q, Ho S-H (2022b) Recent advances of algae-bacteria consortia in aquatic remediation. Crit Rev Environ Sci Technol 53(3):315–339. https://doi.org/10.1080/10643389.2022.2052704

Liew Z, Goudarzi H, Oulhote Y (2018) Developmental exposures to perfluoroalkyl substances (PFASs): an update of associated health outcomes. Curr Environ Health Rep 5(1):1–19. https://doi.org/10.1007/s40572-018-0173-4

Liu J-L, Wong M-H (2013) Pharmaceuticals and personal care products (PPCPs): a review on environmental contamination in China. Environ Int 59:208–224. https://doi.org/10.1016/j.envint.2013.06.012

Liu Y, Wu J, Liu Y, Wu X (2021) Biological process of alkane degradation by Gordonia sihwaniensis. ACS Omega 7(1):55–63. https://doi.org/10.1021/acsomega.1c01708

Liu C-J, Deng S-G, Hu C-Y, Gao P, Khan E, Yu C-P, Ma LQ (2023) Applications of bioremediation and phytoremediation in contaminated soils and waters: CREST publications during 2018–2022. Crit Rev Environ Sci Technol 53(6):723–732. https://doi.org/10.1080/10643389.2023.2168365

Ma M, Gao W, Li Q, Han B, Zhu A, Yang H, Zheng L (2021) Biodiversity and oil degradation capacity of oil-degrading bacteria isolated from deep-sea hydrothermal sediments of the South Mid-Atlantic Ridge. Mar Pollut Bull 171:112770. https://doi.org/10.1016/j.marpolbul.2021.112770

Markande AR, Acharya SR, Nerurkar AS (2013) Physicochemical characterization of a thermostable glycoprotein bioemulsifier from Solibacillus silvestris AM1. Process Biochem 48(11):1800–1808. https://doi.org/10.1016/j.procbio.2013.08.017

Miri S, Naghdi M, Rouissi T, Kaur Brar S, Martel R (2018) Recent biotechnological advances in petroleum hydrocarbons degradation under cold climate conditions: a review. Crit Rev Environ Sci Technol 49(7):553–586. https://doi.org/10.1080/10643389.2018.1552070

Mona S, Kaushik A, Kaushik CP (2011) Biosorption of reactive dye by waste biomass of Nostoc linckia. Ecol Eng 37(10):1589–1594. https://doi.org/10.1016/j.ecoleng.2011.04.005

Mujawar SY, Shamim K, Vaigankar DC, Dubey SK (2018) Arsenite biotransformation and bioaccumulation by Klebsiella pneumoniae strain SSSW7 possessing arsenite oxidase (aioA) gene. Biometals 32(1):65–76. https://doi.org/10.1007/s10534-018-0158-7

Nazari MT, Simon V, Machado BS, Crestani L, Marchezi G, Concolato G, Ferrari V, Colla LM, Piccin JS (2022) Rhodococcus: a promising genus of actinomycetes for the bioremediation of organic and inorganic contaminants. J Environ Manag 323:116220. https://doi.org/10.1016/j.jenvman.2022.116220

Nguyen HT, Yoon Y, Ngo HH, Jang A (2020) The application of microalgae in removing organic micropollutants in wastewater. Crit Rev Environ Sci Technol 51(12):1187–1220. https://doi.org/10.1080/10643389.2020.1753633

OECD (2018) Environmental directorate joint meeting of the chemicals committee and the working party on chemicals, pesticides and biotechnology. In ENV/JM/Mono (Vol.7). OECD.

Paço A, Jacinto J, da Costa JP, Santos PSM, Vitorino R, Duarte AC, Rocha-Santos T (2019) Biotechnological tools for the effective management of plastics in the environment. Crit Rev Environ Sci Technol 49(5):410–441. https://doi.org/10.1080/10643389.2018.1548862

Partovinia A, Rasekh B (2018) Review of the immobilized microbial cell systems for bioremediation of petroleum hydrocarbons polluted environments. Crit Rev Environ Sci Technol 48(1):1–38. https://doi.org/10.1080/10643389.2018.1439652

Patel AK, Singhania RR, Albarico FPJB, Pandey A, Chen C-W, Dong C-D (2022) Organic wastes bioremediation and its changing prospects. Sci Total Environ 824:153889. https://doi.org/10.1016/j.scitotenv.2022.153889

Perelo LW (2010) Review: in situ and bioremediation of organic pollutants in aquatic sediments. J Hazard Mater 177(1–3):81–89. https://doi.org/10.1016/j.jhazmat.2009.12.090

Pinedo-Rivilla C, Aleu J, Collado I (2009) Pollutants biodegradation by fungi. Curr Org Chem 13(12):1194–1214. https://doi.org/10.2174/138527209788921774

Rafeeq H, Afsheen N, Rafique S, Arshad A, Intisar M, Hussain A, Bilal M, Iqbal HMN (2023) Genetically engineered microorganisms for environmental remediation. Chemosphere 310:136751. https://doi.org/10.1016/j.chemosphere.2022.136751

Ranjan A, Jindal T (2021a) Overview of organophosphate compounds. In: Toxicology of organophosphate poisoning, pp 1–25. https://doi.org/10.1007/978-3-030-79128-5_1

Ranjan A, Jindal T (2021b) Toxicology of organophosphate poisoning in human. In: Toxicology of organophosphate poisoning, pp 27–43. https://doi.org/10.1007/978-3-030-79128-5_2

Sajid M, Ahmad Khan MS, Singh Cameotra S, Safar Al-Thubiani A (2020) Biosurfactants: potential applications as immunomodulator drugs. Immunol Lett 223:71–77. https://doi.org/10.1016/j.imlet.2020.04.003

Samardjieva KA, Gonçalves RF, Valentão P, Andrade PB, Pissarra J, Pereira S, Tavares F (2014) Zinc accumulation and tolerance in Solanum nigrum are plant growth dependent. Int J Phytoremediation 17(3):272–279. https://doi.org/10.1080/15226514.2014.898018

Sharma P, Bano A, Singh SP, Dubey NK, Chandra R, Iqbal HMN (2022) Recent advancements in microbial-assisted remediation strategies for toxic contaminants. Cleaner Chem Eng 2:100020. https://doi.org/10.1016/j.clce.2022.100020

Shen M, Song B, Zeng G, Zhang Y, Huang W, Wen X, Tang W (2020) Are biodegradable plastics a promising solution to solve the global plastic pollution? Environ Pollut 263:114469. https://doi.org/10.1016/j.envpol.2020.114469

Singh P, Iyengar L, Pandey A (2007) Bacterial Decolorization and Degradation of Azo Dyes. Int. Biodeterior. Biodegrad, 59:73–84.

Thomas SC, Madaan T, Kamble NS, Siddiqui NA, Pauletti GM, Kotagiri N (2021) Engineered bacteria enhance immunotherapy and targeted therapy through stromal remodeling of tumors. Adv Healthc Mater 11(2). https://doi.org/10.1002/adhm.202101487

Topuz F, Abdulhamid MA, Hardian R, Holtzl T, Szekely G (2022) Nanofibrous membranes comprising intrinsically microporous polyimides with embedded metal–organic frameworks for capturing volatile organic compounds. J Hazard Mater 424:127347. https://doi.org/10.1016/j.jhazmat.2021.127347

Touliabah HE-S, El-Sheekh MM, Ismail MM, El-Kassas H (2022) A review of microalgae- and cyanobacteria-based biodegradation of organic pollutants. Molecules 27(3):1141. https://doi.org/10.3390/molecules27031141

Usmani Z, Sharma M, Lukk T, Karpichev Y, Thakur VK, Kumar V, Allaoui A, Awasthi AK, Gupta VK (2020) Developments in enzyme and microalgae based biotechniques to remediate micropollutants from aqueous systems—a review. Crit Rev Environ Sci Technol 52(10):1684–1729. https://doi.org/10.1080/10643389.2020.1862551

Verasoundarapandian G, Lim ZS, Radziff SBM, Taufik SH, Puasa NA, Shaharuddin NA, Merican F, Wong C-Y, Lalung J, Ahmad SA (2022) Remediation of pesticides by microalgae as feasible approach in agriculture: bibliometric strategies. Agronomy 12(1):117. https://doi.org/10.3390/agronomy12010117

Wang N, Buck RC, Szostek B, Sulecki LM, Wolstenholme BW (2012) 5:3 Polyfluorinated acid aerobic biotransformation in activated sludge via novel "one-carbon removal pathways". Chemosphere 87(5):527–534. https://doi.org/10.1016/j.chemosphere.2011.12.056

Wang J, de Ridder D, van der Wal A, Sutton NB (2020) Harnessing biodegradation potential of rapid sand filtration for organic micropollutant removal from drinking water: a review. Crit Rev Environ Sci Technol:1–33. https://doi.org/10.1080/10643389.2020.1771888

Wang J, de Ridder D, van der Wal A, Sutton NB (2021) Harnessing biodegradation potential of rapid sand filtration for organic micropollutant removal from drinking water: A review. Critical Reviews in Environmental Science and Technology, 51(18):2086–2118. https://doi.org/10.1080/10643389.2020.1771888

Wernke G, Fagundes-Klen MR, Vieira MF, Suzaki PYR, de Souza HKS, Shimabuku QL, Bergamasco R (2018) Mathematical modelling applied to the rate-limiting mass transfer step determination of a herbicide biosorption onto fixed-bed columns. Environ Technol 41(5):638–648. https://doi.org/10.1080/09593330.2018.1508252

Yaashikaa PR, Kumar PS, Saravanan A, Vo D-VN (2021) Advances in biosorbents for removal of environmental pollutants: a review on pretreatment, removal mechanism and future outlook. J Hazard Mater 420:126596. https://doi.org/10.1016/j.jhazmat.2021.126596

Yadav M, Singh G, Jadeja RN (2021) Bioremediation of organic pollutants: a sustainable green approach. In: Sustainable environmental clean-up, pp 131–147. https://doi.org/10.1016/b978-0-12-823828-8.00006-2

Zakaria NN, Convey P, Gomez-Fuentes C, Zulkharnain A, Sabri S, Shaharuddin NA, Ahmad SA (2021) Oil bioremediation in the marine environment of Antarctica: a review and bibliometric keyword cluster analysis. Microorganisms 9(2):419. https://doi.org/10.3390/microorganisms9020419

Zdarta J, Nguyen LN, Jankowska K, Jesionowski T, Nghiem LD (2021) A contemporary review of enzymatic applications in the remediation of emerging estrogenic compounds. Crit Rev Environ Sci Technol 52(15):2661–2690. https://doi.org/10.1080/10643389.2021.1889283

Zhang J, Chen X, Zhou J, Luo X (2020a) Uranium biosorption mechanism model of protonated Saccharomyces cerevisiae. J Hazard Mater 385:121588. https://doi.org/10.1016/j.jhazmat.2019.121588

Zhang Z, Guo H, Sun J, Wang H (2020b) Investigation of anaerobic phenanthrene biodegradation by a highly enriched co-culture, PheN9, with nitrate as an electron acceptor. J Hazard Mater 383:121191. https://doi.org/10.1016/j.jhazmat.2019.121191

Zhang W, Lin Z, Pang S, Bhatt P, Chen S (2020c) Insights into the biodegradation of Lindane (γ-Hexachlorocyclohexane) using a microbial system. Front Microbiol 11. https://doi.org/10.3389/fmicb.2020.00522

Zhang Z, Sarkar D, Biswas JK, Datta R (2022) Biodegradation of per- and polyfluoroalkyl substances (PFAS): a review. Bioresour Technol 344:126223. https://doi.org/10.1016/j.biortech.2021.126223

Zhao S, Liang T, Zhu L, Yang L, Liu T, Fu J, Wang B, Zhan J, Liu L (2019) Fate of 6:2 fluorotelomer sulfonic acid in pumpkin (Cucurbita maxima L.) based on hydroponic culture: uptake, translocation and biotransformation. Environ Pollut 252:804–812. https://doi.org/10.1016/j.envpol.2019.06.020

Zhou H, Gao X, Wang S, Zhang Y, Coulon F, Cai C (2023) Enhanced bioremediation of aged polycyclic aromatic hydrocarbons in soil using immobilized microbial consortia combined with strengthening remediation strategies. Int J Environ Res Public Health 20(3):1766. https://doi.org/10.3390/ijerph20031766

Part III
Biotechnology and Genetic Basis of Extremophile

Chapter 10
Functional Insights of Nutrients Solubilizing Extremophiles for Potential Agriculture Application

Bhalerao Bharat, Khaire Pravin, Borase Dhyaneshwar, Kamble Bhimrao, Arjun Singh, Murugan Kumar, Aniket Gade, and Arunima Mahto

10.1 Introduction

To thrive under hostile or harsh circumstances, the extremophiles have evolved specific modifications, such as 'water cages' in halophiles, smaller pore diameters in acidophiles, and so on (Coker 2019). Extremophiles are classified as polyextremophiles if they can endure a variety of extreme habitats (Gupta et al. 2014).

B. Bharat (✉)
Department of Biochemistry, Post Graduate Institute, Mahatma Phule Krishi Vidyapeeth, Rahuri, Maharashtra, India

K. Pravin
Department of Plant Pathology and Microbiology, Post Graduate Institute, Mahatma Phule Krishi Vidyapeeth, Rahuri, Maharashtra, India

B. Dhyaneshwar
ICAR-Indian Institute of Sugar Research, Biological Control Centre, Pravaranagar, Maharashtra, India

K. Bhimrao
Department of Soil Science, Post Graduate Institute, Mahatma Phule Krishi Vidyapeeth, Rahuri, Maharashtra, India

A. Singh
ICAR-Central Soil Salinity Research Institute-Regional Research Centre, Lucknow, Uttar Pradesh, India

M. Kumar
ICAR-National Bureau of Agriculturally Important Microorganisms (NBAIM), Mau, Uttar Pradesh, India

A. Gade
Department of Microbiology, Nicolaus Copernicus University, Torun, Poland

A. Mahto
National Institute of Plant Genome Research, New Delhi, India

© The Author(s), under exclusive license to Springer Nature Switzerland AG 2024
A. Ranjan et al. (eds.), *Extremophiles for Sustainable Agriculture and Soil Health Improvement*, https://doi.org/10.1007/978-3-031-70203-7_10

Extremolytes are tiny organic compounds that build up within the cells and are either made or ingested by extremophilic bacteria. By creating and maintaining protective water layers, they shield the macromolecules as well as the structure of cells in extremophiles (Becker and Wittmann 2020). Extremophiles can survive under some of the adverse conditions on the planet, including salinity (2–5 M NaCl for halophiles), pH 4 for acidophiles and >9 for alkaliphiles and temperature for psychrophiles –10 °C to 20 °C and thermophiles 60 °C to 115 °C (Saxena et al. 2016; Sahay et al. 2017). Polyextremophiles can thrive in two or more severe or intense environments. Although extremophiles have been prevalent in eubacterial as well as eukaryotic domains, the primitive forms have been previously reported from archaic groups of bacteria. Archael bacterial phyla include *Euryarchaeota* and *Crenarchaeota*, whereas nearly all phyla from eubacteria have lineages of genera having some form of extreme environment tolerance. Due to the long-term viability and catalytic activity of extremozymes, it finds many industrial, biotechnological, and agricultural applications (Hough and Danson 1999; Sarmiento et al. 2015).

Potential strains of archaea, eubacteria, and fungi have been utilized for white and green biotechnology, pharmaceutical, and food industries (Yadav et al. 2019c). Extremophiles have been classified based on their capability to withstand extreme pH (acidophilic- and alkaline-tolerant), piezophiles (pressure), radiation resistance, xerophiles, and so on (Yadav et al. 2015a, b). The polyextremophilic microbiomes can thrive in additional stressful and challenging conditions in the environment.

Due to the extreme adaptation potential of this group, several landmark efforts have been made to decipher its ecology and diversity. Bioprospecting of extremophiles has been done (Kumar et al. 2021; Yadav 2021a, b) for nutrient solubilization (Fig. 10.1) and other plant growth promotion traits (Fig. 10.2 and Table 10.1). The movement of plant nutrients depends heavily on extremophilic microbiomes. Globally, agroecological environments are using these bacteria as bioinoculants to increase production and nutrient availability. From the Indian perspective, several ecological zones experience extreme climatic regimes ranging from drought to cold conditions; a significant portion of lands comes under the category of salinity, sodicity, and heavy metals affected lands, leading to a reduction in agricultural output and productivity and a reduction in the development of plants (Ramegowda and Senthil-Kumar 2015). Abiotic stress is brought on by unfavourable weather conditions, which also lowers crop output. The many different abiotic stresses include low and high temperatures. Low temperatures are one such stress for the plants, as nearly 205 of the soil is frozen on the earth due to ice (nearly 20%) which reduces agricultural production and productivity significantly. Drought has become one of the key issues in the farming sector (Malyan et al. 2016). Agriculture yield is significantly impacted by drought throughout the planet. Whenever microbes have been identified and extensively studied regarding their functions, such as their capacity to resist extremes and their genetic variety, procedures have been developed for their usage in farming. Microbes additionally serve a significant part in managing this stress. As a result, providing plants with a variety of PGP microbiomes might also aid in the relief of drought in dry areas. Another abiotic stress that may ruin crop yield is water flooding stress. This stress is typically concentrated in locations where rice is grown;

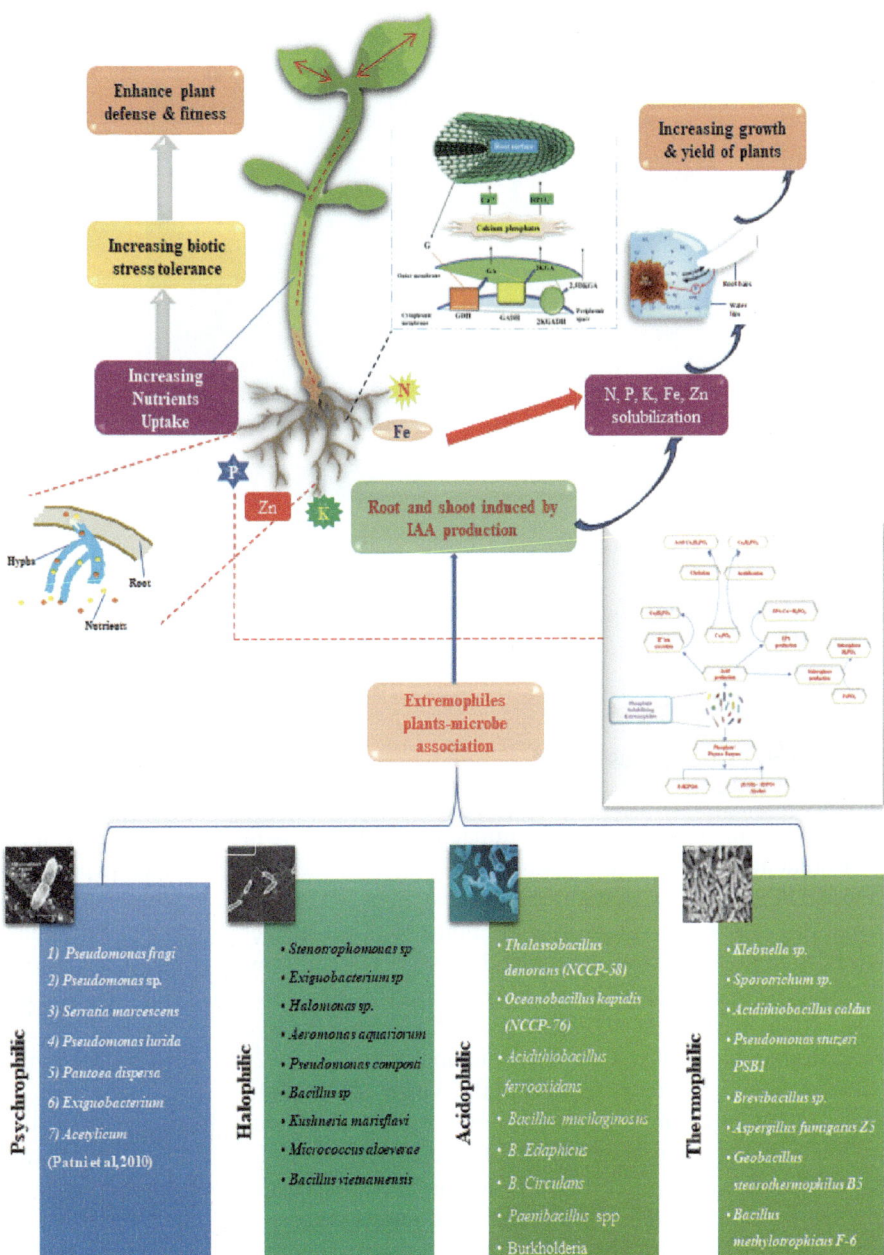

Fig. 10.1 The role of extremophiles in various nutrient solubilization and utilization in plant

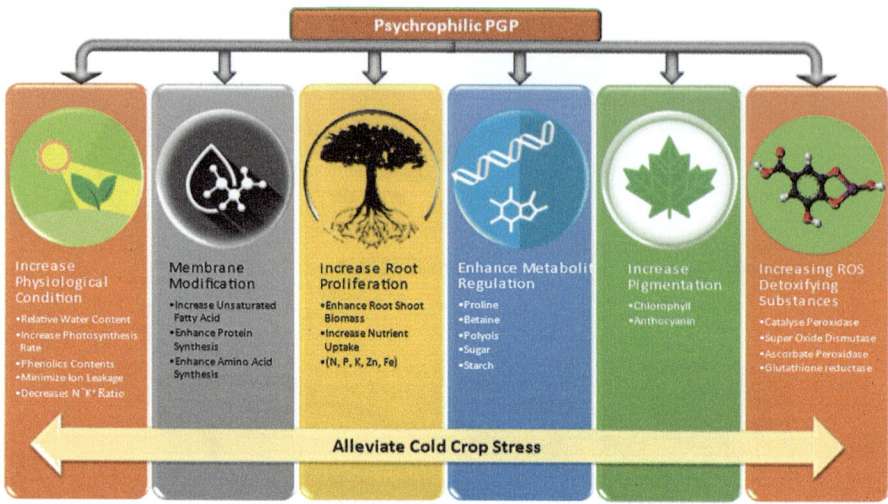

Fig. 10.2 Plant growth promotion by psychrophilic microbes

nearly 25% of paddy fields are annually experiencing erratic floods (Mackill et al. 2012).

10.2 Psychrophiles

One of the primary abiotic factors impacting productivity in agriculture is the harsh environment of cold temperatures. Nearly 20% of the Earth's surface experiences geological features prevalent in permafrost areas. Extremely low temperatures represent distinctive ecosystems that are home to novel species. Studies to explore the diversity of microbial utilized a blend of both classical as well as omics approaches (Yadav 2015; Verma et al. 2015a, b; Frühling et al. 2002; Shivaji et al. 2005; Mayilraj et al. 2006; Kishore et al. 2010). Bacteria known as psychrophiles multiply and flourish in cold settings, typically at temperatures below 20 °C. In 1902, Schmidt-Nielsen coined the term psychrophile which means cold-loving (Ingraham and Stokes 1959). Although they are referred to be 'cold-loving', the majority of them are incapable of developing above 20 °C or below 30 °C. Due to this, they are truly 'cold-tolerant' as opposed to 'cold-loving'. The bacterial groups surviving under low-temperature regimes were categorized into two groups: psychrotrophic (organisms able to thrive between 0 and 30 °C) and psychrophiles (organisms with an optimal growth temperature of 16 °C). Currently, this difference is not made, and any bacteria that can withstand cold temperatures are referred to as psychrophiles. Growth traits studies of microbes isolated from Antarctic biotopes studies revealed that nearly 40–70% of bacteria showed optimum growth between 1 and 30 °C. These individuals were therefore classified as psychrotolerant (Romanovaskaia et al. 2012).

Table 10.1 Significance of extremophilic microbes in agriculture crop growth and production

Type	Name of microbes	Nutrient uptake/functions	Crop	Development of plants as a result	References
Psychrophiles	*Serratia marcescens* SRM (MTCC 8708) and *Pantoea dispersa* 1A	Phosphate solubilization	Wheat	Plant biomass and absorption of nutrients were considerably increased by seed bacterization.	Selvakumar et al. (2008)
	Pseudomonas fragi CS11RH1 (MTCC8984)			The biomass, nutritional absorption, and germination rate in seedlings improved substantially.	Selvakumar et al. (2009b)
	Pseudomonas lurida M2RH3 (MTCC 9245)			Seedling growth and utilization of nutrient metrics were affected by seed bacterization.	Selvakumar et al. (2011)
	Pseudomonas sp. PGERs17 (MTCC9000)	Solubilization of tricalcium phosphate		Seedlings with longer roots and shoots germinated more readily after being bacterized with the isolate.	Mishra et al. (2008)
	Pseudomonas sp. NARs9 (MTCC9002) and *Exiguobacterium acetylicum* 1P (MTCC8707)	Phosphate solubilization		The lengths of the shoots, roots, and germination of seedlings increased.	Mishra et al. (2009)
	Pseudomonas sp.			Plant biochemistry significantly improved. Transportation of N, Fe, and other nutrients by bacterial inoculation increased.	Mishra et al. (2011)
	Pseudomonas sp.	Inorganic phosphatesolubilization	*Deschampsia antarctica* Desv.	Bacterial application encouraged the growth of roots.	Berros et al. (2013)
	Pseudomonas sp.	–	Lentil	Plant biochemical attributes improved.	Bisht et al. (2013)
	Absidia spp.	Potash solubility	–	–	Nenwani et al. (2010)
	Dyadobacter sp.	N	Chickpea, mungbean, urdbean, redgram, and finger millet	Biological N fixation activity improved.	Kumar et al. (2018)
	Pseudomonas jesenii MP1	Phosphorus	Chickpea, common bean, mungbean, urdbean, redgram, and finger millet	Phosphorous availability improved.	Joshi et al. (2019); Suyal et al. (2023)

(continued)

Table 10.1 (continued)

Type	Name of microbes	Nutrient uptake/functions	Crop	Development of plants as a result	References
Acidophilic	*Oceanobacillus kapialis* (NCCP-76) and *Thalassobacillus devorans* (NCCP-58)	Ca^{+2}, K^+, and total N	Rice	The concentrations of Ca^{+2}, K^+, and total N increased significantly.	Shah et al. (2016)
Halophilic	*Stenotrophomonas* sp. and *Exiguobacterium* sp.	Phosphate, N, or ions	Soybean	Availability of phosphate and N improved.	Ana et al. (2022)
	Several strains of *Halomonas* sp.	P solubilization, N fixation		Accumulation of N_2 and PO_4^{3-} levels in shoot and roots improved.	Mukherjee et al. (2019)
	Aeromonas aquariorum, Pseudomonas composti, Bacillus sp.	P solubilization, N fixation		Seed germination enhanced.	Andrades-Moreno et al. (2014)
	Halomonas sp. and *Halobacillus* sp.	P solubilization		Root and dry weight in plants improved.	Desale et al. (2014)
	Bacillus Licheniformis, Bacillus aryabhattai, Bacillus methylotrophicus, B. aryabhattai	N fixation, P solubilization		Longer roots, more lateral roots, and more root hair produced.	Mesa et al. (2015)
	Vibrio neocaledonicus, Pseudarthrobacter oxydans, Thalassospira australica	N fixation, P solubilization		Growth rate enhanced.	Mesa-Marin et al. (2020)
	Consortia 1: *Micrococcus aloeverae, Kushneria marisflavi, Bacillus vietnamensis, Halomonas zincidurans* Consortia 2: *Pseudoalteromonas, Distinct, Vibrio kanaloae, Staphylococcus warneri, P. prydzensis*	P Solubilization, N fixation		Seed vigour increased.	
Thermophilic	*Bacillus* sp.	N uptake and inoculation of S-oxidizing bacteria		N uptake and S transportation improved.	Saharan and Nehra (2011)

The psychrotrophic organisms belong to all three domains, namely, archaea, bacteria, and eukarya. With predominant phyla of *Spirochaetes, Chlamydiae, Euryarchaeota, Ascomycota, Acidobacteria, Verrucomicrobia, Gemmatimonadetes, Planctomycetes, Cyanobacteria, Mucoromycota, Thaumarchaeota, Firmicutes, Basidiomycota, Proteobacteria, Chloroflexi, Actinobacteria, Nitrospirae, Bacteroidetes, and Archaea* (Yadav et al. 2017a, b, c). Psychrophiles have been harnessed for a set of opportunities in biotechnological and agricultural sectors (Table 10.2). The ability of the psychrophilic microbes with plant growth promotional activity helps in the survival of plants under suboptimal temperatures, which has been an important researchable aspect around the world.

Table 10.2 Psychrotrophic microorganism's properties that support plant growth

Psychrotrophic microbes	Phosphorus solubilizing potential (mg L^{-1})	Reference
Aeromonas hydrophila	31.5 ± 1.8	Yadav et al. (2015a)
Arthrobacter methylotrophus	55.9 ± 1.4	Verma et al. (2015a)
Arthrobacter sulfonivorans	25.6 ± 1.2	Yadav et al. (2015b)
Bacillus amyloliquefaciens	54.2 ± 1.5	Verma et al. (2015b)
Bacillus firmus	35.2 ± 3.3	Yadav et al. (2015b)
Bacillus licheniformis	19.2 ± 1.0	Yadav et al. (2016)
Bacillus pumilus	36.1 ± 0.8	Yadav et al. (2015b)
Bacillus subtilis	19.8 ± 0.5	Yadav et al. (2015b)
Cellulosimicrobium cellulans	15.5 ± 1.1	Yadav et al. (2015b)
Desemzia incerta	47.5 ± 1.2	Yadav et al. (2015b)
Paenibacillus tylopili	48.4 ± 2.4	Yadav et al. (2016)
Pantoea dispersa	44.5 ± 0.2	Selvakumar et al. (2008)
Pseudomonas fluorescens	768.3 ± 0.2	Gulati et al. (2008)
Pseudomonas fluorescens	90.2 ± 1.7	Mishra et al. (2011)
Pseudomonas fragi CS11RH1	514.9 ± 0.2	Selvakumar et al. (2009a, b)
Pseudomonas vancouverensis	66.3 ± 0.2	Mishra et al. (2008)
Rahnella sp.	805.0 ± 1	Vyas et al. (2010)
Sanguibacter antarcticus	20.1 ± 0.1	Yadav et al. (2015a)
Sanguibacter suarezii	18.1 ± 0.5	Yadav et al. (2015b)
Stenotrophomonas maltophilia	55.7 ± 0.5	Verma et al. (2015a)

Source: This table was constructed for this manuscript based on information obtained from Yadav et al. (2019a, b, c)

10.2.1 Bioprospecting Agriculturally Important Psychrophilic Microbes

Microorganisms perform a multitude of vital functions within agroecosystems, such as nitrogen (N) fixation, and nutrient solubilization for plant growth promotion. They also support it by producing bioactive compounds for the biocontrol of plant diseases and pests and improving soil fertility. These diverse roles are essential for ensuring the productivity and effectiveness of agricultural systems. The global temperate agricultural ecosystems are categorized by their small seasons of growth, which are separated by times of inadequate temperature. Most microbiological operations will inevitably slow down or, in the worst case, stop altogether in such an environment, which will negatively affect productivity. The transformation of nutrients, where bacteria serve as a vital part, is where the impact is most noticeable. Cold-tolerant bacteria that maintain their activity in inadequate temperature circumstances are essential in this situation because time and temperature are critical factors in both crop development and microbial growth. However, very little work has been done to study the nature and characteristics of these microbes, and there is barely any information on the tolerance mechanisms associated with other critical microorganisms for agriculture.

10.2.2 Psychrophile Microbes for Promoting Plant Growth

Kloepper and Schroth (1978) were the first to characterize the plant-growth-promoting rhizobacteria (PGPR), which are an important part of the rhizosphere microbiome. The microbiome present in the rhizosphere affects the soil's physico-chemical characteristics, thereby supporting the optimal health of the plants and crops. To include various strains that are not rhizospheric in origin, the nomenclature has recently been changed to plant-growth-promoting bacteria (PGPB) (Andrews and Harris 2003). Since numerous physiological processes which support the plants essentially reach an end at inadequate temperatures, the growth and activity of these rhizosphere populations in temperate climates depend mostly on the root zone temperature. Considering the growth of plant stimulation is accomplished by the action of several metabolites, it is crucial in this situation that the root-colonizing bacteria maintain their metabolic versatility at low temperatures. The phytohormones production by PGPR/PGPB in the rhizosphere is also crucial for plant growth promotion (Volkmar and Bremer 1998). These hormones are believed to promote the root structure architecture, which improves the absorption of water as well as minerals via the soil. Siderophore production is another aspect of PGPR that is believed to be helpful for plant growth promotion and improving soil health (Katiyar and Goel 2004). Its antagonistic action towards plant pathogen that affects roots is helpful in the biocontrol of pathogens (Misaghi et al. 1982). *Serratia marcescens* strain SRM and *Pantoea dispersa* strain 1A are psychrotolerant PGPB found in the Northwestern Himalayas of the Indian region, reported to produce IAA at 4 and 15 °C. These bacterial strains were added to the seed before bacterization, greatly increasing plant biomass and absorption of nutrients in wheat seedlings grown in freezing temperatures. The genus *Pseudomonas* contributes significantly to the rhizospheric microbial ecology and frequently aids in plants' growth (Selvakumar et al. 2008a, b).

10.2.3   The Phosphate Solubilization by Psychrophilic Bacteria

Plant growth promotion by phosphate-solubilizing (PS) rhizospheric bacteria is one of the most important techniques. The gluconic acid produced due to the action of membrane-bound enzyme glucose dehydrogenase on glucose has been an important mechanism for mineral phosphate solubilization by microbes (Goldstein 1995). In the second step, gluconic acid is converted into 2,5-diketogluconic acid and 2-ketogluconic acid via Krebs cycle enzymes. Among organic acids, 2-ketogluconic acid has been observed to be better in terms of its PS activity (Kim et al. 2002). Das et al. (2003) tested psychrophilic *Pseudomonas* mutants to assess their PS activities at low temperatures (10 °C), and their findings provided the very first account of PS at low temperatures. When matched to their corresponding wild-type counterparts, they discovered that all of the cold-tolerant mutants were better at PS at 10 °C than they were at 25 °C. It has also been documented that PS by *Pseudomonas* mutants in the psychro-tolerant range of temperature from (4°C) to (28°C) (Katiyar and Goel 2003; Trivedi and Sa 2008). It seems smart to search for naturally produced psychro-tolerant strains in pristine habitats for industrial inoculant production, however, given the ecological stability of mutant strains. The Indian Himalayan region has made the greatest number of advancements in this direction.

The strain of *P. putida* that is antagonistic and cold-tolerant was discovered by Pandey et al. (2006) from a sub-alpine region of the Indian central Himalaya. Phosphate was solubilized by this strain between 4-28 °C. Selvakumar et al. (2009a) showed that a cold-tolerant strain of *P. fragi* was capable of phosphorus solubilization. Given that *P. fragi* is typically linked to the refrigeration-related deterioration of dairy items, this is a unique discovery. This strain greatly increased the rate and percentage of germination, plant biomass, and nutrient assimilations by wheat seedlings in cold environment settings. It also solubilized P at temperatures 4–30 °C. Gulati et al. (2009) discovered an *Acinetobacter rhizosphaerae* strain that is capable of PS in the rhizosphere in the cold regions of the Indian Himalayan area; however, P solubilization at low temperatures by this bacterium was not described. Nineteen fluorescent isolates of *Pseudomonas* from the trans-Himalayan cold deserts were tested by Vyas et al. (2009) for their ability to withstand alkalinity, temperature, calcium (Ca) salts, salinity, as well as desiccation-induced stressors. Selvakumar et al. (2009b) found that the geographical place of origin of cold-tolerant PS *Pseudomonas fragi* CS11RH1 had an impact on their genetic grouping. Repeatitive-element PCR profiles showed that the second cluster was made up of isolates derived from the colder north-facing slopes, whereas the first cluster was made up of isolates from the warmer southern slopes of the Himalayas. The majority of the investigations stated above are exploratory; however, the pursuit of a cold-tolerant PSB inoculant that can be employed successfully in temperate agriculture is what is needed right now.

10.2.4 Nitrogen Fixer

One of the main ways that life on this planet is maintained is through the N fixation processes carried out by symbiotic and asymbiotic bacterial genera; however, cold stress significantly affects it. Depression of nodule competitiveness and nodule activity are two consequences of low temperature on rhizobial activities. By reducing the temperature, *Rhizobium leguminosarum* bv. trifolii produces fewer Nod metabolites, which may have an impact on the nodulation and yield (McKay and Djordjevic 1993). Studies have reported that unfavourable temperature reduces competitiveness in rhizobial bacteria for nodulation, altered nodule function, and also delayed root infection (Lynch and Smith 1994). According to estimates, under temperate circumstances, a week sooner formation of an efficient symbiosis in the cropping period might quadruple the quantity of fixed N and hence boost the productivity of leguminous crops (Sprent 1979). To counteract the stress caused by the cold, it is essential to choose rhizobia strains that are cold-adapted or cold-tolerant. Prevost et al. (1999) made a significant advancement in this regard by choosing psychrotolerant rhizobia from Canadian soils for improved productivity of legumes under cool climates. They used rhizobia linked to native to the arctic and subarctic areas bean plants for this purpose. The candidate rhizobia included *R. leguminosarum* from *Lathyrus* spp., *Oxytropis* spp., and *Mesorhizobium* sp., which was isolated from Astragalus. Psychotropic *Rhizobia* can thrive at 0 ° C. Sainfoin (*Onobrychis viciifolia*), a temperate forage legume, was used to highlight the advantages of cold-adapted *Mesorhizobium* in improving symbiosis in legume. Rhizobial strains isolated from arctics were more infective as compared to temperate rhizobia as they were able to induce more nodulation in sub-zero temperatures. According to biochemical research on cold adaptation, cold-adapted rhizobia produce greater amounts of cold shock proteins than their mesophilic relatives.

It is being researched to exploit cold-adapted *Rhizobia* nodulating field legumes under temperate climates. Nodulation by *Rhizobia* is a summation of the surrounding environment and signals. In contrast to warmer southern climes, *Rhizobia* from North America's temperate regions were able to positively affect soybean nodulation and N fixation (Zhang et al. 2003). In both laboratory as well as field tests, the potential strain of *Sinorhizobium meliloti* showed better nodulation at low temperatures. Prevost et al. (2003) identified a highly effective strain of *Sinorhizobium meliloti* that was specifically adapted for nodulating alfalfa at low temperatures. This strain proved to be the most beneficial in promoting lucern growth during both laboratory and field trials. Notably, it significantly enhanced lucern regrowth after overwintering in cold, anaerobic conditions, such as ice encasement. This study demonstrates that selected rhizobia can cross-adapt to various abiotic challenges typical of temperate climates.

Consequently, the most suitable rhizobia for temperate legumes would need to have strong nodule competitiveness, N-fixing skills, and cold tolerance features. The production and function of membrane-bound Nod factors has been crucial for nodulation and host particularity and would be made possible by such *Rhizobia*'s ability to maintain their cell membrane fluidity at temperatures below freezing. *Azospirillum* is a PGPB that primarily associates with tropical grasses and grain crops. It is a linked symbiotic PGPB. According to Tripathi and Klingmuller (1992), temperature has a significant impact on how well the bacteria can grow, survive, and function. It was previously hypothesized by Kaushik et al. (2001) that the insignificant impact of *Azospirillum* inoculation on crops grown in cold seasons has deterred widespread usage of this bacterium. In order to develop in freezing temperatures, A study with Tn5: lacZ mutants derived from wild type *Azospirillum brasilense* with regards to plant growth promoting activities and root colonization ability. Two *A. brasilense* strains could affect wheat development in field tests when temperatures were not ideal (Kaushik et al. 2002). This represents one of the rare investigations on *Azospirillum* field efficiency at suboptimal temperatures despite the fact that the temperature range where the strains were assessed on plants was not temperate. Cold-tolerant *Azospirillum* has great possibilities for research and development for agriculture under low-temperature conditions. Temperature is a key factor for the decomposition of waste, and at low temperatures, the rate of decomposition reduces; hence, the focus needs to be shifted to the identification of psychrotolerant or psychrophiles microbes that have the potential as PGPR and as a stimulant for soil improvement and bioremediation of metal and organic pollutants under these settings.

10.3 Acidophiles

Acidophiles that thrive below pH value of 3.0 were first reported and identified by Waksman and Joffe in the early 1900s. The bacterium could live in the dilute and weak H_2SO_4 produced by oxidizing elemental S. While there is no precise definition of an acidophile, Johnson (2007) also presented an 'extreme acidophiles' term, which represents microbes that can thrive at pH of 3.0 or below, whereas moderate acidophiles at pH 3.0–5.0, microbes tolerating the pH optima above 5.0 are categorized under acid tolerant. *Picrophilus* can survive pH optimum of 0.7–0, hence surviving the hydronium ion concentration 50–100 times higher than those tolerated even by many extreme acidophiles. Hence are categorized as 'hyper-acidophiles' (grow optimally at pH < 1) though such prokaryotes appear at present to be very rare (Serour and Antranikian 2002).

The mechanisms described, such as the regulation of internal pH, proton pumps, DNA and protein repair systems, biofilm formation, and utilization of various energy sources, are commonly associated with acidophilic microorganisms. The internal pH regulation is done via proton pumps and the GadB-GadC system, reinforced cell membrane impermeability, and DNA/protein repair systems. They utilize chemoorganotrophic metabolism, oxidizing ferrous iron and reducing compounds like inorganic sulfur compounds (ISCs), H_2, and organic matter while employing escape strategies like Quorum Sensing (QS) systems and biofilm formation for adaptation and survival (Quatrini and Johnson 2018). Some acidophiles are also capable of photosynthesis using bacteriochlorophyll pigments, harnessing light energy to drive their metabolic activities. In terms of biodiversity, acidophiles have been derived from all three domains like other extremophiles. Among acidophilic eukaryotes, microalgae, protists, and fungi have also been discovered from low pH environments (Baker et al. 2004) though extreme acidophilic organisms are reported only from archaea and bacteria as described in Table 10.3 (Quatrini and Johnson 2016). Acidophilic bacteria have a high degree of phylogenetic diversity and are covered under *Actinobacteria*, *Aquificae*, *Verrucomicrobia*, *Nitrospirae*, *Proteobacteria*, and *Firmicutes* phyla (Dopson 2016). However, archaeal acidophiles are found under *Thermoplasmatales* and *Sulfolobales* lineage to *Crenarchaeota*

Table 10.3 List of acidophilic extremophiles with their optimal pH and temperature

Phyla	Genera	pH	Temperature	Species
Proteobacteria	*Acidithiobacillus*	2–2.5	30–45 °C	*A. ferridurans, A. ferrivorans, A. ferrooxidans, A. ferriphilus, A. thiooxidans, A. albertensis, and A. caldus*
	Acidiphilium	1–3.5	25–40 °C	*A. acidophilum, A. cryptum, A. Angustum, and A. rubrum. Acidiphilum spp.*
	Acidocella	2.5–3	50–55 °C	*A. facilis, A, aminolytica, Acc. aluminiidurans and A. aromatica*
	Acidicaldus	2.5–3.0	50–55 °C	*A. organivorans*
Nitrospirae	*Leptospirillum*	1–1.6	30–45 °C	*L. ferrooxidans, L. ferriphilum, and L. rubarum.*
Firmicutes	*Sulfobacillus*	1.5–2.5	45–55 °C	*S. thermosulfidooxidans, S. benefacians, S. thermotolerans, S. acidophilus, and S. sibiricus*
	Alicyclobacillus	0.5–6	45–50 °C	*A. disulfidooxidans, A. tolerance, A. aeris, A. ferrooxydans, and A. contaminans*
Actinobacteria	*Acidimicrobium*	2	48 °C	*Acidimicrobium ferrooxidans*

(continued)

Table 10.3 (continued)

Phyla	Genera	pH	Temperature	Species
	Acidithiomicrobium	<3	55 °C	*A. caldus*
	Ferrithrix	1.8	43 °C	*F. thermotolerans*
Aquificae	*Hydrogenobaculum*	3–4	65 °C	*H. acidophilum*
Verrucomicrobia	*Methylacidiphilum*		60 °C	*M. infernorum*
Crenarchaeotes	*Sulfolobus*	2–3	65–85 °C	*S. metallicus, Sb. shibatae, or Sb. tokadaii*
	Metallosphaera	1–6	65–75 °C	*M. sedula, M. prunae, M. hakonensis, M. cuprina, and Ms. tengchongensis*
	Acidianus	1–3	75–90 °C	*A. brierleyi, A. sulfidivorans, A. ambivalens, A. tengchongensis, A. infernus, and A. manzaensis*
	Stygiolobus	2–3	72–102 °C	*S. azoricus*
Euryarchaeota	*Thermoplasma*	0.5–2	60 °C	*T. acidophilum and T. volcanium*
	Acidiplasma	1–1.5	45–55 °C	*A. cupricumulans and A. aeolicum*
	Ferroplasma	1–1.5	35–45 °C	*F. acidophilum and F. acidarmanus*
	Picrophilus	0–0.7	60 °C	*P. torridus and P. oshimae*

Source: This table was constructed for this manuscript based on information obtained from Quast et al. (2013)

and *Euryarchaeota*, respectively (Golyshina et al. 2016). *Crenarchaeota* acidophiles thrive in acidic conditions and often exhibit thermophilic or hyperthermophilic traits, while *Euryarchaeota* tends to favour moderate thermophilic environments.

Ecologically, acidophiles influence biogeochemical cycles of elemental S and iron (Fe) (Druschel et al. 2004), Acidophilic extremophiles can support sustainable agriculture practices by supporting plant health and crop yield in areas with adverse temperature regimes, salinity, drought, and so on (Yadav and Saxena 2018). *Acidithiobacillus* and its associated genera are among the acidophiles that have received the most attention because of their crucial contributions to plant health by S oxidation, P and K solubilization, N fixation and absorption, and solubilization of other nutrients like Zn, Fe, and so on (Kumar et al. 2020). The mentioned organisms are utilized in the reclamation and improvement of high-pH soil, which typically affects crop growth and yield (Bao et al. 2016).

10.3.1   P Solubilization in Crop Plants

Phosphorus is the most crucial nutrient after N, which is crucial for plant growth, crop yield, and quality of produce. To increase soil fertility, a variety of commercial P fertilizers are applied to the soil. However, the majority of the soluble P fertilizer is inaccessible in the soil as insoluble P of aluminium (Al), Ca, and Fe, which are unavailable to plants as nutrients. The microbial-mediated solubilization of insoluble P is brought about by releasing weak organic acids, thereby solubilizing the insoluble P to soluble P, which is then assimilated by the plants (Singh and Amberger 1998). The two basic tactics of P management by microorganisms are the use of P mobilizers and PS bacteria. Another popular approach is to use S-oxidizing bacteria with a reduced form of S in combination with or without a complex P source such as rock phosphate (Besharati et al. 2007; Chi et al. 2006, 2007). *A. thiooxidans* HSS and *A. ferrooxidans* oxidized S to H_2SO_4, which helped in dissolving P from rock mineral fluorapatite. This bacterium was able to solubilize P from 24% to 100% (Maochun et al. 2002). S-oxidizing bacterium *Thiobacillus thiooxidans* has been demonstrated to improve water-soluble P in soil when inoculated with elemental S, rock P, and vermicompost (Aria et al. 2010). In sandy soil, the combined application of arbuscular mycorrhizal fungi and *Acidithiobacillus*, alongside standard doses of S and rock P fertilizers, demonstrated significant improvements in the yield of maize plants, and also, quality and quantity of onion bulbs were improved (Eweda et al. 2015). Inoculation of S-oxidizing *Acidithiobacillus thiooxidans* IW16 with S and struvite (a phosphate mineral) in alkaline soil causes a reduction in pH by formation of H_2SO_4 through oxidation of S, which further solubilized the inorganic P in struvite and also in soil thereby improving P bioavailability to wheat plant (Khan et al. 2019). The utilization of *A. ferrooxidans* for P biosolubilization from poultry bones, fish bones, and phosphorite demonstrated significant effectiveness. Especially, the solubilization factor reached its highest with lower doses of poultry bones (96%) and fish bones (94%), as well as with phosphorite (91%). Subsequently, fertilizers derived from this microbial-mediated solubilization showed P content in terms of P_2O_5 was observed to be 1.6% for poultry bones and 1.9% for fish bones. (Wyciszkiewicz et al. 2017). *Acidithiobacillus* leads to a reduction in pH and promotes changes in phosphate and potassic rocks with a substantial improvement in P and K availability, suggesting an alternative for chemical fertilizers (Silva et al. 2020).

10.3.2   Potassium Solubilization in Crop Plants

Potassium is the third major essential macronutrient after N and P, which plays an important role in the growth, metabolism, development, and yield of the crop plant. The total K content of soils ranges between 3000 and 100,000 kg ha^{-1} in the upper 0.2 m of the soil profile. Approximately 2% of this total K content is present in soil solution in available form, while 98% is bound in minerals in unavailable form (Schroeder 1979; Bertsch and Thomas 1985). The majority of the K is found in feldspar and mica minerals. The K in this crystalline-insoluble state (i.e. unavailable form of K) is not used by plants. These minerals weather (degrade) over extended periods of time, releasing K. However, this procedure is too slow to meet all of the K requirements for field crops. One of the alternative technologies for making K available for plant uptake is microbes-mediated solubilization of K from K-bearing minerals. The bacteria, such as *Bacillus circulans*, *Bacillus edaphicus*, *Pseudomonas*, *Bacillus mucilaginosus*, *Burkholderia*, and *Acidithiobacillus ferrooxidans* and various species of *Paenibacillus* have been reported to release K in an available form from K minerals in soils. These K-solubilizing bacteria facilitate the dissolution of K, Si, and Al from insoluble K minerals such as orthoclase, micas, illite, and so on. They achieve this by secreting organic acids that either directly dissolve the rock-bound K or chelate Si ions, thereby making K available in the soil solution for easy uptake by plants (Sheng et al. 2008; Meena et al. 2015). Recent studies on the solubilization of a K-silicate rock (verdete rock) using *A. thiooxidans* and elemental S have revealed that bacterium-mediated acidification produces 6.6%, 5.8%, 14.1%, and 1.7% solubilization of K, Al, Fe, and Si, respectively, of the total content of these elements in verdete rock (VR) (Matiasa et al. 2019). Research investigating the impact of utilizing biofertilizers produced from phosphate and potash rocks combined with S and treated with *Acidithiobacillus* on the yield of sugarcane revealed significant outcomes. It was observed that the availability of P and K, along with exchangeable Mg and Ca, significantly improved after the application of biofertilizer compared to mineral fertilizers (Stamford et al. 2008). Lima et al. (2007) reported that the application of P and K rock biofertilizers with S and *Acidithiobacillus* in lettuce increases the availability of P and K, specifically for consecutive crops.

10.3.3 S and Plant Nutrition

After N, P, and K, S is the fourth most important nutrient for plants and the sixteenth most important nutrient element crucial for agriculture (Vidyalakshmi and Sridar 2007). The introduction and continuous cultivation of high-yielding varieties causes a deficiency of S, resulting in significant yield loss. Chemical fertilizers help compete with the demand for nutrients; however, their imbalance and improper use result in soil infertility. Sulphur gets into the soil system via microbial-mediated processes, namely, immobilization, mineralization, and redox reactions. Although microbes of all three domains participate in S oxidation, bacteria play a significant role, and among bacteria, *Acidithiobacillus* and related species are important S oxidizers. *Acidithiobacillus* oxidize the reduced form of S from sulfide minerals, releasing S into the soil ecosystem where it can be taken up by plants as sulfate. The oxidation of elemental S by *Thiobacillus thiooxidans* produced sulphate–sulphur in the range of 100–200 µg per ml with a precision of 10 µg (Vogler and Umbreit 1941). Though reports indicated that *Acidithiobacillus* and its related genera were not abundant in soil (Chapman 1990; Lawrence and Germida 2011), a report by Yang et al. (2010) confirmed that the addition of elemental S led to improve the population of *Acidithiobacillus* spp. and allowed S oxidation until 2 weeks. Several studies have found that *Acidithiobacillus* inoculation contributes to S nutrition to crops, resulting in improved yield, and quality. Elemental S, together with *Acidithiobacillus* spp., is employed as a substrate for S oxidation. *Acidithiobacillus* inoculation at 10^4 cells/gram of soil, combined with elemental S treatment, improved soil nutrient levels, nutrient assimilation by plants, and yield in calcareous soils (Besharati 2017). A study reported that S and S-oxidizing bacteria substantially affect certain nutrient deficiencies in calcareous soils. The findings demonstrated S oxidation, pH decrease, and increased sulphate ions (Heydarnezhad et al. 2012). The co-inoculation of biofertilizers containing *Acidithiobacillus thiooxidans* with elemental S at rates of 20 and 30 Kg/ha led to a substantial rise in the uptake of Nitrogen, Phosphorus, and Potassium (NPK) and also S, thereby providing good yield when compared to treatments of S and biofertilizer alone (Pujar et al. 2014).

10.3.4 Role in Zinc and Iron Nutrition to Crop Plants

Soluble zinc (Zn) sources, like fertilizers, have been recommended for application to different crops. As a result, 96–99% of the available Zn is transformed into different, unavailable forms (Saravanan et al. 2003). Microorganisms play a significant role in solubilizing the unavailable form of Zn into the usable form by producing organic acids which regulate the pH of the rhizosphere. Jashni et al. (2017) found that foliar application of Zn and Fe micronutrients combined with S-oxidizing bacterium *Thiobacillus* sp. and a biofertilizer nitrocara increased Zn and Fe accumulation in grains yield and oil content in canola. *Thiobacillus thiooxidans* inoculation with Zn and Fe spraying increased grain yield and quality in maize (Hagh et al. 2016). *Thiobacillus* sp. application combined with ground rubber treatment increased Zn and Fe accumulation in wheat (Asadollahzadeh et al. 2019).

10.4 Soil Reclamations

Restoration of saline–alkali, sodic and calcareous soils as well as soils with heavy metal contamination using *Acidithiobacillus* is considered an innovative, practical, cost-effective, and efficient technique (Yadav et al. 2019a, b, c). A study used a 50 ml culture of *A. thiooxidans* when treated on pH of 7.5 to 8 soil in saline-alkali soil in Jilin, China, it sufficiently lowered the pH of the soil from 7.5 to 7.2 (Bao et al. 2016). A study on the effects of applying a gypsum–S mixture treated with *Acidithiobacillus* to sodic soils found that it reduced soil pH and the level of soil exchangeable Na. The application of Phosphorus and Potassium (PK) rock-based biofertilizer to this reclaimed soil improved cowpea nodulation, dry matter accumulation, and yield and uptake of nutrients (Stamford et al. 2013). In Iran, calcareous soil was reclaimed using elemental S and vermicompost inoculated with *Thiobacillus* sp. Application of these amendments in black seed crops improved P and N uptake and yield and oil content (Seyyedi et al. 2015). Application of S-oxidizing bacteria, particularly when combined with S supplementation and struvite, has been shown to reduce alkaline soil pH by S oxidation (H_2SO_4 production). This reaction releases the locked P from the struvite soil and increases its availability to wheat plants (Khan et al. 2019). In a study, the application of *A. thiooxidans* coupled with S was found to ameliorate the alkalinity of the soils and improve the yield (Li-shu et al. 2013). In contrast, due to growth restrictions *A. thiooxidans* cannot thrive in saline–alkali soil of pH more than 9.0. In the long run, *A. thiooxidans* exhibit significant potential for use in the reclamation of salty soils.

10.5 Thermophiles

Thermophiles are microorganisms that survive in high-temperature conditions, usually above 45 °C. These organisms have attracted interest in various fields, including agriculture, because of their ability to solubilize nutrients and improve plant growth. Thermophilic bacteria like *Bacillus* spp. and *Klebsiella* sp. (Verma et al. 2018; Mukherjee et al. 2020), thermophilic fungi *Sporotrichum* sp. (Singh and Satyanarayana 2010), thermophilic archaea, and thermophilic S oxidizers are some of the most studied thermophiles for their potential in agriculture. This section elaborates on the potential uses of thermophiles in agriculture.

10.5.1 Thermophiles in Phosphorus Nutrition

Thermophiles can play a significant role in phosphorus solubilization and mineralization, contributing to the increased availability of this essential nutrient for plants in agricultural systems. A moderately thermophilic bacterium *Acidithiobacillus caldus* was found to solubilize rock P in the presence of elemental S as an energy source. It was also found that the solubilization rate was influenced by the temperature of incubation (Xiao et al. 2011). Thermotolerant microbes that can solubilize rock P, tricalcium phosphates [$Ca_3(PO_4)_2$], and aluminium phosphates at 25 °C as well as 50 °C were isolated from composts, and their potential was studied (Chang and Yang 2009). A thermotolerant bacterium, *Pseudomonas stutzeri* PSB1, was found to have a significant PS index and improve nodulation in chickpeas (Wasule et al. 2022). A significant increase in plant growth was noticed when inoculated with thermotolerant bacteria. *Bacillus* sp. BISR-HY63 and *Brevibacillus* sp. BISR-HY07 can solubilize rock P and also produce alkaline phosphatase, respectively. Although there is a significant potential for thermophilic and thermotolerant phosphorus solubilizers and mineralizers, there are very few reports and practical applications. More research on thermophiles for phosphorus solubilization and mineralization is essential so that the practical implications of increased nutrient availability, reduced phosphatic fertilizer utilization, and proper disposal of organic phosphorus are met.

10.5.2 Thermophiles in Organic Matter Decomposition and Composting

Any composting process is characterized by four different phases, namely, (i) mesophilic phase where the organic matter is decomposed by mesophilic microbes, (ii) thermophilic phase where due to initial degradation the temperature reaches up to 70 °C, (iii) second mesophilic phase where temperature decreases and degradation continues, and (iv) maturing or curing phase where the produced organics stabilizes (Finore et al. 2023). The role of thermophiles is pronounced in the thermophilic phase of composting where the mesophiles cannot take part in any role in the degradation of organic matter. The presence and activity of thermophiles in the composting process are critical for efficient decomposition, pathogen destruction, and nutrient transformation. Inoculation of thermophiles can speed up the composting process. Inoculation with the thermophilic fungus *Aspergillus fumigatus* Z5 and the thermophilic bacterium *Geobacillus stearothermophilus* B5 was observed to modify key physicochemical factors, enhancing efficiency and maturity during rice straw composting (Wang et al. 2022). Inoculation with thermophilic bacteria such as *Bacillus velezensis* T-B, *Bacillus methylotrophicus* F-6, *and Bacillus haynesii* R-1 in composting of wheat straw and cattle manure prolonged the thermophilic phase and enhanced the degradation of cellulose, hemicellulose, and lignin (Wang et al. 2023). The utilization of thermophiles to hasten the process of compositing and enhance the efficiency of substrate degradation is a major researchable area, and further studies are needed to optimize the processes of composting and organic matter decomposition.

10.5.3 Thermophiles in Nitrogen Nutrition

While thermophiles are not the primary group of N-fixing organisms, some thermophilic bacteria have been found to possess N-fixing capabilities. *Klebsiella* sp. strain PMnew, a thermophilic bacteria from a hot spring has been found to have N fixation ability alongside the capabilities to solubilize phosphorus, produce plant growth hormones like indole compounds and ACC deaminase (Mukherjee et al. 2020). In another study, thermotolerant *Bacillus* sp. and *Weissella* sp. are involved in N transformation and fixation while composting bean dregs (Chen et al. 2022). It is important to understand that while thermophilic N-fixing bacteria exist, they are not as extensively studied or widely used in agricultural practices compared to mesophilic bacteria, such as rhizobia in legume nodules or free-living bacteria in soil. Further research is needed to explore the diversity, physiology, and potential applications of thermophilic N-fixing bacteria, as well as their compatibility with agricultural systems and their ability to enhance N availability in high-temperature environments.

10.5.4 Thermophiles in Abiotic Stress Management

Thermophilic microbes can enhance a plant's tolerance to abiotic stresses, including high temperatures, drought, salinity, and heavy metal toxicity. They produce stress-responsive proteins and enzymes that help plants cope with adverse conditions. Additionally, these microbes may secrete compounds that act as osmoprotectants, reducing water stress in plants. *Bacillus cereus* SA1 was found to protect soybeans from heat stress. Inoculation with this thermotolerant bacilli was found to increase biomass and chlorophyll content in plants challenged with heat stress (Khan et al. 2020). To investigate the ability of bacteria to enhance plant growth in conditions of heat stress, a greenhouse experiment was conducted. Two varieties of tomatoes were inoculated with *B. safensis*, and the results revealed significant improvements in plant growth parameters, antioxidant enzyme activities, and chlorophyll content when compared to non-inoculated plants under heat stress. These findings indicate that the introduction of a heat-tolerant endophytic bacterium with plant growth-promoting properties could serve as an environmentally friendly approach to mitigate the negative effects of heat stress in tomato plants (Mukhtar et al. 2023). Several studies have documented that plants exposed to bacteria capable of producing ACC deaminase exhibit notable enhancements in their well-being under various stress environments, such as flooding, drought, salinity, and heavy metal contamination. Thermotolerant *Bacillus* and *Paenibacillus* strains having ACC deaminase activities were found to protect tomatoes grown in arid regions from drought and heat stress as indicated by their chlorophyll content and shoot and root lengths.

10.6 Halophiles

10.6.1 Agriculturally Important Halophilic Microorganism

Halophiles are extremophilic archaea which thrive in highly saline environments. In another definition for halophilic bacteria by Oren A (2015), they are defined as the groups which require a salt concentration of 0.2 M to attain optimum growth rates. As per the literature surveys, these microbes have been isolated from the Dead Sea and Great Salt Lake, anoxic brine pools of the Red Sea, and elsewhere. Based on their growth optima at various gradients of salts, they have been characterized as moderately halophilic (3–15%) and extremely halophilic (>30%) (Irwin 2020). Taxonomically, the halophiles are part of eubacterial as well as archaeal groups. Under the eubacterial group, they are well-distributed in eight phyla. Under the archaeal group, they are limited to only the 'Nanohaloarchaeota' subphylum, halo-archaea class (Abaramak et al. 2020). Halophilic archaea are known to dominate most of the hypersaline environments on Earth, due to their capability to survive salt concentrations near saturation levels. Among the various strategies to adapt under high salt concentrations, the well-characterized mechanism of osmoadaptation is the most common. In brief, this mechanism works by balancing Na^+/K^+ ions by the accumulation of K^+ ions in the cytoplasm and exudation of Na^+ ions in the surrounding area of the cells. Some halophilic eubacteria and haloarchaea utilize compatible solutes such as betaine and ectoine for osmoregulation. Additionally, haloarchaea has adopted an aerobic lifestyle, which is distinct from other archaea, likely due to gene transfer events with halophilic eubacterial neighbours.

10.6.2   How Does Halophilic Adaptation Benefit a Plant's Fitness Under Salinity?

The benefits of halophilic bacterial adaptations are extended to the plant via its symbiotic associations through its colonization primarily within the endosphere and rhizosphere of the plant system (Fig. 10.3). Associating bacteria primarily enhances plant antioxidant defence mechanisms by upregulating the expression of enzymes such as peroxidase, superoxide dismutase, and catalase. These enzymes play a

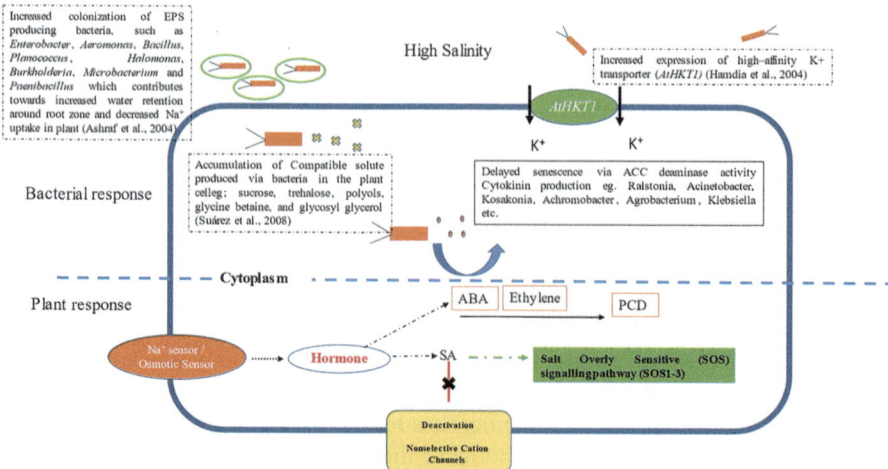

Fig. 10.3 Halotolerant microbes and their role in mitigating salinity-induced stress in plants

crucial role in protecting crop plants from Na^+ and Cl^- toxicity caused by high salt concentrations.

The next mechanism involves the improvement of plant health by increasing nutrient accumulation, bacterization of the corn plant with ACC deaminase positive strains of *Enterobacter aerogenes*, *Pseudomonas syringae*, and *Pseudomonas fluorescens* (Nadeem et al. 2007) not only improved the corn survivability under salinity stress but also lead to increased uptake of phosphorous and potassium. The ability to produce exopolysaccharide (EPS) is quite prevalent among the halophilic bacterial strains. EPS helps in improving the soil structure by improving water retention and reducing the Na^+ uptake in the cells. EPS has free functional groups such as carboxyl, hydroxyl, phosphoryl, and sulfhydryl which bind with the Na^+ ions and reduce its retention in the plant cell system (Nunkaew et al. 2015). In the study conducted on soybeans (Ashraf et al. 2004), EPS-producing *Bacillus subtilis* and *Serratia proteamaculans* strains reduced the uptake of Na^+ ions in the plant system. The plant system facing stress produces senescence hormones such as ethylene as a coping mechanism, which significantly affects plant vigour and health. A specialized group of microbial enzymes known as ACC deaminase inhibits the plant ethylene production pathway by converting 1-aminocyclopropane-1-carboxylate (ACC) to ammonia and alpha-ketobutyrate (Glick 2004), thus delaying the senescence of the plants under stressful conditions. ACC deaminase–producing *Achromobacter piechaudii* when applied to tomato increased the plant's resilience towards salinity stress and further improved its dry and fresh weight (Mayak et al. 2004). Halophilic microbes maintain homeostasis in saline environments through the accumulation of various compounds, including N-acetylated amino acids like proline, peptides, glutamate, and quaternary amines such as carnitine, betaine, and glycine and also sugars like trehalose and sucrose. These compounds, categorized as compatible solutes, aid in osmoregulation not only in bacterial cells but also in plant cells (Creus et al. 2004). Microbial interactions with plants can affect the expression of genes involved in salt stress tolerance (Mahmood et al. 2019). For example, the halotolerant rhizobacterium *Dietzia natronolimnaea* STR1 protected wheat from salt stress by altering the transcriptional machinery responsible for salinity tolerance. This involved the upregulation of genes such as TaABARE, TaOPR1, and TaST related to salt stress resistance, along with modulation of Salt Overly Sensitive (SOS) pathway-related genes. Furthermore, increased expression of antioxidative enzymes and proline content contributed to enhanced salt stress tolerance. The study also noted tissue-specific responses of ion transporters, further influencing salt tolerance mechanisms (Bharti et al. 2016).

Halotolerant plant-growth-promoting bacteria contribute to plant salt tolerance through osmolyte accumulation, ion homeostasis, and exopolysaccharide production. They induce stress resistance genes, antioxidative enzyme synthesis, and phytohormonal modifications, ultimately enhancing plant resilience to salinity (Fig. 10.3) (Bharti et al. 2016; Egamberdieva et al. 2017; Etesami and Maheshwari 2018).

Essential nutrient uptake by extremophiles is still unexplored or limited in literature, and hence, the present chapter explores the minds of young researchers who are working on extremophiles, especially on the insights of nutrient solubilizing extremophilic microbes for potential agriculture applications.

10.7 Conclusion

Extremophiles, including psychrophiles, acidophiles, thermophiles, and halophiles, can survive in extreme environments, making them environmentally resilient, more effective than any other microbes, and cost-effective. They contribute to a better understanding of the fundamental concepts driving biological processes, which inquire about the limits of life and aid in the development of life on Earth. Extremophiles respond readily to available nutrients and are resilient in changing agroclimatic conditions. They have potential for use in agriculture and food security for solubilizing essential nutrients and improving soil fertility, modulating antioxidative/defensive enzymes, and ion transporters in the plants and overall plant health and crop yield. Further investigation must be conducted to discover the most effective strategies to harness the use of extremophiles for environment-friendly agricultural practices.

References

Abaramak G, Kirtel O, Oner ET (2020) Fructanogenic halophiles: a new perspective on extremophiles. In: Physiological and biotechnological aspects of extremophiles. Academic, pp 123–130

Ana SP, Belfiore C, Urbez C, Ferrando A, Miguel AB, Maria EF (2022) Extremophiles as plant probiotics to promote germination and alleviate salt stress in soybean. J Plant Growth Regul 42:946. https://doi.org/10.1007/s00344-022-10605-5

Andrades-Moreno L, Del CI, Parra R, Doukkali B, Redondo-Gomez S, Perez-Palacios P, Caviedes MA, Pajuelo E, Rodriguez-Llorente ID (2014) Prospecting metal-resistant plant-growth promoting rhizobacteria for rhizoremediation of metal contaminated estuaries using spartina densiflora. Environ Sci Pollut Res 21:3713–3721

Andrews JH, Harris RF (2003) The ecology and biogeography of microorganisms on plant surfaces. Annu Rev Phytopathol 38:145–180

Aria MM, Lakzian A, Haghnia GH, Berenji AR, Besharati H, Fotovat A (2010) Effect of Thiobacillus, sulfur, and vermicompost on the water-soluble phosphorus of hard rock phosphate. Bioresour Technol 101:551–554

Asadollahzadeh MJ, Khoshgoftarmanesh AH, Chaney RL (2019) Ability of sulfur-oxidising bacteria to hasten degradation of ground rubber particles in soil for release of zinc as a fertiliser to correct deficiency in wheat. Crop Pasture Sci 70:26–35

Ashraf M, Hasnain S, Berge O, Mahmood T (2004) Inoculating wheat seedlings with exopolysaccharide-producing bacteria restricts sodium uptake and stimulates plant growth under salt stress. Biol Fertil Soils 40:157–162

Baker BJ, Lutz MA, Dawson SC, Bond PL, Banfield JF (2004) Metabolically active eukaryotic communities in extremely acidic mine drainage. Appl Environ Microbiol 70:6264–6271

Bao S, Wang Q, Bao X, Li M, Wang Z (2016) Biological treatment of saline-alkali soil by sulfur-oxidizing bacteria. Bioengineered 7(5):372–375. https://doi.org/10.1080/2165597 9.2016.1226664

Becker J, Wittmann C (2020) Microbial production of extremolytes – high-value active ingredients for nutrition, health care, and well-being. Curr Opin Biotechnol 65:118–128

Berros G, Cabrera-Barjas G, Gidekel M, Gutierrez-Moraga A (2013) Characterization of a novel antarctic plant growth-promoting bacterial strain and its interaction with antarctic hair grass (Deschampsia antarctica Desv). Polar Biology 36:349–362. https://doi.org/10.1007/s00300-012-1264-6

Bertsch PM, Thomas GW (1985) Potassium status of temperate region soils. In: Munson RD (ed) Potassium in agriculture. Soil Science Society of America, Madison, pp 131–162

Besharati H (2017) Effects of sulfur application and Thiobacillus inoculation on soil nutrient availability, wheat yield and plant nutrient concentration in calcareous soils with different calcium carbonate content. J Plant Nutr 40:447–456

Besharati H, Atashnama K, Hatami S (2007) Biosuper as a phosphate fertilizer in a calcareous soil with low available phosphorus. Afr J Biotechnol 6:1325–1329

Bharti N, Pandey SS, Barnawal D, Patel VK, Kalra A (2016) Plant growth promoting rhizobacteria Dietzia natronolimnaea modulates the expression of stress responsive genes providing protection of wheat from salinity stress. Sci Rep 6:1–16

Bisht SC, Mishra PK, Joshi GK (2013) Genetic and functional diversity among root-associated psychrotrophic Pseudomonad isolated from the Himalayan plants. Arch Microbiol 195, 605–615. https://doi.org/10.1007/s00203-013-0908-4

Chang CH, Yang SS (2009) Thermo-tolerant phosphate-solubilizing microbes for multi-functional biofertilizer preparation. Bioresour Technol 100(4):1648–1658

Chapman SJ (1990) Thiobacillus populations in some agricultural soils. Soil Biol Biochem 22:479–482

Chen X, Du G, Wu C, Li Q, Zhou P, Shi J, Zhao Z (2022) Effect of thermophilic microbial agents on nitrogen transformation, N functional genes, and bacterial communities during bean dregs composting. Environ Sci Pollut Res: internat 29(21): 31846–31860. https://doi.org/10.1007/s11356-021-17946-w

Chi R, Xiao C, Gao H (2006) Bioleaching of phosphorus from rock phosphate containing pyrites by Acidithiobacillus ferrooxidans. Min Engin 19(9):979–981. https://doi.org/10.1016/j.mineng.2005.10.003

Chi R, Xiao C, Hang X, Wang C, Wu Y (2007) Bio-decomposition of rock phosphate containing pyrites by Acidithiobacillus ferrooxidans. J Cent S Univ Technol 14:170–175

Coker JA (2019) Recent advances in understanding extremophiles. F1000Res 8:F1000 Faculty Rev-1917

Creus CM, Sueldo RJ, Barassi CA (2004) Water relations and yield in Azospirillum inoculated wheat exposed to drought in the field. Can J Bot 82:273–281

Das K, Katiyar V, Goel R (2003) P solubilization potential of plant growth promoting Pseudomonas mutants at low temperature. Microbiol Res 158:359–362

Desale P, Patel B, Singh S, Malhotra A, Nawani N (2014) Plant growth promoting properties of Halobacillus sp. and Halomonas sp. in presence of salinity and heavy metals. J Basic Microbiol 54:781–791

Dopson M (2016) Physiological and phylogenetic diversity of acidophilic bacteria. In: Quatrini R, Johnson DB (eds) Acidophiles – life in extremely acidic environments. Caister Academic Press

Druschel GK, Baker BJ, Gihring TM, Banfield JF (2004) Acid mine drainage biogeochemistry at Iron Mountain, California. Geochem Trans 5(2):13. https://doi.org/10.1186/1467-4866-5-13

Egamberdieva D, Wirth SJ, Alqarawi AA, AbdAllah EF, Hashem A (2017) Phytohormones and beneficial microbes: essential components for plants to balance stress and fitness. Front Microbiol 8:2104. https://doi.org/10.3389%2Ffmicb.2017.02104

Etesami H, Maheshwari DK (2018) Use of plant growth promoting rhizobacteria (PGPRs) with multiple plant growth promoting traits in stress agriculture: Action mechanisms and future prospects. Eco Envir Saf 156:225–246. https://doi.org/10.1016/j.ecoenv.2018.03.013

Eweda WEE, Hassan EA, Heggo AM, Mohamed AA (2015) Impact of endomycorrhizae and *Acidithiobacillus ferrooxidans* with sulfur and phosphorus nutrition on onion (Allium cepa L.) and maize (Zea mays L.) plants under field conditions. Br Microbiol Res J 9(2):1–15

Finore I, Feola A, Russo L, Cattaneo A, Di Donato P, Nicolaus B, Romano (2023) Thermophilic bacteria and their thermozymes in composting processes: a review. Chem Biol Technol Agric 10(1):7

Frühling A, Schumann P, Hippe H, Straubler B, Stackebrandt E (2002) *Exiguobacterium undae* sp. nov. and *Exiguobacterium antarcticum* sp. nov. Int J Syst Evol Microbiol 52:1171–1176

Glick BR (2004) Bacterial ACC deaminase and the alleviation of plant stress. Adv Appl Microbiol 56:291–312

Goldstein A (1995) Recent progress in understanding the molecular genetics and biochemistry of calcium phosphate solubilization by Gram negative bacteria. Biol Agric Hortic 12:185–193

Golyshina O, Ferrer M, Golyshin PN (2016) Diversity and physiologies of acidophilic archaea. In: Quatrini R, Johnson DB (eds) Acidophiles – life in extremely acidic environments. Caister Academic Press

Gulati A, Rahi P, Vyas P (2008) Characterization of phosphate-solubilizing *fluorescent pseudomonads* from the rhizosphere of seabuckthorn growing in the cold deserts of Himalayas. Curr Microbiol 56:73–79

Gulati A, Vyas P, Rai P, Kasana RC (2009) Plant growth promoting and rhizosphere-competent *Acinetobacter rhizosphaerae* strain BIHB 723 from the cold deserts of the Himalayas. Curr Microbiol 58:371–377

Gupta GN, Srivastava S, Khare SK, Prakash V (2014) Extremophiles: an overview of microorganism from extreme environment. Int J Agric Environ Biotechnol 7(2):371–380

Hagh ED, Mirshekari B, Ardakani MR, Farahvash F, Rejali F (2016) Evaluating maize yield and the quality of response to vermicompost, in Thiobacillus and foliar application of Fe and Zn. Agroecology 8:359–372

Heydarnezhad F, Parisa S, Vahed HS, Hassein B (2012) Influence of elemental sulfur and sulfur oxidizing bacteria on some nutrient deficiency in calcareous soils. Int J Agri Crop Sci 4(12):735–739

Hough DW, Danson MJ (1999) Extremozymes. Curr Opin Chem Biol 3(1):39–46. https://doi.org/10.1016/S1367-5931(99)80008-8

Ingraham JL, Stokes JL (1959) Psychrophilic bacteria. Bacteriol Rev 23:97–108

Irwin JA (2020) Overview of extremophiles and their food and medical applications. In: Physiological and biotechnological aspects of extremophiles. Academic, pp 65–87

Jashni R, Fateh E, Aynehband A (2017) Effect of Thiobacillus and nitrocara biological fertilizers and foliar application of zinc and iron on some qualitative characteristic and remobilization of rapeseed (*Brassica napus* L.). Plant Prod 40:1–14

Johnson DB (2007) Physiology and ecology of acidophilic microorganisms. In: Gerday C, Glansdorff N (eds) Physiology and biochemistry of extremophiles. American Society of Microbiology Press, Washington, DC, pp 257–270

Joshi D, Chandra R, Suyal DC, Kumar S, Reeta GO (2019) Impacts of bioinoculants *Pseudomonas jesenii* MP1 and *Rhodococcus qingshengii* S10107 on chickpea (Cicer arietinum L.) yield and soil nitrogen status. Pedosphere 29(3):388–399

Katiyar V, Goel R (2003) Solubilization of inorganic phosphate and plant growth promotion by cold tolerant mutants of *Pseudomonas fluorescens*. Microbiol Res 158:163–168

Katiyar V, Goel R (2004) Siderophore mediated plant growth promotion at low temperature by mutant of fluorescent pseudomonads. Plant Growth Regul 42:239–244. https://doi.org/10.1023/B:GROW.0000026477.10681.d2

Kaushik R, Saxena AK, Tilak KVBR (2001) Selection and evaluation of *Azospirillum brasilense* strains capable of growing at sub-optimal temperature in rhizocoenosis with wheat. Folia Microbiol 46:327–332

Kaushik R, Saxena AK, Tilak KVBR (2002) Can *Azospirillum* strains capable of growing at a suboptimal temperature perform better in field-grown-wheat rhizosphere. Biol Fertil Soils 35:92–95

Khan A, Jilani G, Zhang D, Akbar S, Malik KM, Shah R, Mujtaba G (2019) *Acidithiobacillus thiooxidans* IW16 and sulfur synergistically with struvite aggrandize the phosphorus bio-availability to wheat in alkaline soil. J Soil Sci Plant Nutr 20:95. https://doi.org/10.1007/s42729-019-00104-0

Khan MA, Asaf S, Khan AL, Jan R, Kang SM, Kim KM, Lee IJ (2020) Thermotolerance effect of plant growth-promoting *Bacillus cereus* SA1 on soybean during heat stress. BMC Microbiol 20(1):1–14

Kim KY, Hwangbo H, Kim YW, Kim HJ, Park KH, Kim YC, Seoung KY (2002) Organic acid production and phosphate solubilization by *Enterobacter intermedium* 60-2G. Korean J Soil Sci Fert 35:59–67

Kishore KH, Begum Z, Pathan AA, Shivaji S (2010) *Paenibacillus glacialis* sp. nov. isolated from the Kafni Glacier of the Himalayas, India. Int J Syst Evol Microbiol 60:1909–1913

Kloepper JW, Schroth MN (1978) Plant growth-promoting rhizobacteria on radishes. In: Proceedings of the 4th international conference on plant pathogenic bacteria. Gibert-Clarey, Tours, pp 879–882

Kumar S, Suyal DC, Bhoriyal M, Goel R (2018) Plant growth promoting potential of psychrotolerant *Dyadobacter* sp. for pulses and finger millet and impact of inoculation on soil chemical properties and diazotrophic abundance. J Plant Nutr 41(8):1035–1046. https://doi.org/10.1080/01904167.2018.1433211

Kumar M, Zeyad MT, Choudhary P, Paul S, Chakdar H, Rajawat MVS (2020) Thiobacillus. In: Amaresan N, Kumar MS, Annapurna K, Kumar K, Sankaranarayanan A (eds) Beneficial microbes in agro-ecology: bacteria and fungi. Academic, pp 545–557

Kumar M, Yadav AN, Saxena R, Paul D, Tomar RS (2021) Biodiversity of pesticides degrading microbial communities and their environmental impact. Biocatal Agric Biotechnol 31:101883. https://doi.org/10.1016/j.bcab.2020.101883

Lawrence JR, Germida JJ (2011) Enumeration of sulfur-oxidizing populations in Saskatchewan agricultural soils. Can J Soil Sci 71:127–136

Lima RCM, Stamford NP, Santos CERS, Dias SHL (2007) Rendimento da alface e atributos químicos de um Latossolo em funcao da aplicacao de biofertilizantes de rochas com fosforo e potassio. Hortic Bras 25:224–229

Li-shu G, Li-ping Y, Bo Y, Ya-bin C, Hao-qiong W, Yan-bo N, Tao Z, A-li D (2013) The effect of *Acidithiobacillus thiooxidans* TT03 to alkaline soil. Heilongjiang Sci 4(5):28–31

Lynch DH, Smith DL (1994) The effects of low temperature stress on two soybean (*Glycine max*) genotypes when combined with Bradyrhizobium strains of varying geographic origin. Physiol Plant 90:105–113

Mackill D, Ismail A, Singh U, Labios R, Paris T (2012) Development and rapid adoption of submergence-tolerant (Sub1) rice varieties. In: Advances in agronomy, vol 115. Elsevier, pp 299–352. https://doi.org/10.1016/B978-0-12-394276-0.00006-8

Mahmood A, Kataoka R, Turgay OC, Yaprak AE (2019) Halophytic microbiome in ameliorating the stress, ecophysiology, abiotic stress responses and utilization of halophytes. Springer, pp 171–194

Malyan SK, Kumar A, Kumar J, Smita KS (2016) Water management tool in rice to combat two major environmental issues: global warming and water scarcity. Environmental concerns of 21st century: Indian and global context. Book Age Publication, New Delhi, pp 46–58

Maochun C, Yongkui Z, Benhe Z, Liyou Q, Bin L (2002) Growth kinetics of Thiobacilli strain HSS and its application in bioleaching phosphate ore. Ind Eng Chem Res 41:1329–1334

Matiasa PC, Mattielloa EM, Santosa WO, Badelb JL, Alvarez VH (2019) Solubilization of a K-silicate rock by *Acidithiobacillus thiooxidans*. Miner Eng 132:69–75

Mayak S, Tirosh T, Glick BR (2004) Plant growth-promoting bacteria that confer resistance to water stress in tomatoes and peppers. Plant Sci 166:525–530

Mayilraj S, Krishnamurthi S, Saha P, Saini HS (2006) *Rhodococcus kroppenstedtii* sp. nov. a novel actinobacterium isolated from a cold desert of the Himalayas, India. Int J Syst Evol Microbiol 56:979–982

McKay IA, Djordjevic MA (1993) Production and excretion of nod metabolites by *Rhizobium leguminosarum* bv. *trifolii* are disrupted by the same environmental factors that reduce nodulation in the field. Appl Environ Microbiol 59:3385–3392

Meena RK, Singh RK, Singh NP, Meena SK, Meena VS (2015) Isolation of low temperature surviving plant growth-promoting rhizobacteria (PGPR) from pea (*Pisum sativum* L.) and documentation of their plant growth promoting traits. Biocatal Agric Biotechnol. https://doi.org/10.1016/j.bcab.2015.08.006

Mesa J, Mateos-Naranjo E, Caviedes MA, Redondo-Gomez S, Pajuelo E, Rodriguez-Llorente ID (2015) Scouting contaminated estuaries: heavy metal resistant and plant growth promoting rhizobacteria in the native metal rhizoaccumulator Spartina maritima. Mar Pollut Bull 90:150–159

Mesa-Marin J, Perez-Romero JA, Redondo-Gomez S, Pajuelo E, Rodriguez-Llorente ID, Mateos-Naranjo E (2020) Impact of plant growth promoting bacteria on Salicornia ramosissima ecophysiology and heavy metal phytoremediation capacity in estuarine soils. Front Microbiol 11:553018

Misaghi IJ, Stowell LJ, Grogan RG, Spearman LC (1982) Fungistatic activity of water-soluble fluorescent pigments of *fluorescent pseudomonads*. Phytopathology 72:33–36

Mishra PK, Mishra S, Selvakumar G, Bisht SC, Bisht JK, Kundu S (2008) Characterisation of a psychrotolerant plant growth promoting *Pseudomonas* sp. strain PGERs17 (MTCC 9000) isolated from North Western Indian Himalayas. Ann Microbiol 58:561–568

Mishra PK, Mishra S, Bisht SC, Selvakumar G, Kundu S (2009) Isolation, molecular characterization and growthpromotion activities of a cold tolerant bacterium Pseudomonas sp. NARs9 (MTCC9002) from the Indian Himalayas. Biol Res 42(3):305–313

Mishra PK, Bisht SC, Ruwari P, Selvakumar G, Joshi GK, Bisht JK (2011) Alleviation of cold stress in inoculated wheat (*Triticum aestivum* L.) seedlings with *psychrotolerant pseudomonads* from NW Himalayas. Arch Microbiol 193:497–513

Mukherjee P, Mitra A, Roy M (2019) Halomonas rhizobacteria of Avicennia marina of Indian Sundarbans promote rice growth under saline and heavy metal stresses through exopolysaccharide production. Front Microbiol 10:1207

Mukherjee T, Banik A, Mukhopadhyay SK (2020) Plant growth-promoting traits of a thermophilic strain of the Klebsiella group with its effect on rice plant growth. Curr Microbiol 77:2613–2622

Mukhtar T, Ali F, Rafique M, Ali J, Afridi MS, Smith D, Chaudhary HJ (2023) Biochemical characterization and potential of *Bacillus safensis* strain SCAL1 to mitigate heat stress in *Solanum lycopersicum* L. J Plant Growth Regul 42(1):523–538

Nadeem S, Zahir ZA, Naveed AM (2007) Preliminary investigations on inducing salt tolerance in maize through inoculation with rhizobacteria containing ACC deaminase activity. Can J Microbiol 53:1141–1149

Nenwani V, Doshi P, Saha T, Rajkumar S (2010) Isolation and characterization of a fungal isolate for phosphate solubilization and plant growth promoting activity. J Yeast Fungal Res 1(1):009–014

Nunkaew T, Kantachote D, Nitoda T, Kanzaki H, Ritchie RJ (2015) Characterization of exopolymeric substances from selected *Rhodopseudomonas palustris* strains and their ability to adsorb sodium ions. Carbohydr Polym 115:334–341

Oren A (2015) Halophilic microbial communities and their environments. Curr Opin Biotechnol 33:119–124

Pandey A, Trivedi P, Palni LMS (2006) Characterization of phosphate solubilizing and antagonistic strain of *Pseudomonas putida* (BO) Isolated from a sub-alpine location in the Indian Central Himalaya. Curr Microbiol 53:102–107

Prevost D, Drouin P, Antoun H (1999) The potential use of cold adapted rhizobia to improve nitrogen fixation in legumes cultivated in temperate regions. In: Margesin R, Schinner F (eds) Biotechnological application of cold-adapted organisms. Springer, Berlin, pp 161–176

Prevost D, Drouin P, Laberge S, Bertrand A, Cloutier J, Levesque G (2003) Cold-adapted rhizobia for nitrogen fixation in temperate regions. Can J Bot 81:1153–1161

Pujar AM, Aravinda Kumar BN, Geeta GS (2014) Response of Sunflower (Helianthus annuus L.) to Graded Levels of Sulphur and Sulphur Oxidising Biofertilizer (Thiobacillus thiooxidans). Biochem. Cell Arch 14:339–342

Quast C, Pruesse E, Yilmaz P, Gerken J, Schweer T, Yarza P, Peplies J, Glockner FO (2013) The SILVA ribosomal RNA gene database project: improved data processing and web-based tools. Nucleic Acids Res 41:590–596

Quatrini R, Johnson DB (2016) Acidophiles: life in extremely acidic environments. Caister Academic Press

Quatrini R, Johnson DB (2018) Microbiomes in extremely acidic environments: functionalities and interactions that allow survival and growth of prokaryotes at low pH. Curr Opin Microbiol 43:139–147

Ramegowda V, Senthil-Kumar M (2015) The interactive effects of simultaneous biotic and abiotic stresses on plants: mechanistic understanding from drought and pathogen combination. J Plant Physiol 176:47–54. https://doi.org/10.1016/j.jplph.2014.11.008

Romanovaskaia VA, Tashirev AB, Gladka GB, Tashireva AA (2012) Temperature range for growth of the Antarctic microorganisms. Mikrobiol Z 74:13–19

Sahay H, Yadav AN, Singh AK, Singh S, Kaushik R, Saxena AK (2017) Hot springs of Indian Himalayas: potential sources of microbial diversity and thermostable hydrolytic enzymes. 3. Biotech 7(2):1–11. https://doi.org/10.1007/s13205-017-0762-1

Saharan B, Nehra V (2011) Plant Growth Promoting Rhizobacteria: A Critical Review. Life Sci Med Res 21:1–30

Saravanan VS, Subramoniam SR, Raj SA (2003) Assessing in vitro solubilization potential of different zinc solubilizing bacterial (ZSB) isolates. Braz J Microbiol 34:121–112

Sarmiento F, Peralta R, Blamey JM (2015) Cold and hot extremozymes: industrial relevance and current trends. Front Bioeng Biotechnol 3:148. https://doi.org/10.3389/fbioe.2015.00148

Saxena AK, Yadav AN, Rajawat M, Kaushik R, Kumar R, Kumar M, Prasanna R, Shukla L (2016) Microbial diversity of extreme regions: an unseen heritage and wealth. Indian J Plant Genet Res 29(3):246–248. https://doi.org/10.5958/0976-1926.2016.00036.X

Schroeder D (1979) Structure and weathering of potassium containing minerals. Proc Congr Int Potash Inst 2:43–63

Selvakumar G, Kundu S, Joshi P, Nazim S, Gupta A, Mishra P (2008) Characterization of a cold-tolerant plant growth-promoting bacterium *Pantoea dispersa* 1A isolated from a sub-alpine soil in the North Western Indian Himalayas. World J Microbiol Biotechnol 24:955–960

Selvakumar G, Kundu S, Joshi P, Gupta AD, Nazim S, Mishra PK, Gupta HS (2008a) Characterization of a cold-tolerant plant growth-promoting bacterium *Pantoea dispersa* 1A isolated from a sub-alpine soil in the North Western Indian Himalayas. World J Microbiol Biotechnol 24:955–960

Selvakumar G, Mohan M, Kundu S, Gupta AD, Joshi P, Nazim S, Gupta HS (2008b) Cold tolerance and plant growth promotion potential of *Serratia marcescens* strain SRM (MTCC 8708) isolated from flowers of summer squash (Cucurbita pepo). Lett Appl Microbiol 46:171–175

Selvakumar G, Joshi P, Mishra PK, Bisht JK, Gupta HS (2009a) Mountain aspect influences the genetic clustering of psychrotolerant phosphate solubilizing *Pseudomonads* in the Uttarakhand Himalayas. Curr Microbiol 59:432–438

Selvakumar G, Joshi P, Nazim S, Mishra PK, Bisht JK, Gupta (2009b) Phosphate solubilization and growth promotion by Pseudomonas fragi CS11RH1 (MTCC 8984) a psychrotolerant bacterium isolated from a high altitude Himalayan rhizosphere. Biologia 64:239–245

Selvakumar G, Joshi P, Suyal P, Mishra PK, Joshi GK, Bisht JK, Bhatt JC, Gupta HS (2011) Pseudomonas lurida M2RH3 (MTCC 9245), a psychrotolerant bacterium from the Uttarakhand Himalayas, solubilizes phosphate and promotes wheat seedling growth. World J Microbiol Biotechnol 27:1129–1135. https://doi.org/10.1007/s11274-010-0559-4

Serour E, Antranikian G (2002) Novel thermoactive glucoamylases from the thermoacidophilic Archaea *Thermoplasma acidophilum*, *Picrophilus torridus* and *Picrophilus oshimae*. Antonie Van Leeuwenhoek 81(1):73–83. https://doi.org/10.1023/A:1020525525490

Seyyedi SM, Moghaddam PR, Khajeh-Hosseini M, Shahandeh H (2015) Influence of phosphorus and soil amendments on black seed (*Nigella sativa* L.) oil yield and nutrient uptake. Ind Crop Prod 77:167–174

Shah G, Jan M, Afreen M, Anees M, Rehman S, Jamil M (2016) Halophilic bacteria mediated phytoremediation of salt-affected soils cultivated with rice. J Geochem Explor 174:59–65. https://doi.org/10.1016/j.gexplo.2016.03.011

Sheng XF, Zhao F, He LY, Qiu G, Chen L (2008) Isolation and characterization of silicate mineral-solubilizing *Bacillus globisporus* Q12 from the surfaces of weathered feldspar. Can J Microbiol 54(12):1064–1068

Shivaji S, Reddy GS, Suresh K, Gupta P, Chintalapati S, Schumann P (2005) *Psychrobacter vallis* sp. nov. and *Psychrobacter aquaticus* sp. nov. from Antarctica. Int J Syst Evol Microbiol 55:757–762

Silva EVN, Stamford NP, Oliveira WS, de Souza Jr VS, Sousa LB, Lira Jr MA (2020) Nutrient availability in phosphate and potassic rocks induced by Acidithiobacillus oxidizing bacteria to produce biofertilizers. AJCS 14(02):250–258

Singh CP, Amberger A (1998) Organic acids and phosphorus solubilisation in straw composted with phosphate rock. Bioresour Technol 63:13–16

Singh B, Satyanarayana T (2010) Applications of phytase of thermophilic mould, *Sporotrichum thermophile*: a review. J Sci Ind Res 69:411–414

Sprent JI (1979) The biology of nitrogen-fixing organisms. McGraw-Hill Book, New York

Stamford NP, Lima RA, Lira Jr MA, Santos CRS (2008) Effectiveness of phosphate and potash rocks with Acidithiobacillus on sugarcane yield and their effects on soil chemical attributes. World J Microbiol Biotechnol 24:2061–2066

Stamford NP, Silva Jr S, Santos CER, Freitas ADS, Lira Jr MA, Barros MFC (2013) Cowpea nodulation, biomass yield and nutrient uptake, as affected by biofertilizers and rhizobia, in a sodic soil amended with Acidithiobacillus. Acta Sci Agron 35(4):453–459

Suyal DC, Khan A, Singh AV, Agarwal A, Pareek N, Sah VK, Goel R (2023) Impact assessment of cold-adapted Pseudomonas jesenii MP1 and Pseudomonas palleroniana N26 on Phaseolus vulgaris yield and soil health. Front Agron 5:2673–3218. https://www.frontiersin.org/articles/10.3389/fagro.2023.1121757

Tripathi AK, Klingmuller W (1992) Temperature sensitivity of nitrogen fixation in *Azospirillum* sp. Can J Microbiol 38:1238–1241

Trivedi P, Sa T (2008) *Pseudomonas corrugata* (NRRL B-30409) mutants increased phosphate solubilization, organic acid production, and plant growth at low temperatures. Curr Microbiol 56:140–144

Verma P, Yadav AN, Khannam KS, Panjiar N, Kumar S, Saxena AK (2015a) Assessment of genetic diversity and plant growth promoting attributes of psychrotolerant bacteria allied with wheat (*Triticum aestivum*) from the Northern Hills zone of India. Ann Microbiol 65:1885–1899

Verma P, Yadav AN, Shukla L, Saxena AK, Suman A (2015b) Alleviation of cold stress in wheat seedlings by *Bacillus amyloliquefaciens* IARIHHS2-30, an endophytic psychrotolerant K-solubilizing bacterium from NW Indian Himalayas. Natl J Life Sci 12:105–110

Verma JP, Jaiswal DK, Krishna PS, Yadav J, Singh V (2018) Characterization and screening of thermophilic *Bacillus* strains for developing plant growth promoting consortium from hot spring of Leh and Ladakh region of India. Front Microbiol 9:1293

Vidyalakshmi R, Sridar R (2007) Isolation and characterization of sulphur oxidizing bacteria. J Cult Collect 5:73–77

Vogler KG, Umbreit WW (1941) The necessity for direct contact in sulfur oxidation by *Thiobacillus thiooxidans*. Soil Sci 51:331–337

Volkmar KM, Bremer E (1998) Effects of seed inoculation with strain of *Pseudomonas fluorescens* on root growth and activity of wheat in well-watered and drought stressed grass-fronted rhizotrons. Can J Plant Sci 78:545–551

Vyas P, Rahi P, Gulati A (2009) Stress tolerance and genetic variability of phosphate-solubilizing *fluorescent Pseudomonas* from the cold deserts of the Trans-Himalayas. Microb Ecol 58:425–434

Vyas P, Joshi R, Sharma KC, Rahi P, Gulati A, Gulati A (2010) Cold-adapted and rhizosphere-competent strain of *Rahnella* sp. with broad-spectrum plant growth-promotion potential. J Microbiol Biotechnol 20:1724–1734

Wang M, Wang X, Wu Y, Wang X, Zhao J, Liu Y, Zhang J (2022) Effects of thermophiles inoculation on the efficiency and maturity of rice straw composting. Bioresour Technol 354:127195

Wang L, Wang T, Xing Z, Zhang Q, Niu X, Yu Y, Chen J (2023) Enhanced lignocellulose degradation and composts fertility of cattle manure and wheat straw composting by Bacillus inoculation. J Environ Chem Eng 11(3):109940

Wasule DL, Shinde RM, Shingote PR, Parlawar ND (2022) Characterization of phosphate solubilizing Pseudomonas stutzeri for nodulation in chickpea. J Plant Nutr:1–13

Wyciszkiewicz M, Saeid A, Malinowski P, Chojnacka K (2017) Valorization of phosphorus secondary raw materials by *Acidithiobacillus ferrooxidans*. Molecules 473(22):1–13. https://doi.org/10.3390/molecules22030473

Xiao CQ, Chi RA, Li WS, Zheng Y (2011) Biosolubilization of phosphorus from rock phosphate by moderately thermophilic and mesophilic bacteria. Miner Eng 24(8):956–958

Yadav AN (2015) Bacterial diversity of cold deserts and mining of genes for low temperature tolerance. Ph.D. thesis, IARI, New Delhi/BIT, Ranchi. https://doi.org/10.13140/RG.2.1.2948.1283/2

Yadav AN (2021a) Beneficial plant-microbe interactions for agricultural sustainability. J Appl Biol Biotechnol 9(1):1–4

Yadav AN (2021b) Microbial biotechnology for bio-prospecting of microbial bioactive compounds and secondary metabolites. J Appl Biol Biotechnol 9(2):1–6

Yadav AN, Saxena AK (2018) Biodiversity and biotechnological applications of halophilic microbes for sustainable agriculture. J Appl Biol Biotechnol 6(1):48–55

Yadav AN, Sachan SG, Verma P, Saxena AK (2015a) Prospecting cold deserts of north western Himalayas for microbial diversity and plant growth promoting attributes. J Biosci Bioeng 119(6):683–693. https://doi.org/10.1016/j.jbiosc.2014.11.006

Yadav AN, Sachan SG, Verma P, Tyagi SP, Kaushik R, Saxena AK (2015b) Culturable diversity and functional annotation of psychrotrophic bacteria from cold desert of Leh Ladakh (India). World J Microbiol Biotechnol 31(1):95–108. https://doi.org/10.1007/s11274-014-1768-z

Yadav AN, Sachan SG, Verma P, Saxena AK (2016) Bioprospecting of plant growth promoting psychrotrophic bacilli from the cold desert of North Western Indian Himalayas. Indian J Exp Biol 54:142–150

Yadav AN, Kumar R, Kumar S, Kumar V, Sugitha T, Singh B (2017a) Beneficial microbiomes: Biodiversity and potential biotechnological applications for sustainable agriculture and human health. J Appl Biol Biotechnol 5:1–13

Yadav AN, Verma P, Kumar V, Sachan SG, Saxena AK (2017b) Extreme cold environments: a suitable niche for selection of novel Psychrotrophic microbes for biotechnological applications. Adv Biotechnol Microbiol 2:1–4

Yadav AN, Verma P, Sachan SG, Saxena AK (2017c) Biodiversity and biotechnological applications of psychrotrophic microbes isolated from Indian Himalayan regions. EC Microbiol 1:48–54

Yadav AN, Verma P, Sachan SG, Kaushik R, Saxena AK (2018) Psychrotrophic microbiomes: Molecular diversity and beneficial role in plant growth promotion and soil health. In: Panpatte DG, Jhala YK, Shelat HN, Vyas RV (eds) Microorganisms for green revolution. Volume 2. Microbes for sustainable agro-ecosystem. Springer, Singapore, pp 197–240

Yadav AN, Gulati S, Sharma D, Singh RN, Rajawat MVS, Kumar R, Dey R, Pal KK, Kaushik R, Saxena AK (2019a) Seasonal variations in culturable archaea and their plant growth promoting attributes to predict their role in establishment of vegetation in Rann of Kutch. Biologia 74(8):1031–1043. https://doi.org/10.2478/s11756-019-00259-2

Yadav AN, Kour D, Rana KL, Yadav N, Singh B, Chauhan VS, Rastegari AA, Hesham AE-L, Gupta VK (2019b) Metabolic engineering to synthetic biology of secondary metabolites production. In: Gupta VK, Pandey A (eds) New and future developments in microbial biotechnology and bioengineering. Elsevier, Amsterdam, pp 279–320. https://doi.org/10.1016/B978-0-444-63504-4.00020-7

Yadav AN, Yadav N, Sachan SG, Saxena AK (2019c) Biodiversity of psychrotrophic microbes and their biotechnological applications. J App Biol Biotech 7(04):99–108. https://doi.org/10.7324/JABB.2019.70415

Yang ZH, Stoven K, Haneklaus S, Singh BR, Schnug E (2010) Elemental sulfur oxidation by Thiobacillus spp. and aerobic heterotrophic sulfur-oxidizing bacteria. Pedosphere 20:71–79

Zhang H, Prithiviraj B, Charles TC, Driscoll BT, Smith DL (2003) Low temperature tolerant *Bradyrhizobium japonicum* strains allowing improved nodulation and nitrogen fixation of soybean in a short season (cool spring) area. Eur J Agron 19:205–213

Chapter 11
Extremophiles and Their Genetic Aspects of Potential Bioactive/Metabolites Beneficial for Promoting Plant Health and Soil Fertility

Bal Krishna, Parkash Verma, Rakesh Kumar, Anil Kumar Singh, Priyanka Upadhyay, Ashutosh Kumar, Talekar Nilesh Suryakant, Birender Singh, Sudeepa Kumari Jha, and Juli Kumari

11.1 Introduction

The human perspective often considers the conditions conducive to human life as the 'normal' state for planet Earth, but this perception is far from the actual reality. Earth is predominantly a cold environment, with approximately 90% of the world's oceans maintaining temperatures no higher than 5 °C (Coker 2019). It is crucial to

B. Krishna (✉) · B. Singh
Department of Plant Breeding and Genetics, Bihar Agricultural University, Sabour, Bhagalpur, Bihar, India

P. Verma
Agricultural Lecturer in Government Boys Senior Secondary School, Alipur, Delhi, India

R. Kumar
Division of Crop Research, ICAR Research Complex for Eastern Region, Patna, Bihar, India

A. K. Singh
Bihar Agricultural University, Sabour, Bhagalpur, Bihar, India

P. Upadhyay · A. Kumar · T. N. Suryakant
Department of Genetics and Plant Breeding, School of Agriculture, Lovely Professional University, Phagwara, Jalandhar, Punjab, India

S. K. Jha
Department of Entomology, Bihar Agricultural University, Sabour, Bhagalpur, Bihar, India

J. Kumari
Education Department, Bihar Government, District Institute of Education & Training, Begusarai, Bihar, India

A. Kumar
Plant Breeding and Genetics, Krishi Vigyan Kendra, Siris, Aurangabad, Bihar, Aurangabad, India

© The Author(s), under exclusive license to Springer Nature Switzerland AG 2024
A. Ranjan et al. (eds.), *Extremophiles for Sustainable Agriculture and Soil Health Improvement*, https://doi.org/10.1007/978-3-031-70203-7_11

recognize that extremophiles are termed 'extreme' solely because they inhabit environments where humans cannot survive without assistance (Coker 2019). Extremophiles are organisms thriving in conditions with extraordinary features, like extreme temperatures, pH levels, salinity, and exposure to metals or chemicals (Nweze et al. 2022). While the majority of extremophiles are unicellular, numerous extremophilic organisms associated with plants have been identified, encompassing bacteria, archaea, and eukarya domains (Nweze et al. 2022). These extreme ecosystems house a distinctive biodiversity of organisms equipped with adaptive traits allowing them to flourish optimally in various extreme environments, adapting to both abiotic and biotic stresses. Polyextremophiles show optimal growth under several environments, while extremophiles have evolved adaptive traits that allow them to thrive optimally under one or more environmental extremes (Saxena et al. 2016). Plant microbiomes are made up of plant-associated extremophilic microorganisms (PAEM) that vary in domain from archaea to eukaryotes. These microbes can colonize various regions of plants, such as the rhizosphere, phyllosphere, and interior tissues. These microorganisms produce phytohormones, solubilize nutrients, and function as antagonists against diseases to help plants grow and adapt to adverse environments, such as temperature, salt, pH fluctuations, and drought stresses. The use of environmentally friendly approaches in agriculture is currently a global trend, and microbes are crucial as plant-growth-promoting microorganisms (PGPM). Potential microbes can be used as biocontrol agents and biofertilizers, which are tolerant to pollution in soil and help maintain soil fertility and health (Fig. 11.1). Additionally, microbes are a source of desirable genes. The goal is to use microbial consortia and apply them along with plants to balance nutrients and environments for soil restoration and improved crop yields, promoting sustainable agriculture practices (Fig. 11.1). This chapter of this book will detail the various adaptations

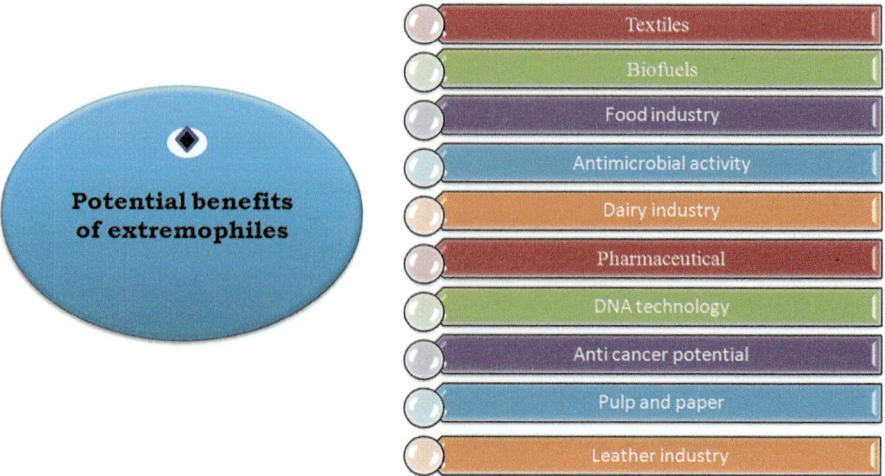

Fig. 11.1 Potential areas of applications of extremophiles

developed by organisms living in extreme environments that allow them to survive and thrive. Additionally, this chapter will highlight some of the significant scientific advancements that have been achieved through studying these unique organisms.

11.2 Extremophile Adjustments for Various Ecological Niches

Most well-described life forms can adapt to 'physiological conditions', which are defined as environments with moderate temperatures (10–37 °C), pH levels around 7, salinities between 0.15 and 0.5 M NaCl, atmospheric pressure of 1 atm, and adequate water availability (Aguilar et al. 1998; Antranikian et al. 2005). Extremophiles may adapt to survive and even thrive in environments that were previously believed to be hostile to them (Rampelotto 2013). Their ability to thrive in such harsh environments has been the subject of extensive research (Orellana et al. 2018). Extremophiles are the subject of applied research to utilize their potential due to the existence of highly complex mechanisms of adaptation and a suite of novel biochemical pathways that sustain unique physiological and metabolic capabilities (Rampelotto 2013).

Extremophiles are divided into groups based on the environments in which they live. According to Berenguer (2011) and Madigan et al. (2000), thermophiles live in an environment where temperatures are normally between 45 and 80 °C, while hyperthermophiles live at temperatures over 80 °C. Opposite to it, psychrophiles can thrive at pH values higher than 9 and can grow at temperatures lower than 5 °C (Fig. 11.2). For halophiles to grow, high concentrations of NaCl between 200 and

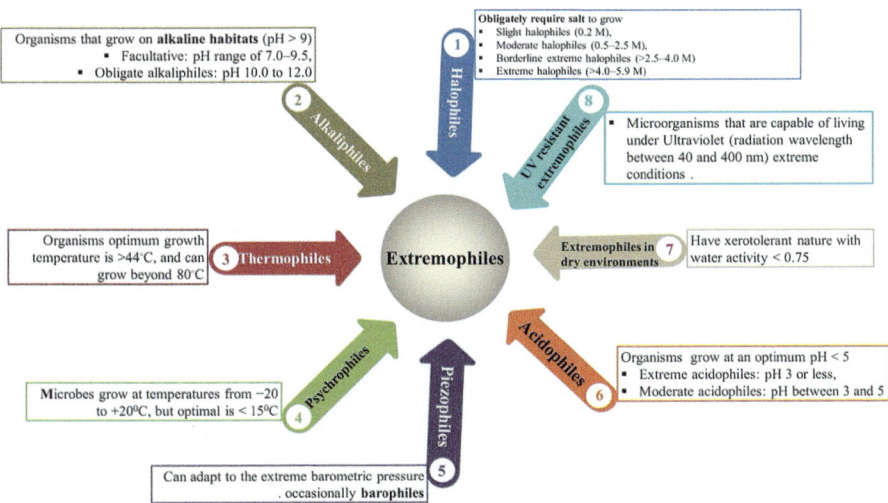

Fig. 11.2 Extremophiles classifications based on various ecological niche

5900 mM are necessary (Edbeib et al. 2016). According to Connon et al. (2007), certain bacteria can thrive in dry conditions with water activity of less than 0.75. UV-resistant and radiotolerant extremophiles are resistant to harmful sun radiation (Gabani et al. 2014). To maintain sustainability on the planet, it is necessary to protect natural ecosystems and restore contaminated areas (Orellana et al. 2018).

11.2.1 Halophiles

According to Margesin and Schinner (2001), halophiles are microorganisms that thrive in high salt concentrations (Margesin and Schinner 2001). According to Edbeib et al. (2016), these halophiles are divided into four groups according to the NaCl concentration needed for their survival and growth (Fig. 11.2). These groups include mild halophiles (0.2 M), moderate halophiles (0.5–2.5 M), borderline extreme halophiles (>2.5–4.0 M), and extreme halophiles (>4.0–5.9 M). They exhibit widespread distribution in diverse saline environments, viz., coastal areas, polar ice, salars, oceans, saline lakes, extreme locations like the Dead Sea, and various geological phenomena and ecosystems that thrive (Oren et al. 1995). Some halophiles have been noted to survive within salt crystals for a limited period (Charlesworth and Burns 2016; Norton and Grant 1988). Microorganisms capable of thriving in the presence or absence of NaCl are termed halotolerant. Extremely halotolerant microorganisms can grow in NaCl concentrations exceeding 2.5 M, according to Margesin and Schinner (2001) (Margesin and Schinner 2001). The salinity levels in agricultural soils are on the rise due to factors such as climate change, the inevitable use of agrochemicals, and irrigation with saline water, as highlighted by Valipour (2014) (Valipour 2014). Salt toxicity poses a significant constraint on crop yield, impacting nearly 20% of cultivated land worldwide (Farhangi-Abriz and Nikpour-Rashidabad 2017; Zhu 2001). The rise in salinity levels has significant implications for food security and highlights the need for strategies to mitigate the effects of salinity on crop productivity. Halophiles have evolved a variety of adaptive features, such as biofilm production, to deal with the high osmotic pressure and desiccation risk associated with saltwater (Fröls 2013), production of compatible solutes (Roberts 2005), and enzyme adaptations to function under ionic stress (Madern et al. 2000). Halophiles have developed several strategies to survive in high-salt environments. One such strategy is to maintain a low intracellular salt concentration, which is achieved through the production or acquisition of compatible solutes, viz., ectoine and betaine (Orellana et al. 2018). Compatible solutes are metabolites that help balance osmotic pressure within cells. Ectoine, glycine betaine, carnitine, glutamate, and proline are some of the commonly described compatible solutes in the literature (Goh et al. 2011), while 2-sulfo trehalose is an example of a more exotic compatible solute found in strongly halo-alkaliphilic archaea (Youssef et al. 2014). To maintain osmotic balance, organisms may employ inorganic solutes like K^+; haloarchaea are well known to employ this tactic (Strahl and Greie 2008). Maintaining an internal salt concentration that is equal to the surrounding environment is another tactic. This calls for the coordinated action of ATP synthase and bacteriorhodopsin, as well as the uptake of Cl⁻ and K⁺ into cells by primary or secondary transporters (Orellana et al. 2018). They commonly produce exopolymer substances (EPS) to form biofilms, which serve as a means of adaptation to their saline environments. These EPS consist of extracellular DNA, lipids, and complex carbohydrates and they can prevent desiccation and reduce UV stresses, which are particularly important to halophilic organisms. In addition to preventing predation, EPS in biofilms can aid in bioremediation by helping to concentrate heavy metals. For example, a particular *Halomonas* isolate demonstrated the effective concentration of lead and cadmium, indicating a possibility of use in the bioremediation of contaminated sites (Amoozegar et al. 2012; Poli et al. 2011; Popescu and Dumitru 2009).

11.2.2 Alkaliphiles

Generally, alkaliphiles grow best in environments with pH values over 9, with a pH of 10 being their ideal growing range (Horikoshi 1999). Facultative alkaliphiles can grow at pH 7.0–9.5, while obligate alkaliphiles, such as *Bacillus krulwichiae*, prefer a pH of 10.0–12.0 (Krulwich and Guffanti 1989) (Fig. 11.2). These extremophiles can tolerate high levels of alkalinity, making them well-suited for inhabiting environments with extreme pH levels, such as soda lakes, which can have a pH range of 10–12 and high salinity (Charlesworth and Burns 2016; Tindall et al. 1984). Despite the harsh conditions, these environments can be highly diverse and support a range of bacteria and archaea, including metabolically active cyanobacteria, thanks to warm temperatures and access to sunlight (Charlesworth and Burns 2016; Duckworth et al. 1996). To survive in such alkaline environments, alkaliphilic bacteria have evolved various molecular mechanisms, including symporter and antiporter systems. The hydro-saline homeostasis and thermodynamic stability of the cell are preserved by the systems, which pump ions into and out of the cell to help establish an electrochemical gradient of Na⁺ and H⁺ (Krulwich 1995; Krulwich et al. 1998; Orellana et al. 2018). Furthermore, alkaliphilic organisms usually modify their cell membranes to maintain a proton motive force inside the cell, which allows them to maintain lower internal pH values than the surrounding environment (Charlesworth and Burns 2016; Krulwich 1995). The cytochrome c-552 is also evident in improving the function of the respiratory system by facilitating electron and H⁺ accumulation (Matsuno and Yumoto 2015).

11.2.3 Thermophiles

Thermophiles are primarily prokaryotic organisms that thrive at optimal growth temperatures exceeding 44 °C (Madigan et al. 2000), with hyperthermophiles capable of growing at even higher temperatures exceeding 80 °C (Fig. 11.2) (Berenguer

2011). These extreme environments, ranging from 40 to 110 °C, are characterized by low nutrient concentrations and high heavy metal (HM) concentrations, including Cd, Cu, Pb, Zn, and Hg (Muñoz et al. 2011). Thermophiles and hyperthermophiles, thriving in environments like hot springs, geothermal waters, fumaroles, mud pots, deep-sea hydrothermal vents, and volcanoes, as well as in engineered settings such as compost facilities and anaerobic reactors, can be found across diverse natural habitats (Ahring 1995; Rastogi et al. 2010; Urbieta et al. 2015a). They have evolved a variety of defence mechanisms in response to the severe heat-induced breakdown of molecules that are vital to their survival, including lipids, enzymes, and nucleic acids. Because of their thermotolerance, thermophile-derived enzymes are very beneficial. One well-known example of such an enzyme is Taq polymerase from *Thermophilus aquaticus*, which made it possible to create the polymerase chain reaction (PCR), an essential tool in molecular biology (Charlesworth and Burns 2016; Saiki et al. 1988). Thermophiles employ several mechanisms to protect their cell organelles in adverse temperatures, including adapting thermophilic proteins by introducing changes to the fundamental structure that enhance stability (Xu et al. 2018). Heat shock proteins (HSPs), such as the chaperones *GroES*, *GroEL*, and *DnaK*, are also involved in assisting protein folding, and thermophiles use compatible solutes to ensure stability to cellular structures and components (Orellana et al. 2018; Urbieta et al. 2015b). Thermophilic organisms have evolved various strategies to protect their cellular components at high temperatures. One such mechanism involves amino acid modifications in the primary structure of thermophilic proteins, which increases their thermal stability (Xu et al. 2018). Heat shock proteins (HSPs), such as the chaperones *DnaK*, *GroEL*, and *GroES*, play a critical role in protein folding and are a primary mechanism in thermophiles to maintain proper protein function (Orellana et al. 2018). Additionally, thermophiles utilize compatible solutes to stabilize cell components, providing an additional layer of protection (Urbieta et al. 2015b).

11.2.4 Psychrophiles

Low temperatures can be just as challenging to live in as high temperatures. Microbes that can thrive in such conditions are called psychrophiles which can grow at temperatures between −20 to 20 °C; however, their ideal growth has been observed at <15 °C (Clarke et al. 2013; Morita 1975) (Fig. 11.2). Psychrophiles can be found in cold environments, viz., mountainous areas, polar regions, some deep-sea niches, and even in high atmospheric zones like stratosphere and mesosphere during certain parts of their life cycle (Joly et al. 2013; Morris et al. 2008). Numerous strategies have been evolved by thermophilic bacteria to enable them to thrive in extremely cold climates. One such strategy to avoid loss of membrane fluidity is an increase in cyclopropane-containing unsaturated fatty acids and short-chain fatty acids in their membranes (D'Amico et al. 2006; Feller and Gerday 2003). They also synthesize high levels of chaperones to protect nucleic acids and protein synthesis and cold-shock proteins (CSPs) (Godin-Roulling et al. 2015; Orellana et al. 2018). Psychrophiles and psychrotrophs also synthesize antifreeze proteins (AFP) which bind to ice to protect the cells and cellular structures at low temperatures (Eskandari et al. 2020; Kristiansen and Zachariassen 2005; Phadtare 2004). AFPs lower the freezing point of water while leaving its melting point unaffected, thus hindering the formation of ice crystals (Eskandari et al. 2020; Gilbert et al. 2005). Additionally, the accumulation of glycine and betaine produced by *Bacillus subtilis* helps prevent the formation of ice crystals and acts as a cold stress protectant enabling bacteria to thrive in cold-stressed environment (Cid et al. 2017; Hoffmann and Bremer 2011; Vega-Celedón et al. 2021). Apart from this approach, the accumulation of cryo-protectants, such as mannitol and polyols, prevents cell damage caused by UV radiation (Orellana et al. 2018). Another approach is the synthesis and secretion of exo-polymeric substances, like those produced by *Colwellia psychrerythraea* 34H (Casillo et al. 2017), which form biofilms, protect against hostile conditions, and facilitate biochemical interactions by trapping nutrients (Nichols et al. 2005). Cold-adapted bacteria also must defend against reactive oxygen species (ROS), which can lead to substantial harm to cells in cold ecosystems (Rizvi et al. 2021). Interestingly, some psychrophiles are capable of driving cloud formation in their high atmosphere environment as part of their life cycle. For example, *Pseudomonas syringae* produces proteins that aid in the formation of ice nucleation, a critical step in cloud formation. This process can have a significant impact on local environments and water availability (Morris et al. 2008).

11.2.5 Piezophiles

Survival in the extreme atmospheric pressures of ocean depths, which can reach up to 110 MPa, with an average of approximately 38 MPa, is significantly challenging. Organisms capable of thriving in such environments are referred to as piezophiles or barophiles (Abe and Horikoshi 2001). Piezophiles exhibit unique adaptations to high pressure, such as the development of densely packed hydrophobic cores in proteins and an inclination for forming multimeric proteins. The challenges associated with culturing piezophiles necessitate the use of complex systems like hydraulic pumps and gas systems to maintain pressures as high as 38 MPa (Y. Zhang et al. 2015). While piezophilic enzymes remain largely untapped in biological research and industrial sectors, they hold promise for various industries. For example, some piezophilic enzymes could potentially contribute to the high-pressure sterilization of foods, given their importance in taste modulation, cheese ripening, the role of proteolytic enzymes, and so on. Further exploration of piezophiles could unlock opportunities in this regard (Y. Zhang et al. 2015).

11.2.6 Acidophiles

Acidophiles are organisms that can survive optimally at a pH lower than 5.0. Acid-tolerant microbes, on the other hand, can survive in low-pH environments despite their optimum pH being above 5 (Johnson 2008). These organisms are found in habitats, viz., hot springs and mine drainage systems (Charlesworth and Burns 2016; Johnson 1995). One of the main approaches that enable acidophiles to survive and grow in low pH environments is active H^+ pumping, which helps maintain a pH of 1 in the cytosol through proton flux systems. In *Thermoplasma acidophilum*, *Bacillus acidocaldarius*, and *Leptospirillum ferriphilum*, researchers have observed the efflux of H^+ through transport pumps of the electron transport chain, alongside the entry of H^+ facilitated by the F0F1-type ATP synthase (Michels and Bakker 1985; Orellana et al. 2018). Another approach utilized by acidophiles to thrive in low pH environments is reducing the plasma membrane permeability to prevent the H^+ influx into the cytoplasm. This is achieved through attaining a positive membrane potential inside the cell generated by K^+ ions, which inhibits proton influx (Christel et al. 2018). Acidophiles also use multiple transporter proteins situated on their plasma membranes to regulate cytosolic pH levels, a crucial adaptation to their surroundings (Baker-Austin and Dopson 2007). Additionally, acidophiles have advanced DNA and protein repair systems compared to neutrophils, which can help mitigate damage caused by low pH conditions. For example, an external pH shift from 3.5 to 1.5 can induce the production of HSPs or chaperones, *Acidithiobacillus ferrooxidans* in the acidophile (Amaro et al. 1991).

11.2.7 Extremophiles in Dry Environments

Climate change has caused an increase in the number of arid environments with high temperatures, low rainfall, and drought (Mukherjee et al. 2018). Desserts are the limit of survivability due to their extreme dryness (Bull and Asenjo 2013; Navarro-González et al. 2003). Xerotolerant extremophiles can survive in these harsh environments where the water activity levels are below 0.75 (Lebre et al. 2017) (Fig. 11.2). Water activity refers to the amount of water available to organisms in the environment and is calculated by comparing the vapour pressure in an environment relative to pure water (Lebre et al. 2017). Xerotolerant microorganisms have evolved two basic strategies to live in these dry environments: environmental stress avoidance and adaptive mechanisms. Spore formation is the process by which cells transform into a non-replicative viable state in order to escape the stress of a dry environment (Crits-Christoph et al. 2013). On the other side, adaptive mechanisms include the synthesis of DNA-repair proteins, the formation of extracellular polymeric materials, the modification of the plasma membrane to hold onto intracellular water, and the prevention of water loss and enhancement of water retention through the accumulation of osmoprotectants like L-glutamate, trehalose, and glycine betaine (Lebre et al. 2017).

11.2.8 UV-Resistant Extremophiles

Ultraviolet (UV) radiation, which has a wavelength between 40 and 400 nm, is a severe abiotic stressor that can cause significant damage to both the community and the cellular levels of living organisms. UV radiation can lead to DNA damage, oxidative stress, protein denaturation, and mutations, resulting in a loss of cell viability (Marizcurrena et al. 2017; Pérez et al. 2017) (Fig. 11.2). However, certain extremophiles, known as UV-resistant organisms, have evolved to withstand these harsh conditions (Gabani et al. 2014). Radiotolerant organisms are particularly resistant to radiation due to their highly efficient DNA repair systems (Charlesworth and Burns 2016; Minton and Daly 1995). Such organisms are now being considered for bioremediation in areas contaminated with radioactive materials (Brim et al. 2003). UV-resistant extremophiles have evolved various approaches to deal with UV stress. These approaches are the synthesis of HSPs/chaperons, advanced DNA repair system, and defence against UV-induced oxidative stress (Pérez et al. 2017). The ability of such microbes to repair DNA damage is linked to their resistance to radiation. For instance, radiotolerant bacteria accumulate high levels of intracellular Mn and low Fe, enabling their tolerance to UV radiation (Paulino-Lima et al. 2016).

11.3 Plant–Microbe Interaction

Plants are seldom found in isolation and are frequently linked to a complex microbiome. The term 'plant microbiome' or 'phytomicrobiome' refers to the group of fungi, bacteria, and archaea that live in nearly all plant tissues (Knack et al. 2015; Smith et al. 2017). Plant–microbe interactions are affected by several environmental factors or abiotic factors (Kumar Meena et al. 2015; Meena et al. 2014). Low temperature is a general constraint for plant growth, development, and production (Hatfield and Prueger 2015). Studies have highlighted beneficial microorganisms, including those that colonize crop plants, and help them tolerate various stressors such as lack of nutrients (Mylona et al. 1995; Yazdani et al. 2009), drought (Lim and Kim 2013) and salinity (Subramanian et al. 2016). In this context, psychrophiles have a benefit over their mesophilic microbes in supporting plant growth and productivity, especially in regions affected by low temperatures (Rizvi et al. 2021). Furthermore, these cold-adaptive bacterial formulations have the potential to replace agrochemicals currently used to optimize plant-microbe interactions. However, there is relatively less research on the effects of soil acidity and alkalinity on plant–microbe interactions (Rajasreelatha and Thippeswamy 2022).

11.4   Role of Extremophiles on Abiotic Stresses

Soil microbial diversity is largely controlled and affected by soil texture, particle size, plant roots, mineral content, and agricultural practices Hartman and Tringe 2019). The rhizosphere, where most microorganisms are concentrated, is modulated by root exudates of the plants. The use of extremophiles as PGPR or biostimulant could be an environmentally friendly, cost-effective, and sustainable approach to combat abiotic stresses in plants. However, changes in the plant's exudation pattern could lead to unexpected interactions between the same plant genotype and microbial communities, potentially causing genetic changes in microorganisms and loss of their ability to colonize the rhizosphere (Enebe and Babalola 2018; Hartman and Tringe 2019; Van Oosten et al. 2017). The use of ammonium-based fertilizers, sulfur, urea, and legume cultivation are major factors contributing to soil acidity in agricultural soils. The salts present in these fertilizers strongly impact soil acidification through processes like nitrification and fixation (Rajasreelatha and Thippeswamy 2022). Compared to non-legumes, legumes cause greater soil acidification due to their excessive uptake of cations compared to anions in the process of N_2 fixation (Tang 1998; Tang et al. 1999). Another reason for soil acidification is crop growth, which leads to the localized acidification of soil as plants take up nutrients. Additionally, organic acids are naturally secreted by plant roots, contributing to soil acidification. Acidophilic microorganisms like *L. ferriphilum* possess a diverse range of genes which modulate cell membrane synthesis and are believed to aid in tolerance to acidic environments. Furthermore, archaea like *Ferroplasma acidiphilum* and *Sulfolobus solfataricus* have tetrapetic lipids in their cell membranes, protecting acidic pH levels (Orellana et al. 2018). Salinity of the soil poses a major threat to agriculture as it impacts soil, soil microorganisms, and plants throughout their life cycle. Soil alkalinity can occur naturally or because of human activities, such as the hydrolysis of silicates, aluminosilicates, and carbonate-containing mixtures, leading to the release of OH^- ions and increasing soil pH levels (Rajasreelatha and Thippeswamy 2022). Additionally, drought can cause soil alkalinity by limiting water availability, preventing the flushing of soluble salts and allowing their accumulation on the soil surface. PGPMs have been shown to possess a remarkable ability to mitigate saline stress due to their adaptability, resistance, and diverse mechanisms involved in the process (Rajasreelatha and Thippeswamy 2022). Endophytic PGPM that form closer associations with plants are also effective in promoting plant growth (R. de Souza et al. 2015). The use of halophilic microorganisms for the reclamation of saline soil is an eco-friendly, safer, and more efficient approach, as they have the potential to eliminate salt from the soil (Arora et al. 2013). Water availability is crucial for plant growth, but drought or flooding can cause abiotic stress, limiting crop production (Danish et al. 2020; İpek et al. 2019). The use of PGPM has been shown to enhance drought stress tolerance in plants, as evidenced in *Mentha pulegium* L. through the use of *Azospirillum brasilense* and *Azotobacter chroococcum* (Asghari et al. 2020), as well as in *Cymbopogon citratus* and *Zea mays* with *Pseudomonas* sp. and *Azotobacter* sp. (Danish et al. 2020;

Mirzaei et al. 2020). Inoculating microbes which impart drought tolerance in plants, viz., *Azotobacter*, *Bacillus*, *Burkholderia*, *Flavobacterium*, *Pseudomonas*, *Methylobacterium*, *Serratia*, and so on have been proven effective in the management of adverse effects of drought in plants. This microbial intervention not only improves plant health and growth but also improves overall yield. Flood tolerance can be improved with the help of *Klebsiella variicola* and *Azospirillum* sp., which induce endogenous hormonal regulation leading to the formation of adventitious roots, as reported in *Zea mays* and *Glycine max* (Czarnes et al. 2020; Kim et al. 2017). Agriculture is highly vulnerable to extreme temperatures which can negatively affect seed germination, seedling growth, and crop yields, resulting in significant economic losses. As the climate changes, high and low temperatures are becoming an increasingly significant threat to global agriculture (İpek et al. 2019; Mukhtar et al. 2020). Inoculation of beneficial extremophilic microorganisms is an effective strategy for improving plant growth and mitigating the adverse effects of extreme abiotic stress. For instance, *Pseudomonas putida* was used to enhance the heat tolerance of *Triticum* sp. (Ali et al. 2011), while *Bacillus cereus* improved the heat tolerance of *Solanum lycopersicum* (Mukhtar et al. 2020). Several extremophilic bacteria, such as *Pseudomonas*, *Serratia*, *Staphylococcus*, *Exiguobacterium*, *Rahnella*, *Stenotrophomonas*, *Leucobacter*, and *Flavobacterium*, have been found to influence plant growth promotion in extremely cold environments (Araya et al. 2020; Dolkar et al. 2018; Kadioglu et al. 2018; Mishra et al. 2011; Verma et al. 2015; Vyas et al. 2010). These bacteria have the potential to synthesize cold-resistant enzymes and mitigate the effects of cold stress on plant growth in agroecosystems, thus optimizing crop production (Feller 2013; Furhan 2020; Kadioglu et al. 2018; Subramanian et al. 2016; Yadav et al. 2016b). Under low-temperature conditions, *Burkholderia* sp. has been shown to increase the tolerance of *Vitis vinifera* to low temperatures by modifying carbohydrate metabolism and increasing crop yield (Fernandez et al. 2012). Similarly, cold-tolerant P-solubilizing *Pseudomonas simiae*, recovered from the *Seabuckthorn rhizosphere*, has been documented to promote plant growth at low-temperature conditions (Dolkar et al. 2018; Rizvi et al. 2021). In order to ensure the resilience of extremophilic PGPM against the various environmental changes encountered by crop plants, it is imperative to isolate these microorganisms from the rhizosphere environments exposed to different environmental conditions, including both high and low temperatures (Etesami 2020). Extremophiles can also aid in HM remediation for plant growth and health. Heavy metal contamination in the soil can have adverse effects on plant growth, as these metals can accumulate in plant tissues and interfere with vital physiological processes. However, certain extremophiles possess the capability to reduce the harmful effects of heavy metals on plants. Studies have shown that psychrophiles possess natural adaptations that allow them to tolerate heavy metals in the environment (M.-J. De Souza et al. 2006; González-Aravena et al. 2016; Tomova et al. 2015). *Pseudomonas putida* strain ATH-43 tolerating Cd, Cu, Cr, and Se, and *Psychrobacter* sp. ATH-62 is tolerant to Hg and tellurium (Te) metals (Rodríguez-Rojas et al. 2016). In addition, thermophilic pure cultures and consortia have demonstrated high potential for bioremediation of HM-polluted groundwater as well as surface water, with capabilities such as immobilization of radionuclides, biosorption and HMs (Sar et al. 2013), removal of HMs (Ilyas et al. 2014), and persistent organic pollutants, viz., polyaromatic hydrocarbons, polychlorinated biphenyls, and pesticides degradation. *Serratia odorifera* and *Pseudomonas fluorescens* strains isolated from the As metal-contaminated river of the Atacama Desert demonstrated resistance to high concentrations of As ranging from 800 to 1000 mM (Campos et al. 2009; Escalante et al. 2009). Acidophilic bacteria, including *Leptospirillum ferrooxidans* and *At. ferrooxidans* have demonstrated their potential in the removal of arsenite (AsO_4^{3-}) from polluted sites. Furthermore, in microcosm experiments, a consortium of acidophilic bacteria *A. ferrooxidans*, *A. thiooxidans* and *L. ferrooxidans* has also exhibited Fe-oxidizing potential useful to remove Hg and Ni from port sediments (Beolchini et al. 2009). The drainage systems of acid mines have also been reported to have toxic aliphatic organic compounds which could be mineralized by acidophiles to mitigate the impact of mining on surrounding water systems (Gemmell and Knowles 2000).

11.5 Extremophiles in Biotic Stress

Extremophiles have developed unique mechanisms to adapt and thrive in harsh conditions. Such adaptation may also aid them in mitigating biotic stress, caused by living organisms, viz., pests, pathogens, or competitors. Certain cold-tolerant phosphate solubilizing bacteria (PSB) species, such as *Flavobacterium*, *Pseudomonas*, and *Serratia*, produce ACC deaminase, a phytoenzyme which can reduce plant ethylene levels synthesized under biotic/abiotic stress responsible for inducing chlorosis, senescence, and abscission in plants. Consequently, the reduction in ethylene levels caused by ACC deaminase helps alleviate the harmful effects of biotic stress or pests on plants (Dubois et al. 2018; Etesami 2020; Vega-Celedón et al. 2021). Decreasing levels of ethylene in plants allows them to grow normally even under stress, and studies have shown that cold-active PGPB can protect plants from phytopathogens by the synthesis of enzymes like ACC deaminase, chitinases, and proteases (Hu et al. 2015; Melo et al. 2016; Rizvi et al. 2021; Verma et al. 2015). These enzymes are helpful in breaking down and degrading fungal cell walls, thereby restricting the fungal pathogen of the plants (Rizvi et al. 2021). To ensure the sustainability of agro-ecosystems, it is important to apply cold-adapted PGPB to inhibit the phytopathogens infestation and succession of diseases (Rizvi et al. 2021; Torracchi C. et al. 2020). Studies have exhibited that certain psychrophilic *Pseudomonas*, e.g., can secrete chitinases and proteases that inhibit the growth of fungal phytopathogens like *R. solani*, *Fusarium* sp., *P. capsica*, *A. solani* (Tapia-Vázquez et al. 2020). A chitinase-producing *Pseudomonas* sp. has also been isolated from marine sediments, which suppress the growth of *F. oxysporum* f. sp. cucumerinum and *V. dahlia* (K. Liu et al. 2019). In addition, natural products from thermophilic environments have been discovered, viz., antimicrobial and antiquorum sensing compounds from cyanobacterial mats in Oman, and novel antinematode compounds like thermolides from *Talaromyces thermophilus* (Dobretsov et al. 2011; Guo et al. 2012).

11.6 Extremophiles in Soil Ecosystem

Extremophilic organisms exhibit high diversity in soil environments, as noticed in studies by Pikuta et al. (2007) (Pikuta et al. 2007). Until a decade ago, the presence and importance of Archaea in soil were largely unknown, as these microorganisms were typically isolated from extreme environments that were challenging to study using traditional microbiological methods. However, recent surveys of soil samples from Darmstadt, Germany, have revealed the relative abundance of crenarchaeotal rDNA to be between 0.5% and 3% in bulk soil, with 0.16% and 0.17% in the rhizosphere of sandy and agricultural ecosystems, respectively. These findings indicate that *crenarchaeota* constitutes a significant component of the terrestrial ecosystem (Ochsenreiter et al. 2003; Pikuta et al. 2007). Additionally, crenarchaeota have been found to associate with plants, viz., tomato, maize, rice, and pine seedlings, revealing their ecological importance in the soil ecosystem. Crenarchaeotal small subunit rRNA gene sequences have also been identified in rhizosphere of various plants, e.g., rice (Großkopf et al. 1998), tomato and maize (Chelius and Triplett 2001; Simon et al. 2005), and the pine seedlings mycorrhiza (Bomberg et al. 2003). An epifluorescence microscopy and culture-independent methods revealed that *Crenarchaeota* colonizes young as well as ageing plant roots at surprisingly higher rates (Simon et al. 2000). These microorganisms were found to be particularly proficient at colonizing senescent roots, suggesting that they are crucial in root development and ecology (Pikuta et al. 2007). Studies have inferred that the associations between *Crenarchaeota* and plants in natural environments were likely mediated by environmental factors and displayed the complex interactions between microbes and plants (Sliwinski and Goodman 2004). Nicol et al. (2004) reported that archaea can also be found in upland pastures (Nicol et al. 2004). The crenarchaeal community in managed pastures where input of inorganic N fertilizer was used, differed from the unmanaged upland pastures, where N sources were mainly from grazing sheep (Pikuta et al. 2007). These soil microcosms showed stable archaeal communities that were least influenced by changes in nitrogen sources or pH levels. These findings suggest that mesophilic soil crenarchaeotes have important ecological functions and are significantly associated with plant roots. *Euryarchaeota*, another group of archaea, has been found in various types of soil. Most of the research on the archaeal community of *Euryarchaeota* has been conducted in rice paddy field soil, especially in flooded rice paddies (Pikuta et al. 2007). In flooded environments, archaea are key methanogens due to the absence of oxygen (Liesack et al. 2000). They use acetate or H_2–CO_2 as substrates for methanogenesis, and they can also endure dry seasons.

Bacteria indirectly support plant growth by biocontrolling pathogens through the production of antimicrobial substances or by providing induced systemic resistance to the plants. This can be achieved by releasing, chitinases, antimicrobial compounds, siderophores, fluorescent pigments, or cyanide (Verma et al. 2017). Biocontrol systems offer eco-friendly and cost-effective solutions to maintain soil health and enhance consistency. Effective biocontrol agents must endure diverse conditions, including pH, temperature, and ion variations. Apart from that, they should also ideally have the potential to deter nematodes and insects through direct and indirect mechanisms. They should produce antimicrobial substances, compete for resources, and induce systemic resistance in plants. Recent research highlights bacterial wilt disease management using antagonistic bacteria. Various bacterial species like *Bacillus*, *Alcaligenes*, and *Pseudomonas* demonstrate inhibition of plant pathogens, offering promising characteristics for sustainable agriculture (Gholami et al. 2014; Purnawati 2014; Verma et al. 2017; Yadav et al. 2015). Iron is a crucial nutrient for all living organisms, and bacterial siderophores can effectively transport and solubilize Fe metal. Microorganisms that produce siderophores can efficiently transport the Fe-siderophore complex under Fe-limited conditions through specific proteins' expression. Microbial siderophore production aids in restraining plant pathogen propagation and can additionally boost plant growth via PGPB.

Biosurfactants represent promising alternatives to chemical surfactants in agriculture, as they are of significant application as adjuvants in the formulation of herbicides and pesticides. They contribute to soil bioremediation and exhibit antimicrobial properties, facilitating the biocontrol of phytopathogens. This dual role of siderophores and biosurfactants underscores their significance in sustainable agricultural practices. Their ability to enhance plant defence mechanisms highlights their potential for eco-friendly pest management strategies (Raddadi et al. 2015; Sachdev and Cameotra 2013). Biosurfactants can also improve the quality and structure of arid-zone soil by hydrophilising soils, improving wettability, and reducing water infiltration thereby achieving an expansion in sustainable agriculture practices in arid conditions.

11.7 Abiotic Stress Resistance Genes in Extremophiles

Debaryomyces hansenii is a type of yeast that is commonly found in salty environments and is known to thrive under extreme pH and high temperatures (Almagro et al. 2000; S.-H. Zhang 2016). Research involving *S. cerevisiae* mutants holding a genomic library from *D. hansenii* revealed multiple genes linked to salt tolerance. Among these, *DhGZF3* encodes a GATA transcription factor similar to *Dal80* and *Gzf3* in *S. cerevisiae*. Despite exhibiting negative transcriptional regulation in *S. cerevisiae*, *DhGZF3* underwent functional analysis within *D. hansenii* (García-Salcedo et al. 2006; S.-H. Zhang 2016). Studies have shown that among extremophiles, genes conferring the resistance to abiotic stress in plants tend to be more resilient than those from non-extremophiles (S.-H. Zhang 2016). For instance, *EhHOG*, a *MAPK* kinase gene found in *E. herbariorum*, was found to outperform wild-type HOG1 under conditions of high salt and freezing-thawing even though it is like *HOG1* homologs found in other fungi, such as *S. cerevisiae*, *A. nidulans*, and *Schizosaccharomyces pombe* (Jin et al. 2005). During plant evolution, gene transfer from unrelated species has been observed, and such genes can provide plants with beneficial traits. For example, *Candida tropicalis*, a halophytic yeast, contains the *CtHSR1* gene, which confers the capability to adapt to adverse environments, and this gene has been successfully transferred to plants (Martínez et al. 2015). Fungi, such as *Trichoderma harzianum*, have been used for agricultural biological control and have moderate levels of stress tolerance. However, these fungi also possess genes responsible for stress resistance that could be transferred to plants to enhance their resistance to stressors (S.-H. Zhang 2016). *ThHog1*, *HSP70*, and *Thkel1* are some of the genes identified from *T. harzianum* that are known to provide salinity tolerance or other stress resistance (Delgado-Jarana et al. 2006; Hermosa et al. 2011; Montero-Barrientos et al. 2008, 2010; S.-H. Zhang 2016). Furthermore, genes from extremophilic fungi may confer even greater stress resistance to plants. Therefore, identifying and characterizing stress-resistance genes from various sources is important and necessary. Additionally, the significant applications of such genes in transgenic plants are improved by their multiple functions. For instance, *Trichoderma* HSP70 has been shown to enhance tolerance to osmotic, salt, and oxidative stresses in transgenic *Arabidopsis* (Montero-Barrientos et al. 2010). The ribosomal protein subunits RPL44 and RPS3aE, which are highly conserved, suggest them as potential candidates for producing crops with enhanced tolerance to stress (Liang et al. 2015; X.-D. Liu et al. 2014). The genes, though not exclusive to extremophilic fungi, exhibit high conservation and seem to promote the survival of transgenic cells or organisms during stressful conditions, indicating potential for genetic engineering applications. Research conducted on *Actinobacter* from thermal springs in Tengchong, China, has identified gene clusters for non-ribosomal peptide (NRPs) and polyketide (PKs), hinting at the opportunity of discovering novel compounds in thermophilic environments (L. Liu et al. 2016). Various genera, including *Psychrobacter*, *Octadecabacter*, *Glaciecola*, *Terriglobus*, and *Photobacter*, have been found to harbour novel genes, as reported by Bowman and Deming in 2014 (Bowman and Deming 2014). Microorganisms utilizing hydrocarbons as only carbon source, such as *Oleispira antarctica* near the Antarctic coast, possess a rich set of alkane degradation genes and a protein system acting as a cold barrier, preventing low-temperature effects from hindering hydroxylase reactions (Kube et al. 2013). Moreover, *aioA* gene sequences similar to Chloroflexi have been identified in the microbial community structure of El Tatio, suggesting potential diversity in As-resistant bacteria or resistance mechanisms. Halophilic strains, like *Shewanella* sp. CC-1 and Asc-3 from the low-pH salt flats of Ascotán can precipitate As metal. A survey of metal resistance in halophilic and halotolerant strains isolated from the Atacama Desert has revealed their ability to tolerate HMs such as Cu, Cd, Ni, Zn, and Co. Noteworthy examples include *Thalassobacillus devorans*, which exhibits significant resistance to Cd and Ni (Moreno et al. 2012). The whole genome sequence of *Streptomyces* sp. H-KF8 has identified 49 genes associated with HM tolerance (Undabarrena et al. 2017). Research suggests that halotolerant bacteria play a role in protecting plants by improving the synthesis of extracellular hydrolytic enzymes (Rohban et al. 2009). Furthermore, these bacteria enhance ACC deaminase activity, effectively reducing plant ethylene levels which increase due to salinity stress (Siddikee et al. 2010). Additionally, they improve indole-3-acetic acid (IAA) levels, supporting plants in nutrient uptake during salt stress (Vacheron et al. 2013). Table 11.1 provides examples of potential extremophiles employed in various crops.

11.8   Conclusion

Environmental stressors, such as soil salinity, drought, and temperature extremes, pose significant challenges to plant growth and crop yield worldwide. Extremophilic and extremotolerant microorganisms have the potential to play a major role in agriculture, as they can thrive in harsh environments. To investigate the potential of such microbes, extensive investigations have been carried out, given their vast diversity and potential applications in specific fields. However, the peculiarities of these organisms often render culture conditions ineffective; making it necessary to study those further to better understand their behaviour and mechanisms. The deficiency resistance to abiotic stresses in germplasms limits the success of conventional plant breeding, but recent research focusing on molecular aspects and genetic basis in extremophiles offers hope for addressing such issues. The publication of whole genome sequences of extremophiles promises to reveal more promising resistant genes, with considerable economic potential in agriculture, food and beverages, and feed production. While only a few extremozymes are currently being used in agriculture, there is significant potential for further application in the development of a biobased economy. In addition, the plant–microbe interaction significantly influences the selection and enrichment of specific bacterial types in the rhizosphere, endosphere, or episphere. Microorganisms associated with crops play a crucial role in agriculture, promoting plant growth, enhancing plant nutrition through biological N_2 fixation, and providing additional benefits such as increased crop yield, control of pathogens, contaminant elimination, and the production of nitrogen or novel substances. Stimulating microbial growth can occur through various approaches, viz., N_2 fixation, nutrient bioavailability, phytohormones (e.g., IAA and cytokines) production, plant pathogens biocontrol, competing for nutrients, siderophore production, and induction systemic resistance. The future exploration of extremophilic microbes certainly relies on research advancement and field trials to deepen our understanding of these organisms. The introduction of innovative methods like metagenomics has expanded the research scope to encompass non-culturable

Table 11.1 Extremophiles conferring optimal growth and tolerance to various stresses in crops

Sl. no.	Crops	Extremophiles	Remarks	References
1.	Wheat	*Aspergillus niger*	Enhanced growth	Alori et al. (2017)
		Serratia sp.	Improved growth and nutrient absorption	Swarnalakshmi et al. (2013)
		Bacillus subtilis and *Azotobacter chroococcum*	Enhanced wheat productivity	Kumar et al. (2014)
		Pantoea, *Pseudomonas*, *Enterobacter*, and *Serratia*	Citric, fumaric acids, gluconic succinic, and oxalic	Rfaki et al. (2020)
		Bacilli strains	Gluconic, lactic, citric, malic, propionic acid, and succinic acid	Azaroual et al. (2020)
		Pseudomonas	Markedly extended root and shoot lengths	Rizvi et al. (2021)
		Pseudomonas fragi CS11RH1 (MTCC 8984)	Increased biomass, nutrient uptake, germination percentage, and germination rate	Selvakumar et al. (2009)
		Pseudomonas vancouverensis	Improved germination, shoot length, and root length	Rizvi et al. (2021)
		Pantoea agglomerans	Enhanced drought tolerance through rhizosphere soil aggregation via EPS	Grover et al. (2011)
		Azospirillum sp.	Enhanced drought stress resistance through improved water relations	Creus et al. (2004)
		Microbacterium sp., *B. amyloliquefaciens*, *B. insolitus*, and *P. syringae*	Salinity due to limited influx of Na^+	Ashraf et al. (2004)
2.	Rice	*Bacillus thuringiensis*	Elevated shoot length	Raj (2014)
		Burkholderia, *Bacillus*, and *Paenibacillus* sp.	Citric, tartaric, succinic, gluconic oxalic, acetic acid, and formic acid	Chawngthu et al. (2020)
		Arthrobacter nitroguajacolicus strain YB4 and *Pseudomonas koreensis*	Effectively boosted biomass and phosphorus uptake	Rizvi et al. (2021)
3.	Maize	*Burkholderia cepacia*	Enhanced plant growth	Zhao et al. (2014)
4.	Chickpea	*Bacillus* sp. strain AZ17 and *Pseudomonas* sp. strain AZ5	Acetic, oxalic, and gluconic acids, as well as acetic, citric, and lactic acids	Zaheer et al. (2019)

(continued)

Table 11.1 (continued)

Sl. no.	Crops	Extremophiles	Remarks	References
5.	Tomato	*Curtobacterium* sp. BmP22c (BC3) and *Mixture of Pseudomonas* sp. TmR5a	Facilitated a 90% increase in germination and notably extended root lengths	Vega-Celedón et al. (2021)
		Pseudomonas	Elevated germination and development of plantlets	Tapia-Vázquez et al. (2020)
		Pseudomonas simiae	Improved plant growth and heightened fruit yield	Rizvi et al. (2021)
		Enterobacter cloacae, *Pseudomonas putida*, and *P. putida*	Flooding by Synthesis of ACC-deaminase	Grover et al. (2011)
		Achromobacter piechaudii	Salt and drought attributed to the synthesis of ACC-deaminase	Mayak et al. (2004)
		Burkholderia sp. and *Methylobacterium oryzae*	Mitigates Ni and Cd toxicity and reduce uptake and translocation	Madhaiyan et al. (2007)
6.	Chickpeas, peas, and maize	*P. palleroniana* N-26, *Lysinibacillus macroides* ST-30, and *P. jessenii* MP-1	Markedly enhanced germination efficiency	Tomer et al. (2017)
7.	Lentil	*Pseudomonas* spp.	Plant growth, crop yield, and P uptake was improved	Rizvi et al. (2021)
8.	Pea	*Variovorax paradoxus*	Drought due to the synthesis of ACC-deaminase	Grover et al. (2011)
		Pseudomonas sp.	Drought by a reduction in ethylene production	Arshad et al. (2008)
9.	Barley, Pea, and Corn	*Rahnella* sp.	Remarkably heightened growth in all crops; micro-plot testing of the PPSB inoculum similarly resulted in a substantial increase in pea growth and yield	Vyas et al. (2010)
10.	Sorghum	*Pseudomonas* sp. AMK-P6	Enhanced heat tolerance through the induction of HSPs and enhanced physiology of the plants	Ali et al. (2009)

microorganisms. Additionally, new cloning techniques will facilitate the rapid and large-scale expression of extremophilic proteins in easily cultivable microorganisms. This knowledge will be crucial for further advancements in the field of extremophiles and their utilization across various biotechnological domains.

References

Abe F, Horikoshi K (2001) The biotechnological potential of piezophiles. Trends Biotechnol 19(3):102–108. https://doi.org/10.1016/S0167-7799(00)01539-0

Aguilar A, Ingemansson T, Magnien E (1998) Extremophile microorganisms as cell factories: support from the European Union. Extremophiles 2(3):367–373. https://doi.org/10.1007/s007920050080

Ahring B (1995) Methanogenesis in thermophilic biogas reactors. Antonie Van Leeuwenhoek 67:91–102

Ali SZ, Sandhya V, Grover M, Kishore N, Rao LV, Venkateswarlu B (2009) Pseudomonas sp. strain AKM-P6 enhances tolerance of sorghum seedlings to elevated temperatures. Biol Fertil Soils 46(1):45–55. https://doi.org/10.1007/s00374-009-0404-9

Ali SZ, Sandhya V, Grover M, Linga VR, Bandi V (2011) Effect of inoculation with a thermotolerant plant growth promoting *Pseudomonas putida* strain AKMP7 on growth of wheat (*Triticum* spp.) under heat stress. J Plant Interact 6(4):239–246. https://doi.org/10.1080/17429145.2010.545147

Almagro A, Prista C, Castro S, Quintas C, Madeira-Lopes A, Ramos J, Loureiro-Dias MC (2000) Effects of salts on Debaryomyces hansenii and Saccharomyces cerevisiae under stress conditions. Int J Food Microbiol 56(2–3):191–197. https://doi.org/10.1016/S0168-1605(00)00220-8

Alori ET, Glick BR, Babalola OO (2017) Microbial phosphorus solubilization and its potential for use in sustainable agriculture. Front Microbiol 8. https://doi.org/10.3389/fmicb.2017.00971

Amaro AM, Chamorro D, Seeger M, Arredondo R, Peirano I, Jerez CA (1991) Effect of external pH perturbations on in vivo protein synthesis by the acidophilic bacterium Thiobacillus ferrooxidans. J Bacteriol 173(2):910–915. https://doi.org/10.1128/jb.173.2.910-915.1991

Amoozegar MA, Ghazanfari N, Didari M (2012) Lead and cadmium bioremoval by Halomonas sp., an exopolysaccharide-producing halophilic bacterium. Prog Biol Sci 2(1):1–11

Antranikian G, Vorgias CE, Bertoldo C (2005) Extreme environments as a resource for microorganisms and novel biocatalysts. pp 219–262. https://doi.org/10.1007/b135786

Araya MA, Valenzuela T, Inostroza NG, Maruyama F, Jorquera MA, Acuña JJ (2020) Isolation and characterization of cold-tolerant hyper-ACC-degrading bacteria from the rhizosphere, endosphere, and phyllosphere of Antarctic vascular plants. Microorganisms 8(11):1788. https://doi.org/10.3390/microorganisms8111788

Arora S, Trivedi R, Rao GG (2013) Bioremediation of coastal and inland salt affected soils using halophyte plants and halophilic soil microbes

Arshad M, Shaharoona B, Mahmood T (2008) Inoculation with pseudomonas spp. containing ACC-deaminase partially eliminates the effects of drought stress on growth, yield, and ripening of pea (Pisum sativum L.). Pedosphere 18(5):611–620. https://doi.org/10.1016/S1002-0160(08)60055-7

Asghari B, Khademian R, Sedaghati B (2020) Plant growth promoting rhizobacteria (PGPR) confer drought resistance and stimulate biosynthesis of secondary metabolites in pennyroyal (Mentha pulegium L.) under water shortage condition. Sci Hortic 263:109132. https://doi.org/10.1016/j.scienta.2019.109132

Ashraf M, Hasnain S, Berge O, Mahmood T (2004) Inoculating wheat seedlings with exopolysaccharide-producing bacteria restricts sodium uptake and stimulates plant growth under salt stress. Biol Fertil Soils 40(3). https://doi.org/10.1007/s00374-004-0766-y

Azaroual SE, Hazzoumi Z, El Mernissi N, Aasfar A, Meftah Kadmiri I, Bouizgarne B (2020) Role of inorganic phosphate solubilizing bacilli isolated from Moroccan phosphate rock mine and rhizosphere soils in wheat (Triticum aestivum L) phosphorus uptake. Curr Microbiol 77(9):2391–2404. https://doi.org/10.1007/s00284-020-02046-8

Baker-Austin C, Dopson M (2007) Life in acid: pH homeostasis in acidophiles. Trends Microbiol 15(4):165–171. https://doi.org/10.1016/j.tim.2007.02.005

Beolchini F, Dell'Anno A, Rocchetti L, Vegliò F, Danovaro R (2009) Biohydrometallurgy as a remediation strategy for marine sediments contaminated by heavy metals. Adv Mater Res 71–73:669–672. https://doi.org/10.4028/www.scientific.net/AMR.71-73.669

Berenguer J (2011) Thermophile. In: Encyclopedia of astrobiology. Springer, Berlin/Heidelberg, pp 1666–1667. https://doi.org/10.1007/978-3-642-11274-4_1583

Bomberg M, Jurgens G, Saano A, Sen R, Timonen S (2003) Nested PCR detection of archaea in defined compartments of pine mycorrhizospheres developed in boreal forest humus microcosms. FEMS Microbiol Ecol 43(2):163–171. https://doi.org/10.1111/j.1574-6941.2003. tb01055.x

Bowman JS, Deming JW (2014) Alkane hydroxylase genes in psychrophile genomes and the potential for cold active catalysis. BMC Genomics 15(1):1120. https://doi.org/10.1186/1471-2164-15-1120

Brim H, Venkateswaran A, Kostandarithes HM, Fredrickson JK, Daly MJ (2003) Engineering *Deinococcus geothermalis* for bioremediation of high-temperature radioactive waste environments. Appl Environ Microbiol 69(8):4575–4582. https://doi.org/10.1128/AEM.69.8.4575-4582.2003

Bull AT, Asenjo JA (2013) Microbiology of hyper-arid environments: recent insights from the Atacama Desert, Chile. Antonie Van Leeuwenhoek 103(6):1173–1179. https://doi.org/10.1007/s10482-013-9911-7

Campos VL, Escalante G, Yañez J, Zaror CA, Mondaca MA (2009) Isolation of arsenite-oxidizing bacteria from a natural biofilm associated to volcanic rocks of Atacama Desert, Chile. *J Basic Microbiol* 49(S1):S93–S97. https://doi.org/10.1002/jobm.200900028

Casillo A, Parrilli E, Sannino F, Mitchell DE, Gibson MI, Marino G, Lanzetta R, Parrilli M, Cosconati S, Novellino E, Randazzo A, Tutino ML, Corsaro MM (2017) Structure-activity relationship of the exopolysaccharide from a psychrophilic bacterium: a strategy for cryoprotection. Carbohydr Polym 156:364–371. https://doi.org/10.1016/j.carbpol.2016.09.037

Charlesworth J, Burns P (2016) Extremophilic adaptations and biotechnological applications in diverse environments. AIMS Microbiol 2(3):251–261. https://doi.org/10.3934/microbiol.2016.3.251

Chawngthu L, Hnamte R, Lalfakzuala R (2020) Isolation and characterization of rhizospheric phosphate solubilizing bacteria from Wetland Paddy Field of Mizoram, India. *Geomicrobiol J* 37(4):366–375. https://doi.org/10.1080/01490451.2019.1709108

Chelius MK, Triplett EW (2001) The diversity of archaea and bacteria in association with the roots of Zea mays L. Microb Ecol 41(3):252–263. https://doi.org/10.1007/s002480000087

Christel S, Herold M, Bellenberg S, El Hajjami M, Buetti-Dinh A, Pivkin IV, Sand W, Wilmes P, Poetsch A, Dopson M (2018) Multi-omics reveals the lifestyle of the acidophilic, mineral-oxidizing model species Leptospirillum ferriphilum [T]. Appl Environ Microbiol 84(3). https://doi.org/10.1128/AEM.02091-17

Cid FP, Inostroza NG, Graether SP, Bravo LA, Jorquera MA (2017) Bacterial community structures and ice recrystallization inhibition activity of bacteria isolated from the phyllosphere of the Antarctic vascular plant Deschampsia Antarctica. Polar Biol 40(6):1319–1331. https://doi.org/10.1007/s00300-016-2036-5

Clarke A, Morris GJ, Fonseca F, Murray BJ, Acton E, Price HC (2013) A low temperature limit for life on earth. PLoS One 8(6):e66207. https://doi.org/10.1371/journal.pone.0066207

Coker JA (2019) Recent advances in understanding extremophiles. F1000Research 8:1917. https://doi.org/10.12688/f1000research.20765.1

Connon SA, Lester ED, Shafaat HS, Obenhuber DC, Ponce A (2007) Bacterial diversity in hyperarid Atacama Desert soils. J Geophys Res Biogeosci 112(G4). https://doi.org/10.1029/2006JG000311

Creus CM, Sueldo RJ, Barassi CA (2004) Water relations and yield in *Azospirillum*-inoculated wheat exposed to drought in the field. Can J Bot 82(2):273–281. https://doi.org/10.1139/b03-119

Crits-Christoph A, Robinson CK, Barnum T, Fricke WF, Davila AF, Jedynak B, McKay CP, DiRuggiero J (2013) Colonization patterns of soil microbial communities in the Atacama Desert. Microbiome 1(1):28. https://doi.org/10.1186/2049-2618-1-28

Czarnes S, Mercier P, Lemoine DG, Hamzaoui J, Legendre L (2020) Impact of soil water content on maize responses to the plant growth-promoting rhizobacterium *Azospirillum lipoferum* CRT1. J Agron Crop Sci 206(5):505–516. https://doi.org/10.1111/jac.12399

D'Amico S, Collins T, Marx J, Feller G, Gerday C, Gerday C (2006) Psychrophilic microorganisms: challenges for life. EMBO Rep 7(4):385–389. https://doi.org/10.1038/sj.embor.7400662

Danish S, Zafar-Ul-Hye M, Hussain S, Riaz M, Qayyum MF (2020) Mitigation of drought stress in maize through inoculation with drought tolerant ACC deaminase containing PGPR under axenic conditions. Pak J Bot 52(1):10.30848/PJB2020-1(7)

De Souza M-J, Nair S, Loka Bharathi PA, Chandramohan D (2006) Metal and antibiotic-resistance in psychrotrophic bacteria from Antarctic Marine waters. Ecotoxicology 15(4):379–384. https://doi.org/10.1007/s10646-006-0068-2

de Souza R, Ambrosini A, Passaglia LMP (2015) Plant growth-promoting bacteria as inoculants in agricultural soils. Genet Mol Biol 38(4):401–419. https://doi.org/10.1590/S1415-475738420150053

Delgado-Jarana J, Sousa S, González F, Rey M, Llobell A (2006) ThHog1 controls the hyperosmotic stress response in Trichoderma harzianum. Microbiology 152(6):1687–1700. https://doi.org/10.1099/mic.0.28729-0

Dobretsov S, Abed RMM, Al Maskari SMS, Al Sabahi JN, Victor R (2011) Cyanobacterial mats from hot springs produce antimicrobial compounds and quorum-sensing inhibitors under natural conditions. J Appl Phycol 23(6):983–993. https://doi.org/10.1007/s10811-010-9627-2

Dolkar D, Dolkar P, Stobdan T, Katiyar AK (2018) Tomato growth promotion induced by stress tolerant phosphate solubilizing Pseudomonas simiae in arid trans-Himalaya. Defence Life Sci J 3(2):105. https://doi.org/10.14429/dlsj.3.12565

Dubois M, Van den Broeck L, Inzé D (2018) The pivotal role of ethylene in plant growth. Trends Plant Sci 23(4):311–323. https://doi.org/10.1016/j.tplants.2018.01.003

Duckworth AW, Grant WD, Jones BE, Steenbergen R (1996) Phylogenetic diversity of soda lake alkaliphiles. FEMS Microbiol Ecol 19(3):181–191. https://doi.org/10.1111/j.1574-6941.1996.tb00211.x

Edbeib MF, Wahab RA, Huyop F (2016) Halophiles: biology, adaptation, and their role in decontamination of hypersaline environments. World J Microbiol Biotechnol 32(8):135. https://doi.org/10.1007/s11274-016-2081-9

Enebe MC, Babalola OO (2018) The influence of plant growth-promoting rhizobacteria in plant tolerance to abiotic stress: a survival strategy. Appl Microbiol Biotechnol 102(18):7821–7835. https://doi.org/10.1007/s00253-018-9214-z

Escalante G, Campos VL, Valenzuela C, Yañez J, Zaror C, Mondaca MA (2009) Arsenic resistant bacteria isolated from arsenic contaminated river in the Atacama Desert (Chile). Bull Environ Contam Toxicol 83(5):657–661. https://doi.org/10.1007/s00128-009-9868-4

Eskandari A, Leow TC, Rahman MBA, Oslan SN (2020) Antifreeze proteins and their practical utilization in industry, medicine, and agriculture. Biomol Ther 10(12):1649. https://doi.org/10.3390/biom10121649

Etesami H (2020) Plant–microbe interactions in plants and stress tolerance. In: Plant life under changing environment. Elsevier, pp 355–396. https://doi.org/10.1016/B978-0-12-818204-8.00018-7

Farhangi-Abriz S, Nikpour-Rashidabad N (2017) Effect of lignite on alleviation of salt toxicity in soybean (Glycine max L.) plants. Plant Physiol Biochem 120:186–193. https://doi.org/10.1016/j.plaphy.2017.10.007

Feller G (2013) Psychrophilic enzymes: from folding to function and biotechnology. Scientifica 2013:1–28. https://doi.org/10.1155/2013/512840

Feller G, Gerday C (2003) Psychrophilic enzymes: hot topics in cold adaptation. Nat Rev Microbiol 1(3):200–208. https://doi.org/10.1038/nrmicro773

Fernandez O, Theocharis A, Bordiec S, Feil R, Jacquens L, Clément C, Fontaine F, Barka EA (2012) *Burkholderia phytofirmans* PsJN acclimates grapevine to cold by modulating carbohydrate metabolism. Mol Plant-Microbe Interact 25(4):496–504. https://doi.org/10.1094/MPMI-09-11-0245

Fröls S (2013) Archaeal biofilms: widespread and complex. Biochem Soc Trans 41(1):393–398. https://doi.org/10.1042/BST20120304

Furhan J (2020) Adaptation, production, and biotechnological potential of cold-adapted proteases from psychrophiles and psychrotrophs: recent overview. J Genet Eng Biotechnol 18(1):36. https://doi.org/10.1186/s43141-020-00053-7

Gabani P, Prakash D, Singh V, O. (2014) Bio-signature of ultraviolet-radiation-resistant extremophiles from elevated land. Am J Microbiol Res 2(3):94–104. https://doi.org/10.12691/ajmr-2-3-3

García-Salcedo R, Casamayor A, Ruiz A, González A, Prista C, Loureiro-Dias MC, Ramos J, Ariño J (2006) Heterologous expression implicates a GATA factor in regulation of nitrogen metabolic genes and ion homeostasis in the halotolerant yeast *Debaryomyces hansenii*. Eukaryot Cell 5(8):1388–1398. https://doi.org/10.1128/EC.00154-06

Gemmell RT, Knowles CJ (2000) Utilisation of aliphatic compounds by acidophilic heterotrophic bacteria. The potential for bioremediation of acidic wastewaters contaminated with toxic organic compounds and heavy metals. FEMS Microbiol Lett 192(2):185–190. https://doi.org/10.1111/j.1574-6968.2000.tb09380.x

Gholami M, Khakvar R, Niknam G (2014) Introduction of some new endophytic bacteria from *Bacillus* and *Streptomyces* genera as successful biocontrol agents against *Sclerotium rolfsii*. Arch Phytopathol Plant Protect 47(1):122–130. https://doi.org/10.1080/03235408.2013.805043

Gilbert JA, Davies PL, Laybourn-Parry J (2005) A hyperactive, Ca $^{2+}$ -dependent antifreeze protein in an Antarctic bacterium. FEMS Microbiol Lett 245(1):67–72. https://doi.org/10.1016/j.femsle.2005.02.022

Godin-Roulling A, Schmidpeter PAM, Schmid FX, Feller G (2015) Functional adaptations of the bacterial chaperone trigger factor to extreme environmental temperatures. Environ Microbiol 17(7):2407–2420. https://doi.org/10.1111/1462-2920.12707

Goh F, Jeon YJ, Barrow K, Neilan BA, Burns BP (2011) Osmoadaptive strategies of the archaeon *Halococcus hamelinensis* isolated from a hypersaline Stromatolite environment. Astrobiology 11(6):529–536. https://doi.org/10.1089/ast.2010.0591

González-Aravena M, Urtubia R, Del Campo K, Lavín P, Wong CMVL, Cárdenas CA, González-Rocha G (2016) Antibiotic and metal resistance of cultivable bacteria in the Antarctic sea urchin. Antarct Sci 28(4):261–268. https://doi.org/10.1017/S0954102016000109

Großkopf R, Stubner S, Liesack W (1998) Novel Euryarchaeotal lineages detected on Rice roots and in the anoxic bulk soil of flooded Rice microcosms. Appl Environ Microbiol 64(12):4983–4989. https://doi.org/10.1128/AEM.64.12.4983-4989.1998

Grover M, Ali SZ, Sandhya V, Rasul A, Venkateswarlu B (2011) Role of microorganisms in adaptation of agriculture crops to abiotic stresses. World J Microbiol Biotechnol 27(5):1231–1240. https://doi.org/10.1007/s11274-010-0572-7

Guo J-P, Zhu C-Y, Zhang C-P, Chu Y-S, Wang Y-L, Zhang J-X, Wu D-K, Zhang K-Q, Niu X-M (2012) Thermolides, potent nematocidal PKS-NRPS hybrid metabolites from thermophilic fungus *Talaromyces thermophilus*. J Am Chem Soc 134(50):20306–20309. https://doi.org/10.1021/ja3104044

Hartman K, Tringe SG (2019) Interactions between plants and soil shaping the root microbiome under abiotic stress. Biochem J 476(19):2705–2724. https://doi.org/10.1042/BCJ20180615

Hatfield JL, Prueger JH (2015) Temperature extremes: effect on plant growth and development. Weather Clim Extremes 10:4–10. https://doi.org/10.1016/j.wace.2015.08.001

Hermosa R, Botella L, Keck E, Jiménez JÁ, Montero-Barrientos M, Arbona V, Gómez-Cadenas A, Monte E, Nicolás C (2011) The overexpression in Arabidopsis thaliana of a Trichoderma harzianum gene that modulates glucosidase activity, and enhances tolerance to salt and osmotic stresses. J Plant Physiol 168(11):1295–1302. https://doi.org/10.1016/j.jplph.2011.01.027

Hoffmann T, Bremer E (2011) Protection of *Bacillus subtilis* against cold stress via compatible-solute acquisition. J Bacteriol 193(7):1552–1562. https://doi.org/10.1128/JB.01319-10

Horikoshi K (1999) Alkaliphiles: some applications of their products for biotechnology. Microbiol Mol Biol Rev 63(4):735–750. https://doi.org/10.1128/MMBR.63.4.735-750.1999

Hu H, Yan F, Wilson C, Shen Q, Zheng X (2015) The ability of a cold-adapted Rhodotorula mucilaginosa strain from Tibet to control blue mold in pear fruit. Antonie Van Leeuwenhoek 108(6):1391–1404. https://doi.org/10.1007/s10482-015-0593-1

Ilyas S, Lee J, Kim B (2014) Bioremoval of heavy metals from recycling industry electronic waste by a consortium of moderate thermophiles: process development and optimization. J Clean Prod 70:194–202. https://doi.org/10.1016/j.jclepro.2014.02.019

İpek M, Arıkan Ş, Pırlak L, Eşitken A (2019) Sustainability of crop production by PGPR under abiotic stress conditions. In: Plant growth promoting Rhizobacteria for agricultural sustainability. Springer, Singapore, pp 293–314. https://doi.org/10.1007/978-981-13-7553-8_15

Jin Y, Weining S, Nevo E (2005) A MAPK gene from Dead Sea fungus confers stress tolerance to lithium salt and freezing–thawing: prospects for saline agriculture. Proc Natl Acad Sci 102(52):18992–18997. https://doi.org/10.1073/pnas.0509653102

Johnson DB (1995) Acidophilic microbial communities: candidates for bioremediation of acidic mine effluents. Int Biodeterior Biodegradation 35(1–3):41–58. https://doi.org/10.1016/0964-8305(95)00065-D

Johnson DB (2008) Biodiversity and interactions of acidophiles: key to understanding and optimizing microbial processing of ores and concentrates. Trans Nonferrous Metals Soc China 18(6):1367–1373. https://doi.org/10.1016/S1003-6326(09)60010-8

Joly M, Attard E, Sancelme M, Deguillaume L, Guilbaud C, Morris CE, Amato P, Delort A-M (2013) Ice nucleation activity of bacteria isolated from cloud water. Atmos Environ 70:392–400. https://doi.org/10.1016/j.atmosenv.2013.01.027

Kadioglu GB, Koseoglu MS, Ozdal M, Sezen A, Ozdal OG, Algur OF (2018) Isolation of cold tolerant and ACC deaminase producing plant growth promoting rhizobacteria from high altitudes. Roman Biotechnol Lett 23(2):13479–13486

Kim A-Y, Shahzad R, Kang S-M, Seo C-W, Park Y-G, Park H-J, Lee I-J (2017) IAA-producing Klebsiella variicola AY13 reprograms soybean growth during flooding stress. J Crop Sci Biotechnol 20(4):235–242. https://doi.org/10.1007/s12892-017-0041-0

Knack JJ, Wilcox LW, Delaux P-M, Ané J-M, Piotrowski MJ, Cook ME, Graham JM, Graham LE (2015) Microbiomes of Streptophyte algae and bryophytes suggest that a functional suite of microbiota fostered plant colonization of land. Int J Plant Sci 176(5):405–420. https://doi.org/10.1086/681161

Kristiansen E, Zachariassen KE (2005) The mechanism by which fish antifreeze proteins cause thermal hysteresis. Cryobiology 51(3):262–280. https://doi.org/10.1016/j.cryobiol.2005.07.007

Krulwich TA (1995) Alkaliphiles:'basic' molecular problems of pH tolerance and bioenergetics. Mol Microbiol 15(3):403–410. https://doi.org/10.1111/j.1365-2958.1995.tb02253.x

Krulwich TA, Guffanti AA (1989) Alkalophilic bacteria. Ann Rev Microbiol 43(1):435–463. https://doi.org/10.1146/annurev.mi.43.100189.002251

Krulwich TA, Ito M, Hicks DB, Gilmour R, Guffanti AA (1998) pH homeostasis and ATP synthesis: studies of two processes that necessitate inward proton translocation in extremely alkaliphilic Bacillus species. Extremophiles 2(3):217–222. https://doi.org/10.1007/s007920050063

Kube M, Chernikova TN, Al-Ramahi Y, Beloqui A, Lopez-Cortez N, Guazzaroni M-E, Heipieper HJ, Klages S, Kotsyurbenko OR, Langer I, Nechitaylo TY, Lünsdorf H, Fernández M, Juárez S, Ciordia S, Singer A, Kagan O, Egorova O, Alain Petit P et al (2013) Genome sequence and functional genomic analysis of the oil-degrading bacterium Oleispira Antarctica. Nat Commun 4(1):2156. https://doi.org/10.1038/ncomms3156

Kumar Meena R, Kumar Singh R, Pal Singh N, Kumari Meena S, Singh Meena V (2015) Isolation of low temperature surviving plant growth—promoting rhizobacteria (PGPR) from pea (Pisum sativum L.) and documentation of their plant growth promoting traits. Biocatal Agric Biotechnol 4(4):806–811. https://doi.org/10.1016/j.bcab.2015.08.006

Kumar S, Bauddh K, Barman SC, Singh RP (2014) Amendments of microbial biofertilizers and organic substances reduces requirement of urea and DAP with enhanced nutrient availability and productivity of wheat (Triticum aestivum L.). Ecol Eng 71:432–437. https://doi.org/10.1016/j.ecoleng.2014.07.007

Lebre PH, De Maayer P, Cowan DA (2017) Xerotolerant bacteria: surviving through a dry spell. Nat Rev Microbiol 15(5):285–296. https://doi.org/10.1038/nrmicro.2017.16

Liang X, Liu Y, Xie L, Liu X, Wei Y, Zhou X, Zhang S (2015) A ribosomal protein AgRPS3aE from halophilic aspergillus glaucus confers salt tolerance in heterologous organisms. Int J Mol Sci 16(2):3058–3070. https://doi.org/10.3390/ijms16023058

Liesack W, Schnell S, Revsbech NP (2000) Microbiology of flooded rice paddies. FEMS Microbiol Rev 24(5):625–645. https://doi.org/10.1111/j.1574-6976.2000.tb00563.x

Lim J-H, Kim S-D (2013) Induction of drought stress resistance by multi-functional PGPR bacillus licheniformis K11 in pepper. Plant Pathol J 29(2):201–208. https://doi.org/10.5423/PPJ.SI.02.2013.0021

Liu X-D, Xie L, Wei Y, Zhou X, Jia B, Liu J, Zhang S (2014) Abiotic stress resistance, a novel moonlighting function of ribosomal protein RPL44 in the halophilic fungus Aspergillus glaucus. Appl Environ Microbiol 80(14):4294–4300. https://doi.org/10.1128/AEM.00292-14

Liu L, Salam N, Jiao J-Y, Jiang H-C, Zhou E-M, Yin Y-R, Ming H, Li W-J (2016) Diversity of culturable thermophilic actinobacteria in Hot Springs in Tengchong, China and studies of their biosynthetic gene profiles. Microb Ecol 72(1):150–162. https://doi.org/10.1007/s00248-016-0756-2

Liu K, Ding H, Yu Y, Chen B (2019) A cold-adapted Chitinase-producing bacterium from Antarctica and its potential in biocontrol of plant pathogenic fungi. Mar Drugs 17(12):695. https://doi.org/10.3390/md17120695

Madern D, Ebel C, Zaccai G (2000) Halophilic adaptation of enzymes. Extremophiles 4(2):91–98. https://doi.org/10.1007/s007920050142

Madhaiyan M, Poonguzhali S, Sa T (2007) Metal tolerating methylotrophic bacteria reduces nickel and cadmium toxicity and promotes plant growth of tomato (Lycopersicon esculentum L.). Chemosphere 69(2):220–228. https://doi.org/10.1016/j.chemosphere.2007.04.017

Madigan MT, Martinko JM, Parker J (2000) Brock biology of microorganisms.9th edn. Prentice Hall, New Jersey

Margesin R, Schinner F (2001) Potential of halotolerant and halophilic microorganisms for biotechnology. Extremophiles 5(2):73–83. https://doi.org/10.1007/s007920100184

Marizcurrena JJ, Morel MA, Braña V, Morales D, Martinez-López W, Castro-Sowinski S (2017) Searching for novel photolyases in UVC-resistant Antarctic bacteria. Extremophiles 21(2):409–418. https://doi.org/10.1007/s00792-016-0914-y

Martínez F, Arif A, Nebauer SG, Bueso E, Ali R, Montesinos C, Brunaud V, Muñoz-Bertomeu J, Serrano R (2015) A fungal transcription factor gene is expressed in plants from its own promoter and improves drought tolerance. Planta 242(1):39–52. https://doi.org/10.1007/s00425-015-2285-5

Matsuno T, Yumoto I (2015) Bioenergetics and the role of soluble cytochromes c for alkaline adaptation in gram-negative Alkaliphilic *Pseudomonas*. Biomed Res Int 2015:1–14. https://doi.org/10.1155/2015/847945

Mayak S, Tirosh T, Glick BR (2004) Plant growth-promoting bacteria that confer resistance to water stress in tomatoes and peppers. Plant Sci 166(2):525–530. https://doi.org/10.1016/j.plantsci.2003.10.025

Meena VS, Maurya BR, Verma JP (2014) Does a rhizospheric microorganism enhance K+ availability in agricultural soils? Microbiol Res 169(5–6):337–347. https://doi.org/10.1016/j.micres.2013.09.003

Melo I, Souza W, Silva L, Santos S, Assalin M, Zucchi T, Queiroz S (2016) Antifungal activity of pseudomonas frederiksbergensis CMAA 1323 isolated from the Antarctic Hair Grass Deschampsia Antarctica. Br Microbiol Res J 14(3):1–11. https://doi.org/10.9734/BMRJ/2016/25314

Michels M, Bakker EP (1985) Generation of a large, protonophore-sensitive proton motive force and pH difference in the acidophilic bacteria Thermoplasma acidophilum and bacillus acidocaldarius. J Bacteriol 161(1):231–237. https://doi.org/10.1128/jb.161.1.231-237.1985

Minton KW, Daly MJ (1995) A model for repair of radiation-induced DNA double-strand breaks in the extreme radiophile Deinococcus radiodurans. BioEssays 17(5):457–464. https://doi.org/10.1002/bies.950170514

Mirzaei M, Ladan Moghadam A, Hakimi L, Danaee E (2020) Plant growth promoting rhizobacteria (PGPR) improve plant growth, antioxidant capacity, and essential oil properties of lemongrass (Cymbopogon citratus) under water stress. Iran J Plant Physiol 10(2)

Mishra PK, Bisht SC, Ruwari P, Selvakumar G, Joshi GK, Bisht JK, Bhatt JC, Gupta HS (2011) Alleviation of cold stress in inoculated wheat (Triticum aestivum L.) seedlings with psychrotolerant Pseudomonads from NW Himalayas. Arch Microbiol 193(7):497–513. https://doi.org/10.1007/s00203-011-0693-x

Montero-Barrientos M, Hermosa R, Nicolás C, Cardoza RE, Gutiérrez S, Monte E (2008) Overexpression of a Trichoderma HSP70 gene increases fungal resistance to heat and other abiotic stresses. Fungal Genet Biol 45(11):1506–1513. https://doi.org/10.1016/j.fgb.2008.09.003

Montero-Barrientos M, Hermosa R, Cardoza RE, Gutiérrez S, Nicolás C, Monte E (2010) Transgenic expression of the Trichoderma harzianum hsp70 gene increases Arabidopsis resistance to heat and other abiotic stresses. J Plant Physiol 167(8):659–665. https://doi.org/10.1016/j.jplph.2009.11.012

Moreno ML, Piubeli F, Bonfá MRL, García MT, Durrant LR, Mellado E (2012) Analysis and characterization of cultivable extremophilic hydrolytic bacterial community in heavy-metal-contaminated soils from the Atacama Desert and their biotechnological potentials. J Appl Microbiol 113(3):550–559. https://doi.org/10.1111/j.1365-2672.2012.05366.x

Morita RY (1975) Psychrophilic bacteria. Bacteriol Rev 39(2):144–167

Morris CE, Sands DC, Vinatzer BA, Glaux C, Guilbaud C, Buffière A, Yan S, Dominguez H, Thompson BM (2008) The life history of the plant pathogen Pseudomonas syringae is linked to the water cycle. ISME J 2:321–334

Mukherjee S, Mishra A, Trenberth KE (2018) Climate change and drought: a perspective on drought indices. Curr Clim Chang Rep 4(2):145–163. https://doi.org/10.1007/s40641-018-0098-x

Mukhtar T, Rehman S u, Smith D, Sultan T, Seleiman MF, Alsadon AA, Amna, Ali S, Chaudhary HJ, Solieman THI, Ibrahim AA, Saad MAO (2020) Mitigation of heat stress in Solanum lycopersicum L. by ACC-deaminase and exopolysaccharide producing Bacillus cereus: effects on biochemical profiling. Sustainability 12(6):2159. https://doi.org/10.3390/su12062159

Muñoz PA, Flores PA, Boehmwald FA, Blamey JM (2011) Thermophilic bacteria present in a sample from Fumarole Bay, Deception Island. Antarct Sci 23(6):549–555. https://doi.org/10.1017/S0954102011000393

Mylona P, Pawlowski K, Bisseling T (1995) Symbiotic nitrogen fixation. Plant Cell:869–885. https://doi.org/10.1105/tpc.7.7.869

Navarro-González R, Rainey FA, Molina P, Bagaley DR, Hollen BJ, de la Rosa J, Small AM, Quinn RC, Grunthaner FJ, Cáceres L, Gomez-Silva B, McKay CP (2003) Mars-like soils in the Atacama Desert, Chile, and the dry limit of microbial life. Science 302(5647):1018–1021. https://doi.org/10.1126/science.1089143

Nichols CAM, Guezennec J, Bowman JP (2005) Bacterial exopolysaccharides from extreme marine environments with special consideration of the Southern Ocean, sea ice, and Deep-Sea hydrothermal vents: A review. Mar Biotechnol 7(4):253–271. https://doi.org/10.1007/s10126-004-5118-2

Nicol GW, Webster G, Glover LA, Prosser JI (2004) Differential response of archaeal and bacterial communities to nitrogen inputs and pH changes in upland pasture rhizosphere soil. Environ Microbiol 6(8):861–867. https://doi.org/10.1111/j.1462-2920.2004.00627.x

Norton CF, Grant WD (1988) Survival of halobacteria within fluid inclusions in salt crystals. Microbiology 134(5):1365–1373. https://doi.org/10.1099/00221287-134-5-1365

Nweze JE, Nweze JA, Gupta S (2022) Application of extremophiles in sustainable agriculture, pp 233–250. https://doi.org/10.4018/978-1-7998-9144-4.ch011

Ochsenreiter T, Selezi D, Quaiser A, Bonch-Osmolovskaya L, Schleper C (2003) Diversity and abundance of Crenarchaeota in terrestrial habitats studied by 16S RNA surveys and real time PCR. Environ Microbiol 5(9):787–797. https://doi.org/10.1046/j.1462-2920.2003.00476.x

Orellana R, Macaya C, Bravo G, Dorochesi F, Cumsille A, Valencia R, Rojas C, Seeger M (2018) Living at the frontiers of life: extremophiles in Chile and their potential for bioremediation. Front Microbiol 9. https://doi.org/10.3389/fmicb.2018.02309

Oren A, Gurevich P, Gemmell RT, Teske A (1995) Halobaculum gomorrense gen. nov., sp. nov., a novel extremely halophilic archaeon from the Dead Sea. Int J Syst Bacteriol 45(4):747–754. https://doi.org/10.1099/00207713-45-4-747

Paulino-Lima IG, Fujishima K, Navarrete JU, Galante D, Rodrigues F, Azua-Bustos A, Rothschild LJ (2016) Extremely high UV-C radiation resistant microorganisms from desert environments with different manganese concentrations. J Photochem Photobiol B Biol 163:327–336. https://doi.org/10.1016/j.jphotobiol.2016.08.017

Pérez V, Hengst M, Kurte L, Dorador C, Jeffrey WH, Wattiez R, Molina V, Matallana-Surget S (2017) Bacterial survival under extreme UV radiation: a comparative proteomics study of Rhodobacter sp., isolated from high altitude wetlands in Chile. Front Microbiol 8. https://doi.org/10.3389/fmicb.2017.01173

Phadtare S (2004) Recent developments in bacterial cold-shock response. Curr Issues Mol Biol 6(2):125–136. https://doi.org/10.21775/cimb.006.125

Pikuta EV, Hoover RB, Tang J (2007) Microbial extremophiles at the limits of life. Crit Rev Microbiol 33(3):183–209. https://doi.org/10.1080/10408410701451948

Poli A, Di Donato P, Abbamondi GR, Nicolaus B (2011) Synthesis, production, and biotechnological applications of exopolysaccharides and polyhydroxyalkanoates by archaea. Archaea 2011:1–13. https://doi.org/10.1155/2011/693253

Popescu G, Dumitru L (2009) Biosorption of some heavy metals from media with high salt concentrations by halophilic *Archaea*. Biotechnol Biotechnol Equip 23(Suppl 1):791–795. https://doi.org/10.1080/13102818.2009.10818542

Purnawati A (2014) Endophytic bacteria as biocontrol agents of tomato bacterial wilt disease. J Trop Life Sci 4:33–36

Raddadi N, Cherif A, Daffonchio D, Neifar M, Fava F (2015) Biotechnological applications of extremophiles, extremozymes and extremolytes. Appl Microbiol Biotechnol 99(19):7907–7913. https://doi.org/10.1007/s00253-015-6874-9

Raj DP (2014) Molecular characterization of Phosphate Solubilizing Bacteria (PSB) and Plant Growth Promoting Rhizobacteria (PGPR) from pristine soils. https://www.researchgate.net/publication/266393420

Rajasreelatha V, Thippeswamy M (2022) The role of plant growth promoting extremophilic microbiomes under stressful environments. J Stress Physiol Biochem 18(2):2022

Rampelotto P (2013) Extremophiles and extreme environments. Life 3(3):482–485. https://doi.org/10.3390/life3030482

Rastogi G, Bhalla A, Adhikari A, Bischoff KM, Hughes SR, Christopher LP, Sani RK (2010) Characterization of thermostable cellulases produced by Bacillus and Geobacillus strains. Bioresour Technol 101(22):8798–8806. https://doi.org/10.1016/j.biortech.2010.06.001

Rfaki A, Zennouhi O, Aliyat FZ, Nassiri L, Ibijbijen J (2020) Isolation, selection and characterization of root-associated rock phosphate solubilizing bacteria in Moroccan wheat (*Triticum aestivum* L.). Geomicrobiol J 37(3):230–241. https://doi.org/10.1080/01490451.2019.1694106

Rizvi A, Ahmed B, Khan MS, Umar S, Lee J (2021) Psychrophilic bacterial phosphate-biofertilizers: A novel extremophile for sustainable crop production under cold environment. Microorganisms 9(12):2451. https://doi.org/10.3390/microorganisms9122451

Roberts MF (2005) Organic compatible solutes of halotolerant and halophilic microorganisms. Saline Syst 1(1):5. https://doi.org/10.1186/1746-1448-1-5

Rodríguez-Rojas F, Tapia P, Castro-Nallar E, Undabarrena A, Muñoz-Díaz P, Arenas-Salinas M, Díaz-Vásquez W, Valdés J, Vásquez C (2016) Draft genome sequence of a multi-metal resistant bacterium pseudomonas putida ATH-43 isolated from Greenwich Island, Antarctica. Front Microbiol 7. https://doi.org/10.3389/fmicb.2016.01777

Rohban R, Amoozegar MA, Ventosa A (2009) Screening and isolation of halophilic bacteria producing extracellular hydrolyses from Howz Soltan Lake, Iran. *J Ind Microbiol Biotechnol* 36(3):333–340. https://doi.org/10.1007/s10295-008-0500-0

Sachdev DP, Cameotra SS (2013) Biosurfactants in agriculture. Appl Microbiol Biotechnol 97(3):1005–1016. https://doi.org/10.1007/s00253-012-4641-8

Saiki RK, Gelfand DH, Stoffel S, Scharf SJ, Higuchi R, Horn GT, Mullis KB, Erlich HA (1988) Primer-directed enzymatic amplification of DNA with a thermostable DNA polymerase. Science 239(4839):487–491. https://doi.org/10.1126/science.2448875

Sar P, Kazy SK, Paul D, Sarkar A (2013) Metal bioremediation by thermophilic microorganisms. In: Thermophilic microbes in environmental and industrial biotechnology. Springer, Dordrecht, pp 171–201. https://doi.org/10.1007/978-94-007-5899-5_6

Saxena A, Yadav AN, Rajawat M, Kaushik R, Kumar R, Kumar M, Prasanna R, Shukla L (2016) Microbial diversity of extreme regions: an unseen heritage and wealth. Ind J Plant Genet Res 29(3):246–248. https://doi.org/10.5958/0976-1926.2016.00036.X

Selvakumar G, Joshi P, Nazim S, Mishra PK, Bisht JK, Gupta HS (2009) Phosphate solubilization and growth promotion by Pseudomonas fragi CS11RH1 (MTCC 8984), a psychrotolerant bacterium isolated from a high altitude Himalayan rhizosphere. Biologia 64(2):239–245. https://doi.org/10.2478/s11756-009-0041-7

Siddikee MA, Chauhan PS, Anandham R, Han G-H, Sa T (2010) Isolation, characterization, and use for plant growth promotion under salt stress, of ACC deaminase-producing halotolerant bacteria derived from coastal soil. J Microbiol Biotechnol 20(11):1577–1584. https://doi.org/10.4014/jmb.1007.07011

Simon HM, Dodsworth JA, Goodman RM (2000) Crenarchaeota colonize terrestrial plant roots. Environ Microbiol 2(5):495–505. https://doi.org/10.1046/j.1462-2920.2000.00131.x

Simon HM, Jahn CE, Bergerud LT, Sliwinski MK, Weimer PJ, Willis DK, Goodman RM (2005) Cultivation of mesophilic soil Crenarchaeotes in enrichment cultures from plant roots. Appl Environ Microbiol 71(8):4751–4760. https://doi.org/10.1128/AEM.71.8.4751-4760.2005

Sliwinski MK, Goodman RM (2004) Spatial heterogeneity of Crenarchaeal assemblages within mesophilic soil ecosystems as revealed by PCR-single-stranded conformation polymorphism profiling. Appl Environ Microbiol 70(3):1811–1820. https://doi.org/10.1128/AEM.70.3.1811-1820.2004

Smith DL, Gravel V, Yergeau E (2017) Editorial: signaling in the phytomicrobiome. Front Plant Sci 8. https://doi.org/10.3389/fpls.2017.00611

Strahl H, Greie J-C (2008) The extremely halophilic archaeon Halobacterium salinarum R1 responds to potassium limitation by expression of the K+-transporting KdpFABC P-type ATPase and by a decrease in intracellular K+. Extremophiles 12(6):741–752. https://doi.org/10.1007/s00792-008-0177-3

Subramanian S, Souleimanov A, Smith DL (2016) Proteomic studies on the effects of lipo-chitooligosaccharide and Thuricin 17 under unstressed and salt stressed conditions in Arabidopsis thaliana. Front Plant Sci 7. https://doi.org/10.3389/fpls.2016.01314

Swarnalakshmi K, Prasanna R, Kumar A, Pattnaik S, Chakravarty K, Shivay YS, Singh R, Saxena AK (2013) Evaluating the influence of novel cyanobacterial biofilmed biofertilizers on soil fertility and plant nutrition in wheat. Eur J Soil Biol 55:107–116. https://doi.org/10.1016/j.ejsobi.2012.12.008

Tang C (1998) Factors affecting soil acidification under legumes I. Effect of potassium supply. Plant Soil 199(2):275–282. https://doi.org/10.1023/A:1004361205533

Tang C, Unkovich MJ, Bowden JW (1999) Factors affecting soil acidification under legumes. III. Acid production by N_2-fixing legumes as influenced by nitrate supply. New Phytol 143(3):513–521. https://doi.org/10.1046/j.1469-8137.1999.00475.x

Tapia-Vázquez I, Sánchez-Cruz R, Arroyo-Domínguez M, Lira-Ruan V, Sánchez-Reyes A, del Rayo Sánchez-Carbente M, Padilla-Chacón D, Batista-García RA, Folch-Mallol JL (2020) Isolation and characterization of psychrophilic and psychrotolerant plant-growth promoting microorganisms from a high-altitude volcano crater in Mexico. Microbiol Res 232:126394. https://doi.org/10.1016/j.micres.2019.126394

Tindall BJ, Ross HNM, Grant WD (1984) Natronobacterium gen. nov. and Natronococcus gen. nov., two new genera of haloalkaliphilic archaebacteria. Syst Appl Microbiol 5(1):41–57. https://doi.org/10.1016/S0723-2020(84)80050-8

Tomer S, Suyal DC, Shukla A, Rajwar J, Yadav A, Shouche Y, Goel R (2017) Isolation and characterization of phosphate solubilizing bacteria from Western Indian Himalayan soils. 3. Biotech 7(2):95. https://doi.org/10.1007/s13205-017-0738-1

Tomova I, Stoilova-Disheva M, Lazarkevich I, Vasileva-Tonkova E (2015) Antimicrobial activity and resistance to heavy metals and antibiotics of heterotrophic bacteria isolated from sediment and soil samples collected from two Antarctic islands. Front Life Sci 8(4):348–357. https://doi.org/10.1080/21553769.2015.1044130

Torracchi C. JE, Morel MA, Tapia-Vázquez I, Castro-Sowinski S, Batista-García RA, Yarzábal R LA (2020) Fighting plant pathogens with cold-active microorganisms: biopesticide development and agriculture intensification in cold climates. Appl Microbiol Biotechnol 104(19):8243–8256. https://doi.org/10.1007/s00253-020-10812-8

Undabarrena A, Ugalde JA, Seeger M, Cámara B (2017) Genomic data mining of the marine actinobacteria *Streptomyces* sp. H-KF8 unveils insights into multi-stress related genes and metabolic pathways involved in antimicrobial synthesis. PeerJ 5:e2912. https://doi.org/10.7717/peerj.2912

Urbieta MS, Donati ER, Chan K-G, Shahar S, Sin LL, Goh KM (2015a) Thermophiles in the genomic era: biodiversity, science, and applications. Biotechnol Adv 33(6):633–647. https://doi.org/10.1016/j.biotechadv.2015.04.007

Urbieta MS, González-Toril E, Bazán ÁA, Giaveno MA, Donati E (2015b) Comparison of the microbial communities of hot springs waters and the microbial biofilms in the acidic geothermal area of Copahue (Neuquén, Argentina). Extremophiles 19(2):437–450. https://doi.org/10.1007/s00792-015-0729-2

Vacheron J, Desbrosses G, Bouffaud M-L, Touraine B, Moënne-Loccoz Y, Muller D, Legendre L, Wisniewski-Dyé F, Prigent-Combaret C (2013) Plant growth-promoting rhizobacteria and root system functioning. Front Plant Sci 4. https://doi.org/10.3389/fpls.2013.00356

Valipour M (2014) Drainage, waterlogging, and salinity. Arch Agron Soil Sci 60(12):1625–1640. https://doi.org/10.1080/03650340.2014.905676

Van Oosten MJ, Pepe O, De Pascale S, Silletti S, Maggio A (2017) The role of biostimulants and bioeffectors as alleviators of abiotic stress in crop plants. Chem Biol Technol Agric 4(1):5. https://doi.org/10.1186/s40538-017-0089-5

Vega-Celedón P, Bravo G, Velásquez A, Cid FP, Valenzuela M, Ramírez I, Vasconez I-N, Álvarez I, Jorquera MA, Seeger M (2021) Microbial diversity of psychrotolerant bacteria isolated from wild Flora of Andes Mountains and Patagonia of Chile towards the selection of plant growth-promoting bacterial consortia to alleviate cold stress in plants. Microorganisms 9(3):538. https://doi.org/10.3390/microorganisms9030538

Verma P, Yadav AN, Khannam KS, Panjiar N, Kumar S, Saxena AK, Suman A (2015) Assessment of genetic diversity and plant growth promoting attributes of psychrotolerant bacteria allied with wheat (Triticum aestivum) from the northern hills zone of India. Ann Microbiol 65(4):1885–1899. https://doi.org/10.1007/s13213-014-1027-4

Verma P, Yadav AN, Kumar V, Singh DP, Saxena AK (2017) Beneficial plant-microbes interactions: biodiversity of microbes from diverse extreme environments and its impact for crop improvement. In: Plant-microbe interactions in agro-ecological perspectives. Springer, Singapore, pp 543–580. https://doi.org/10.1007/978-981-10-6593-4_22

Vyas P, Joshi R, Sharma KC, Rahi P, Gulati A, Gulati A (2010) Cold-adapted and rhizosphere-competent strain of Rahnella sp. with broad-spectrum plant growth-promotion potential. J Microbiol Biotechnol 20(12):1724–1734. https://doi.org/10.4014/jmb.1007.07030

Xu L, Wu Y-H, Zhou P, Cheng H, Liu Q, Xu X-W (2018) Investigation of the thermophilic mechanism in the genus Porphyrobacter by comparative genomic analysis. BMC Genomics 19(1):385. https://doi.org/10.1186/s12864-018-4789-4

Yadav AN, Verma P, Kumar M, Pal KK, Dey R, Gupta A, Padaria JC, Gujar GT, Kumar S, Suman A, Prasanna R, Saxena AK (2015) Diversity and phylogenetic profiling of niche-specific Bacilli from extreme environments of India. Ann Microbiol 65(2):611–629. https://doi.org/10.1007/s13213-014-0897-9

Yadav AN, Sachan SG, Verma P, Saxena AK (2016b) Bioprospecting of plant growth promoting psychrotrophic Bacilli from the cold desert of north western Indian Himalayas. Indian J Exp Biol 54(2):142–150

Yazdani M, Bahmanyar MA, Pirdashti H, Esmaili MA (2009) Effect of phosphate solubilisation microorganisms (PSM) and plant growth promoting rhizobacteria (PGPR) on yield and yield components of corn (Zea mays L.). World Acad Sci Eng Technol 49:90–92

Youssef NH, Savage-Ashlock KN, McCully AL, Luedtke B, Shaw EI, Hoff WD, Elshahed MS (2014) Trehalose/2-sulfotrehalose biosynthesis and glycine-betaine uptake are widely spread mechanisms for osmoadaptation in the Halobacteriales. ISME J 8(3):636–649. https://doi.org/10.1038/ismej.2013.165

Zaheer A, Malik A, Sher A, Mansoor Qaisrani M, Mehmood A, Ullah Khan S, Ashraf M, Mirza Z, Karim S, Rasool M (2019) Isolation, characterization, and effect of phosphate-zinc-solubilizing bacterial strains on chickpea (Cicer arietinum L.) growth. Saudi J Biol Sci 26(5):1061–1067. https://doi.org/10.1016/j.sjbs.2019.04.004

Zhang S-H (2016) The genetic basis of abiotic stress resistance in extremophilic fungi: the genes cloning and application. pp 29–42. https://doi.org/10.1007/978-3-319-42852-9_2

Zhang Y, Li X, Bartlett DH, Xiao X (2015) Current developments in marine microbiology: high-pressure biotechnology and the genetic engineering of piezophiles. Curr Opin Biotechnol 33:157–164. https://doi.org/10.1016/j.copbio.2015.02.013

Zhao K, Penttinen P, Zhang X, Ao X, Liu M, Yu X, Chen Q (2014) Maize rhizosphere in Sichuan, China, hosts plant growth promoting Burkholderia cepacia with phosphate solubilizing and antifungal abilities. Microbiol Res 169(1):76–82. https://doi.org/10.1016/j.micres.2013.07.003

Zhu J-K (2001) Plant salt tolerance. Trends Plant Sci 6(2):66–71. https://doi.org/10.1016/S1360-1385(00)01838-0

Chapter 12
Biotechnology and Genomics Exploration of Halotolerant Microbes: Application for Improving the Fertility of Saline Soil

Smita Kumari and Balaram Mohapatra

12.1 Introduction

In the aspect of global challenge, strategies have been initiated towards the solution of fertility of saline soil to yield the productivity of crops. These challenges arise in the matter of food security and soil quality deterioration which is due to the impact of a growing population and environmental and climate transformation. Due to climate change, some extremophilic plants and microorganisms are emerging as biotechnological tools by enhancing the endurable capacity of crops from extremely harsh conditions to avoid any reasonable loss in agriculture. Extremophiles are those that can thrive only in harsh conditions. Thus, the environments that are not suitable to survive for most organisms are defined as extreme environments. These include most of the oceanic depth, high mountains, volcanoes, saline lakes, rivers, thrilling desert conditions, extreme cold and hot temperate, highly acidic, radiation, pressure and alkaline environments. Since our planet is mostly shielded from extreme conditions, still it is enriched with diverse creatures, especially plants and microorganisms. Microorganisms can easily proliferate, survive and have specific applications and benefits to overcome environmental issues. Extremophilic microorganisms have great potential in biotechnological applications as they can produce key bioactive molecules like extremozymes and extremolytes. These extremozymes are very supportive in the agriculture and food industries.

For a few decades, agriculture productivity has been declining because of environmental stress such as salinity, pH, temperature and alkalinity (Saghafi et al.

S. Kumari (✉)
Department of Basic and Applied Sciences, School of Engineering and Sciences, G D Goenka University, Sohna, Haryana, India

B. Mohapatra
Environment Biotechnology Division, Gujarat Biotechnology University, Gandhinagar, Gujarat, India

2019). As per the Food and Agriculture Organization (Habib et al. 2008), most of the land is unproductive due to elevated levels of soil salinity. Therefore, researchers are paying more attention towards this issue because of the increased salinity of the soil that prevents the optimal growth and function of flora and, consequently, affects crop production. However, scientists have been resolving the issue through genetically modified organisms or plant breeding methods, but this technology is not easy to apply in all types of fields as it is expensive for regular agriculture practice (Imam et al. 2016).

Nowadays, plant-growth-promoting rhizobacteria (PGPR) are explored extensively to understand the morphological and physiochemical properties and their genomic organization, function and application in resolving the environmental stress conditions arising in agricultural matters. Later, PGPR became an effective biotechnological tool for mitigating stress in agriculture (Goswami et al. 2016). Hence, evolving PGPR, cultivation of genetically modified crops and organisms and applying plant growth regulators are suggested to enhance the plant health and yield of crops (Ghorbanpour 2013; Glick 2014). Rhizobacteria of halophyte plants have a salt tolerance ability that can facilitate root colonization and exhibit a critical role in biotic and abiotic stress modulation, thereby improving the growth and yield of the crops. Several reported rhizobacteria like *Bacillus*, *Lysobacter*, *Serratia*, *Burkholderia*, *Enterobacter* and *Pseudomonas* have effectively promoted plant growth and defended plants by biocontrol and biostimulation mechanisms (Khan et al. 2016; Islam et al. 2020).

The specific goals of this chapter are to emphasize on diversity of PGPR at the morphology and genetic level and also briefly discuss the interaction of plant–soil–bacteria. Further, how the genomics approach of PGPR reduces soil stress to overcome soil infertility is analyzed. In addition, it discusses the importance of biotechnological tools and advanced technology which resolve the issue of crop yield in soil salinity. Moreover, deciphering the mechanistic approaches of PGPR plant growth-promoting rhizobacteria (PGPR) by creating biofilms, extracellular polymeric compounds (EPS) or exopolysaccharides, and volatile organic compounds (VOC) such as geosmin, dimethyl disulphide etc., in improving the salt-stressed soil for critically enhancing the soil quality and plant growth. In the end, the chapter focused on the prospects and challenges met in handling salt stress.

12.2 Morphology and Diversity of PGPR

12.2.1 Diversity, Morphology and Effectiveness of PGPR

Plant-growth-promoting rhizobacteria is believed to be the potential alternative to agrochemicals to enhance soil fertility and regulate growth hormones in plants. The word PGPR was first used for fluorescent *Pseudomonas* which contributes to dual function by increasing the growth and regulating the pathogen (Agbodjato et al.

2015). Rhizospheres are that environment of roots where rhizobacteria colonize and impart several important functions, cycling minor plant nutrients, fixing atmospheric nitrogen and PGPR hormone production for sustainable cultivation. The potency of soil and yield of crops are essentially regulated by the mutualistic interaction of innate soil microorganisms and plants. In addition, the main factor for the interaction of rhizobacteria and plants that enhance soil fertility is when the root signal for the exudate's secretion at the rhizosphere zone and soil organic matter takes place. Hence, PGPR serves as a biological fertilizer to maintain the sustainability of the soil.

Researchers have isolated rhizobacteria from different soil, and their experiment results in diversity in their population which depends on key factors for instance O_2, pH and nutrient concentrations. Owing to the varied concentrations of soil nutrients available and the root zone of plants, a high population density of *Rhizobium* growth is feasible (Biswas and Gresshoff 2014; Ju et al. 2019). The root exudates of the plant usually contain vitamins amino acids, sugar compounds, organic acids, glycosides, nucleotide compounds, anions and their enzymes which greatly certainly influence the density of the bacterial strain in the soil system (Dakora and Phillips 2002). In addition to these compounds, plant metabolites released in the soil can also boost the growth of rhizospheric bacteria which results in increased soil fertility (Purwaningsih et al. 2021). Bacteria also assist in increasing the productivity of crops by providing compounds such as iron by releasing siderophores which are not produced by plants and a few specific enzymes like protease enable to hydrolyze of protein into peptides and amino acids that are essential for optimal plant health and growth (Baig et al. 2010). Several bacteria genera like *Burkholderia*, *Agrobacterium*, *Pseudomonas alcaligenes*, *Klebsiella*, *Serratia*, *Enterobacter* and *Enterobacter* have been abundantly identified and is suggested that these genera could effectively augment plant health using several approaches. They fix nitrogen, solubilize phosphorus and potassium to improve soil nutrient availability and also produce compounds such as non-ribosomal peptides and siderophores (Farina et al. 2012; Agbodjato et al. 2015).

Application of the significant PGPR has been deliberated by scientists and researchers and detected the mechanism and its beneficial outcome on agricultural stressed soil and crop growth. Inoculation of PGPR specifically in the rhizosphere of alfalfa has significantly changed the microbial communities; increased the diversity of bacteria taxa such as *Sphingomonas*, *Arthrobacter* and *Bacillus*; and is favourable to the growth of alfalfa in the agriculture field (Tang et al. 2014). Alfalfa is a very noteworthy legume known as the 'king of pastures' which is high-quality grass that can tolerate alkali and salt stress (Zhang et al. 2017). When we inoculate strains like *Sphingomonas* A55, *Sinorhizobium* A15, *Enterobacter* P24 and *Bacillus* A28, they enhance the growth of maize. These strains also increase the diversity of maize rhizosphere (Hao et al. 2020).

12.2.2 Genetic Diversity

The strain of several PGPRs due to variability in their genes has opened the possibility of controlling the diseased rhizomes in different soil. Genetic variability in the *Fusarium oxysporum* and *Fusarium solani* from diseased ginger rhizome were examined by using amplified DNA amplicon. These fusarium strains' infection was inhibited by applying different mixtures of effective PGPR like *Bacillus subtilis* and *Burkholderia cepacia* upon overcoming the disease of rhizome rot and healthy soil with increased rhizome production up to 85%. This happened due to the rise in the defence enzyme activity such as polyphenol oxidase, β-1,3-glucanase and chitinase (Farina et al. 2012). Implementation of engineered PGPR has high possibilities for sustainable soil fertility and crop productivity. Strains were isolated to ensure the characteristics related to PGPR for the early growth of maize using the technique amplified ribosomal DNA restriction analysis (ARDRA) along with various restriction endonucleases. Among two different phyla, 89% of genera were *Bordetella*, *Stenotrophomonas*, *Cupriavidus*, *Ochrobactrum*, *Achromobacter*, *Agrobacterium* and *Pseudoxanthomonas*, while 11% of genera were *Flavobacterium* and *Chryseobacterium*. These genera have effectively enhanced the growth of root and shoot with a further increase in the mass of maize. Besides this, some of them could fix nitrogen and adequately produce indole-3-acetic acid which has contributed major biofertilizer of maize crops (Youseif 2018). PGPR like various genera of *Bacillus*, *Paenibacillus* and *Pseudomonas* are also studied for cucumber seedlings and other vegetables which greatly influence plant growth, act as a biocontrol and deal with environmental stress (Kim 2011). The genetic and functional diversity of abundant PGPR species *Pseudomonas* associated with *Oryza sativa* L. has unique properties to produce indole acetic acid, denitrification ability and antifungal metabolites, which are suitable for sustainable rice production (Ramesh Kumar et al. 2014).

12.2.3 Advantage of Plant Microbes' Interaction in Enhancing Soil Fertility

The interaction of microorganisms with plants is an interesting mechanism. However, the interaction is positive and negative. Here, the key focus is on the positive interaction which reflects the beneficial outcomes for plants. Microorganisms colonize the adjacent surface area of the plant and near the root locality. In this way, plants provide an environment or habitat and essential nutrients from the decaying parts for the microorganisms. In turn, rhizospheric microorganisms secrete specific compounds which let the plant grow and defend during stress conditions (Schirawski and Perlin 2018). *Streptomyces* species is a very well-known soil bacteria that colonizes the roots, rhizosoil and aerial plant parts and actively produces antibiotics and volatile compounds to save plants (Vurukonda et al. 2018). Nowadays genetically modified bacteria are significantly important in the mitigation of saline soil and make the soil fertile (Fig. 12.1).

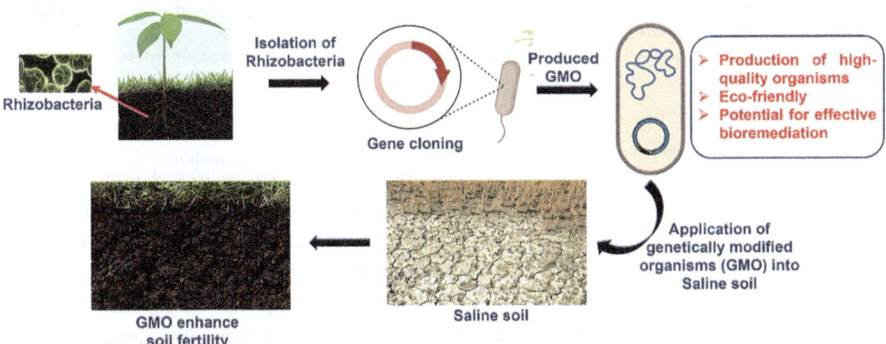

Fig. 12.1 Graphic representation to reflect the effect of genetically modified rhizobacteria in saline soil

12.3 Advanced Biotechnological Tools and Their Application

The application of PGPR at the lab scale or the pilot scale is easy to overcome crop diseases, growth retardation, reduction in enzyme activities, metabolite production and reduction of nitrogen fixation. However, the major challenges to commercializing the benefits of these rhizospheric bacteria are application and performance in the original field, lack of appropriate design while packing and loss of efficiency over time (Bashan et al. 2013). Thus, scientists have rigorously worked to find out the advanced but sustainable research techniques to overcome the problems. They scientifically improve the genetic make-up of the PGPR by applying different omics methods such as genomics, proteomics, transcriptomics, metabolomics and so on and other advanced methods (Fig. 12.2) used like next-generation sequencing (NGS) for sequencing of the whole genome (Cramer et al. 2011). This breakthrough has brought a revolution in genotyping.

Further, development at the genomics level for salinity tolerance is quite a difficult task as the multigenicity in the plant create difficulties in understanding the genomes at structural and functional level. However, genomics-related technology and evidence led to the generation of innovative resources to produce efficient tolerant genotypes. Various advanced tools and techniques such as identification of single nucleotide polymorphism (SNP) variations, their functional effects and genome-wide analyses (GWM) have manipulated the genomics information and revealed the understanding of salt tolerance mechanism by plant breeding method (Afzal et al. 2022). The depth of knowledge of plant cellular biology reflects a deep awareness of the traits associated with salinity tolerance and increases the functional and structural genomics approaches that are appropriate for the detection of specific traits associated with quantitative trait loci (QTLs) genes (Saradadevi et al. 2021).

High-saline soil has both ionic and osmotic stress, and this tends to increase biosynthesis of the abscisic acid (ABA), further modulating the ABA pathway.

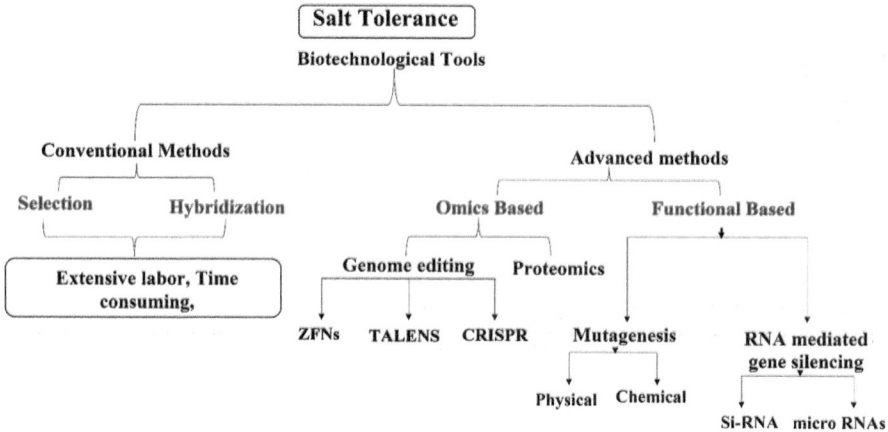

Fig. 12.2 Flow chart to represent advanced tools for developing PGPR

Under this salt-stressed condition, various genes are induced because this ABA pathway is active in rice crops. Genes that are studied in rice in salt stress conditions are: phytoene synthase gene, OsPSY3 and 9-cis-epoxy carotenoid dioxygenases genes (OsNCED3, OsNCED4 and OsNCED5), protein kinases and so on are involved through ABA-dependent pathway (Kumar et al. 2013). Thus, the rise in ABA biosynthesis regulates several genes in many agriculture crops during the high salinity in the soil. Moreover, there are two advanced tools known as transgenic plants and genome editing; both are genetically modified tools, but genome editing tools use specific DNA or nucleases to obtain traits. This genome editing has opened a great possibility for the demanding crops of different traits. Scientists have developed several tools such as zinc finger nucleases (ZFNs), TALENS and CRISPR/Cas9 for gene editing that are for specific editing. TALENs and ZFNs are known as first-generation gene technology, but this is time-consuming extensive labour and hard to accomplish a specific target compared to CRISPR/Cas9 which consumes less time and is effectively applied in tobacco and rice (Afzal et al. 2022). However, CRISPR/Cas 9 technology uses two important components, that is, guide RNA and Cas9. The significant feature of these two components is that they can aim for multiple sites within the genome (Shelake et al. 2022). The application of CRISPR in salt stress or abiotic conditions can be considered into three categories based on the target DNA. In the plant genome, CRISPR is divided into structural genes, regulatory genes and cis-regulatory elements (CREs) of regulatory genes (Zafar et al. 2019).

Besides genomics-based tools, other highly specific techniques like proteomics and NGS are involved in reducing salinity stress in soil. Proteomics is considered the best tool in different abiotic stress environments because the changes that are examined in protein will vary in multiple numbers. Those proteins which are related to the antioxidant system, sugar metabolism and signal transduction suggest a key role in resistance to salinity in plants by regulating metabolic processes and protein stability (Mansour and Hassan 2021). Crops produce several key enzymes and antioxidant proteins that assist in damage instigated by salt stress and remove reactive oxygen species (ROS) in *Zea mays* (Chen et al. 2019). ROS has a major role in controlling water and osmotic balance in growing crops like *Hordeum vulgare* (Szypulska et al. 2017). Proteomics techniques used for the identification of different proteins that are accumulated during stress conditions in photosynthesis, sugar metabolism and redox reactions and to study young seedlings in *Brassica napus* and pigeon pea are isobaric tags for relative and absolute quantitation (iTRAQ), iodoacetyl tandem mass tag (iodoTMT) and liquid chromatography-mass spectrometry (LC-MS) (Hussain et al. 2019; Yu et al. 2021; Jain et al. 2021). Nowadays, awareness of advanced tools and understanding at the genetic level for functional characterization is very important for effective agriculture practice. Another important approach that led to deciphering the genomic organization of the crops to develop a wide source of genomes is known as NGS. Whole genome analyses have played a key role in setting a benchmark for halophytes and allele identification in several studied crops, including chickpeas, rice and maize (Hoyos-Villegas et al. 2017). Additionally, rice genes like auxin efflux that are expressed during salt stress situations are identified by genome-wide analysis, and this data is possible only by whole genome sequencing data (Islam et al. 2022). Therefore, the time has come to practice these advanced techniques to identify the specific alleles for breeding purposes and further develop various salt-tolerant crops.

12.4 Alleviating Soil Salinity: Role of PGPR/B

Salinity in the soil is primarily caused by (a) modified agricultural practices, namely, the use of saline water for irrigation and fertilization; (b) intrusion of sea/river water into vulnerable areas, such as coastal arid and semi-arid regions; (c) altered precipitation patterns; or (d) an increase in sea level because of climate change (Zhang et al. 2020; Kumar et al. 2022). The rhizospheric microbial community and physiological development such as stomatal movement, premature senescence, photosynthetic rate, oxidative stress response and so on are all impacted by the high osmotic stress (Mahawar and Shekhawat 2019; Dubey et al. 2022). The overall crop production is affected even though plants are known to use strategies like an increased synthesis of osmoprotectants, phytohormones, upregulation of antioxidant activities, regulation of Na^+/K^+ homeostasis, cellular compartmentalization and altering various morphological responses (Zhao et al. 2020; Arif et al. 2020). Recent studies have demonstrated that a variety of halotolerant microorganisms can coexist with plants in soil or rhizospheric niches to increase stress tolerance, nutrient uptake and plant growth, thereby increasing crop yield. In this perspective, the application/use of halophilic or halotolerant PGPB/R is found to be a suitable alternative. Some key attributes of such salt-adapted plant growth promoting (PGP) microbes are efficient (de)mineralization of minerals/nutrients (P, K, Zn, Si, etc.), efficient uptake/transport, fixation of N_2, production of ammonia, biocontrol activities, modulation of phytohormones (IAA, ACC deaminase and gibberellic acid), Fe uptake by siderophore, induced systematic resistance, biotransformation of toxic compounds, and so on (Kumar et al. 2017; Pattnaik et al. 2021). The involvement of PGPR is attributed to possessing three-way mechanisms: (a) PGP survival under a hyperosmotic environment, (b) introducing salinity tolerance in plants and (c) improving soil quality. However, studies have shown that the assemblage of such microbes at the soil/rhizosphere sites is dependent on several factors, namely, plant genotype, the composition of root exudates, substrate availability, metabolic dependency, soil composition, mineral facies, local hydro-geological parameters, the genomic-metabolic repertoire of the community and so on (Pattnaik et al. 2021). Hence a detailed insight into such halophilic and halotolerant PGP microbes at the OMICS level is required.

12.5   Key PGP Mechanisms of Halotolerant Bacteria: Effects on Crop Growth and Yield

Halotolerant PGP microbes of diverse taxa (*Rhizobium, Arthrobacter, Flavobacterium, P. alcaligenes, Azospirillum, Burkholderia, Bacillus,* etc.) exert their beneficial effects in multiple ways. Various key plant growth beneficial activities of halotolerant PGP bacteria on different crops are listed in Table 12.1. PGPB/R typically forms biofilms and creates extracellular polymeric compounds, also known as exopolysaccharides (EPSs), in saline agroecosystems (Haque et al. 2022). Kumar Arora et al. (2020) and Sunita et al. (2020) highlighted that the EPS production (40–90% of weight) by PGPR is dependent on the bacterial growth phase and the extent of external stimuli, like nutrient availability/medium's composition, pH, temperature and so on. EPS forms the rhizosheath, or plaque, around the roots of plants, improving nutrient availability and uptake while also acting as a physical barrier against pathogens and increased ions due to salinity. According to Ramasamy and Mahawar (2023), it has also been linked to improved humification, formation of soil aggregation, water retention, quorum sensing, nodulation and antioxidants for lowering oxidative damage and biofilm that shields plant cells from drying. A study examined the importance of halotolerant *Enterobacter cloacae* and *Bacillus drentensis*, which produce EPS, in promoting the plant health and growth of mung beans grown in saline-stressed soil, by improving nutrient availability and water uptake (Mahmood et al. 2016). It has been demonstrated that the EPS-producing endophyte *Pantoea alhagi* NX-11 reduces oxidative stress and promotes rice development by increasing antioxidant activity (Sun et al. 2020). In order to boost antioxidant activity in *Solanum lycopersicum* under salt stress, a combined mixture of EPS-producing bacteria and silicon dioxide (SiO_2) nanoparticles has been attempted. Mukherjee et al. (2019) described that rice growth was stimulated and plants acquired resistance to osmotic stress by the EPS synthesized by *Halomonas* sp. EX01 in salt stress experimental settings. EPS lowers the risk of plant toxicity due to ions by expressing the HKT1/K^+ transporter and restricting Na^+ influx (Zhang et al. 2008). A study observed that the EPS-producing *Pseudomonas* sp. improved the yield of sunflowers in a high-saline soil (EC > 10 dS/m). The findings

Table 12.1 Various key plant growth beneficial activities of halotolerant PGP bacteria on different crops at varied salinity conditions

S. No.	Crop plant	Halotolerant PGP microbes	Salt Conc. (NaCl)	Activity of PGP microbes	Effects on crop plant	Level of study	References
1.	Barley (*Hordeum vulgare* L.)	*Bacillus mojavensis* S1, *Pseudomonas fluorescens* S3	200 mM	Osmoprotective	Alleviation of Na+ concentration in, stimulation of root development enhanced water and nutrient uptake	Pot	Mahmoud et al. (2020)
2.	Sunflower (*Helianthus annuus* L.)	*Bacillus aryabhattai* RS341, *Bacillus licheniformis* RS656, *Pseudomonas*	125 mM	EPS	Plant growth promotion and biocontrol against phytopathogenic fungus	Pot & field	Tewari et al. (2016)
3.	Maize (*Zea mays* L.)	*Bacillus pumilus* (STR2), *Halomonas desiderata* (STR8), *Exiguobacterium*	500 mM	EPS and Osmoprotective	Improved plant growth, influenced indigenous microbial communities	Pot	Bharti et al. (2015)
4.	White clover (*Trifolium repens* L.)	*Bacillus subtilis* strain GB03	150 mM	VOCs	Decreased Na+ accumulation, increase in chlorophyll content, leaf osmotic potential, cell membrane integrity	Pot	Han et al. (2014)
5.	Mint (*Mentha arvensis* L.)	*Bacillus pumilus* STR2, *Halomonas desiderata* STR8, *Exiguobacterium*	500 mM	EPS	Improved nutrient uptake and antioxidant machinery	Pot	Bharti et al. (2014)
6.	Ginseng (*Panax ginseng* L.)	*Paenibacillus yonginensis* DCY84T	300 mM	Osmoprotective	Enhanced nutrient availability, induction of defence-related systems ion transport, antioxidant enzymes, total sugars, ABA and root hair formation	Pot	Sukweenadhi et al. (2018)
7.	Canola (*Brassica napus* L.)	*Brevibacterium iodinum* RS16, *Micrococcus yunnanensis* RS222	100 mM	EPS	Increased vigour index, fresh weight and growth hormones; production of stress alleviating enzymes	Pot	Hong et al. (2017)
8.	Lotus (*Lotus japonicus* cv Gifu	*Bacillus amyloliquefaciens* RHF6	400 mM	Osmoprotective, multiple PGP traits	Increased seedling, root growth and stress management	Pot	Castaldi et al. (2023)

(continued)

Table 12.1 (continued)

S. No.	Crop plant	Halotolerant PGP microbes	Salt Conc. (NaCl)	Activity of PGP microbes	Effects on crop plant	Level of study	References
9.	Maize (*Zea mays* L.)	*Bacillus safensis* PM22	100–300 mM	Osmoprotective, multiple PGP traits	photosynthetic pigment, carotenoid, leaf relative water content, stress tolerance	Pot	Atif Azeem et al. (2022)
10.	Tomato (*Solanum lycopersicum* L.)	*Bacillus spizizenii*	150 mM	Enzymes, osmoprotection	Photosynthetic pigment, membrane integrity (MI) and phenol peroxidase (POX)	Pot	Masmoudi et al. (2021a)
11.	Tomato (*Solanum lycopersicum* L.)	*Bacillus velezensis* FMH2	300–700 mM	Enzymes, osmoprotection	Photosynthetic pigment, membrane integrity (MI) and phenol peroxidase (POX)	Pot	Masmoudi et al. (2021b)
12.	Soybean (*Glycine max* L.)	*Arthrobacter woluwensis* (AK1), *Microbacterium oxydans* (AK2)	200 mM	VOCs	Increased level of antioxidant enzymes and K+ uptake; reduced Na+ ion concentration in plant tissue	Pot	Khan et al. (2019)
13.	Sunflower (*Helianthus annuus* L.)	*Arthrobacter aurescens*, *Bacillus megaterium*, *Bacillus aryabhattai*, *Pseudomonas ae ruginosa*	500 mM	EPS	Production of salicylic acid (SA), enhancement in plant growth parameters and reduction in disease incidence	Field	Tewari and Arora (2018)
14.	Tomato (*Solanum lycopersicum* L.)	*Pseudomonas* sp. UW4	800 mM	Osmoprotective	Increased root and shoot length, total dry weight and chlorophyll content	Pot	Orozco-Mosqueda et al. (2019)
15.	Peanut (*Arachis hypogaea* L.)	*Klebsiella*, *Pseudomonas*, *Agrobacterium* and *Ochrobactrum*	100 mM	VOCs	Nutrients uptake, ion homeostasis and defence against ROS	Pot	Sharma et al. (2016)
16.	Maize (*Zea mays* L.)	*Pseudomonas fluorescens* 002	150 mM	VOCs	Improved plant growth	Pot and field	Zerrouk et al. (2016)

No.	Plant	Bacteria	Concentration	Mechanism	Effect	Study type	Reference
17.	Soybean (*Glycine max* L.)	*Pseudomonas simiae*	100 mM	VOCs	Decrease root Na+ accumulation and increase in proline and chlorophyll content	Pot	Vaishnav et al. (2016)
18.	Soybean (*Glycine max* L.)	*Pseudomonas pseudoalcaligenes*	100 mM	Osmoprotective, multiple PGP traits	Enhanced biomass, relative water content and osmolytes	Pot	Yasmin et al. (2020)
19.	Soybean (*Glycine max* L.)	*Pseudomonas aeruginosa* GS-33	200 mM	Multiple PGP traits	Enhanced biomass and biocontrol activity		Patil et al. (2016)
20.	Pigeon pea (*Cajanus cajan* L.)	*Rhizobium* sp. IC3123	16 mM	EPS	Enhanced germination percentage, pod number, seed yield, protein content and nodule formation (per plant)	Pot and field	Tewari and Sharma (2020)
21.	Quinoa (*Chenopodium quinoa* L.)	*Enterobacter* sp. MN17 and *Bacillus* sp. MN54	400 mM	EPS	Improved plant-water relationship	Pot	Yang et al. (2016)
22.	Tomato (*Solanum lycopersicum* L.)	*Enterobacter* sp. EJ01	200 mM	VOCs	Enhanced plant growth and increase in salt-stress tolerance	Pot	Kim et al. (2014)
23.	Thale cress (*Arabidopsis thaliana* L.)	*Paraburkholderia phytofirmans* PsJN	150 mM	VOCs	Improved plant growth	Pot	Ledger et al. (2016)
24.	Tomato (*Solanum lycopersicum* L.)	*Leclercia adecarboxylata* MO1	120 mM	VOCs	Enhancement in soluble sugars, i.e., glucose, sucrose, fructose; organic acids, i.e., citric acid, malic acid; amino acids, i.e., serine, glycine, methionine, threonine, and proline in the fruits	Pot	Kang et al. (2019)
25.	Chickpea (*Cicer arietinum* L.)	*Halomonas variabilis* HT1 and *Planococcus rifietoensis* RT4	200 mM	EPS	Increased plant growth, improved soil nutrient status	Pot	Qurashi and Sabri (2012)

also observed a decline in the *Macrophomina phaseolina*'s prevalence of charcoal rot disease (Tewari and Arora, 2018). By inhibiting Na$^+$ inflow into the stele of plants, *Bacillus amyloliquefaciens*, *Microbacterium* spp., *Bacillus insolitus* and *Pseudomonas syringae* have been shown by Ashraf et al. (2004) to play a favourable effect in promoting wheat growth. Similar to this, treating maize with EPS-producing *Azotobacter chrococcum* in salty circumstances reduced saline stress by raising the K$^+$/Na$^+$ ratio; causing altered absorption of Na$^+$, K$^+$, Ca^{2+} and Mg^{2+}; increasing the amount of chlorophyll; and causing polyphenols and proline to accumulate (Rojas-Tapias et al. 2012). An EPS bioformulation developed from *Alcaligenes* sp. has exhibited success in lowering osmotic stress and improving rice growth in saline environments (Fatima et al. 2020). All in all, it is generally accepted that EPS-producing PGP microorganisms can be used as bioinoculants to enhance soil quality, nitrogen uptake in saline extremities and rhizosphere colonization.

Furthermore, under conditions of salinity stress, PGPR generates lipophilic metabolites (with a greater vapour pressure) and low molecular weight volatile organic compounds (VOCs) (Sunita et al. 2020). The most studied VOCs are geosmin, dimethyl disulphide, acetoin and 2,3-butanediol. These have been linked to various processes such as composting, soil formation, cell expansion, auxin-mediated homeostasis and systemic and drought resistance (Meldau et al. 2013). VOCs aid in the plant health, growth and adaptation of plants under stress by promoting the phytohormones, osmoprotectants and siderophores production in PGPR/B. They also enhance the HKT1/K$^+$ transporter's expression and microbial motility to improve plant–PGPR interaction and modulate the virulence factors (Bhat et al. 2020). According to Gutierrez-Luna et al. (2010), *Bacillus* strains were found to create 22 different types of VOCs, which were found to promote the growth of both primary and lateral roots in *Arabidopsis thaliana*. Even when *A. thaliana* was briefly exposed to volatile organic compounds, root-associated *Microbacterium* spp. showed similar effects, and these effects were tissue-specific (Cordovez et al. 2018). Even at extremely high salt concentrations (150 mM), VOCs such as 4-nitro guaiacol and quinoline generated by halophilic *Pseudomonas simiae* have been shown to promote soybean growth (Vaishnav et al. 2016). *B. subtilis* was found to downregulate gene expression in relation to the HKT1/K+ transporter (inhibition of Na$^+$ inflow) (Zhang et al. 2008). Bhattacharyya and Lee (2017) discovered that *Alcaligenes faecalis* JBCS1294 produced a cocktail of VOCs (butyric acid, propionic acid and benzoic acid) that improved salt tolerance and plant development via altering phytohormones and inorganic ion transport. Further systematic research is required to better comprehend the novel physiological roles that remain unknown.

Halotolerant PGP microorganisms have been demonstrated to produce and accumulate suitable osmolytes/osmoprotectants in host crop plants, reducing osmotic stress, maintaining high turgor pressure and sustaining ion balance. PGPR have been evident in producing compounds, namely, amino acids and their derivatives, polyols and non-reducing sugars. It has also been evident that these compounds are either acquired/transported from the nearby environment or synthesized in response to salinity stress (Sunita et al. 2020; Ramasamy and Mahawar 2023).

12.6   Exploring Genomes of Halotolerant PGPR

Bacillus fortis SSB21 bioinoculation has been demonstrated to enhance synthesis of proline and expression of genes associated with stress in capsicum during salinity stress, including pepper osmotin-like protein 1 (CaOSM1), pepper pathogen-induced protein gene (CAPIP2), pepper class II basic chitinase (CAChi2) and ketoacyl-ACP reductase (CaKR1) (Yasin et al. 2018). As per Sukweenadhi et al. (2018), seed priming of *Panax ginseng* seeds with *Paenibacillus yonginensis* DCY84T has been demonstrated to increase chlorophyll, polyamine, proline, total soluble sugar and abscisic acid (ABA) during salinity stress. Drought, high salinity and heavy metals are salinity stressors that are known to induce proline accumulation (genes: *pro*B, *pro*A and *pro*C) by PGPR that works as ROS scavengers, control cytosolic acidity and stabilize protein structures (Krasensky and Jonak 2012, Kaur and Asthir 2015). According to Pan et al. (2019), halotolerant PGP bacteria are crucial for preserving ion homeostasis (K/Na); osmolyte accumulation, which includes the build-up of proteins and soluble carbohydrates; and increased N, P, K, Ca and Mg bioavailability. It has been demonstrated that the microbial synthesis of glycine betaine lowers osmotic stress and preserves the general integrity of the plant cell. Research has demonstrated that two enzymes, namely, glycine betaine aldehyde dehydrogenase and type III alcohol dehydrogenase, are crucial for initiating the synthesis of glycine betaine in *B. subtilis* (Kappes et al. 1999). In a study, *Acacia gerrardii* plants treated with *B. subtilis* BERA71 exhibited improved glycine betaine and other osmolytes (Hashem et al. 2018). *Bacillus* sp. HL3RS14 was also reported to improve glycine betaine levels in maize and encourage plant growth in salinity-stressed environments (Mukhtar et al. 2019).

In transgenic plants, with *codA* (choline oxidase) gene from *Arthrobacter globiformis* was reported to convert choline into glycine betaine (Giri 2011). According to Moghaieb et al. (2011), the synthesis of ectoine by PGPR via the *ect*A, *ect*B and *ect*C genes is said to counteract nitrate problems in plant roots. According to a report, *Chromohalobacter salexigens* KT989776, a halophilic bacteria, improved germination in flax seed and reduced salt buildup, peroxidase and phenoloxidase activity evident to ectoine production (Elsakhawy et al. 2019). It is well-known that PGPR aids in the production of trehalose, a crucial osmoprotectant that plants cannot produce when they are under salinity stress. The main enzymes that are elevated during drought and salinity stress are alpha-trehalose-phosphate synthase, trehalose synthase and trehalose-6-phosphate phosphatase (treY/treZ and treS) (Vilchez et al. 2018). According to recent research, halotolerant PGPR also modulates or upregulates phytohormone synthesis (indole acetic acid), changing shape and preventing the accumulation of excess ionic ions (Bhat et al. 2020). Increased IAA production is thought to aid in improved water uptake and nutrient availability by reducing the growth of taproots, promoting root hair elongation and increasing lateral roots, according to in vitro research (Grover et al. 2021; Sarkar et al. 2022).

Application of *Pseudomonas stutzeri*, *Pseudomonas putida* and *Stenotrophomonas maltophilia* on *Coleus forskohlii* enabled the plant to improve the phytohrmones cytokinin, gibberellic acid and IAA (Patel and Saraf 2017).

Similar to this, under salinity stress, *Pseudomonas* sp. improved the synthesis of cytokinin in *Zea mays* and gibberellin in *Glycine max* (Sandhya et al. 2010; Kang et al. 2014). In addition to growth hormones, PGPR may synthesize ABA and ethylene and modulate their gene expression (Bhat et al. 2020). According to Bharti et al. (2016), TaST in *Triticum aestivum* was activated by the halotolerant *Dietzia natronolimnaea* STR1 which led to 12-oxophytodienoate reductase 1 (TaOPR1) and ABA response elements (TaABARE) involving ABA signalling.

12.7   Conclusion

The practice of useful rhizobacteria to maintain salt stress conditions and further develop plant growth and yield seems to be an eco-friendly approach for sustainable agricultural production. The present chapter focussed on the recent use of advantageous rhizobacteria towards the maintenance of soil quality, plant diversity, physiology, genetic development and approaches to combat issues in the crop field. The approach of halotolerant PGPR to combat salt stress in their genomes and its effects on plant health, growth and crop yield have been discussed appropriately. The application of advanced biotechnological tools, NGS, editing tools and several multiomics technologies provides an in-depth understanding at the molecular level of PGPR and plants for the improvement of beneficial rhizobacteria has been also discussed in the current chapter. Even though wide research has been conducted and reported on the PGPR for several crops, the variation in the strain performance in the original field is detrimental for commercial purposes. Furthermore, research into potential microbial consortia, their optimization in different field conditions and planning to commercialize PGPR and effective consortia remain in the early phases. The reported studies are at the primary stages of the omics era for crop development, and therefore, the upcoming research seems promising. Thus, further studies are desired to appropriately elucidate the mechanisms of combat of crops under salt stress conditions. Seemingly, despite remarkable progress in combatting soil salinity by applying omics, various other biotechnological tools and mechanisms it remains a vast challenge. Overall, the concentration should be to uphold a continuous development in agricultural yield, which may accomplish the food demand and security globally.

References

Afzal M, Hindawi SES, Alghamdi SS et al (2022) Potential breeding strategies for improving salt tolerance in crop plants. J Plant Growth Regul 42:3365–3387. https://doi.org/10.1007/s00344-022-10797-w

Agbodjato NA, Noumavo PA, Baba-Moussa F, Salami HA, Sina H, Sèzan A, Baba-Moussa L (2015) Characterization of potential plant growth promoting rhizobacteria isolated from

Maize (Zea mays L.) in central and Northern Benin (West Africa). Appl Environ Soil Sci 2015(1):901656.

Arif Y, Singh P, Siddiqui H et al (2020) Salinity induced physiological and biochemical changes in plants: an omic approach towards salt stress tolerance. Plant Physiol Biochem 156:64–77. https://doi.org/10.1016/j.plaphy.2020.08.042

Ashraf M, Hasnain S, Berge O, Mahmood T (2004) Inoculating wheat seedlings with exopolysaccharide-producing bacteria restricts sodium uptake and stimulates plant growth under salt stress. Biol Fertil Soils 40. https://doi.org/10.1007/s00374-004-0766-y

Azeem MA, Shah FH, Ullah A, Ali K, Jones DA, Khan MEH, Ashraf A (2022) Biochemical characterization of halotolerant Bacillus safensis pm22 and its potential to enhance growth of maize under salinity stress. Plants, 11(13), 1721.

Baig KS, Arshad M, Zahir ZA, Cheema MA, (2010) Short communication comparative efficacy of qualitative and quantitative methods for rock phosphate solubilization with phosphate solubilizing rhizobacteria. Soil Environ 29(1):82–86.

Bashan Y, de Bashan LE, Prabhu SR, Hernandez J-P (2013) Advances in plant growth-promoting bacterial inoculant technology: formulations and practical perspectives (1998–2013). Plant Soil 378:1–33. https://doi.org/10.1007/s11104-013-1956-x

Bharti N, Barnawal D, Awasthi A, Yadav A, Kalra A (2014) Plant growth promoting rhizobacteria alleviate salinity induced negative effects on growth, oil content and physiological status in Mentha arvensis. Acta Physiol Plant 36:45–60.

Bharti N, Barnawal D, Maji D, Kalra A (2015) Halotolerant PGPRs prevent major shifts in indigenous microbial community structure under salinity stress. Microb Ecol 70:196–208.

Bharti N, Pandey SS, Barnawal D et al (2016) Plant growth promoting rhizobacteria Dietzia natronolimnaea modulates the expression of stress responsive genes providing protection of wheat from salinity stress. Sci Rep 6. https://doi.org/10.1038/srep34768

Bhat MA, Kumar V, Bhat MA et al (2020) Mechanistic insights of the interaction of Plant Growth-Promoting Rhizobacteria (PGPR) with plant roots toward enhancing plant productivity by alleviating salinity stress. Front Microbiol 11. https://doi.org/10.3389/fmicb.2020.01952

Bhattacharyya D, Lee YH (2017) A cocktail of volatile compounds emitted from Alcaligenes faecalis JBCS1294 induces salt tolerance in Arabidopsis thaliana by modulating hormonal pathways and ion transporters. J Plant Physiol 214:64–73. https://doi.org/10.1016/j.jplph.2017.04.002

Biswas B, Gresshoff P (2014) The role of symbiotic nitrogen fixation in sustainable production of biofuels. Int J Mol Sci 15:7380–7397. https://doi.org/10.3390/ijms15057380

Castaldi S, Valkov VT, Ricca E, Chiurazzi M, Isticato R (2023) Use of halotolerant Bacillus amyloliquefaciens RHF6 as a bio-based strategy for alleviating salinity stress in Lotus japonicus cv Gifu. Microbiol Res 268:127274.

Chen F, Fang P, Peng Y et al (2019) Comparative proteomics of salt-tolerant and salt-sensitive maize inbred lines to reveal the molecular mechanism of salt tolerance. Int J Mol Sci 20:4725. https://doi.org/10.3390/ijms20194725

Cordovez V, Schop S, Hordijk K et al (2018) Priming of plant growth promotion by volatiles of root-associated microbacterium spp. Appl Environ Microbiol 84. https://doi.org/10.1128/aem.01865-18

Cramer GR, Urano K, Delrot S et al (2011) Effects of abiotic stress on plants: a systems biology perspective. BMC Plant Biol 11:163. https://doi.org/10.1186/1471-2229-11-163

Dakora FD, Phillips DA (2002) Root exudates as mediators of mineral acquisition in low-nutrient environments. In: Food security in nutrient-stressed environments: Exploiting plants' genetic capabilities, pp 201–213. https://doi.org/10.1007/978-94-017-1570-6_23

Dubey S, Khatri S, Bhattacharjee A, Sharma S (2022) Multiple passaging of rhizospheric microbiome enables mitigation of salinity stress in Vigna Radiata. Plant Growth Regul 97:537–549. https://doi.org/10.1007/s10725-022-00820-1

Elsakhawy TA, Fetyan N, Ghazi AA (2019) The potential use of ectoine produced by a moderately halophilic bacteria Chromohalobacter salexigens KT989776 for enhancing germination and primary seedling of flax "Linum usitatissimum L." under salinity conditions. Biotechnol J Int 23(3):1–12.

Farina R, Beneduzi A, Ambrosini A et al (2012) Diversity of plant growth-promoting rhizobacteria communities associated with the stages of canola growth. Appl Soil Ecol 55:44–52. https://doi.org/10.1016/j.apsoil.2011.12.011

Fatima T, Mishra I, Verma R, Arora NK (2020) Mechanisms of halotolerant plant growth promoting Alcaligenes sp involved in salt tolerance and enhancement of the growth of rice under salinity stress. 3 Biotech 10. https://doi.org/10.1007/s13205-020-02348-5

Ghorbanpour M (2013) Role of plant growth promoting rhizobacteria on antioxidant enzyme activities and tropane alkaloids production of Hyoscyamus Niger under water deficit stress. Turk J Biol. https://doi.org/10.3906/biy-1209-12

Giri J. (2011) Glycinebetaine and abiotic stress tolerance in plants. Plant Signal Behav 6(11):1746–1751.

Glick BR (2014) Bacteria with ACC deaminase can promote plant growth and help to feed the world. Microbiol Res 169:30–39. https://doi.org/10.1016/j.micres.2013.09.009

Goswami D, Thakker JN, Dhandhukia PC (2016) Portraying mechanics of plant growth promoting rhizobacteria (PGPR): a review. Cogent Food Agric 2. https://doi.org/10.1080/23311932.2015.1127500

Grover M, Bodhankar S, Sharma A et al (2021) PGPR mediated alterations in root traits: way toward sustainable crop production. Front Sustain Food Syst 4. https://doi.org/10.3389/fsufs.2020.618230

Gutiérrez-Luna FM, López-Bucio J, Altamirano-Hernández J et al (2010) Plant growth-promoting rhizobacteria modulate root-system architecture in Arabidopsis thaliana through volatile organic compound emission. Symbiosis 51:75–83. https://doi.org/10.1007/s13199-010-0066-2

Habib MAB, Parvin M, Huntington TC, Hasan MR (2008) Food and agriculture organization of the united nations; Rome: (2008) A Review on Culture, Production and Use of Spirulina as Food for Humans and Feeds for Domestic Animals and Fish. [Google Scholar].

Han QQ, Lü XP, Bai JP, Qiao Y, Paré PW, Wang SM, Wang ZL (2014) Beneficial soil bacterium Bacillus subtilis (GB03) augments salt tolerance of white clover. Front Plant Sci 5:525.

Hao Z, Li K, Sha Y, Wang E, Sui X, Mi G, Tian C, Chen W (2020) Effectsof growth-promoting rhizobacteria on maize growth and rhizosphere microbial community under conservation tillage in Northeast China. Microb Biotechnol 14(2):535–550.

Haque MM, Biswas MS, Mosharaf MK et al (2022) Halotolerant biofilm-producing rhizobacteria mitigate seawater-induced salt stress and promote growth of tomato. Sci Rep 12:5599. https://doi.org/10.1038/s41598-022-09519-9

Hashem A, Alqarawi AA, Radhakrishnan R et al (2018) Arbuscular mycorrhizal fungi regulate the oxidative system, hormones and ionic equilibrium to trigger salt stress tolerance in Cucumis sativus L. Saudi J Biol Sci 25:1102–1114. https://doi.org/10.1016/j.sjbs.2018.03.009

Hong BH, Joe MM, Selvakumar G, Kim KY, Choi JH, Sa TM (2017) Influence of salinity variations on exocellular polysaccharide production, biofilm formation and flocculation in halotolerant bacteria. J Environ Biol 38(4):657.

Hoyos-Villegas V, Song Q, Kelly JD (2017) Genome-wide association analysis for drought tolerance and associated traits in common bean. Plant Genome 10. https://doi.org/10.3835/plantgenome2015.12.0122

Hussain S, Zhu C, Bai Z et al (2019) iTRAQ-based protein profiling and biochemical analysis of two contrasting rice genotypes revealed their differential responses to salt stress. Int J Mol Sci 20:547. https://doi.org/10.3390/ijms20030547

Imam J, Singh PK, Shukla P (2016) Plant microbe interactions in post genomic era: perspectives and applications. Front Microbiol 7. https://doi.org/10.3389/fmicb.2016.01488

Islam MR, Naveed SA, Zhang Y et al (2022) Identification of candidate genes for salinity and anaerobic tolerance at the germination stage in rice by genome-wide association analyses. Front Genet 13. https://doi.org/10.3389/fgene.2022.822516

Islam MT, Rahman MM, Pandey P et al (2020) Bacilli and agrobiotechnology: Phytostimulation and biocontrol. Springer Nature.

Jain N, Farhat S, Kumar R et al (2021) Alteration of proteome in germinating seedlings of piegonpea (Cajanus cajan) after salt stress. Physiol Mol Biol Plants 27:2833–2848. https://doi.org/10.1007/s12298-021-01116-w

Ju W, Liu L, Fang L et al (2019) Impact of co-inoculation with plant-growth-promoting rhizobacteria and rhizobium on the biochemical responses of alfalfa-soil system in copper contaminated soil. Ecotoxicol Environ Saf 167:218–226. https://doi.org/10.1016/j.ecoenv.2018.10.016

Kang S-M, Khan AL, Waqas M et al (2014) Plant growth-promoting rhizobacteria reduce adverse effects of salinity and osmotic stress by regulating phytohormones and antioxidants in Cucumis sativus. J Plant Interact 9:673–682. https://doi.org/10.1080/17429145.2014.894587

Kang SM, Shahzad R, Bilal S, Khan AL, Park YG, Lee KE, Lee IJ (2019) Indole-3-acetic-acid and ACC deaminase producing Leclercia adecarboxylata MO1 improves Solanum lycopersicum L. growth and salinity stress tolerance by endogenous secondary metabolites regulation. BMC microbiology, 19:1–14.

Kappes RM, Kempf B, Kneip S, Boch J, Gade J, Meier Wagner J, Bremer E (1999) Two evolutionarily closely related ABC transporters mediate the uptake of choline for synthesis of the osmoprotectant glycine betaine in Bacillus subtilis. Mol Microbiol 32(1):203–216.

Kaur G, Asthir B (2015) Proline: a key player in plant abiotic stress tolerance. Biol Plant 59:609–619. https://doi.org/10.1007/s10535-015-0549-3

Khan N, Bano A, Babar MA (2016) The root growth of wheat plants, the water conservation and fertility status of sandy soils influenced by plant growth promoting rhizobacteria. Symbiosis 72:195–205. https://doi.org/10.1007/s13199-016-0457-0

Khan MA, Asaf S, Khan AL, Adhikari A, Jan R, Ali S, Lee IJ (2019) Halotolerant rhizobacterial strains mitigate the adverse effects of NaCl stress in soybean seedlings. BioMed research international, 2019(1):9530963.

Kim W-I (2011) Genetic diversity of cultivable plant growth-promoting Rhizobacteria in Korea. J Microbiol Biotechnol 21:777–790. https://doi.org/10.4014/jmb.1101.01031

Kim K, Jang YJ, Lee SM, Oh BT, Chae JC, Lee KJ (2014) Alleviation of salt stress by Enterobacter sp. EJ01 in tomato and Arabidopsis is accompanied by up-regulation of conserved salinity responsive factors in plants. Mol Cells 37(2):109–117.

Krasensky J, Jonak C (2012) Drought, salt, and temperature stress-induced metabolic rearrangements and regulatory networks. J Exp Bot 63:1593–1608. https://doi.org/10.1093/jxb/err460

Kumar Arora N, Fatima T, Mishra J et al (2020) Halo-tolerant plant growth promoting rhizobacteria for improving productivity and remediation of saline soils. J Adv Res 26:69–82. https://doi.org/10.1016/j.jare.2020.07.003

Kumar K, Amaresan N, Madhuri K (2017) Alleviation of the adverse effect of salinity stress by inoculation of plant growth promoting rhizobacteria isolated from hot humid tropical climate. Ecol Eng 102:361–366. https://doi.org/10.1016/j.ecoleng.2017.02.023

Kumar K, Kumar M, Kim S-R et al (2013) Insights into genomics of salt stress response in rice. Rice 6. https://doi.org/10.1186/1939-8433-6-27

Kumar P, Choudhary M, Halder T et al (2022) Salinity stress tolerance and omics approaches: revisiting the progress and achievements in major cereal crops. Heredity 128:497–518. https://doi.org/10.1038/s41437-022-00516-2

Ledger T, Rojas S, Timmermann T, Pinedo I, Poupin MJ, Garrido T, Donoso R (2016) Volatilemediated effects predominate in Paraburkholderia phytofirmans growth promotion and salt stress tolerance of Arabidopsis thaliana. Front Microbiol 7:1838.

Mahawar L, Shekhawat GS (2019) EsHO 1 mediated mitigation of NaCl induced oxidative stress and correlation between ROS, antioxidants and HO 1 in seedlings of Eruca sativa: underutilized oil yielding crop of arid region. Physiol Mol Biol Plants 25:895–904. https://doi.org/10.1007/s12298-019-00663-7

Mahmood S, Daur I, Al-Solaimani SG et al (2016) Plant growth promoting rhizobacteria and silicon synergistically enhance salinity tolerance of mung bean. Front Plant Sci 7. https://doi.org/10.3389/fpls.2016.00876

Mahmoud OMB, Hidri R, Talbi-Zribi O, Taamalli W, Abdelly C, Djébali N (2020) Auxin and proline producing rhizobacteria mitigate salt-induced growth inhibition of barley plants by enhancing water and nutrient status. S Afr J Bot 128:209–217.

Mansour MMF, Hassan FAS (2021) How salt stress-responsive proteins regulate plant adaptation to saline conditions. Plant Mol Biol 108:175–224. https://doi.org/10.1007/s11103-021-01232-x

Masmoudi F, Tounsi S, Dunlap CA, Trigui M (2021a) Halotolerant Bacillus spizizenii FMH45 promoting growth, physiological, and antioxidant parameters of tomato plants exposed to salt stress. Plant Cell Reports, 40(7):1199–1213.

Masmoudi F, Tounsi S, Dunlap CA, Trigui M (2021b) Endophytic halotolerant Bacillus velezensis FMH2 alleviates salt stress on tomato plants by improving plant growth and altering physiological and antioxidant responses. Plant Physiology and Biochemistry, 165:217–227.

Meldau DG, Meldau S, Hoang LH et al (2013) Dimethyl disulfide produced by the naturally associated bacterium bacillus sp B55 promotes Nicotiana attenuata growth by enhancing sulfur nutrition. Plant Cell 25:2731–2747. https://doi.org/10.1105/tpc.113.114744

Moghaieb REA, Nakamura A, Saneoka H, Fujita K (2011) Evaluation of salt tolerance in ectoine-transgenic tomato plants (Lycopersicon esculentum) in terms of photosynthesis, osmotic adjustment, and carbon partitioning. GM Crops 2:58–65. https://doi.org/10.4161/gmcr.2.1.15831

Mukherjee P, Mitra A, Roy M (2019) Halomonas Rhizobacteria of Avicennia marina of Indian Sundarbans promote rice growth under saline and heavy metal stresses through exopolysaccharide production. Front Microbiol 10. https://doi.org/10.3389/fmicb.2019.01207

Mukhtar S, Malik K, Mehnaz S (2019) Microbiome of halophytes: diversity and importance for plant health and productivity. Microbiol Biotechnol Lett 47:1–10. https://doi.org/10.4014/mbl.1804.04021

Orozco-Mosqueda MDC, Duan J, DiBernardo M, Zetter E, Campos-García J, Glick BR, Santoyo G (2019) The production of ACC deaminase and trehalose by the plant growth promoting bacterium Pseudomonas sp. UW4 synergistically protect tomato plants against salt stress. Front Microbiol 10:1392.

Pan J, Peng F, Xue X, You Q, Zhang W, Wang T, Huang C (2019) The growth promotion of two salttolerant plant groups with PGPR inoculation: a meta-analysis. Sustainability 11(2):378.

Patel T, Saraf M (2017) Exploration of novel plant growth promoting bacteria Stenotrophomonas maltophilia MTP42 isolated from the rhizospheric soil of Coleus Forskohlii. Int J Curr Microbiol Appl Sci 6:944–955. https://doi.org/10.20546/ijcmas.2017.611.111

Patil S, Paradeshi J, Chaudhari B (2016) Suppression of charcoal rot in soybean by moderately halotolerant Pseudomonas aeruginosa GS-33 under saline conditions. J Basic Microbiol 56(8):889–899.

Pattnaik S, Mohapatra B, Gupta A (2021) Plant growth-promoting microbe mediated uptake of essential nutrients (Fe, P, K) for crop stress management: microbe–soil–plant continuum. Front Agron 3. https://doi.org/10.3389/fagro.2021.689972

Purwaningsih S, Agustiyani D, Antonius S (2021) Diversity, activity, and effectiveness of rhizobium bacteria as plant growth promoting rhizobacteria (PGPR) isolated from Dieng, central Java. Iran J Microbiol. https://doi.org/10.18502/ijm.v13i1.5504

Qurashi AW, Sabri AN (2012) Bacterial exopolysaccharide and biofilm formation stimulate chickpea growth and soil aggregation under salt stress. Braz J Microbiol 43:1183–1191.

Ramasamy KP, Mahawar L (2023) Coping with salt stress-interaction of halotolerant bacteria in crop plants: a mini review. Front Microbiol 14. https://doi.org/10.3389/fmicb.2023.1077561

Ramesh Kumar N, Krishnan M, Kandeepan C, Kayalvizhi, N, (2014) Molecular and functional diversity of PGPR fluorescent Pseudomonas isolated from rhizosphere of rice (Oryza sativa L.). Int J Adv Biotechnol Res 5(3):490–505.

Rojas-Tapias D, Moreno-Galván A, Pardo-Díaz S et al (2012) Effect of inoculation with plant growth-promoting bacteria (PGPB) on amelioration of saline stress in maize (Zea mays). Appl Soil Ecol 61:264–272. https://doi.org/10.1016/j.apsoil.2012.01.006

Saghafi D, Delangiz N, Lajayer BA, Ghorbanpour M (2019) An overview on improvement of crop productivity in saline soils by halotolerant and halophilic PGPRs. 3 Biotech 9. https://doi.org/10.1007/s13205-019-1799-0

Sandhya V, SkZ A, Grover M et al (2010) Effect of plant growth promoting Pseudomonas spp. on compatible solutes, antioxidant status and plant growth of maize under drought stress. Plant Growth Regul 62:21–30. https://doi.org/10.1007/s10725-010-9479-4

Saradadevi R, Mukankusi C, Li L, Amongi W, Mbiu JP, Raatz B, Cowling WA (2021) Multivariate genomic analysis and optimal contributions selection predicts high genetic gains in cooking time, iron, zinc, and grain yield in common beans in East Africa. The Plant Genome 14(3):e20156.

Sarkar D, Rakshit A, Parewa HP et al (2022) Bio-priming with compatible rhizospheric microbes enhances growth and micronutrient uptake of red cabbage. Land 11:536. https://doi.org/10.3390/land11040536

Schirawski J, Perlin M (2018) Plant–microbe interaction 2017—The good, the bad and the diverse. Int J Mol Sci 19:1374. https://doi.org/10.3390/ijms19051374

Shelake RM, Kadam US, Kumar R et al (2022) Engineering drought and salinity tolerance traits in crops through CRISPR-mediated genome editing: targets, tools, challenges, and perspectives. Plant Commun 3:100417. https://doi.org/10.1016/j.xplc.2022.100417

Sharma S, Kulkarni J, Jha B (2016) Halotolerant rhizobacteria promote growth and enhance salinity tolerance in peanut. Front Microbiol 7:1600.

Sukweenadhi J, Balusamy SR, Kim Y-J et al (2018) A growth-promoting bacteria, Paenibacillus yonginensis DCY84T enhanced salt stress tolerance by activating defense-related Systems in Panax ginseng. Front Plant Sci 9. https://doi.org/10.3389/fpls.2018.00813

Sun L, Di D-W, Li G et al (2020) Transcriptome analysis of rice (Oryza sativa L.) in response to ammonium resupply reveals the involvement of phytohormone signaling and the transcription factor OsJAZ9 in reprogramming of nitrogen uptake and metabolism. J Plant Physiol 246–247:153137. https://doi.org/10.1016/j.jplph.2020.153137

Sunita K, Mishra I, Mishra J et al (2020) Secondary metabolites from halotolerant plant growth promoting rhizobacteria for ameliorating salinity stress in plants. Front Microbiol 11. https://doi.org/10.3389/fmicb.2020.567768

Szypulska E, Jankowski K, Weidner S (2017) ABA pretreatment can limit salinity-induced proteome changes in growing barley sprouts. Acta Physiol Plant 39. https://doi.org/10.1007/s11738-017-2490-x

Tang X, Mu X, Shao H et al (2014) Global plant-responding mechanisms to salt stress: physiological and molecular levels and implications in biotechnology. Crit Rev Biotechnol 35:425–437. https://doi.org/10.3109/07388551.2014.889080

Tewari S, Arora NK, Miransari M (2016) Plant growth promoting rhizobacteria to alleviate soybean growth under abiotic and biotic stresses. In Abiotic and biotic stresses in soybean production (pp. 131–155). Academic Press.

Tewari S, Sharma S (2020) Rhizobial-metabolite based biocontrol of fusarium wilt in pigeon pea. Microb Pathog 147:104278.

Tewari S, Arora NK (2018) Role of salicylic acid from Pseudomonas aeruginosa PF23EPS+ in growth promotion of sunflower in saline soils infested with phytopathogen Macrophomina phaseolina. Environ Sustain 1:49–59. https://doi.org/10.1007/s42398-018-0002-6

Vaishnav A, Kumari S, Jain S et al (2016) PGPR-mediated expression of salt tolerance gene in soybean through volatiles under sodium nitroprusside. J Basic Microbiol 56:1274–1288. https://doi.org/10.1002/jobm.201600188

Vílchez JI, Tang Q, Kaushal R et al (2018) Genome sequence of bacillus megaterium strain YC4-R4, a plant growth-promoting rhizobacterium isolated from a high-salinity environment. Genome Announc 6. https://doi.org/10.1128/genomea.00527-18

Vurukonda SSKP, Giovanardi D, Stefani E (2018) Plant growth promoting and biocontrol activity of Streptomyces spp. as endophytes. Int J Mol Sci 19:952. https://doi.org/10.3390/ijms19040952

Yang A, Akhtar SS, Iqbal S, Amjad M, Naveed M, Zahir ZA, Jacobsen SE (2016) Enhancing salt tolerance in quinoa by halotolerant bacterial inoculation. Funct Plant Biol 43(7):632–642.

Yasin NA, Akram W, Khan WU et al (2018) Halotolerant plant-growth promoting rhizobacteria modulate gene expression and osmolyte production to improve salinity tolerance and growth in Capsicum annum L. Environ Sci Pollut Res 25:23236–23250. https://doi.org/10.1007/s11356-018-2381-8

Yasmin H, Naeem S, Bakhtawar M, Jabeen Z, Nosheen A, Naz R, Hassan MN (2020) Halotolerant rhizobacteria Pseudomonas pseudoalcaligenes and Bacillus subtilis mediate systemic tolerance in hydroponically grown soybean (Glycine max L.) against salinity stress. PLoS One, 15(4):e0231348.

Youseif SH (2018) Genetic diversity of plant growth promoting rhizobacteria and their effects on the growth of maize plants under greenhouse conditions. Ann Agric Sci 63:25–35. https://doi.org/10.1016/j.aoas.2018.04.002

Yu L, Iqbal S, Zhang Y et al (2021) Proteome-wide identification of S-sulphenylated cysteines in Brassica napus. Plant Cell Environ 44:3571–3582. https://doi.org/10.1111/pce.14160

Zafar SA, Zaidi SS-A, Gaba Y et al (2019) Engineering abiotic stress tolerance via CRISPR/ Cas-mediated genome editing. J Exp Bot 71:470–479. https://doi.org/10.1093/jxb/erz476

Zhang G, Bai J, Tebbe CC et al (2020) Salinity controls soil microbial community structure and function in coastal estuarine wetlands. Environ Microbiol 23:1020–1037. https://doi.org/10.1111/1462-2920.15281

Zhang H, Li X, Nan X, et al (2017) Alkalinity and salinity tolerance during seed germination and early seedling stages of three alfalfa (Medicago sativa L.) cultivars. Legume Res. https://doi.org/10.18805/lr.v0i0.8401

Zhao C, Zhang H, Song C et al (2020) Mechanisms of plant responses and adaptation to soil salinity. The Innovation 1:100017. https://doi.org/10.1016/j.xinn.2020.100017

Zerrouk IZ, Benchabane M, Khelifi L, Yokawa K, Ludwig-Müller J, Baluska F (2016) A Pseudomonas strain isolated from date-palm rhizospheres improves root growth and promotes root formation in maize exposed to salt and aluminum stress. J Plant Physiol 191:111–119.

Zhang H, Kim MS, Sun Y, Dowd SE, Shi H, Paré PW (2008) Soil bacteria confer plant salt tolerance by tissue-specific regulation of the sodium transporter HKT1. Mol Plant Microbe Interact 21(6):737–744.

Chapter 13
Biotechnological Insights into Cold-Stressed, Adapted Microorganisms for Plant Health and Soil Improvement

Vishnu Mishra, Jawahar Singh, and Vishal Varshney

13.1 Introduction

Cold temperature stress is a prominent environmental challenge that microorganisms encounter due to their extensive distribution on Earth (Mishra et al. 2011; Puranik et al. 2022). The impact of temperature on life processes is significant, as the majority of these processes are influenced by temperature. In particular, life tends to slow down or cease entirely when exposed to unfavorable cold temperatures (Collins and Margesin 2019). Certain regions of the world face extremely cold temperatures, slower biological processes, and limited availability of nutrients and water (Collins and Margesin 2019; Jha et al. 2021). These conditions pose significant challenges to the survival of native organisms. In these areas, the soil remains frozen for extended periods, posing challenges to agricultural activities as well as the production and productivity of crops (Nikrad et al. 2016; Yadav et al. 2019a, b). Additionally, the low temperature also leads to cellular damage and the production of reactive oxygen species (ROS) in plants (Thakur and Nayyar 2013). These factors negatively impact various biochemical and physiological processes at the molecular level, ultimately affecting plant health and fitness (Mishra et al. 2011;

V. Mishra
National Institute of Plant Genome Research, New Delhi, India

Department of Plant and Soil Sciences, Delaware Biotechnology Institute, University of Delaware, Newark, DE, USA

J. Singh
National Institute of Plant Genome Research, New Delhi, India

Laboratorio de Genomica Funcional de Leguminosas, Facultad de Estudios Superiores Iztacala, Universidad Nacional Autonoma de Mexico, Tlalnepantla, Mexico

V. Varshney (✉)
Departmnet of Botany, Govt. Shaheed GendSingh College, Charama, Chhattisgarh, India

© The Author(s), under exclusive license to Springer Nature Switzerland AG 2024
A. Ranjan et al. (eds.), *Extremophiles for Sustainable Agriculture and Soil Health Improvement*, https://doi.org/10.1007/978-3-031-70203-7_13

Puranik et al. 2022). Since more than 80% of the Earth's biosphere experiences temperatures below 5 °C for a substantial duration, annually, this has resulted in the emergence of microorganisms that can endure cold temperatures in diverse environments (Puranik et al. 2022). Based on their cardinal temperatures, these microorganisms can be categorized as either psychrophiles (cold lovers) or psychrotolerants (cold-tolerant). Psychrophiles can thrive at temperatures below 150 °C (Puranik et al. 2022). They are primarily found in permanent frigid habitats, such as polar ice caps, extremely high altitudes, and the depths of the ocean. On the other hand, psychrotolerant organisms are capable of thriving at temperatures above 20 °C, but they also possess the ability to withstand lower temperatures (Puranik et al. 2022). These cold-adapted microorganisms belong to various groups, including fungi, bacteria, viruses, algae, nematodes, protozoans, and actinomycetes (Ciobanu et al. 2014; Margesin et al. 2016; Zachariah et al. 2016; Li et al. 2020; Puranik et al. 2022). These populations exhibit significant variability, resulting in rich and varied microflora within the soil (Puranik et al. 2022).

According to numerous studies, Actinobacteria, Firmicutes, Chloroflexi, Armatimonadetes, and Thermi have been identified as dominant bacterial phyla in various environments. Among these, the phylum Actinobacteria, specifically the *Arthrobacter* species, has been commonly observed in both the Antarctic and Alpine regions (Shin et al. 2020). Different studies have reported the prevalence of *Arthrobacter* species in these specific environments (Shin et al. 2020). Certain bacteria, such as *Thiobacillus*, demonstrate that they not only survive but even metabolize chemotrophically in cold environments, highlighting their remarkable adaptation to these conditions (Harrold et al. 2016; Puranik et al. 2022). Cold-tolerant fungi from various phyla, such as Ascomycota, Deuteromycota, Zygomycota, and Basidiomycota, have been identified in Antarctica. These fungi include *Aspergillus*, *Monodictys*, *Cladosporium*, *Lecanicillium*, *Botrytis*, *Alternaria*, *Geomyces*, *Rhizopus*, *Mucor*, and *Penicillium*. These fungi exhibit adaptations that allow them to thrive in cold environments (Anand and Sharma 2022; Puranik et al. 2022; Rai and Sharma 2022; Sharma et al. 2022; Suresh and Dass 2022). These fungi can exist as saprobes (decomposers of organic matter), symbionts (living in mutually beneficial relationships with other organisms), parasites (harming host organisms), and pathogens of both plants and animals. Furthermore, these cold-tolerant fungi play crucial roles in various ecosystems, performing essential functions that contribute to the ecological balance of the region (Suresh and Dass 2022). Moreover, research has demonstrated that several genera of archaea possess the capacity to thrive in low-temperature environments. Some of these cold-adapted genera include *Methanosaeta*, *Methanolobus*, *Methanosarcina*, *Methanococcoides*, and *Methanogenium* (Chaya et al. 2019; Anand and Sharma 2022; Kumar et al. 2022a, b; Pathania et al. 2022; Puranik et al. 2022; Sharma et al. 2022). These microorganisms have adapted to survive in extremely cold environments (Puranik et al. 2022). Even with these challenges, the investigation of habitats with low temperatures has led to the discovery of psychrotrophic microbiomes. These microbiomes can be utilized for agricultural practices even when the temperature is extremely low. These cold-stressed, adapted microorganisms serve a vital role in maintaining the production and productivity of agroecosystems (Puranik et al. 2022). They play a significant role in promoting plant health and soil improvement through various mechanisms. These mechanisms include nutrient cycling and nitrogen fixation (N_2 fixation), disease suppression, symbiotic relationships, plant growth promotion, organic matter decomposition, soil aggregation, and bioremediation (Mishra et al. 2011; Nath Yadav et al. 2017a, b; Yadav 2017; Yadav et al. 2017; Jha et al. 2021; Kumar et al. 2022a, b; Puranik et al. 2022; Rai and Sharma 2022). Various molecular and biotechnological techniques have been developed to isolate, characterize, and understand the functions of cold-adapted microorganisms. These techniques include phospholipid fatty acid (PLFA) analysis; nucleic acid techniques, including PCR, Rt-PCR, DNA microarray, sequencing, Southern or Northern blot, and so on; gene clone libraries and fluorescence in situ hybridization; DNA fingerprinting techniques; multiomics approaches; genome or gene-editing techniques; and stable isotope probing (SIP) (Anand and Sharma 2022; Dasila et al. 2022; Puranik et al. 2022; Sehgal and Chaturvedi 2022). These biotechnological tools have unlocked new opportunities for advancing beneficial microbiomes in soil, which can promote plant growth and effectively control soilborne pathogens. It is important to understand microbial diversity and how it could be used in agriculture, as it can provide valuable insights into indicators for measuring plant growth, yield, and soil health.

13.2 Diversity of Cold-Stressed, Adapted Microorganisms in Their Ecosystems

Earth is primarily a frigid, aquatic planet, where about 85% of the ocean's waters have temperatures of 5 °C or lower. Nearly 20% of the Earth's surface is covered by permafrost soils, glaciers, ice sheets, polar sea ice, and snow (Puranik et al. 2022). Artificial habitats, like refrigeration and freezer systems, contribute only a small portion of the potential environments for organisms that are adapted to cold temperatures (Casanueva et al. 2010; Yadav et al. 2018a, b, c). Ecological diversity is the variety of different organisms and their interactions within an ecosystem. Microorganisms that have adapted to cold environments are classified into a group known as cold-stressed, adapted microorganisms (Mishra et al. 2011; Puranik et al. 2022). These microorganisms are a distinct group of organisms that have evolved specific traits and mechanisms that allow them to survive and even thrive in cold environments (Yadav et al. 2016, Yadav et al. 2017, Yadav et al. 2018a, b, c; Nath Yadav et al. 2017a, b, Nath Yadav et al. 2019; Yadav 2017; Kumar et al. 2022a, b). There are a total of 17 distinct phyla that have been identified, associated with various domains: archaea, bacteria, and eukarya (Yadav et al. 2018a, b, c). The overall percentages of different phyla are shown in Fig. 13.1.

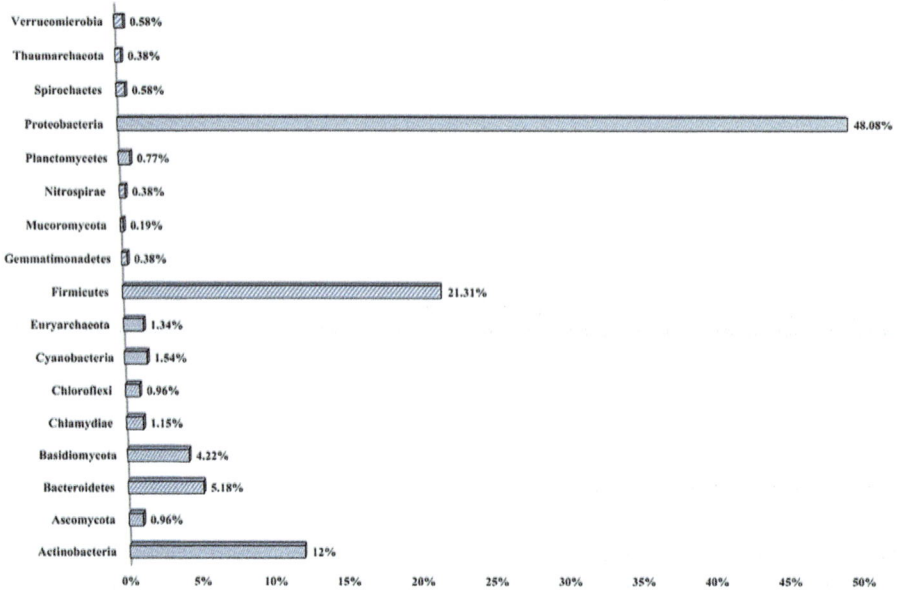

Fig. 13.1 Distribution of various phyla of psychrotrophic microbes, isolated from a wide range of cold environments

13.3   Impact of Cold Temperature on Microbial Cells

Cold temperatures have a significant impact on microbial cells, affecting their growth and survival. At very low temperatures, membranes lose their fluidity and are damaged by ice crystal formation, while chemical reactions and diffusion slow considerably (Rizzo and Lo Giudice 2022). Proteins also become excessively rigid, making them unable to efficiently catalyze reactions and potentially undergo denaturation. On the other hand, heat denatures proteins and nucleic acids (Rizzo and Lo Giudice 2022). Antifreeze proteins and solutes that lower the freezing temperature of the cytoplasm are common, and additionally, the lipids found in membranes often possess unsaturated bonds, which helps enhance their fluidity (Fig. 13.2). Cold-tolerant microorganisms have been reported to undergo phospholipid and fatty acid modifications, leading to enhanced membrane fluidity in colder environments. However, not all microorganisms exhibit similar adaptations, and some may remain unchanged in response to lower temperatures (Berry and Foegeding 1997). Enzymes with flexible conformational structures are required to function at low temperatures to compensate for lower reaction rates (Berry and Foegeding 1997). Cold shock response and low-temperature adaptation mechanisms, such as modulation in fatty acid to maintain membrane fluidity, modulation in carotene/carotenoids, enzyme adaptation, alteration in genes and transcription factors, alteration in clod shock protein and proline accumulation, elevated levels of ROS, changes in pH and temperature, compatible solutes and cryoprotectants, and so on (Fig. 13.2) (De Maayer

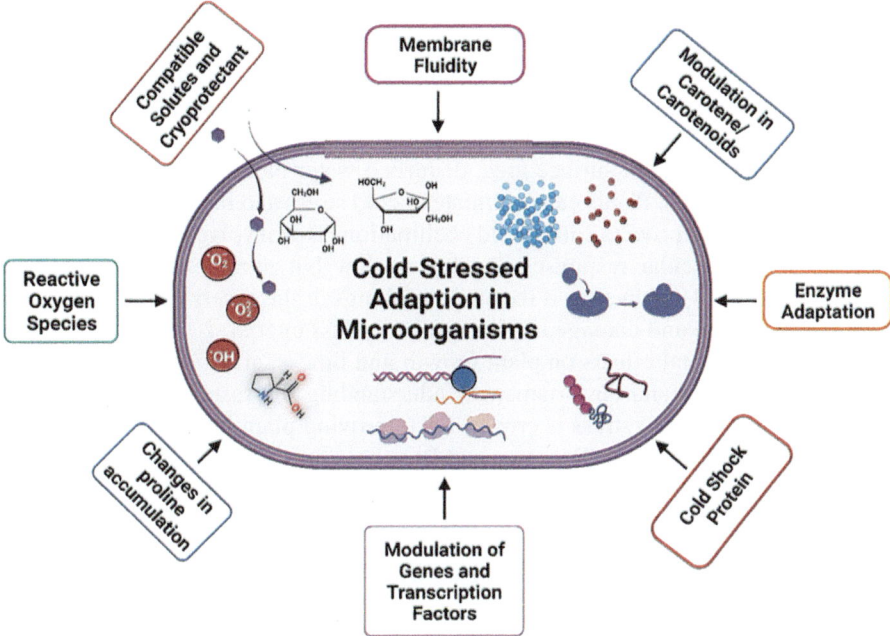

Fig. 13.2 Cold-stressed adaptation in microorganisms. Cold-stressed, adapted microorganisms employ diverse approaches and characteristics to uphold cellular functionality and structural integrity at low-temperature environments such as modulation in fatty acid to maintain membrane fluidity, modulation in carotene/carotenoids, enzyme adaptation, modulation in genes and transcription factors, alteration in clod shock protein and proline accumulation, elevated levels of ROS, compatible solutes, and cryoprotectants that play a crucial role in providing cryoprotection

et al. 2014). Overall, cold temperature has a significant impact on microbial cells, affecting their growth, survival, and adaptation mechanisms. Understanding the responses and management strategies for cold stress is crucial for improving microbial tolerance and productivity in cold environments (Hassan et al. 2020).

13.4   Impact of Cold Stress on Plant Growth and Overall Fitness

Cold stress has significant effects on plant growth and fitness. It can lead to various morphological alterations in plants, such as leaf chlorosis, wilting, and necrosis, resulting in stunted growth and reduced productivity (Adhikari et al. 2022; Kameniarová et al. 2022). Cold stress can also delay and reduce germination, limit root proliferation and surface area, and hinder nutrient and water uptake, further impacting plant growth (Hassan et al. 2021). The reproductive stage of plants is also susceptible to cold stress, leading to the shedding of reproductive structures (Boinot et al. 2022; Kameniarová et al. 2022). The inhibitory effect of cold stress on plant growth is well-known, and it can override the impact of light quality on growth (Kameniarová et al. 2022). Suboptimal temperatures can inhibit metabolic activity and negatively affect cold resilience and chilling tolerance in plants (Hoermiller et al. 2017; Hurry 2017). Additionally, prolonged exposure to cold stress can result in diminished root–shoot surface area, disturbed water balance, and leaf chlorosis (Adhikari et al. 2022). Plants can acclimate to cold stress and increase their freezing tolerance through a process called cold acclimation. This involves a series of physiological and molecular responses to adapt to low but nonfreezing temperatures (Hassan et al. 2021). However, if the stress prolongs or the severity increases, it can lead to destruction and damage to plant growth (Hassan et al. 2021). Overall, cold stress has detrimental effects on plant growth and fitness, affecting various aspects of plant physiology and development. Understanding the responses and management strategies for cold stress is crucial for improving plant tolerance and productivity in cold environments (Banerjee and Roychoudhury 2019).

13.5 Importance of Cold-Stressed, Adapted Microorganisms in Plant Health and Soil Improvement

Low temperature creates extreme conditions for plants and significantly reduces agricultural productivity and growth of commercially important crop plants worldwide (Chattopadhyay and Jagannadham 2001; Di Pietro et al. 2013; Yadav et al. 2019a, b). Several investigations unveil the role of psychrotrophic microbes in the promotion of plant growth through different modes of mechanisms, particularly biological N_2 fixation. These microbes can solubilize micronutrients like zinc, phosphorus, potassium, and so on under extremely low temperatures and ultimately improve soil health and plant growth (Table 13.1) (Yadav et al. 2019a, b). Additionally, they also help the plants in combating cold environmental conditions. There are several studies are available that provide the wider applications of psychrophilic microbes in mitigating stressful conditions (Table 13.1). For instance, genera like *Bacillus*, *Kluyvera*, *Pseudomonas*, *Arthrobacter*, *Serratia*, and so on help in the mobilization of phosphorus to plants and in adaptation and mitigation under extreme cold habits (Kumar et al. 2015, Kumar et al. 2016; Nath Yadav et al. 2017a, b, Nath Yadav et al. 2019; Yadav et al. 2018a, b, c, Yadav et al. 2019a, b). One study identified the cold-tolerant *Pseudomonas* strains TRTRP2 and RT6RP from the rhizoplane of wild grass in the Rudraprayag district of Uttarakhand that showed maximum growth at 4–30 °C (Selvakumar et al. 2013). The bacterial strains' phosphate solubilization kinetics reveal a nonlinear regression of the rate of phosphorus solubilization that fit best in the power model and had a decreasing trend over three district temperatures. In a pot experiment, plants took up more phosphorus when lentil seeds were bacterized with *Pseudomonas* strains as the only source of phosphorus as opposed to when rock phosphorus was applied more

Table 13.1 Diverse functions of psychrophilic microbes in advancing crop productivity

S. no.	Psychrophilic microbes	Functions	References
1.	*Pseudomonas vancoverensis*	Increase germination rates	Mishra et al. (2008)
2.	*Pantoea dispersa*	Increase the plant growth and nutrient uptake	Selvakumar et al. (2008)
3.	*Pseudomonas* spp.	Increase chlorophyll and anthocyanin content; increase root and shoot length; improve seed germination; increase nutrient uptake, especially phosphorus; restrict the growth of pathogenic fungi; reduce antioxidant activity	Selvakumar et al. (2013), Suyal et al. (2014), Subramanian et al. (2016), Ghorbanpour et al. 2018, Tiryaki et al. (2019)
4.	*Azospirillum brasilense*	Dry weight	Turan et al. (2012)
5.	*Pseudomonas* sp. Da-bac TI-8	Promote phosphate solubilization from diverse sources and foster growth in *Deschampsia. antarctica*	Berríos et al. (2013), Rai and Sharma (2022)
6.	*Bacillus* spp.	Act as a biocontrol agent to many Phytopathogenic spp.	Wu et al. (2019)
7.	*Burkholderia phytofirmans*	Affect carbohydrate metabolism	Fernandez et al. (2012)
8.	*Serratia* spp.	Improve crop growth and productivity	Kang et al. (2015)
9.	*Trichoderma harzianum*	Increase plant growth and development and the relative water content	Ghorbanpour et al. 2018
10.	*Penicillium* sp. (GBPI_P155)	Carotenoid pigment and antimicrobial properties	Pandey et al. (2018), Rai and Sharma (2022)
11.	*Microbial consortium*	Increase the germination and the enzymatic activity	Kakar et al. (2016)
12.	*Mortierella* sp.	Insecticidal attributes effective against both houseflies and wax moths	Edgington et al. (2014), Rai and Sharma (2022)
13.	*Sphingomonas faeni*	Improve the root length, shoot length, and antioxidant enzymatic activities	Srinivasan et al. (2017)
14.	*Thermobifida halotolerans* YIM 90462	Synthesize β-glucosidase and involve in the hydrolysis of sugarcane bagasse	Yin et al. (2021), Rai and Sharma (2022)
15.	*Pseudomonas koreensis* P2	Solubilize phosphorus	Srivastava et al. (2019), Rai and Sharma (2022)
16.	*Exiguobacterium acetylicum* 1P (MTCC 8707)	Synthesize IAA, siderophores, and hydrogen cyanide (HCN) and solubilize phosphate at 4°C, coupled with stimulating growth in wheat seedling	Selvakumar et al. (2010), Rai and Sharma (2022)
17.	*Pseudomonas putida* (B0)	Promote plant biomass, ß-1,3-glucanase, and salicylic acid (SA), siderophore, and produce chitinase and HCN	Pandey et al. (2006), Puranik et al. (2022)
18	*Pseudomonas* spp.	Promote lentil shoot and root length and root and shoot biomass	Bisht et al. (2013), Puranik et al. (2022)

(Table 13.1) (Selvakumar et al. 2013). Likewise, microbes like *Serratia, Rhizobium, Herbaspirillum,   Azospirillum,   Azoarcus,   Bacillus,   Enterobacter, Gluconoacetobacter*, and *Arthrobacter* have been reported to fix atmospheric nitrogen (N₂) under low-temperature conditions (Verma et al. 2015; Nath Yadav et al. 2017a, b; Kumar et al. 2019; Rana et al. 2019; Sharma et al. 2019). The use of psychrophilic microbes acts as a sustainable way to boost plant development and crop output under normal as well as abiotic stress conditions. They can be utilized as biofertilizers and/or bioinoculants in a single form or in combination with other microbial consortiums (Table 13.1). They might help maintain crop productivity under diverse stress conditions.

Further, there are several psychrotrophic microbes like *Azospirillum, Alcaligenes, Achromobacter, Acinetobacter, Enterobacter, Burkholderia, Rhizobium, Ralstonia*, and so on that produce stress-induced plant hormones aiding in the adaptation of plants under severely harsh conditions (Khalid et al. 2006; Xu et al. 2014; Verma et al. 2015; Cai et al. 2017; Yadav et al. 2018a, b, c). For example, they produce ACC deaminase and control the level of ethylene in plants supporting plant growth and development under cold-stressed environments. By using the precursor 1-amin ocyclopropane-1-carboxylate (ACC) of ethylene generated by the plants, psychrotrophic microorganisms can reduce the amount of c2H2 in the plants. Qin et al. (2017) identified a psychrotrophic *Pseudochrobactrum kiredjianiae* A4 from cave soil. The strain was examined for many characteristics that encourage plant development. The strain produced indoleacetic acid (IAA) and siderophores and had activity for ACC deaminase. Additionally, under in vitro circumstances, the stain prevented the growth and development of *Magnaporthe grisea, Fusarium graminearum, Fusarium oxysporum, Botrytis cinerea*, and *Rhizoctonia cerealis*. In the presence of *R. cerealis*, the isolate enhanced the physiological parameters and decreased the enzymatic activities of wheat (Qin et al. 2017). The results have shown that the A4-inoculated therapy reduced pathogenic stress in wheat plants. Moreover, the impact of a combination of *Serratia* sp. XY21, *Bacillus cereus* AR156, and *Bacillus subtilis* SM21 on tomato seedling chilling tolerance was examined. The study showed a sixfold improvement in survival rates for treated tomato seedlings compared to untreated ones. Furthermore, the commencement of the freezing stress increased the buildup of malondialdehyde (MDA) and H₂O₂ (Wang et al. 2016).

Additionally, certain groups of psychrophilic microbes produce bioactive compounds like fluorescent pigment, chitinases, Fe-chelating compounds, 1,3-glucanase, and antibiotics which can inhibit the negative effects of other harmful pathogens and promote the plant growth and development indirectly (Rana et al. 2019; Sharma et al. 2019). There are numerous studies available that examine the role of microbial bioresources that are inhibitory to plant diseases as biocontrol agents. Recently, *Trichoderma harzianum* AK20G strain has been proven as a beneficial biocontrol agent, promoting the tolerance in tomato plants under chilling stress. Further in this study, the tomato plants were pretreated with *Trichoderma harzianum* AK20G strain before subjecting them to low temperatures to investigate the physiological, biochemical, and molecular responses of the plants over different periods

(Ghorbanpour et al. 2018). Results showed that treating plants significantly reduced the negative impacts of cold stress, as seen by improvements in photosynthesis and growth rates. In treated plants, there was a noticeable decrease in the rate of lipid peroxidation and electrolyte leakage as well as an increase in leaf water content and proline accumulation (Ghorbanpour et al. 2018).

Overall, the possible use of psychrophilic microbes as bioinoculants/biofertilizers as plant growth promoters with diverse multifunctional plant growth promoting (PGP) attributes, such as N_2 fixation, solubilization of micronutrients and making them available to plants and soil, induction of plant stress hormones under extreme conditions, and production of bioactive compounds in combating other harmful pathogens and diseases, can be helpful and productive to maintain the current rate of crop and food production and soil nutrient enrichment, thereby increasing the agricultural productivity through sustainable means (Table 13.1). Numerous studies have reported the positive impact of psychrotrophic microbes in reducing cold stress and increasing the plant yield in a variety of commercial crops.

13.6 Molecular Aspects and Biotechnological Applications of Cold-Stressed, Adapted Microorganisms

The research on psychrotrophic bacteria is of immense importance and possesses potential applications in diverse fields because they can help know about the mysteries involving the functioning of life at extremely low temperatures. From varied investigations, the researchers have been able to dissect the causes of unique characteristics of psychrotrophic microbes at the genome and molecular level (Fig. 13.3). The genomes of psychrophilic microbes contain information that codes proteins for cold shock and antifreeze, membrane fluidity maintenance, hydrolytic activities, making of cryoprotectants, and so on (Methé et al. 2005; Hallam et al. 2006; Yadav et al. 2018a, b, c, Yadav et al. 2019a, b; Nath Yadav et al. 2019). To date, the genomes of several psychrophilic and psychrotolerant microbes have been sequenced and are available that provide information about the roles and regulations of different genes responsible for their special characteristics and the mechanisms involved in adapting to low-temperature conditions. From the same, it has also been evaluated that in addition to genes needed for basic metabolic and physiological processes, there are genes related to survival under low-temperature conditions, including the genes for different secondary metabolites synthesis, cold shock proteins (CSPs), chaperone protein synthesis, DNA repair system, and so on (Methé et al. 2005; Hallam et al. 2006; Yadav et al. 2018a, b, c, Yadav et al. 2019a, b; Nath Yadav et al. 2019). The important players in providing immunity and surviving low temperatures are CSPs, cold acclimation proteins (Caps), antifreeze proteins (AFPs), and antifreezing compounds. CSPs are ubiquitously present proteins that help protect the microbes from low temperatures. Caps, another class of cold stress proteins, help in the adaptation of microbes to cold stresses (Yadav et al. 2019a, b). AFPs have been reported in

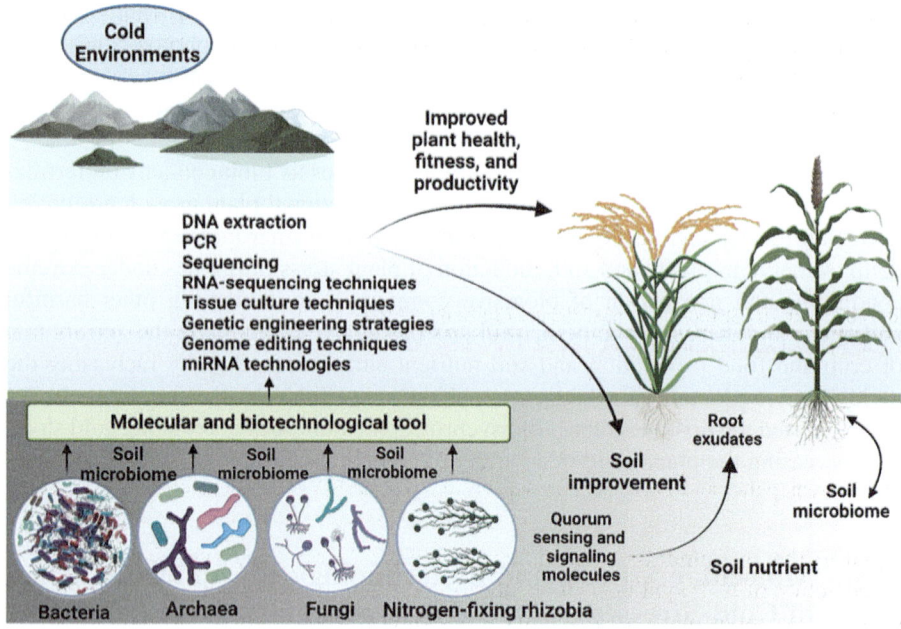

Fig. 13.3 A schematic illustration of the implementation of biotechnological tools to harness cold-stressed, adapted microorganisms for plant health and soil improvement

psychrotrophic bacteria, particularly in Antarctic Lake microbes (Gilbert et al. 2004). Collaborating with antifreezing compounds like organic acids, sugars, cryoprotectants, and so on helps organisms survive at subfreezing temperatures. Psychrotolerant microbes like *Arthrobacter*, *Sporosarcina*, and *Pseudomonas* spp. protect their cytoplasmic components by synthesizing and accumulating antifreezing compounds, proteins, cryoprotectants, and sugars. The adaptation to cold stress or extremely low temperatures includes several changes in bacteria due to the downshift of temperatures through the accumulation of osmolytes like glycine, mannitol, membrane fluidity, antifreezing compound production, pigment biosynthesis, and so on (De Maayer et al. 2014). Based on available genome data modern omics technologies coupled with bioinformatic tools, CRISPR/cas9 and miRNAs technologies can be used to provide a detailed and deep understanding of the adaptation of psychrophilic microbes at low-temperature conditions (Yadav et al. 2020; Anand and Sharma 2022; Puranik et al. 2022; Sehgal and Chaturvedi 2022) (Fig. 13.3).

13.7   Conclusion

Microorganisms inhabiting frigid environments have evolved a diverse array of sophisticated cellular adaptations to endure the inhospitable conditions. Scientists have started to uncover and comprehend several of these adaptive features and how they work, but there are still unanswered questions in various areas. These microorganisms have been observed to enhance plant growth in cold conditions through processes like N_2 fixation, phosphorus solubilization, modulation in phytohormone, production of siderophores and ACC deaminase, and biocontrol. They achieve this by utilizing different mechanisms to combat the effects of cold stress. These mechanisms include modifying their cell membrane and fluidity, regulating energy metabolism, neutralizing ROS, and producing compatible solutes, CSP, and extracellular polymeric substances. This enables them to adapt and thrive even in subzero temperatures. Based on the current knowledge of cold-adapted microorganisms and their role in plant health and soil improvement, we suggest the following areas for further research: (1) comprehensive study of cold-adapted microorganisms in plant associations for remediating organic contamination in soils: this involves a detailed exploration of the various aspects of cold-adapted microorganisms and their interactions with plants. Researchers should investigate how these microorganisms can effectively degrade organic pollutants present in soils, thereby improving soil health and plant growth. This research can shed light on the specific mechanisms employed by cold-adapted microorganisms to break down organic contaminants under low-temperature conditions. (2) Utilization of cutting-edge engineering strategies (e.g., biotechnology) to enhance the functionality of cold-adapted microorganisms in plant systems: by employing advanced techniques like biotechnology, researchers can enhance the performance and efficiency of cold-adapted microorganisms in remediating organic pollutant contamination in soils. This may involve genetically modifying these microorganisms to optimize their pollutant-degrading capabilities to improve their survival, colonization, and activity in plant systems. (3) Investigation of the remediation of co-contaminated soils with both organic and metal pollutants: Many soils are co-contaminated with both organic compounds and heavy metals, which pose a significant challenge for remediation. Further research should focus on understanding how cold-adapted microorganisms can be utilized to remediate such co-contaminated soils effectively. This includes studying their ability to tolerate and degrade both organic pollutants and heavy metals simultaneously, as well as investigating potential synergistic interactions between cold-adapted microorganisms and plants in the remediation process. By addressing these research areas, we can deepen our understanding of the capabilities of cold-adapted microorganisms in plant associations and their potential for improving soil health and remediating contaminated soils. These findings can contribute to the development of innovative strategies for sustainable agriculture, environmental remediation, and soil management practices.

Acknowledgments We acknowledge https://www.biorender.com/ for designing the figures.

Author Contribution VM, JS, and VV outlined the original draft. VM, JS, and VV wrote the article and created the tables and figs. VV, VM, and JS actively engaged in the discussion and worked together to revise and edit the article. All authors have reviewed and approved for publication.

Declaration of Interests All authors claim there are no conflicts of interest.

References

Adhikari L, Baral R, Paudel D, Min D, Makaju SO, Poudel HP, Acharya JP, Missaoui AM (2022) Cold stress in plants: strategies to improve cold tolerance in forage species. Plant Stress 4:100081

Anand A, Sharma A (2022) Use of proteomics and transcriptomics to identify proteins for cold adaptation in microbes. In: Goel R, Soni R, Suyal DC, Khan M (eds) Survival strategies in cold-adapted microorganisms. Springer Singapore, Singapore, pp 285–319

Banerjee A, Roychoudhury A (2019) Cold stress and photosynthesis. Photosynthesis, Productivity and Environmental Stress Wiley:27–37

Berríos G, Cabrera G, Gidekel M, Gutiérrez-Moraga A (2013) Characterization of a novel antarctic plant growth-promoting bacterial strain and its interaction with antarctic hair grass (Deschampsia Antarctica Desv). Polar Biol 36:349–362

Berry ED, Foegeding PM (1997) Cold temperature adaptation and growth of microorganisms. J Food Prot 60:1583–1594

Bisht S, Mishra P, K Joshi G (2013) Genetic and functional diversity among root-associated psychrotrophic Pseudomonad's isolated from the Himalayan plants. Arch Microbiol, 195, 605

Boinot M, Karakas E, Koehl K, Pagter M, Zuther E (2022) Cold stress and freezing tolerance negatively affect the fitness of Arabidopsis thaliana accessions under field and controlled conditions. Planta 255:39

Cai Q, Ye X, Chen B, Zhang B (2017) Complete genome sequence of Exiguobacterium sp. strain N4-1P, a psychrophilic bioemulsifier producer isolated from a cold marine environment in North Atlantic Canada. Genome Announc 5

Casanueva A, Tuffin M, Cary C, Cowan DA (2010) Molecular adaptations to psychrophily: the impact of 'omic' technologies. Trends Microbiol 18:374–381

Chattopadhyay MK, Jagannadham MV (2001) Maintenance of membrane fluidity in Antarctic bacteria. Polar Biol 24:386–388

Chaya A, Kurosawa N, Kawamata A, Kosugi M, Imura S (2019) Community structures of bacteria, archaea, and eukaryotic microbes in the freshwater glacier lake Yukidori-Ike in Langhovde, East Antarctica. Diversity (Basel) 11

Ciobanu M-C, Burgaud G, Dufresne A, Breuker A, Rédou V, Ben Maamar S, Gaboyer F, Vandenabeele-Trambouze O, Lipp JS, Schippers A, Vandenkoornhuyse P, Barbier G, Jebbar M, Godfroy A, Alain K (2014) Microorganisms persist at record depths in the subseafloor of the Canterbury Basin. ISME J 8:1370–1380

Collins T, Margesin R (2019) Psychrophilic lifestyles: mechanisms of adaptation and biotechnological tools. Appl Microbiol Biotechnol 103:2857–2871

Dasila H, Maithani D, Suyal DC, Debbarma P (2022) Cold-adapted microorganisms: survival strategies and biotechnological significance. In: Goel R, Soni R, Suyal DC, Khan M (eds) Survival strategies in cold-adapted microorganisms. Springer Singapore, Singapore, pp 357–378

De Maayer P, Anderson D, Cary C, Cowan DA (2014) Some like it cold: understanding the survival strategies of psychrophiles. EMBO Rep 15:508–517

Di Pietro F, Brandi A, Dzeladini N, Fabbretti A, Carzaniga T, Piersimoni L, Pon CL, Giuliodori AM (2013) Role of the ribosome-associated protein PY in the cold-shock response of Escherichia coli. Microbiology 2:293–307

Edgington S, Thompson E, Moore D, Hughes KA, Bridge P (2014) Investigating the insecticidal potential of Geomyces (Myxotrichaceae: Helotiales) and Mortierella (Mortierellacea: Mortierellales) isolated from Antarctica. Springerplus 3:289

Fernandez O, Theocharis A, Bordiec S, Feil R, Jacquens L, Clément C, Fontaine F, Barka EA (2012) Burkholderia phytofirmans PsJN acclimates grapevine to cold by modulating carbohydrate metabolism. Mol Plant-Microbe Interact 25:496–504

Ghorbanpour A, Salimi A, Ghanbary MAT, Pirdashti H, Dehestani A (2018) The effect of Trichoderma harzianum in mitigating low temperature stress in tomato (Solanum lycopersicum L.) plants. Sci Hortic 230:134–141

Gilbert JA, Hill PJ, Dodd CER, Laybourn-Parry J (2004) Demonstration of antifreeze protein activity in Antarctic lake bacteria. Microbiology (N Y) 150:171–180

Hallam SJ, Konstantinidis KT, Putnam N, Schleper C, Watanabe YI, Sugahara J, Preston C, De La Torre J, Richardson PM, DeLong EF (2006) Genomic analysis of the uncultivated marine crenarchaeote Cenarchaeum symbiosum. Proc Natl Acad Sci USA 103:18296–18301

Harrold ZR, Skidmore ML, Hamilton TL, Desch L, Amada K, van Gelder W, Glover K, Roden EE, Boyd ES (2016) Aerobic and anaerobic thiosulfate oxidation by a cold-adapted, subglacial chemoautotroph. Appl Environ Microbiol 82:1486–1495

Hassan N, Anesio AM, Rafiq M, Holtvoeth J, Bull I, Haleem A, Shah AA, Hasan F (2020) Temperature driven membrane lipid adaptation in glacial psychrophilic bacteria. Front Microbiol 11:533104

Hassan MA, Xiang C, Farooq M, Muhammad N, Yan Z, Hui X, Yuanyuan K, Bruno AK, Lele Z, Jincai L (2021) Cold stress in wheat: Plant acclimation responses and management strategies. Front Plant Sci 12:676884

Hoermiller II, Naegele T, Augustin H, Stutz S, Weckwerth W, Heyer AG (2017) Subcellular reprogramming of metabolism during cold acclimation in Arabidopsis thaliana. Plant Cell Environ 40:602–610

Hurry V (2017) Metabolic reprogramming in response to cold stress is like real estate, it's all about location. Plant Cell Environ 40:599–601

Jha Y, Kulkarni A, Subramanian RB (2021) Psychrotrophic soil microbes and their role in alleviation of cold stress in plants. In: Yadav AN (ed) Soil microbiomes for sustainable agriculture: functional annotation. Springer International Publishing, Cham, pp 267–286

Kakar KU, Ren XL, Nawaz Z, Cui ZQ, Li B, Xie GL, Hassan MA, Ali E, Sun GC (2016) A consortium of rhizobacterial strains and biochemical growth elicitors improve cold and drought stress tolerance in rice (Oryza sativa L.). Plant Biol 18:471–483

Kameniarová M, Černý M, Novák J, Ondrisková V, Hrušková L, Berka M, Vankova R, Brzobohatý B (2022) Light quality modulates plant cold response and freezing tolerance. Front Plant Sci 13:887103

Kang SM, Khan AL, Waqas M, You YH, Hamayun M, Joo GJ, Shahzad R, Choi KS, Lee IJ (2015) Gibberellin-producing Serratia nematodiphila PEJ1011 ameliorates low temperature stress in Capsicum annuum L. Eur J Soil Biol 68:85–93

Khalid A, Akhtar MJ, Mahmood MH, Arshad M (2006) Effect of substrate-dependent microbial ethylene production on plant growth. Microbiology (N Y) 75:231–236

Kumar R, Singh D, Swarnkar MK, Singh AK, Kumar S (2015) Complete genome sequence of Arthrobacter sp. ERGS1:01, a putative novel bacterium with prospective cold active industrial enzymes, isolated from East Rathong glacier in India. J Biotechnol 214:139–140

Kumar R, Singh D, Swarnkar MK, Singh AK, Kumar S (2016) Complete genome sequence of Arthrobacter alpinus ERGS4:06, a yellow pigmented bacterium tolerant to cold and radiations isolated from Sikkim Himalaya. J Biotechnol 220:86–87

Kumar M, Kour D, Yadav AN, Saxena R, Rai PK, Jyoti A, Tomar RS (2019) Biodiversity of methylotrophic microbial communities and their potential role in mitigation of abiotic stresses in plants. Biologia (Bratisl) 74:287–308

Kumar S, Joshi D, Pandey SC, Debbarma P, Suyal DC, Chaubey AK, Soni R (2022a) Structure and functions of rice and wheat microbiome. In: Goel R, Soni R, Suyal DC, Khan M (eds) Survival strategies in cold-adapted microorganisms. Springer Singapore, Singapore, pp 343–356

Kumar S, Sravani B, Korra T, Behera L, Datta D, Dhakad PK, Yadav MK (2022b) Psychrophilic microbes: Biodiversity, beneficial role and improvement of cold stress in crop plants. In: New and future developments in microbial biotechnology and bioengineering: sustainable agriculture: microorganisms as biostimulants, pp 177–198

Li L, Ren M, Xu Y, Jin C, Zhang W, Dong X (2020) Enhanced glycosylation of an S-layer protein enables a psychrophilic methanogenic archaeon to adapt to elevated temperatures in abundant substrates. FEBS Lett 594:665–677

Margesin R, Zhang DC, Frasson D, Brouchkov A (2016) Glaciimonas frigoris sp. Nov., a psychrophilic bacterium isolated from ancient Siberian permafrost sediment, and emended description
of the genus Glaciimonas. Int J Syst Evol Microbiol 66:744–748

Methé BA, Nelson KE, Deming JW, Momen B, Melamud E, Zhang X, Moult J, Madupu R, Nelson
WC, Dodson RJ, Brinkac LM, Daugherty SC, Durkin AS, DeBoy RT, Kolonay JF, Sullivan
SA, Zhou L, Davidsen TM, Wu M, Huston AL, Lewis M, Weaver B, Weidman JF, Khouri H,
Utterback TR, Feldblyum TV, Fraser CM (2005) The psychrophilic lifestyle as revealed by the
genome sequence of Colwellia psychrerythraea 34H through genomic and proteomic analyses.
Proc Natl Acad Sci USA 102:10913–10918

Mishra PK, Mishra S, Selvakumar G, Bisht SC, Bisht JK, Kundu S, Gupta HS (2008) Characterisation
of a psychrotolerant plant growth promoting Pseudomonas sp. strain PGERs17 (MTCC 9000)
isolated from North Western Indian Himalayas. Ann Microbiol 58:561–568

Mishra PK, Joshi P, Bisht SC, Bisht JK, Selvakumar G (2011) Cold-tolerant agriculturally important microorganisms. In: Maheshwari DK (ed) Plant growth and health promoting bacteria.
Springer, Berlin Heidelberg, pp 273–296

Nath Yadav A, Kumar R, Kumar S, Kumar V, Sugitha T, Singh B, Singh Chauahan V, Singh
Dhaliwal H, Kumar Saxena A (2017a) Beneficial microbiomes: Biodiversity and potential biotechnological applications for sustainable agriculture and human health. https://doi.
org/10.7324/JABB.2017.50607

Nath Yadav A, Verma P, Kumar V, Ghosh Sachan S, Kumar Saxena A (2017b) Extreme cold environments: a suitable niche for selection of novel psychrotrophic microbes for biotechnological
applications. Adv Biotechnol Microbiol 2(2):1–4

Nath Yadav A, Kumar R, Kumar S, Kumar V, Sugitha T, Singh B, Singh Chauahan V, Singh
Dhaliwal H, Kumar Saxena A (2019) Biodiversity of psychrotrophic microbes and their biotechnological applications. J Appl Biol Biotechnol 7:99–108

Nikrad MP, Kerkhof LJ, Häggblom MM (2016) The subzero microbiome: microbial activity in
frozen and thawing soils. FEMS Microbiol Ecol 92:fiw081

Pandey A, Trivedi P, Kumar B, Palni LMS (2006) Characterization of a phosphate solubilizing
and antagonistic strain of Pseudomonas putida (B0) isolated from a sub-alpine location in the
Indian Central Himalaya. Curr Microbiol 53:102–107

Pandey N, Jain R, Pandey A, Tamta S (2018) Optimisation and characterisation of the orange pigment produced by a cold adapted strain of Penicillium sp. (GBPI_P155) isolated from mountain ecosystem. Mycology 9:81–92

Pathania S, Solanki P, Putatunda C, Bhatia RK, Walia A (2022) Adaptation to cold environment:
the survival strategy of psychrophiles. In: Goel R, Soni R, Suyal DC, Khan M (eds) Survival
strategies in cold-adapted microorganisms. Springer Singapore, Singapore, pp 87–111

Puranik S, Singh SK, Shukla L (2022) An insight to cold-adapted microorganisms and their importance in agriculture. In: Goel R, Soni R, Suyal DC, Khan M (eds) Survival strategies in cold-
adapted microorganisms. Springer Singapore, Singapore, pp 379–411

Qin Y, Fu Y, Kang W, Li H, Gao H, Vitalievitch KS, Liu H (2017) Isolation and identification of
a cold-adapted bacterium and its characterization for biocontrol and plant growth-promoting
activity. Ecol Eng 105:362–369

Rai AK, Sharma H (2022) Cold-adapted microorganisms and their potential role in plant growth.
In: Goel R, Soni R, Suyal DC, Khan M (eds) Survival strategies in cold-adapted microorganisms. Springer Singapore, Singapore, pp 321–342

Rana KL, Kour D, Sheikh I, Dhiman A, Yadav N, Yadav AN, Rastegari AA, Singh K, Saxena AK
(2019) Endophytic fungi: biodiversity, ecological significance, and potential industrial applications, pp 1–62

Rizzo C, Lo Giudice A (2022) Life from a snowflake: diversity and adaptation of cold-loving bacteria among ice crystals. Crystals (Basel) 12:312

Sehgal P, Chaturvedi P (2022) Omic technologies and cold adaptations. In: Goel R, Soni R, Suyal
DC, Khan M (eds) Survival strategies in cold-adapted microorganisms. Springer Singapore,
Singapore, pp 253–284

Selvakumar G, Kundu S, Joshi P, Nazim S, Gupta AD, Mishra PK, Gupta HS (2008) Characterization of a cold-tolerant plant growth-promoting bacterium Pantoea dispersa 1A isolated from a subalpine soil in the North Western Indian Himalayas. World J Microbiol Biotechnol 24:955–960

Selvakumar G, Kundu S, Joshi P, Nazim S, Gupta AD, Gupta HS (2010) Growth promotion of wheat seedlings by Exiguobacterium acetylicum 1P (MTCC 8707) a cold tolerant bacterial strain from the Uttarakhand Himalayas. Indian J Microbiol 50:50–56

Selvakumar G, Joshi P, Suyal P, Mishra PK, Joshi GK, Venugopalan R, Kumar Bisht J, Bhatt JC, Gupta HS (2013) Rock phosphate solubilization by psychrotolerant Pseudomonas spp. and their effect on lentil growth and nutrient uptake under polyhouse conditions. Ann Microbiol 63:1353–1362

Sharma S, Kour D, Rana KL, Dhiman A, Thakur S, Thakur P, Thakur S, Thakur N, Sudheer S, Yadav N, Yadav AN, Rastegari AA, Singh K (2019) Trichoderma: biodiversity, ecological significances, and industrial spplications, pp 85–120

Sharma S, Chaturvedi U, Sharma K, Vaishnav A, Singh HB (2022) An overview of survival strategies of psychrophiles and their applications. In: Goel R, Soni R, Suyal DC, Khan M (eds) Survival strategies in cold-adapted microorganisms. Springer Singapore, Singapore, pp 133–151

Shin Y, Lee B-H, Lee K-E, Park W (2020) Pseudarthrobacter psychrotolerans sp. nov., a cold-adapted bacterium isolated from Antarctic soil. Int J Syst Evol Microbiol 70:6106–6114

Srinivasan R, Mageswari A, Subramanian P, Maurya VK, Sugnathi C, Amballa C, Sa T, Gothandam K (2017) Exogenous expression of ACC deaminase gene in psychrotolerant bacteria alleviates chilling stress and promotes plant growth in millets under chilling conditions. IJEB 55(07)

Srivastava AK, Saxena P, Sharma A, Srivastava R, Jamali H, Bharati AP, Yadav J, Srivastava AK, Kumar M, Chakdar H, Kashyap PL, Saxena AK (2019) Draft genome sequence of a cold-adapted phosphorous-solubilizing Pseudomonas koreensis P2 isolated from Sela Lake, India. 3 Biotech 9:256

Subramanian P, Kim K, Krishnamoorthy R, Mageswari A, Selvakumar G, Sa T (2016) Cold stress tolerance in psychrotolerant soil bacteria and their conferred chilling resistance in tomato (Solanum lycopersicum Mill.) under low temperatures. PLoS One 11:e0161592

Suresh AJ, Dass RS (2022) Cold-adapted fungi: evaluation and comparison of their habitats, molecular adaptations and industrial applications. In: Goel R, Soni R, Suyal DC, Khan M (eds) Survival strategies in cold-adapted microorganisms. Springer Singapore, Singapore, pp 31–61

Suyal DC, Shukla A, Goel R (2014) Growth promotory potential of the cold adapted diazotroph Pseudomonas migulae S10724 against native green gram (Vigna radiata (L.) Wilczek). 3 Biotech 2014 4 6(4):665–668

Thakur P, Nayyar H (2013) Facing the cold stress by plants in the changing environment: sensing, signaling, and defending mechanisms. In: Tuteja N, Singh Gill S (eds) Plant acclimation to environmental stress. Springer, New York, pp 29–69

Tiryaki D, Aydın İ, Atıcı Ö (2019) Psychrotolerant bacteria isolated from the leaf apoplast of cold-adapted wild plants improve the cold resistance of bean (Phaseolus vulgaris L.) under low temperature. Cryobiology 86:111–119

Turan M, Gulluce M, Şahin F (2012) Effects… - Google Scholar. Available at https://scholar.google.co.in/scholar?hl=en&as_sdt=0%2C5&q=Turan%2C+M.%2C+Gulluce%2C+M.%2C+%C5%9Eahin%2C+F.%2C+2012.+Effects+of+plant-growth-promoting+rhizobacteria+on+yield%2C+psychrotolerant++K-solubilizing+bacterium+from+NW+Indian+Himalayas.+Natl+J+Life+Sci+12%2C+105%E2%80%93110.&btnG=. Accessed 11 July 2023

Verma P, Yadav AN, Shukla L, Saxena AK, Suman A (2015) Alleviation of cold stress in wheat seedlings by Bacillus amyloliquefaciens IARI-HHS2–30, an endophytic psychrotolerant K-solubilizing bacterium from NW Indian Himalayas. Natl J Life Sci 12:105–110

Wang C, Wang C, Gao YL, Wang YP, Guo JH (2016) A consortium of three plant growth-promoting rhizobacterium strains acclimates lycopersicon esculentum and confers a better tolerance to chilling stress. J Plant Growth Regul 35:54–64

Wu H, Gu Q, Xie Y, Lou Z, Xue P, Fang L, Yu C, Jia D, Huang G, Zhu B, Schneider A, Blom J, Lasch P, Borriss R, Gao X (2019) Cold-adapted Bacilli isolated from the Qinghai–Tibetan Plateau are able to promote plant growth in extreme environments. Environ Microbiol 21:3505–3526

Xu M, Sheng J, Chen L, Men Y, Gan L, Guo S, Shen L (2014) Bacterial community compositions of tomato (Lycopersicum esculentum Mill.) seeds and plant growth promoting activity of ACC deaminase producing Bacillus subtilis (HYT-12-1) on tomato seedlings. World J Microbiol Biotechnol 30:835–845

Yadav AN (2017) Agriculturally important microbiomes: biodiversity and multifarious PGP attributes for amelioration of diverse abiotic stresses in crops for sustainable agriculture. Biomed Res Int 1:1–4

Yadav AN, Sachan SG, Verma P, Kaushik R, Saxena AK (2016) Cold active hydrolytic enzymes production by psychrotrophic Bacilli isolated from three sub-glacial lakes of NW Indian Himalayas. J Basic Microbiol 56:294–307

Yadav AN, Verma P, Sachan SG, Saxena AK (2017) Biodiversity and biotechnological applications of psychrotrophic microbes isolated from Indian Himalayan regions. EC Microbiol Eco 1:48–54

Yadav AN, Kumar V, Dhaliwal HS, Prasad R, Saxena AK (2018a) Microbiome in crops: diversity, distribution, and potential role in crop improvement. In: New and future developments in microbial biotechnology and bioengineering: crop improvement through microbial biotechnology, pp 305–332

Yadav AN, Verma P, Kumar S, Kumar V, Kumar M, Kumari Sugitha TC, Singh BP, Saxena AK, Dhaliwal HS (2018b) Actinobacteria from rhizosphere: molecular diversity, distributions, and potential biotechnological applications. In: New and future developments in microbial biotechnology and bioengineering: Actinobacteria: diversity and biotechnological applications, pp 13–41

Yadav AN, Verma P, Sachan SG, Kaushik R, Saxena AK (2018c) Psychrotrophic microbiomes: molecular diversity and beneficial role in plant growth promotion and soil health. In: Panpatte DG, Jhala YK, Shelat HN, Vyas RV (eds) Microorganisms for green revolution: Volume 2: Microbes for sustainable agro-ecosystem. Springer Singapore, Singapore, pp 197–240

Yadav AN, Kour D, Sharma S, Sachan SG, Singh B, Chauhan VS, Sayyed RZ, Kaushik R, Saxena AK (2019a) Psychrotrophic microbes: biodiversity, mechanisms of adaptation, and biotechnological implications in alleviation of cold stress in plants, pp 219–253

Yadav AN, Yadav N, Ghosh Sachan S, Saxena A (2019b) Biodiversity of psychrotrophic microbes and their biotechnological applications. Biotechnol Adv

Yadav S, Sarkar Das S, Kumar P, Mishra V, Sarkar AK (2020) Chapter 3 - Tweaking microRNA-mediated gene regulation for crop improvement. In: Tuteja N, Tuteja R, Passricha N, Saifi SK (eds) Advancement in crop improvement techniques. Woodhead Publishing, pp 45–66

Yin Y-R, Sang P, Xiao M, Xian W-D, Dong Z-Y, Liu L, Yang L-Q, Li W-J (2021) Expression and characterization of a cold-adapted, salt- and glucose-tolerant GH1 β-glucosidase obtained from Thermobifida halotolerans and its use in sugarcane bagasse hydrolysis. Biomass Convers Biorefin 11

Zachariah S, Kumari P, Das SK (2016) Psychrobacter pocilloporae sp. Nov., isolated from a coral, pocillopora eydouxi. Int J Syst Evol Microbiol 66:5091–5098

Chapter 14
Thermophiles and Their Diverse Function in Agricultural and Biotechnological Applications

Himanshi Aggarwal, Divya Chaudhary ⓘ, Jaagiriti Tyagi ⓘ,
Naveen Chandra Joshi ⓘ, Sakshi Arora ⓘ, Vaibhav Mishra ⓘ,
and Manoj Kumar ⓘ

14.1 Introduction

Unicellular microorganisms were the first inhabitant when the Earth came into existence, and among the deluge of microorganisms present today, thermophiles are the most adapting to their harsh environmental conditions. This strong adaptability is because of their unique genetic makeup which codes for a few proteins and enzymes that are stable at extreme temperatures and are not denatured by the extremities of the environmental factors. This flexibility in the genome attracts metabolic engineers to stretch the possibilities of employing thermophiles as potential candidates in the field of metabolic engineering (Brininger et al. 2018; Zeldes et al. 2015). The thermophiles display a discrepancy in the ratio of the uncharged and charged amino acids in the superficial proteins; the charged amino acids make up for a significant portion, thereby asserting protein stability at extremely high temperatures (Taylor 2010). The net amino acid composition of a typical thermophilic cellular protein shows a steady increase in the quantity of isoleucine, lysine, tyrosine and glutamine to form stronger bonds at the side chain which lowers the net dielectric constant of the protein and provides high thermostability. This unique protein conformation increases the stability of thermophiles at elevated temperatures and pressures (Taylor 2010).

Enzymes are biological catalysts with a high level of catalytic activity and selectivity for their substrate. The majority of enzymes are only active under normal physiological settings and become ineffective during industrial activities that operate in severe environments, necessitating the look for enzymes that can resist extreme environmental conditions. Thermozymes are enzymes derived from

H. Aggarwal · D. Chaudhary · J. Tyagi · N. C. Joshi (✉) · S. Arora · V. Mishra · M. Kumar
Amity Institute of Microbial Technology, Amity University Noida,
Noida, Uttar Pradesh, India
e-mail: ncjoshi@amity.edu

© The Author(s), under exclusive license to Springer Nature Switzerland AG 2024   317
A. Ranjan et al. (eds.), *Extremophiles for Sustainable Agriculture and Soil
Health Improvement*, https://doi.org/10.1007/978-3-031-70203-7_14

microbes that thrive at increased temperatures and are promising candidates for high-temperature catalytic reactions. Thermophiles are microorganisms that can thrive in environments with higher-than-normal temperatures. In contrast to mesophilic enzymes, the robustness of thermophilic enzymes protects them from becoming inactive when exposed to high temperatures. Thermozymes provide unique benefits, including increased enzymatic performance, chemical resistance to denaturants, and thermostability, due to lower costs, less pollution, and reduced hydrolysis speed at extreme temperatures (Akram et al. 2018; Arora et al. 2015; Han et al. 2019). Recent advances have elucidated the characteristics of thermophiles for innovative industrial uses, such as xylanase, cellulase, pullulanase, and protease, which play a crucial role in the pulp, paper, food, waste treatment, and pharmaceutical industries (Kumar et al. 2019). Presently, site-directed mutagenesis is employed for the engineering of enzymes, which involves altering the gene in such a way that a codon for a certain amino acid is either inserted or replaced. This strategy utilises targeted mutagenesis and random mutagenesis mostly (Sunden et al. 2015). On top of that, de novo production of a gene encoding for thermostable enzymes may be employed. With the help of recombinant DNA technology, the altered gene is incorporated into appropriate cloning and expression vectors. *Saccharomyces* spp., *Escherichia coli*, and *Bacillus* spp. are the most often employed expression hosts for the production and purification of recombinant enzymes (Liu et al. 2013). Despite their evolutionary distance from the host, thermozymes frequently preserve their thermal capabilities when cloned and produced in mesophilic hosts. This suggests that these traits are genetically encoded (Atalah et al. 2019). Through the process of site-directed mutagenesis, the thermal stability of the enzyme xylanase purified from *Geobacillus thermodenitrificans* was increased by a factor of 13, allowing the enzyme to survive at temperatures up to 75 °C (Irfan et al. 2018).

In nearly all living organisms existing on the planet Earth, cellular integrity and membrane permeability play an active role in maintaining stability at extreme temperatures and pressure. When a particular microorganism gets exposed to unfavourable conditions, its survival in such conditions depends upon the presence of various enzymes which regulate some specific membrane lipids, that is, fatty acyl ester, that helps regain or build up the cell integrity and membrane permeability in adverse environmental conditions (Guan et al. 2017). Witnessing their membrane permeability and charged amino acids on their cell walls, it is quite clear that thermophiles have extended their application for producing enzymes (Irwin 2020). The most distinctive properties that separate thermophiles from other extremophiles in the pharmaceutical industries with less culture and immense product harvesting are their tremendous activity and high stability (Fig. 14.1). The most important enzyme emanated by thermophiles is amylase which has been frequently used in the biotechnological application for a long time. Amylase acts as a thermostable enzyme which breaks down α-1,4-glycosidic bonds of starch, yielding dextrin, maltose, and glucose in the presence of water (Mesbah and Wiegel 2018; Wang et al. 2019b; Zhang et al. 2017). Amylases are promising candidates for producing maltose, high fructose syrups, and maltotetraose syrups (Bedade et al. 2018). Physical factors such as temperature and pH have a massive impact on the stability of amylase (Monteiro De

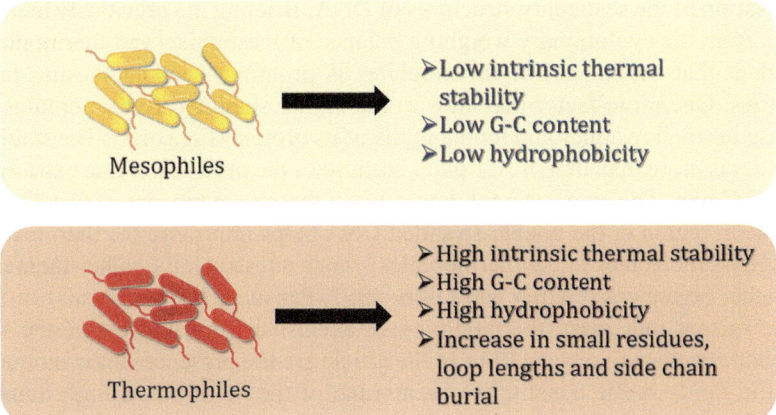

Fig. 14.1 Schematic representation of evolutionary differences in thermophiles in comparison to mesophiles

Souza 2010). It exhibits its maximal growth in the range of 45–60°C with 55°C being the optimum temperature (Pande et al. 2015). In the case of pH, neutral pH (pH 7.0) is the optimum medium pH for the maximal increase of cells in the exponential phase (Kohli et al. 2016).

14.2 Thermophiles and Their Uniqueness

Thermophiles have been playing enormous and numerous roles in the betterment of various strategies and technologies. Their temperature tolerance stands out to be the most significant factor that separates them from other microorganisms. But in this recent era, it has been found that in addition to temperature tolerance, many other factors contribute to the uniqueness of thermophiles. One such is the unique lipids on the outer cell membrane; these lipids contribute to cell membrane permeability, cell stability, and cellular flexibility of these thermophiles (Mehta et al. 2016). The uniqueness of these lipids is based on their ability to retain the synthesis of elongated and branched fatty acid chains at higher temperatures (Siliakus et al. 2017). The second important factor which adds to the uniqueness of thermophiles is the elevated number of intramolecular salt bridges due to the dominance of charged amino acids on the surface proteins (Ahmed et al. 2022). The reason behind the dominance of charged amino acids is a truncation of the N and C terminus of the thermostable protein. Shortening or deletion of the exposed loop regions and increment in the proline residues enhances protein stability (Thompson and Eisenberg, 1999 n.d.). The third and most important factor that contributes to thermophiles' uniqueness is the high G + C content of these organisms (Wu et al. 2012). G + C content of all the organisms plays an essential role in stabilising the structural organisation of the secondary structures of DNA. Briefing the previously mentioned factors from the evolutionary weighing balance of mesophiles and thermophiles, it is evident that the impact of thermophiles is prominent and increasing in most industries. One more factor that adds to the higher stability of thermophiles is the increase in small residues and loop lengths of its protein (Fig. 14.1). The stability of G:C pairs is more than that of A:T pairs because of the presence of one extra H-bond in the G:C pair. The experimental data suggest that the increment of G + C content in the stem region of the double-stranded DNA helps improvise the thermostability of the RNA molecules in the thermophiles, which subsequently makes them unique from other organisms (Galtier and Lobry 1997; Paz et al. 2004).

The factors that make thermophiles so exclusive from the rest of the known microbial species as voiced earlier in the article are the presence of the unique thermophilic lipids in the constitutive membrane of the organism, a high number of intramolecular salt bridges in the surface proteins, exceptional chemical stability of the membranes, and the augmented G + C content (Koga 2012). These properties beef up the application of thermophiles not only in the field of genetic and metabolic engineering but also in the fermentation industries. A few conveniences provided by the process of thermophilic fermentation include the limiting flux of carbon source in the media to form new cell mass; the explanation behind this is that most of the cell's internal energy gets invested in the maintenance of equilibrium of the cellular thermodynamics which is continuously being distorted due to the higher temperature in the cell's external environment; thus, a maximum of the cellular energy produced gets channelised towards keeping the cell functional, and there is barely any energy left for the cell to undergo mitosis, which leads to maximisation of the product formation which is economically benefitting. As the temperature rises, the concentration of the dissolved oxygen is also observed to get reduced, and the liquid rapidly transforms into the gaseous phase and vaporises leading to the reduced solubility of the gases in the liquid, as a result of which the maintenance of anaerobic conditions becomes effortless (Hendriks et al. 2018).

The thermostability imitated by the thermophiles is because of the presence of three main factors in the thermostable protein. The factors are higher ionic charge distributions, more disulphide bonds, and immense hydrogen bonding (Fig. 14.2) (Kumar and Tsai 2000). Hydrogen bonds contribute to protein stability via intramolecular interactions, whereas disulphide bridges are responsible for increased conformational stability (Xu et al. 2020). Numerous mutagenesis experiments in which disulphide bonds were introduced into enzymes provide evidence that supports the notion that disulphide bridges have a stabilising role (Li et al. 2005). In the case of hyperthermophiles, optimising electrostatic interactions with an increase in the number of salt bridges has been shown to improve thermostability (Karshikoff and Ladenstein 2001). On the surface of proteins, extensive networks of ion pairs are formed. Thus, ionic interactions and salt bridges are of crucial relevance for enhancing protein stability (Sinha and Khare 2013). When compared to mesophiles, thermophilic proteins contain a higher concentration of charged amino acids like arginine, glutamate, and lysine (Liang et al. 2005).

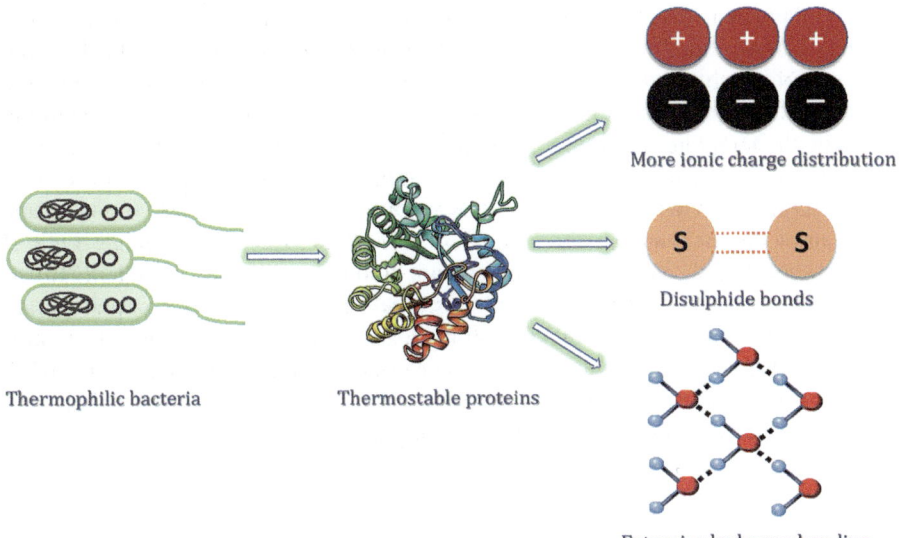

More ionic charge distribution

Disulphide bonds

Extensive hydrogen bonding

Thermophilic bacteria

Thermostable proteins

Fig. 14.2 Molecular interactions contributing to thermophilic extensive properties

14.3   Thermophiles in Agriculture

The agricultural sector is expanding annually to accommodate the constantly rising global population. To avoid environmental pollution, there is an utmost need to utilise the organic matter present in agricultural waste. Agricultural waste is a major point of concern in agronomy and composting is an effective method to deal with it. The composting is a biological process, wherein organic waste is recycled into value-added products that could be applied to fields as organic fertilisers for crop production and could be used as livestock feed. Such organic fertilisers generated could reduce the dependence on chemical fertilisers (Mengqi et al. 2021). The composting process is influenced by many factors like temperature (45–55 °C), pH (5–7), 50–60% moisture content, and C/N ratio (25–35%). In the agriculture sector, crop productivity is increased by the application of humic substances that are a by-product of the composting process, it enhances the physio-chemical, biological properties of agricultural soil (Ho et al. 2022). So there is a great importance of composting in agriculture to increase crop productivity. Agricultural waste mainly consists of lignocellulosic origin that contains materials like lignin, cellulose, and hemicellulose. Bio fabrication is a process, wherein the fungus mainly converts the agricultural waste into nutrients in the form of mycelium-based composites (MBC) for their growth production. Lignocellulosic waste is the major waste produced via agricultural practices. However, in recent years, the amount of this waste generated has increased across the world exponentially (Chuen et al. 2021). The average increase in waste is 5–10% every year (Wang et al. 2016; Xue et al. 2016). If such wastes are not treated properly, they may pose serious health issues to human society; therefore, researchers are focusing more on looking after such problems. Scientists have developed various effective strategies that utilise agricultural waste and convert it to high-value products like enzymes, biofuels, antioxidants, antibiotics, and so on. Such strategies reduce the waste load and environmental pollution (Sadh et al. 2018). In addition, agricultural waste could be utilised to produce biodegradable composite materials that are eco-friendly as well. Fungi are well known for their high efficiency in degrading agricultural waste like straw husks, wood chips, sugar bagasse, and so on.

Microorganisms like fungi and bacteria are ubiquitous in nature. Microorganisms produce enzymes and secondary metabolites that help them survive, reproduce, and adapt to extreme environmental conditions, as seen in thermophilic fungi. In agricultural management practices, the thermophilic fungus is utilized for its capacity to enhance biomass production, generate green fuel, and aid in stress management. Lignin and certain components are resistant to decomposition, but some fungal strains have the potential to degrade them. Certain thermophilic microorganisms accelerate the decomposition processes by their high ability to degrade the organic compounds. During composting, these microorganisms produce hydrolytic enzymes like phosphatase, cellulase, xylanase, chitinase, and so on. The agricultural waste contains chitin the second most abundant polymer (Yang et al. 2020). Chitinases are enzymes that degrade the chitin present in chitinaceous agricultural waste (Akram et al. 2022). Apart from that, they are promising agents for controlling plant diseases. The thermostability is the key factor for such enzymes, so various bacterial and fungal strains are being explored to obtain such enzymes. Through recombinant DNA technology, new ways have been opened for sustainable agriculture by the development of pathogen-resistant crops, thereby minimising crop yield. Bacterial genera like *Pseudomonas*, *Bacillus*, *Rhodococcus*, and *Amycolatopsis* are lignocellulosic degrading bacteria that have gained recent attention. Bacteria tend to degrade compounds that have low molecular weight. Such bacteria act both aerobically and anaerobically in mass but with high efficiency under aerobic conditions (Hemati et al. 2021).

14.4 Applications of Thermophiles

Thermophiles have diverse applications, ranging from bioremediation to the industrial production of commercially important biocatalysts. Beyond aiding in bioremediation and industrial processes, thermophiles also contribute to fields such as bioleaching, plastic biodegradation, biofuel production, and crude oil recovery (Bhattacharya and Gupta 2022; Kochhar et al. 2022). Figure 14.3 depicts the applications of thermophiles in various fields. In further discussion, we will be advocating the need for reinstatement of mesophiles in industries with more apt thermophiles.

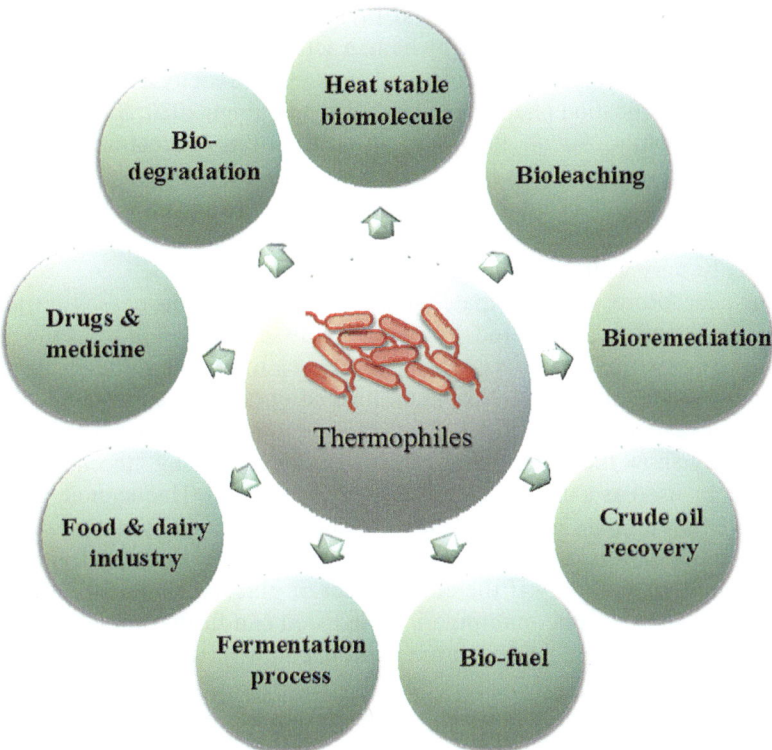

Fig. 14.3 Application diversity of thermophiles

14.4.1 Thermophiles in Invigoration of the Environment

In today's world, the greatest challenge facing humanity is the worsening state of the environment. Six out of the seventeen UN Sustainable Development Goals (SDGs) implemented in the year 2015 directly address ecological protection. Thermophiles have shown a promising approach towards combating two of the seven pollution types, which are soil and water pollution (Margesin and Schinner, 2001). Regions of extreme febricities typically exhibit hydrocarbon contamination, providing a niche for bio transforming or biomineralizing extremophiles. Oil-polluted desert soil was found to be predominantly enhanced with thermophiles capable of biodegrading petroleum-based contaminants of soil (Tahri et al. 2013). The petroleum contamination in and around Middle Eastern countries demonstrated a unique self-cleaning action that led to the inquisition for the presence of some biological activity, which was later found to be the petroleum-utilising activity of thermophilic bacteria, *Bacillus stearothermophilus* (Sorkhoh et al. 1993).

A consortium of thermophilic *Pseudomonas* demonstrated the assimilation of aromatic hydrocarbons such as benzothiazole, phenols, lindane (hexachlorocyclo-hexane), aniline, and toluene. Co-metabolic degradation of BTEX by thermophilic bacteria was also probed (C.-I. Chen and Taylor 1997; Zhou et al. 2013). Expulsion of heavy metals from the biosphere by human activities may be efficiently immobilised or precipitated, thereby reducing the heavy metal toxicity and mobility via the route of biomagnification phenomena (Rahman and Singh 2020; Sikdar and Kundu 2018). Thermophilic *Bacilli* flourished in a strontium concentration as high as 100 mg/l, thereby demonstrating the detoxification of heavy metal-contaminated samples (Chaalal and Islam 2001). The transition of humanity from traditional crude oil to much more environmentally sustainable alternatives is also a furore which can be very well backed by the thermophilic bacteria.

In the near future, it is anticipated that all automotive industries will transition to hydrogen fuel cell-powered engines. These engines will use the energy released from the fusion of oxygen and hydrogen to propel vehicles, producing deionized water as the only by-product, instead of harmful compounds like carbon dioxide and nitrogen oxides. The microorganisms procured from the hot springs array show the exceptional capabilities of producing "thermophilic H_2," which can be used as a cheap biologically produced fuel (Haque et al. 2022; Lee et al. 2011; Ören et al. 2022). A study conducted on *Clostridium thermocellum* and *Thermoanaerobacterium thermosaccharolyticum* theoretically produced hydrogen and ethanol only if the end product of glucose would be butyrate and acetate, obviously genetic manipulations induced can further increase the metabolic kinetics and yield of the fermentation pathway (Liu et al. 2008; Wang et al. 2019a).

Thermophiles have been playing a great role in the reduction of metals since their importance in industries is being known. They nearly reduce all kinds of metals that are essential for the flora and fauna of the atmosphere, that is, Fe (III), Mn (IV), Cr (VI), U (VI), Tc (VII), Co (III), Mo (VI), Au (I), Au (III), and Hg (II) (Sar et al. 2013; Urbieta et al. 2015). During their growth, they undergo metal reductions that aid them in accepting electrons, which, in turn, aids in rapid metabolic growth and multiplication. For example, *Petrothermobacter organivorans* utilise metal ions like Fe (III) and Mn (IV) as electron acceptors which, in turn, increase their metabolic rates (Tamazawa et al. 2017). Thermophiles also play a great role in the biogeochemical cycle as the reduction of the elements employs a heavy ecological impact on the atmosphere (Offre et al. 2013). Most of the chemical elements in the atmosphere are taken up by the biotic components in the atmosphere through the abiotic components, which make up the biogeochemical cycle of the atmosphere. The biogeochemical cycle comprises the gaseous phase, which includes the carbon cycle, nitrogen cycle, and the sedimentary cycle, which includes the phosphorus cycle and the sulphur cycle (Basu et al. 2021). The metal-containing compounds are mostly insoluble, and microorganisms have developed numerous approaches to bypass their insoluble nature and reduce them into soluble compounds. These approaches include the utilisation of chelating amalgams, the implementation of endogenous and exogenous carriers of electrons, and interconnection with the electrons (Slobodkin 2005).

14.4.2 Industrial Application of Thermophiles

Thermophiles are way more potent candidates for industrial fermentation processes since cell biomass production is lower than that of any other microorganism. This aspect is very intriguing fermentation microbiologists since it reduces downstream processing costs and amount of cell mass discarded in rummage. The nutrient media is solely utilised for the fructification of the product and not for the biomass, which is ultimately discarded. The ambient temperature for thermophiles is way intolerant for mesophiles, and most of the pathogenic microorganisms are also not obligate thermophiles; therefore, the upstreaming costs such as sterilisation to maintain the axenic conditions of the fermentation are also abated.

To have an understanding of why thermostable enzymes are preferred over regular enzymes, let us take an example of the most extensively utilised enzyme amongst all known biocatalysts, which is alpha-amylase (Elleuche et al. 2014; Gurung et al. 2013; Tiwari et al. 2015 n.d.). The major confrontation during any fermentation procedure is to maintain the integrity of reaction broth. Although mesophiles require specific temperature and pH to optimally produce enzymes, the thermostable enzymes procured from *Bacillus subtilis, B. amyloliquifaciens, B. thermooleovorans* and *Anoxybacillus beppuensis* are more tolerant to alkaline pH (Kikani and Singh 2012, 2015). The acid-stable amylases are also of great importance to the industry (starch industry), as they provide the flexibility to work at low pH cardinal (Ahmad et al. 2022 n.d.). The amylases sourced from mesophilic organisms usually tend to denature and lose their catalytic activity as soon as they stray from the neutral pH, which is induced due to starch hydrolysis.

This, in turn, leads to the lower turnover of the enzyme, questioning the efficiency. Now, as we are already aware as soon as an organism confronts a highly stressful situation, it generates two responses: "adaption" or "escape"; when an organism adapts, it starts producing the structurally more stress-tolerant compounds which can resist the environmental stressors (Karita et al. 2003). Now, there are a plethora of factors which support the stability of enzymes under such extremities; they include inherent stability and acerbity as well as flexibility which also contributes to the defended function of the enzyme (Sharma et al. 2019).

14.4.3 Thermophiles in Biofuel Production

We all know that fossil fuels have always been the propelling force to humanity, but this fact is also true that the levels of crude and fossilised matter are also depleting at a high rate. Under these circumstances, there is a sure need to switch to a cheaper and more sustainable source of energy: biologically generated fuels or biofuels. Now that is where the thermophiles come to rescue (Srivastava et al. 2018). The process of biofuel production becomes much more convenient with the employment of thermophiles (Abdel-Fattah et al. 2018). Now, the process of forming fermentation products such as ethanol, butanol, and even methane is more complex than it seems. We have the fundamental knowledge of the fermentation processes, we lack the necessary information to comprehend their interrelation with other chemical pathways. Well, this can be controlled by some degree of bioengineering, which is the science of modification of an organism's genetic makeup to alter its metabolism to produce one particular product of interest. For example, to increase the production of ethanol, *Thermoanaerobacterium saccharolyticum* in preliminary attempts was engineered by knocking off the genes translating the enzyme lactate dehydrogenase (ldh) (Joe Shaw et al. 2012). The concurrent deletion of the gene for enzyme acetate kinase and phosphotransacetylase theoretically yields only ethanol without any side products, thereby ensuring a 100% yield. Eliminating the genes responsible for the production of the enzymes for alternate pathways ensures that only one product of interest is generated, thereby naturally inhibiting or suppressing the production of any unwanted by-products. The ethanol production also depends on the type of carbon source present in the growth media. *Thermoanaerobacter mathranii*, for instance, is recognised to give maximum yield when mannitol is employed followed by xylose and glucose (Michael Scully and Orlygsson 2019). *Clostridium thermocellum* was subjected to the same treatment for the removal of the enzymes participating in alternate pathways, but it was contemplated that along with ethanol production, significant amounts of amino acids were also produced in the fermentation (Holwerda et al. 2020).

14.4.4   Applications of Thermophiles in the Dairy Industry

Usually, the employment of thermophilic microorganisms in food and dairy products is limited to indicator organisms which help as a determinant of contamination in the raw materials. This information aids in determining the load of microbial population in the raw materials and in the designing of the most efficient sterilisation process for attaining the axenic conditions required for the fermentation process without harming the physical or chemical composition of the raw materials (Flint et al. 2020). Even in post-sterilisation, the presence of these organisms in the fermentation media questions the efficiency of the fermentation process; an ideal example of this would be *Anoxybacillus flavithermus*, which is a potent indicator organism for contamination in the dairy industry (Burgess et al. 2010; Delaunay et al. 2021; Sadiq et al. 2016). This spoilage caused by these pathogens leads to not only the loss of product but also a substantial monetary loss. Thermophiles are now being intensively promoted as the probiotic concoction which essentially lays the foundation for a healthier gut microbiota by inoculating desired microbes, which in turn leads to the colonisation of the healthy bacteria thereby preventing the brooding of pathogenic microbes as well as increased absorption of amino acids and vitamins (Boro et al. 2022; Vargas-Ramella et al. 2021).

14.4.5 Application in Molecular Biology and Genetic Engineering

Genome editing has become a versatile technique nowadays. Various genome editing tools have been designed for genetic modifications in the case of mesophiles but not in the case of thermophiles. Thermophiles are studied in industrial biotechnology for various aspects. It is very challenging to develop genetic tools for thermophiles due to their optimal growth at higher temperatures because genetic markers and selection substrates are mostly thermolabelile. Emerging uses for the thermophilic enzymes from archaea in a variety of environmental settings include the removal of organophosphates, the elimination of the presence of arsenate and phosphate from the disinfection of water, and the reduction of nitrogen from effluent (Straub et al. 2018). An important enzyme Taq polymerase is used in the polymerase chain reaction (PCR) synthesised from the *Thermus aquaticus* bacterium because the constituents would ordinarily get denatured at high temperatures (Singh et al. 2014). Archaeal DNA polymerases are important from the perspective of molecular replication pathways; they have mostly served as laboratory reagents maintaining accuracy in PCR experiments, often in conjunction with bacterial DNA polymerases (Bergquist et al. 2014). The Pfu polymerase synthesised from *Pyrococcus furiosus* was utilised as a substitute for the bacterial Taq polymerase with increased accuracy (Lundberg et al. 1991). A triple mutant of important biotechnological protein SsoPox which is a bifunctional phosphotriesterase and lactonase first found in the thermophile *Sulfolobus solfataricus* showed 300 times more activity against paraoxon (Del Giudice et al. 2016). This mutant shows great potential for eliminating chemical warfare agents and agricultural produce in addition to being able to degrade paraoxon on apple surfaces and cotton tissues without the use of detergents. Furthermore, it has been demonstrated that mutated forms of SsoPox maintain activity despite sterilisation, immobilisation, and solvent exposure, widening the range of operating conditions and functions that this enzyme can withstand (Rémy et al. 2016). Insoluble steryl glucosides (SGs) are pollutants produced as biodiesel residues that may lead to the biodiesel failing quality inspections. Within two hours, *Thermococcus litoralis* eliminated the majority of SGs, indicating rather than distillation, enzymatic treatment might be employed to minimise the contaminants in biodiesel (Peiru et al. 2015). Adalsteinsson et al. (2021) have designed a CRISPR-Cas9-based thermostable genome editing system for thermophiles. This system is CaldoCas9, along with guide RNA, exhibits nuclease activity at elevated temperatures. A protospacer adjacent motif (PAM) was identified, indicating a preference for the 5'-NNNNGNMA sequence. Knockout mutants of *Thermus thermophilus* were generated using this vector having 90% efficiency.

14.4.6 Biomining

As per the research done with different extremophile microbes, the method of biomining came across. It is also called bioleaching; this method mainly comprises the employment of various acid-tolerant microbes in the elimination of sulphides and oxides from different metals as these are extracted from the ground surface (Vera et al. 2013). The ordinary activity of heap leaching comprises the mingling and blending of mined metals with some elevated volatile chemicals, for example, cyanide (Coker 2016). This method of bioleaching is a cautious technique for mining metals from the ores of metals (Villares et al. 2017). Additionally, it is also more efficient for the ecological system. The heap leaching process brings up the potential for runoff and spills, which can pose a threat to the environment through the toxins that drip into the ground surface, degrading the soil quality and disturbing the microbial population in the soil (Candeias et al. 2019). Biomining can also be used to reduce pollution because the right conditions can be set quickly with the help of thermophiles and bacteria that can live in acidic conditions (Donati et al. 2016). *Acidithiobacillus* is a highly acidophilic chemolithoautotrophic bacterium that has been shown to contribute to the extraction of metals from metal bioremediation settings; its effectiveness in biomining might be improved by genetic engineering techniques (J. Chen et al. 2022).

This method is safer and environmentally sustainable although it is also able to extract much additional metal. The heap leaching process has a nearly 55% metal extraction rate, while bioleaching has a rapid rate of 85%. Till today, gold, nickel, copper, zinc, and silver are the important metals which are successfully extracted through this method (Podar and Reysenbach 2006; Vera et al. 2013). The three previously mentioned examples are some of the chief utilisation of extremophiles in biotechnology practices; however, they are not the least. Various other utilizations are in the process of their analysis and examination such as lipase production, protease generation, making of glycosyl hydrolase, and sugar synthesis (Dalmaso et al. 2015). Therefore, these secondary biotechnological applications usually include the production of significant bioactive chemicals that may be used for primary purposes in a very sustainable manner (Bugge et al. 2016).

14.5 Recent Advancements in Extremozymes Discovery with Multi-omics Approaches

Bioinformatics, as well as algorithms, play an important role in designing in situ mutagenesis and the creation of different combinations of alleles during meiosis to enhance protein stability for its potential application in commercial biorefinery activities (Annamalai et al. 2016). These modern ways have been demonstrated and proved to assist in the development of enzymes based on extremophiles in the field of biotechnology. As above 99% of microbes in the ecological surrounding consist of microbes which are unable to be grown by culturing techniques, hence referred to as 'uncultured microbes', these techniques cannot acquire the target enzyme from the huge enzyme pool in nature. Hence, omics technology has a promising potential to serve as an important tool for the isolation of novel enzymes from the environment (Juerges and Hansjürgens, 2018).

With the help of new bioinformatic analysis, the complete DNA sequence of psychrophilic bacteria can be studied and examined such as *Pseudomonas* sp. MPC6. The toxic pathway of the metabolism of harmful aromatic compounds can be easily studied as well as the synthetase system of polycyclic aromatic hydrocarbon can be spotted and recognised (Orellana-Saez et al. 2019). Recent advancements involve combining computational and structure-driven research with evolutionary-directed methods, which has significantly improved the discovery of novel extremozymes with promising industrial and commercial applications (Ibrahim and Ma 2017). Data related to the genome for greater than 100 thermophiles are now accessible in common shared popular databases. It also helped in studying the genomes of various important extremophiles such as *Anaerobranca, Pyrococcus* and *Thermus* genera. Additionally, genomic data of some other important members of thermophiles such as *Thermotoga, Thermoplasma* have also been examined elaborately, and it came up with some important enzymes which have essential biotechnological roles to play in biorefinery industries (Chettri et al. 2021).

A multi-omics analysis discloses and provides some thermal tolerant mechanisms and techniques shown by thermophiles such as *Thermus filiformis*, including oxidative stress produced by high temperature, which results in the hindrance of genes that play a major role in glycolysis and Krebs cycle; glucose metabolism is attained chiefly via the hexose monophosphate shunt and the accumulation of various antioxidant enzymes which have to neutralise to free radical (Mandelli et al. 2017). In numerous research studies, transcriptional engineering has proved to serve as an important technique to enhance recombinant bacteria. It will alter the pH and ion flux and will offer more efficient enzymatic action (Zhu et al. 2020).

14.6 Conclusion

Considering the literature discussed earlier, one can easily infer that thermophiles when looked upon for industrial applications are best suited. The reason behind thermophiles being a promising candidate is backed up by the extra stability and rigidity imparted to the organism via its genetic advancements over mesophiles. Mesophiles, when employed in an industrial set-up, are found to require a lot of pampering to retrieve the maximum output. On the other hand, recent studies and investigations have demonstrated that thermophiles have comparatively a flexible cardinal range. Thermophiles are not only employed in basal enzyme production at the industrial level but are also coming forth as active participants in the relay of environment conservation by efficiently recycling resources and in some cases providing means of replacing scarce resources.

Much work has been performed on thermophilic genetic makeup which makes them unique from their mesophilic counterparts. Also, it has been extensively studied that thermophiles are potential candidates in various fields of life, such as life's evolution or its utilisation in food or the health industry. Its structural makeup, amino acid stability, lipid organisation, and other genomic aspects have paved its way to becoming a universal user in areas of research and technology, industries, and day-to-day life. We need to focus more on how we can make more use of it in the industrial field. An extensive study of the thermophiles on the genetic level can clear the picture and can be useful in the bioengineering of heat-stable biomolecules. As discussed in the chapter, thermophiles are highly capable of deflating the cost of the fermentation process. We know that fermentation has a paramount role in the multiplicity of industries. Not much research has been done on thermophiles after the 1990s.

As we know, non-renewable resources are depleting, and excessive exploitation of non-renewable resources also leads to environmental retrogression. So thermo-tolerant microorganisms can be exploited to produce lignocellulosic-based ethanol because biofuel is the need of an hour. This type of production of biofuels can benefit the environment in the coming time. A lot of research is going on to develop this alternative source of fuel on a mass level using thermophiles. Currently, thermophiles have promising uses in the dairy sector even beyond their role as contaminant indicator organisms. They are extensively promoted as probiotics establishing the basis for a more beneficial gut microbiota. Fundamentals for maximizing the research opportunity in thermophilic species are genetic techniques that enable the substitution, insertion, and deletion of genetic material. Nevertheless, the present state of tools utilised for genome editing of thermophiles is mostly undeveloped because of the degradation of molecular components at high temperatures. Notwithstanding this, significant advancements have been achieved in the last ten years in the development of genome editing technologies for thermophiles. Expanding the range of nucleic acid sequences is crucial for optimizing the uses of thermostable Cas9 enzymes, which may extend beyond the thermophiles.

References

Abdel-Fattah AM, El-Gamal MS, Ismail SA, Emran MA, Hashem AM (2018) Biodegradation of feather waste by keratinase produced from newly isolated Bacillus licheniformis ALW1. J Genet Eng Biotechnol 16(2):311–318. https://doi.org/10.1016/j.jgeb.2018.05.005

Adalsteinsson BT, Kristjansdottir T, Merre W, Helleux A, Dusaucy J, Tourigny M, Fridjonsson O, Hreggvidsson GO (2021) Efficient genome editing of an extreme thermophile, Thermus thermophilus, using a thermostable Cas9 variant. Sci Rep 11(1). https://doi.org/10.1038/s41598-021-89029-2

Ahmad A, Rahamtullah, Mishra R (2022) Structural and functional adaptation in extremo-philic microbial α-amylases. Biophys Rev 14(2):499–515. https://doi.org/10.1007/s12551-022-00931-z

Ahmed Z, Zulfiqar H, Tang L, Lin H (2022) A statistical analysis of the sequence and structure of thermophilic and non-thermophilic proteins. Int J Mol Sci 23(17). https://doi.org/10.3390/ijms231710116

Akram F, Haq I, ul, Imran, W., & Mukhtar, H. (2018) Insight perspectives of thermostable endoglucanases for bioethanol production: a review. In: Renewable energy, vol 122. Elsevier Ltd, pp 225–238. https://doi.org/10.1016/j.renene.2018.01.095

Akram F, Jabbar Z, Aqeel A, Haq IU, Tariq S, Malik K (2022) A contemporary appraisal on impending industrial and agricultural applications of thermophilic-recombinant chitinolytic enzymes from microbial sources. Mol Biotechnol 64(10):1055–1075. https://doi.org/10.1007/s12033-022-00486-0

Annamalai M, Dhinesh B, Nanthagopal K, SivaramaKrishnan P, Isaac JoshuaRamesh Lalvani J, Parthasarathy M, Annamalai K (2016) An assessment on performance, combustion and emission behavior of a diesel engine powered by ceria nanoparticle blended emulsified biofuel. Energy Convers Manag 123:372–380. https://doi.org/10.1016/j.enconman.2016.06.062

Arora R, Behera S, Kumar S (2015) Bioprospecting thermophilic/thermotolerant microbes for production of lignocellulosic ethanol: a future perspective. In: Renewable and sustainable energy reviews, vol 51. Elsevier Ltd, pp 699–717. https://doi.org/10.1016/j.rser.2015.06.050

Atalah J, Cáceres-Moreno P, Espina G, Blamey JM (2019) Thermophiles and the applications of their enzymes as new biocatalysts. In: Bioresource technology, vol 280. Elsevier Ltd, pp 478–488. https://doi.org/10.1016/j.biortech.2019.02.008

Basu S, Kumar G, Chhabra S, Prasad R (2021) Role of soil microbes in biogeochemical cycle for enhancing soil fertility. In: New and future developments in microbial biotechnology and bioengineering. Elsevier, pp 149–157. https://doi.org/10.1016/B978-0-444-64325-4.00013-4

Bedade D, Deska J, Bankar S, Bejar S, Singhal R, Shamekh S (2018) Fermentative production of extracellular amylase from novel amylase producer, tuber maculatum mycelium, and its characterization. Prep Biochem Biotechnol 48(6):549–555. https://doi.org/10.1080/10826068.2018.1476876

Bergquist PL, Morgan HW, Saul D (2014) Send orders for reprints to reprints@benthamscience.net selected enzymes from extreme thermophiles with applications in biotechnology

Bhattacharya A, Gupta A (2022) Current trends in applicability of thermophiles and thermozymes in bioremediation of environmental pollutants. In: Microbial Extremozymes. Elsevier, pp 161–176. https://doi.org/10.1016/B978-0-12-822945-3.00013-0

Boro M, Bhadra S, Verma AK (2022) Prebiotics and probiotics in regulation of metabolic disorders. In: Prebiotics and probiotics in disease regulation and management. Wiley, pp 239–269. https://doi.org/10.1002/9781394167227.ch9

Brininger C, Spradlin S, Cobani L, Evilia C (2018) The more adaptive to change, the more likely you are to survive: protein adaptation in extremophiles. In: Seminars in cell and developmental biology, vol 84. Elsevier Ltd., pp 158–169. https://doi.org/10.1016/j.semcdb.2017.12.016

Bugge MM, Hansen T, Klitkou A (2016) What is the bioeconomy? A review of the literature. Sustainability (Switzerland) 8(7). MDPI. https://doi.org/10.3390/su8070691

Burgess SA, Lindsay D, Flint SH (2010) Thermophilic bacilli and their importance in dairy processing. Int J Food Microbiol 144(2):215–225. https://doi.org/10.1016/j.ijfoodmicro.2010.09.027

Candeias C, Ávila P, Coelho P, Teixeira JP (2019) Mining activities: health impacts. In: Encyclopedia of environmental health. Elsevier, pp 415–435. https://doi.org/10.1016/B978-0-12-409548-9.11056-5

Chaalal O, Islam MR (2001) Integrated management of radioactive strontium contamination in aqueous stream systems. J Environ Manag 61(1):51–59. https://doi.org/10.1006/jema.2000.0399

Chen C-I, Taylor RT (1997) Thermophilic biodegradation of BTEX by two consortia of anaerobic bacteria. Appl Microbiol Biotechnol 48(1):121–128. https://doi.org/10.1007/s002530051026

Chen J, Liu Y, Diep P, Mahadevan R (2022) Genetic engineering of extremely acidophilic Acidithiobacillus species for biomining: progress and perspectives. J Hazard Mater 438:129456. https://doi.org/10.1016/j.jhazmat.2022.129456

Chettri D, Verma AK, Sarkar L, Verma AK (2021) Role of extremophiles and their extremozymes in biorefinery process of lignocellulose degradation. Extremophiles 25(3):203–219). Springer Japan. https://doi.org/10.1007/s00792-021-01225-0

Chuen NL, Ghazali MSM, Hassim MFN, Bhat R, Ahmad A (2021) Agro-waste-derived silica nanoparticles (Si-NPs) as biofertilizer. In: Valorization of agri-food wastes and by-products. Academic Press, pp 881–897. https://doi.org/10.1016/B978-0-12-824044-1.00029-5

Coker JA (2016) Extremophiles and biotechnology: current uses and prospects. F1000Res 5. Faculty of 1000 Ltd. https://doi.org/10.12688/f1000research.7432.1

Dalmaso GZL, Ferreira D, Vermelho AB (2015) Marine extremophiles a source of hydrolases for biotechnological applications. Marine Drugs 13(4):1925–1965). MDPI AG. https://doi.org/10.3390/md13041925

Del Giudice I, Coppolecchia R, Merone L, Porzio E, Carusone TM, Mandrich L, Worek F, Manco G (2016) An efficient thermostable organophosphate hydrolase and its application in pesticide decontamination. Biotechnol Bioeng 113(4):724–734. https://doi.org/10.1002/bit.25843

Delaunay L, Cozien E, Gehannin P, Mouhali N, Mace S, Postollec F, Leguerinel I, Mathot AG (2021) Occurrence and diversity of thermophilic sporeformers in French dairy powders. Int Dairy J 113. https://doi.org/10.1016/j.idairyj.2020.104889

Donati ER, Castro C, Urbieta MS (2016) Thermophilic microorganisms in biomining. World J Microbiol Biotechnol 32(11). Springer Netherlands. https://doi.org/10.1007/s11274-016-2140-2

Elleuche S, Schröder C, Sahm K, Antranikian G (2014) Extremozymes-biocatalysts with unique properties from extremophilic microorganisms. Curr Opin Biotechnol 29(1):116–123). Elsevier Ltd. https://doi.org/10.1016/j.copbio.2014.04.003

Flint S, Bremer P, Brooks J, Palmer J, Sadiq FA, Seale B, Teh KH, Wu S, Md Zain SN (2020) Bacterial fouling in dairy processing. Int Fairy J 101). Elsevier Ltd. https://doi.org/10.1016/j.idairyj.2019.104593

Galtier N, Lobry JR (1997) Relationships between genomic G+C content, RNA secondary structures, and optimal growth temperature in prokaryotes. J Mol Evol 44(6):632–636. https://doi.org/10.1007/PL00006186

Guan N, Li J, Shin HD, Du G, Chen J, Liu L (2017) Microbial response to environmental stresses: from fundamental mechanisms to practical applications. Appl Microbiol Biotechnol 101(10):3991–4008). Springer Verlag. https://doi.org/10.1007/s00253-017-8264-y

Gurung N, Ray S, Bose S, Rai V (2013) A broader view: microbial enzymes and their relevance in industries, medicine, and beyond. BioMed Res Int 2013. https://doi.org/10.1155/2013/329121

Han H, Ling Z, Khan A, Virk AK, Kulshrestha S, Li X (2019) Improvements of thermophilic enzymes: from genetic modifications to applications. Bioresour Technol 279:350–361). Elsevier Ltd. https://doi.org/10.1016/j.biortech.2019.01.087

Haque S, Singh R, Pal DB, Faidah H, Ashgar SS, Areeshi MY, Almalki AH, Verma B, Srivastava N, Gupta VK (2022) Thermophilic biohydrogen production strategy using agro industrial wastes: current update, challenges, and sustainable solutions. Chemosphere 307:136120. https://doi.org/10.1016/j.chemosphere.2022.136120

Hemati A, Aliasgharzad N, Khakvar R, Khoshmanzar E, Lajayer BA, van Hullebusch ED (2021) Role of lignin and thermophilic lignocellulolytic bacteria in the evolution of humification indices and enzymatic activities during compost production. Waste Manag 119:122–134. https://doi.org/10.1016/j.wasman.2020.09.042

Hendriks ATWM, van Lier JB, de Kreuk MK (2018) Growth media in anaerobic fermentative processes: the underestimated potential of thermophilic fermentation and anaerobic digestion. Biotechnol Adv 36(1):1–13). Elsevier Inc. https://doi.org/10.1016/j.biotechadv.2017.08.004

Ho TTK, Le TH, Tran CS, Nguyen PT, Thai VN, Bui XT (2022) Compost to improve sustainable soil cultivation and crop productivity. Case Stud Chem Environ Eng 6:100211. https://doi.org/10.1016/j.cscee.2022.100211

Holwerda EK, Olson DG, Ruppertsberger NM, Stevenson DM, Murphy SJL, Maloney MI, Lanahan AA, Amador-Noguez D, Lynd LR (2020) Metabolic and evolutionary responses of clostridium thermocellum to genetic interventions aimed at improving ethanol production. Biotechnol Biofuels 13(1). https://doi.org/10.1186/s13068-020-01680-5

Ibrahim NE, Ma K (2017) Industrial applications of thermostable enzymes from Extremophilic microorganism. Curr Biochem Eng 4(2). https://doi.org/10.2174/2212711904666170405123414

Irfan M, Gonzalez CF, Raza S, Rafiq M, Hasan F, Khan S, Shah AA (2018) Improvement in thermostability of xylanase from Geobacillus thermodenitrificans C5 by site directed mutagenesis. Enzym Microb Technol 111:38–47. https://doi.org/10.1016/j.enzmictec.2018.01.004

Irwin JA (2020) Overview of extremophiles and their food and medical applications. In: Physiological and biotechnological aspects of extremophiles. Elsevier, pp 65–87. https://doi.org/10.1016/b978-0-12-818322-9.00006-x

Joe Shaw A, Covalla SF, Miller BB, Firliet BT, Hogsett DA, Herring CD (2012) Urease expression in a Thermoanaerobacterium saccharolyticum ethanologen allows high titer ethanol production. Metab Eng 14(5):528–532. https://doi.org/10.1016/j.ymben.2012.06.004

Juerges N, Hansjürgens B (2018) Soil governance in the transition towards a sustainable bioeconomy—a review. J Clean Prod 170:1628–1639. https://doi.org/10.1016/j.jclepro.2016.10.143

Karita S, Nakayama K, Goto M, Sakka K, Kim W-J, Ogawa S (2003) A novel cellulolytic, anaerobic, and thermophilic bacterium, Moorella sp. Strain F21. Biosci Biotechnol Biochem 67(1). https://academic.oup.com/bbb/article/67/1/183/5944363

Karshikoff A, Ladenstein R (2001) Ion pairs and the thermotolerance of proteins from hyperthermophiles: a 'traffic rule' for hot roads. Trends Biochem Sci 26(9):550–557. https://doi.org/10.1016/S0968-0004(01)01918-1

Kikani BA, Singh SP (2012) The stability and thermodynamic parameters of a very thermostable and calcium-independent α-amylase from a newly isolated bacterium, Anoxybacillus beppuensis TSSC-1. Process Biochem 47(12):1791–1798. https://doi.org/10.1016/j.procbio.2012.06.005

Kikani BA, Singh SP (2015) Enzyme stability, thermodynamics and secondary structures of α-amylase as probed by the CD spectroscopy. Int J Biol Macromol 81:450–460. https://doi.org/10.1016/j.ijbiomac.2015.08.032

Kochhar N, Kavya IK, Shrivastava S, Ghosh A, Rawat VS, Sodhi KK, Kumar M (2022) Perspectives on the microorganism of extreme environments and their applications. Curr Res Microb Sci 3). Elsevier Ltd. https://doi.org/10.1016/j.crmicr.2022.100134

Koga Y (2012) Thermal adaptation of the archaeal and bacterial lipid membranes. Archaea 2012. https://doi.org/10.1155/2012/789652

Kohli I, Tuli R, Singh VP (2016) Purification and characterization of maltose forming thermostable alkaline α-Amylase from Bacillus gibsonii S213. Int J Adv Res 4(1). http://www.megasoftware.net/

Kumar S, Tsai C-J (2000) Factors enhancing protein thermostability and starch processing, production of high fructose corn syrup. Protein Eng 13(3). Xiao and Honig

Kumar S, Dangi AK, Shukla P, Baishya D, Khare SK (2019) Thermozymes: adaptive strategies and tools for their biotechnological applications. Bioresour Technol 278:372–382. Elsevier Ltd. https://doi.org/10.1016/j.biortech.2019.01.088

Lee DJ, Show KY, Su A (2011) Dark fermentation on biohydrogen production: pure culture. Bioresour Technol 102(18):8393–8402. https://doi.org/10.1016/j.biortech.2011.03.041

Li WF, Zhou XX, Lu P (2005) Structural features of thermozymes. Biotechnol Adv 23(4):271–281). Elsevier Inc. https://doi.org/10.1016/j.biotechadv.2005.01.002

Liang HK, Huang CM, Ko MT, Hwang JK (2005) Amino acid coupling patterns in thermophilic proteins. Proteins Struct Funct Genet 59(1):58–63. https://doi.org/10.1002/prot.20386

Liu Y, Yu P, Song X, Qu Y (2008) Hydrogen production from cellulose by co-culture of Clostridium thermocellum JN4 and Thermoanaerobacterium thermosaccharolyticum GD17. Int J Hydrog Energy 33(12):2927–2933. https://doi.org/10.1016/j.ijhydene.2008.04.004

Liu L, Liu Y, Shin HD, Chen RR, Wang NS, Li J, Du G, Chen J (2013) Developing Bacillus spp. as a cell factory for production of microbial enzymes and industrially important biochemicals in the context of systems and synthetic biology. Appl Microbiol Biotechnol 97(14):6113–6127. https://doi.org/10.1007/s00253-013-4960-4

Lundberg KS, Shoemaker DD, Adams MWW, Short JM, Sorge JA, Mathur EJ (1991) High-fidelity amplification using a thermostable DNA polymerase isolated from Pyrococcus furiosus. Gene 108(1):1–6. https://doi.org/10.1016/0378-1119(91)90480-Y

Mandelli F, Couger MB, Paixão DAA, Machado CB, Carnielli CM, Aricetti JA, Polikarpov I, Prade R, Caldana C, Paes Leme AF, Mercadante AZ, Riaño-Pachón DM, Squina FM (2017) Thermal adaptation strategies of the extremophile bacterium Thermus filiformis based on multi-omics analysis. Extremophiles 21(4):775–788. https://doi.org/10.1007/s00792-017-0942-2

Margesin R, Schinner F (2001) Biodegradation and bioremediation of hydrocarbons in extreme environments. Appl Microbiol Biotechnol 56(5–6):650–663. https://doi.org/10.1007/s002530100701

Mehta R, Singhal P, Singh H, Damle D, Sharma AK (2016) Insight into thermophiles and their wide-spectrum applications. 3 Biotech 6(1):1–9). Springer Verlag. https://doi.org/10.1007/s13205-016-0368-z

Mengqi Z, Shi A, Ajmal M, Ye L, Awais M (2021) Comprehensive review on agricultural waste utilization and high-temperature fermentation and composting. Biomass Convers Biorefinery:1–24. https://doi.org/10.1007/s13399-021-01438-5

Mesbah NM, Wiegel J (2018) Improvement of activity and thermostability of agar-entrapped, thermophilic, haloalkaliphilic amylase AmyD8. Catal Lett 148(9):2665–2674. https://doi.org/10.1007/s10562-018-2493-2

Michael Scully S, Orlygsson J (2019) Progress in second generation ethanol production with thermophilic bacteria. In: Fuel ethanol production from sugarcane. IntechOpen. https://doi.org/10.5772/intechopen.78020

Monteiro De Souza P (2010) Application of microbial-amylase in industry-a review. Braz J Microbiol 41:850–861

Offre P, Spang A, Schleper C (2013) Archaea in biogeochemical cycles. Ann Rev Microbiol 67:437–457. https://doi.org/10.1146/annurev-micro-092412-155614

Orellana-Saez M, Pacheco N, Costa JI, Mendez KN, Miossec MJ, Meneses C, Castro-Nallar E, Marcoleta AE, Poblete-Castro I (2019) In-depth genomic and phenotypic characterization of the antarctic psychrotolerant strain pseudomonas sp. MPC6 reveals unique metabolic features, plasticity, and biotechnological potential. Front Microbiol 10. https://doi.org/10.3389/fmicb.2019.01154

Ören İ, Çalkaya A, Han H, Keskin N, Karaoğlan Z, Mıynat ME, Görgül İ, Argun H (2022) Dark fermentative hydrogen gas production from molasses using hot spring microflora. Int J Hydrog Energy 47(34):15370–15382. https://doi.org/10.1016/j.ijhydene.2022.03.149

Pande, A., Singh, S., Samad, J., Saurabh, K., & Haider, Z. A. (2015). Studies on potential of finger millet (Eleusine Coracana Gaertn. L.) amylases for industrial applications. Int J Biotechnol 4(4), 20–29. https://doi.org/10.18488/journal.57/2015.4.4/57.4.20.29

Paz A, Mester D, Baca I, Nevo E, Korol A (2004) Adaptive role of increased frequency of polypurine tracts in mRNA sequences of thermophilic prokaryotes. PNAS March 2(9) https://www.pnas.org/cgi/doi/10.1073pnas.0308594100

Peiru S, Aguirre A, Eberhardt F, Braia M, Cabrera R, Menzella HG (2015) An industrial scale process for the enzymatic removal of steryl glucosides from biodiesel. Biotechnol Biofuels 8(1):223. https://doi.org/10.1186/s13068-015-0405-x

Podar M, Reysenbach AL (2006) New opportunities revealed by biotechnological explorations of extremophiles. Curr Opin Biotechnol 17(3):250–255. https://doi.org/10.1016/j.copbio.2006.05.002

Rahman Z, Singh VP (2020) Bioremediation of toxic heavy metals (THMs) contaminated sites: concepts, applications and challenges. Environ Sci Pollut Res 27(22):27563–27581. Springer. https://doi.org/10.1007/s11356-020-08903-0

Rémy B, Plener L, Poirier L, Elias M, Daudé D, Chabrière E (2016) Harnessing hyperthermostable lactonase from Sulfolobus solfataricus for biotechnological applications. Sci Rep 6. https://doi.org/10.1038/srep37780

Sadh PK, Duhan S, Duhan JS (2018) Agro-industrial wastes and their utilization using solid state fermentation: a review. Bioresour Bioprocess 5(1):1–15

Sadiq FA, Li Y, Liu TJ, Flint S, Zhang G, He GQ (2016) A RAPD based study revealing a previously unreported wide range of mesophilic and thermophilic spore formers associated with milk powders in China. Int J Food Microbiol 217:200–208. https://doi.org/10.1016/j.ijfoodmicro.2015.10.030

Sar P, Kazy SK, Paul D, Sarkar A (2013) Metal bioremediation by thermophilic microorganisms. In: Thermophilic microbes in environmental and industrial biotechnology: biotechnology of thermophiles. Springer, Netherlands, pp 171–201. https://doi.org/10.1007/978-94-007-5899-5_6

Sharma V, Ayothiraman S, Dhakshinamoorthy V (2019) Production of highly thermo-tolerant laccase from novel thermophilic bacterium Bacillus sp. PC-3 and its application in functionalization of chitosan film. J Biosci Bioeng 127(6):672–678. https://doi.org/10.1016/j.jbiosc.2018.11.008

Sikdar S, Kundu M (2018) A review on detection and abatement of heavy metals. ChemBioEng Rev 5(1):18–29. Wiley-Blackwell. https://doi.org/10.1002/cben.201700005

Siliakus, M. F., van der Oost, J., & Kengen, S. W. M. (2017). Adaptations of archaeal and bacterial membranes to variations in temperature, pH and pressure. In Extremophiles (Vol. 21, Issue 4, pp. 651–670). Springer Tokyo. https://doi.org/10.1007/s00792-017-0939-x

Singh J, Birbian N, Sinha S, Goswami A (2014) A critical review on PCR, its types and applications. Int J Adv Res BiolSci 1(7):65–80. www.ijarbs.com

Sinha R, Khare SK (2013) Thermostable proteases. In: Thermophilic microbes in environmental and industrial biotechnology: biotechnology of thermophiles. Springer, Cham, pp 859–880. https://doi.org/10.1007/978-94-007-5899-5_32

Slobodkin AI (2005) Thermophilic microbial metal reduction. Transl Mikrobiol 74(5)

Sorkhoh NA, Ibrahim AS, Ghannoum MA, Radwan SS (1993) Applied Microbiology Biotechnology High-temperature hydrocarbon degradation by Bacillus stearothermophilus from oil-polluted Kuwaiti desert. Appl Microbiol Biotechnol 39

Srivastava N, Srivastava M, Mishra PK, Gupta VK, Molina G, Rodriguez-Couto S, Manikanta A, Ramteke PW (2018) Applications of fungal cellulases in biofuel production: advances and limitations. Renew Sustain Energy Rev 82:2379–2386. Elsevier Ltd. https://doi.org/10.1016/j.rser.2017.08.074

Straub CT, Counts JA, Nguyen DMN, Wu CH, Zeldes BM, Crosby JR, Conway JM, Otten JK, Lipscomb GL, Schut GJ, Adams MWW, Kelly RM (2018) Biotechnology of extremely thermophilic archaea. FEMS Microbiol Rev 42(5):543–578). Oxford University Press. https://doi.org/10.1093/femsre/fuy012

Sunden F, Peck A, Salzman J, Ressl S, Herschlag D (2015) Extensive site-directed mutagenesis reveals interconnected functional units in the alkaline phosphatase active site. elife 4. https://doi.org/10.7554/eLife.06181

Tahri N, Bahafid W, Sayel H, el Ghachtouli N (2013) Biodegradation: involved microorganisms and genetically engineered microorganisms. Biodegradation. InTech. https://doi.org/10.5772/56194

Tamazawa S, Mayumi D, Mochimaru H, Sakata S, Maeda H, Wakayama T, Ikarashi M, Kamagata Y, Tamaki H (2017) Petrothermobacter organivorans gen. nov., sp. nov., a thermophilic, strictly anaerobic bacterium of the phylum Deferribacteres isolated from a deep subsurface oil reservoir. Int J Syst Evol Microbiol 67(10):3982–3986. https://doi.org/10.1099/ijsem.0.002234

Taylor TJ (2010) Discrimination of thermophilic and mesophilic proteins. http://www.biomedcentral.com/1472-6807/10/S1/S5

Thompson MJ, Eisenberg D.(1999) Transproteomic evidence of a loop-deletion mechanism for enhancing protein thermostability. J Mol Biol. Jul 9;290(2):595-604. https://doi.org/10.1006/jmbi.1999.2889. Erratum in: J Mol Biol 1999 Oct 1;292(4):946. PMID: 10390356.

Tiwari SP, Srivastava R, Singh CS, Shukla K, Singh RK, Singh P, Singh R, Singh NL, Sharma R (2015) Amylases: an overview with special reference to alpha amylase. www.mutagens.co.in

Urbieta MS, Donati ER, Chan KG, Shahar S, Sin LL, Goh KM (2015) Thermophiles in the genomic era: biodiversity, science, and applications. Biotechnol Adv 33(6):633–647. Elsevier Inc. https://doi.org/10.1016/j.biotechadv.2015.04.007

Vargas-Ramella M, Pateiro M, Maggiolino A, Faccia M, Franco D, de Palo P, Lorenzo JM (2021) Buffalo milk as a source of probiotic functional products. Microorganisms 9(11). MDPI. https://doi.org/10.3390/microorganisms9112303

Vera M, Schippers A, Sand W (2013) Progress in bioleaching: fundamentals and mechanisms of bacterial metal sulfide oxidation-part A. Appl Microbiol Biotechnol 97(17):7529–7541. Springer Verlag. https://doi.org/10.1007/s00253-013-4954-2

Villares M, Işıldar A, van der Giesen C, Guinée J (2017) Does ex ante application enhance the usefulness of LCA? A case study on an emerging technology for metal recovery from e-waste. Int J Life Cycle Assess 22(10):1618–1633. https://doi.org/10.1007/s11367-017-1270-6

Wang B, Dong F, Chen M, Zhu J, Tan J, Fu X, Wang Y, Chen S (2016) Advances in recycling and utilization of agricultural wastes in China: based on environmental risk, crucial pathways, influencing factors, policy mechanism. Procedia Environ Sci 31:12–17. https://doi.org/10.1016/j.proenv.2016.02.002

Wang F, Wang M, Zhao Q, Niu K, Liu S, He D, Liu Y, Xu S, Fang X (2019a) Exploring the relationship between Clostridium thermocellum JN4 and Thermoanaerobacterium thermosaccharolyticum GD17. Front Microbiol 10. https://doi.org/10.3389/fmicb.2019.02035

Wang Y, Pan S, Jiang Z, Liu S, Feng Y, Gu Z, Li C, Li Z (2019b) A novel maltooligosaccharide-forming amylase from Bacillus stearothermophilus. Food Biosci 30. https://doi.org/10.1016/j.fbio.2019.100415

Wu H, Zhang Z, Hu S, Yu J (2012) On the molecular mechanism of GC content variation among eubacterial genomes. Biol Direct 7. https://doi.org/10.1186/1745-6150-7-2

Xu Z, Cen Y-K, Zou S-P, Xue Y-P, Zheng Y-G (2020) Recent advances in the improvement of enzyme thermostability by structure modification. Crit Rev Biotechnol 40(1):83–98. https://doi.org/10.1080/07388551.2019.1682963

Xue L, Zhang P, Shu H, Wang R, Zhang S (2016) Agricultural waste. Water Environ Res 88(10):1334–1369. https://doi.org/10.2175/106143016X14696400495019

Yang Y, Shen H, Qiu J (2020) Bio-inspired self-bonding nanofibrillated cellulose composite: a response surface methodology for optimization of processing variables in binderless biomass materials produced from wheat-straw-lignocelluloses. Ind Crop Prod 149:112335. https://doi.org/10.1016/j.indcrop.2020.112335

Zeldes BM, Keller MW, Loder AJ, Straub CT, Adams MWW, Kelly RM (2015) Extremely thermophilic microorganisms as metabolic engineering platforms for production of fuels and industrial chemicals. Front Microbiol 6(NOV). Frontiers Research Foundation. https://doi.org/10.3389/fmicb.2015.01209

Zhang Q, Han Y, Xiao H (2017) Microbial α-amylase: a biomolecular overview. In: Process biochemistry, vol 53. Elsevier Ltd, pp 88–101. https://doi.org/10.1016/j.procbio.2016.11.012

Zhou MY, Wang GL, Li D, Zhao DL, Qin QL, Chen XL, Chen B, Zhou BC, Zhang XY, Zhang YZ (2013) Diversity of both the cultivable protease-producing bacteria and bacterial extracellular proteases in the coastal sediments of King George Island, Antarctica. PLoS One 8(11). https://doi.org/10.1371/journal.pone.0079668

Zhu D, Adebisi WA, Ahmad F, Sethupathy S, Danso B, Sun J (2020) Recent development of Extremophilic bacteria and their application in biorefinery. In: Frontiers in bioengineering and biotechnology, vol 8. Frontiers Media S.A. https://doi.org/10.3389/fbioe.2020.00483

Chapter 15
Thermophilic Microbes: Their Role in Plant Growth Promotion and Mitigation of Biotic Stress

Sumit Kumar, Mehjebin Rahman, Mateti Gayithri, Anjali, Ali Chenari Bouket, R. Naveenkumar, Anuj Ranjan, Vishnu D. Rajput, Tatiana Minkina, and Rupesh Kumar Singh

15.1 Introduction

In nature, plants are generally sessile and throughout their life cycles are constantly faced with environmental variations and numerous biotic and abiotic stress challenges (Miryeganeh 2021; Sinha et al. 2022; Lal et al. 2023). Biotic stresses, such

S. Kumar
Department of Mycology and Plant Pathology, Institute of Agricultural Sciences, Banaras Hindu University, Varanasi, Uttar Pradesh, India

Department of Plant Pathology, B.M. College of Agriculture, Khandwa, Rajmata Vijayaraje Scindia Krishi Vishwa Vidyalaya, Gwalior, Madhya Pradesh, India

M. Rahman · M. Gayithri
Department of Plant Pathology, Assam Agricultural University, Jorhat, Assam, India

Anjali
Department of Plant Pathology, Punjab Agricultural University, Ludhiana, India

A. C. Bouket
East Azarbaijan Agricultural and Natural Resources Research and Education Centre, Plant Protection Research Department, Agricultural Research, Education and Extension Organization (AREEO), Tabriz, Iran

R. Naveenkumar
Division of Plant Pathology, School of Agricultural Sciences, Karunya Institute of Technology and Sciences, Karunya Nagar, Coimbatore, Tamil Nadu, India

A. Ranjan (✉) · V. D. Rajput · T. Minkina
Academy of Biology and Biotechnology, Southern Federal University, Rostov-On-Don, Russia
e-mail: randzhan@sfedu.ru

R. K. Singh
Centre of Molecular and Environmental Biology, Department of Biology, University of Minho, Braga, Portugal

© The Author(s), under exclusive license to Springer Nature Switzerland AG 2024
A. Ranjan et al. (eds.), *Extremophiles for Sustainable Agriculture and Soil Health Improvement*, https://doi.org/10.1007/978-3-031-70203-7_15

as herbivorous insects, parasitic plants, weeds, and plant pathogens, may cause devastating effects on crop plants (Gull et al. 2018; Kumar et al. 2022; Kumar and Nautiyal 2022). Unpredictable climate changes incite different abiotic stresses that mainly comprise salinity, drought, acidity, low and high temperatures, heavy metal contamination, and low fertility of soil (Tuteja et al. 2011; Phour and Sindhu 2022). These stresses cause undesirable effects on the morphological, physiological, and biochemical traits of plants that lead to reduced growth and development. Among abiotic stresses, high temperatures are recognized as a major problem to food security and its production worldwide. High temperatures may cause a manifold detrimental effect on all stages of plants from germination to maturity (Khan et al. 2013; Tiwari and Yadav 2019; Alsajri et al. 2022). Figure 15.1 illustrates the adverse side effects of high temperatures on crop plants through the blocking of various processes, e.g., water uptake and translocation capacity, essential nutrient absorption, biosynthesis of primary and secondary metabolites, and photosynthesis. Moreover, the continuously rising ambient temperature is one of the most limiting factors for the global human population (Stillman 2019; Abbass et al. 2022). According to various climate model predictions, the global average temperature could rise to 2–5.4 °C, depending on regions, by the end of this century, and it is also predicted that enhancing changes in the minimum night temperature will be greater than those in the maximum day temperature (Intergovernmental Panel on Climate Change [IPCC], 2019).

What is extreme? According to an anthropocentric point of view, the term "extreme" is described as the conditions that are not linked to "normal" and

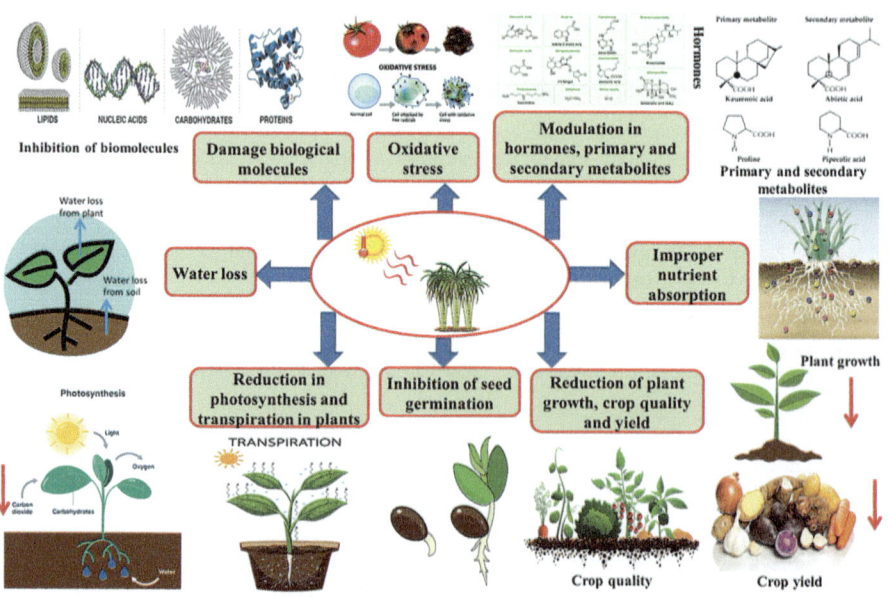

Fig. 15.1 Major adverse effects of high temperatures on different plant activities

"typical" ones under which our species thrives. In another way, extreme conditions are defined as those conditions that are not suitable for living or surviving most life forms (Rothschild and Mancinelli 2001; Merino et al. 2019; Kumar et al. 2022). Conditions include hyper-natural environments such as arid deserts and rocks, high mountains, volcanoes, salt lakes, the deep ocean, and a high atmosphere. Because more than 80% of the earth's surface is covered by hyper-environments, only microbes can survive, colonize, and proliferate easily in these circumstances. Microorganisms that have the ability to flourish in these extreme environments or habitats are known as extremophiles, or extreme-loving microbes (Labes 2018; Yadav et al. 2021; Somayaji et al. 2022). The word "extremophiles" was first used by MacElroy in 1974 to describe a diverse group of microorganisms that are able to survive easily in extreme environments. Extremophiles have well-developed defense mechanisms that help them to survive in extreme environments, e.g., extreme temperature, salinity, pressure, pH, radiation, heavy metals, and nutrient concentrations (Satyanarayana et al. 2013; Arora and Panosyan 2019). Several ecologists reported that extremophilic microbes mainly belong to all three life domains, including bacteria, archaea, and eukarya (Gupta et al. 2014; Jiang et al. 2022; Lhoste et al. 2023). The microorganisms are widely distributed and belong to functionally diverse groups that could be classified as acidophiles (low pH), alkaliphiles (high pH), halophiles (high salt concentration), osmophiles (high organic solute concentration), oligophiles (low organic solute concentration), piezophiles (high pressure), xerophiles (very dry conditions), thermophiles (high temperature), and psychrophiles (low temperature) (Cavicchioli et al. 2011; Neifar et al. 2015; de la Haba et al. 2022; Millan Arias et al. 2023). Some extremophilic microbes thrive in more than one type of extreme environment, known as polyextremophiles, for instance, thermoacidophiles (Nemoto et al. 2003; Colman et al. 2017) and haloalkaliphiles (Raval et al. 2015; Harirchi et al. 2020).

A wide range of temperatures is known from 0 to approximately 3×10^9 K, but only a small fraction of the range is compatible with living forms (Mehta and Satyanarayana 2013). Temperatures cause several challenges, from structural destruction brought on by the development of ice crystals at one extreme (below zero side) to the denaturation of proteins and cell components (very high temp.) (Chen et al. 2020). However, a wide range of microbes identified that are capable of surviving these obstacles (Becerra et al. 2007; Milojevic et al. 2022). The term "thermophile" derives from two Greek words: "thermotita," which means heat, and "philia," which means love. Thermophiles are extremophile-type organisms that are generally found at high temperatures, from 40 to 120 °C (Rasuk et al. 2016). Thermophiles are common in nature and can be found in a variety of settings, including hot springs, pasteurized milk, and newly fallen snow. The natural environments for anaerobic metabolism thermophiles range from terrestrial volcanic sites (like solfatara fields) with temperatures barely above ambient to subterranean hydrothermal environments at temperatures above 300 °C, as well as surface soils heated by the Sun up to 65 °C. In addition, there are hot settings that have been created by humans, such as slag heaps, industrial processes, water heaters, and compost piles (which can reach temperatures of up to 100 °C) (Holden et al. 1998;

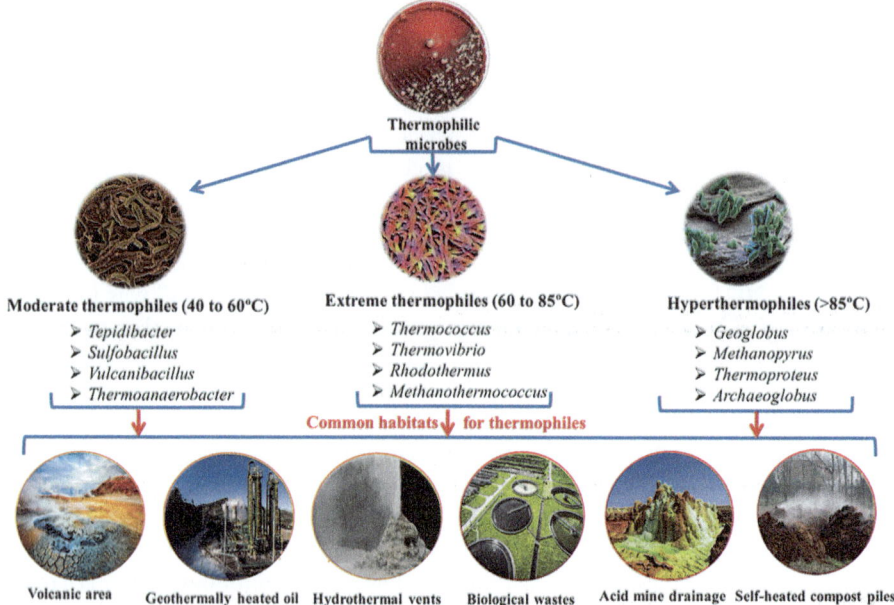

Fig. 15.2 Types and some common habitats of heat-loving organisms

Suddin et al. 2019). Some common habitats are harboring some thermophilic microorganisms (illustrated in Fig. 15.2). According to Brock (2012), thermophiles are organisms that are capable of living at temperatures at or near those of the taxonomic group of which they are a part. Thermophilic microbes belong to a heterogeneous microbiota of heat-loving microorganisms that require temperatures higher than 50 °C for optimum growth (Slobodkin et al. 2013; Vavitsas et al. 2022). Mehta and Satyanarayana et al. (2013) classified thermophiles based on temperature into three main categories: moderate, extreme, and hyperthermophiles, which are explained in Table 15.1.

Heat-loving microbes are recognized for their wide significance in various aims, including medicine, industry, mining, bioremediation, and agriculture. The thermophile population diversity in different high-temperature environments in India, Russia, Iceland, the United States, Korea, and New Zealand has been recently investigated and identified through 16S-rRNA gene sequences from culture-dependent and culture-independent methods (Kecha et al. 2006; Adiguzel et al. 2009; Ghati et al. 2013; Lee et al. 2022; Priyadharshini et al. 2023). Among extremophiles, thermophiles have attracted more attention from research groups because of their ability to produce anti-heating proteins, antibiotics, and several important thermostable enzymes such as amylase, chitinase, pectinase, lipase, cellulase, and proteases (Wongwilaiwalin et al. 2010; Xia et al. 2014; Arbab et al. 2021). Because of their unique features, these enzymes efficiently catalyze cellular reactions at high temperatures, which is significant from an industrial and biotechnological point of view.

Table 15.1 Classification of thermophile microbes proposed by Mehta and Satyanarayana et al. (2013)

S. No.	Category	Optimum temperature (°C)	Microorganisms
1.	Moderate thermophile	40–60 °C	*Clostridium, Exiguobacterium, Tepidibacter, Lebetimonas, Hydrogenimonas, Mohella, Caminibacter, Nautilia, Desulfonauticus, Caminicella, Sulfurivirga, Thermoplasma, Sulfobacillus, Vulcanibacillus, Marinotoga, Acidimicrobium, Desulfovibrio, Caldithrix, Hydrogenobacter, Thermoanaerobacter*
2.	Extreme thermophile	60–85 °C	*Thermococcus, Methanotorris, Palaeococcus, Methanocaldococcus, Thermovibrio, Aeropyrum, Thermosipho, Caloranaerobacter, Methanothermococcus, Deferribacter, Thermodesulfobacterium, Kosmotoga, Thermodesulfatator, Rhodothermus, Petrotoga, Desulfurobacterium, Persephonella, Balnearium, Acidianus, Oceanithermus, Marinithermus, Vulcanithermus, Metallosphaera, Carboxydobrachium, Thermosulfidibacter, Thermaerobacter*
3.	Hyperthermophile	>85 °C	*Geoglobus, Geogemma, Acidianus, Archaeoglobus, Pyrococcus, Methanopyrus, Sulfodobus, Thermoproteus, Ignisphaera, Methanothermus, Ignicoccus*

For instance, Taq-DNA polymerase produced by *Thermus aquaticus* is extensively used in microbiology labs worldwide (Counts et al. 2017). Besides, thermotolerant microbes could produce anti-heating proteins that help microbes adapt to high temperatures. The specialized form of anti-heating protein produced by thermophiles is called thermozymes. Various adaptation mechanisms adopted by thermophiles that help in living under high temperatures include modifications in the cell membrane, protein modifications, genomic modifications, and modifications in DNA and RNA (Gong et al. 2020; Somayaji et al. 2022).

Extensive usage of thermophilic microbes in agriculture provides support to plant growth and development through plant growth-promoting activities, resulting in sustainable agriculture (Salazar-Badillo et al. 2017). These microbes have different growth-promoting mechanisms such as siderophore production, nitrogen fixation, phosphate solubilization, hydrogen cyanide production, 1-aminocyclopropane-1-carboxylate (ACC) deaminase production, growth-inducing hormone production, and antagonism activities that stimulate plant growth and development, aid in the uptake and translocation of essential nutrients, and mitigate biotic and abiotic stress (Verma et al. 2018; Mukherjee et al. 2020; Liu et al. 2007; Rajashree and Borkar 2018; Rahman et al. 2023). Nowadays, these microbes are used as biofertilizers, bioinoculates, bioremediation agents, and biocontrol agents, which play a significant role in enhancing agricultural sustainability.

Recently, several thermophilic microorganisms have been isolated from different springs of hot regions and belong to different genera, e.g., *Bacillus*, *Trichoderma*, *Pseudomonas*, and *Burkholderia* (Camargo et al. 2022; Shah et al. 2023; Peng et al. 2023). The application of thermophilic microorganisms in agriculture fields is unexploited, or very little information is available. Here, we focus on the identification of thermophilic microbes and a detailed discussion of their important role in plant growth promotion and sustainable agriculture. Further, we discuss their pivotal role in the mitigation of plant diseases based on the research literature to date.

15.2   Mechanisms of Adaptation in Thermophilic Microbes

Thermophilic microorganisms adapt by making several modifications at the molecular, morphological, and physiological levels. Stress reactions are the collective term for the genetic and functional changes that acute environmental damage causes in a cell (Helmann et al. 2001; Tam et al. 2006). Many alterations are made in a variety of biomolecules, including proteins, lipids, and nucleic acids for the survival of these microorganisms at high temperatures (Wani et al. 2022; Sarma et al. 2023). Extensive research has been conducted on proteins to study their changes in the adaptation of thermophilic microbes (Liu and LiCata 2013). It is known that, at higher temperatures, proteins and enzymes often denature. Protein molecules are made to tolerate high temperatures by various processes in thermophiles, such as covalent bond changes and the addition of amino acids. Two fundamental mechanisms have been observed for avoiding denaturation of proteins at high temperatures, including enhancing the number of interactions among the constituent amino acids and use of disulfide bonding (Tse and Ma 2016; Berezovsky and Shakhnovich 2005) and forming more compact three-dimensional protein structures (Fig. 15.3) to stabilize the proteins (Berezovsky and Shakhnovich 2005).

15.2.1   Adaptations by Modifications in Structural Elements

Additionally, thermophilic microbes contain distinctive biomolecules in their structural elements (Pati et al. 2023). They have abundant saturated fatty acid content in the cytoplasmic membrane. The thermophilic bacterial membrane is entirely made up of ester-lipids up to a temperature of 80 °C, but ether bonds are necessary for hyperthermophilic growth once the temperature reaches 80 °C (Huber et al. 1992). *Thermotoga maritima* has *n*-fatty acids, diabolic acid, and a glycerol ether lipid in its membrane. The heat-resistant lipid of thermophilic archaea consists of diphytanyl glycerol ether or its dimer, di (diphytanyl) glycerol tetra ether. They also frequently have heat-stable lipid monolayer membranes to maintain the fluidity and permeability of the cell membrane at high temperatures. The predominant thermophiles and hyperthermophiles have obligated heterotrophic modes of nutrition.

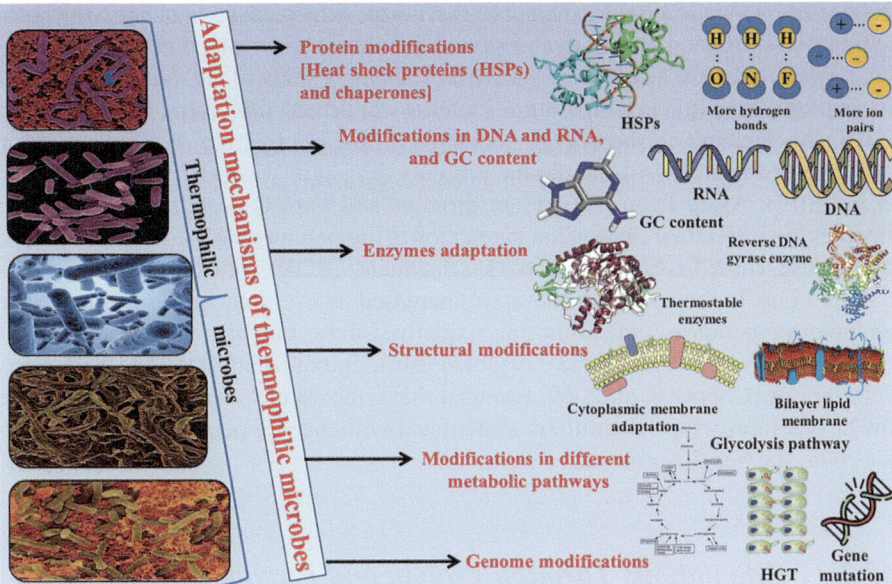

Fig. 15.3 Different types of adaptation mechanisms used by thermophiles

They prefer a combination of carbohydrates and/or polypeptides as sources of carbon and energy. Many of the thermophilic and hyperthermophiles are chemosynthetic. The oxidation of inorganic and organic substances provides them with energy for cellular function and similar to their mesophilic relatives, rely on sugar (carbohydrate) metabolism for energy production.

15.2.2 Adaptations Through Changes in Glycolysis Pathway

The main components of thermophilic bacteria are proteins that are related to the glycolysis pathway. When the temperature rises, the transcriptional regulators that are part of the gene clusters for sulfur respiration and glycolysis come into action. Under aerobic conditions, two-thirds of the glucose consumed by thermophilic bacteria is transferred to the glycolysis pathway, while the remaining part is used for pentose phosphate metabolism. They provide thermophiles with instant energy to combat heat stress. Their physiological response is represented by these thermopile proteins. Genes of upregulation of the glycolysis pathway, including glyceraldehyde-3-phosphate dehydrogenase (GAPDH), phosphoglycerate kinase (PGK), and phosphogluconolactonase (PGL), in response to rising temperatures at mRNA and protein levels were found in many thermophiles, including *Thermoanaerobacter tengcongensis, Thermus thermophilus, Geobacillus thermoglucosidasius,* and *Thermotoga maritima* (Browning and Busby 2004; Gao et al.

2004). *Thermoanaerobacter tengcongensis* was subjected to a quantitative proteomics investigation, which demonstrated that the genes involved in sulfur or aerobic respiration, were also downregulated. Proteomics study of *Geobacillus* sp. NTU 03 exposed to a quick temperature increase confirmed this event (Shih and Pan 2011). An increase in the energy generated with carbohydrate metabolism and a decrease in the ability of membrane proteins to respire were also observed (Chen et al. 2012). Aside from antioxidant proteins and VapBC proteins, other protein families also alter their expression in reaction to temperature variations. Thioredoxin peroxidase, rubredoxin, and superoxide dismutase (SOD) are examples of antioxidant proteins in thermophiles that have increased as a consequence of a rise in the ambient temperature, indicating that oxidative stress may also have occurred (Li et al. 2010; Trauger et al. 2008). A global transcriptome analysis of the *Sulfolobus solfataricus* (Cooper et al. 2009) revealed that temperature variations may activate the largest family type II antitoxin system and virulence-associated proteins in prokaryotes, the VapBC proteins.

15.2.3   Adaptations Through Protein and Enzyme Modification

According to studies by Feder and Hofmann (1999) and Hendrick and Hartl (1993), thermophiles feature two special proteins viz., heat-shock proteins (HSPs) and chaperons that are specifically designated to help proteins fold when they fail to achieve the desired conformation at high temperature. HSPs are activated during a heat-shock reaction (Buchner 1996; Trent 1996) and chaperones execute the folding of an unfolded or nascent polypeptide chain into a full protein shape and stop the bulk synthesis of denatured proteins during high temperatures. They merely aid in the folding processes and do not incorporate themselves into the formed proteins. Molecular chaperones appear to be relatively common and have highly conserved sequences. The ability of molecular chaperons to stop incorrect protein aggregation is one of their key roles. Molecular chaperons can refold partially denatured proteins in the cell before proteases identify them as poorly folded and remove them, in addition to folding freshly synthesized proteins. Molecular chaperones are successfully extracted from *Sulfolobus shibatae* and *S. solfataricus*. Amazingly, the chaperone Hsp60 complex misfolds and denatures proteins and enzymes into chromosomes, which are cellular protein degradation systems. Therefore, the aggregation of unfolded proteins at high temperatures preserves cell homeostasis (Tse and Ma 2016). Many enzymes exhibit distinctive features to give proteins greater stiffness and resistance to chemical denaturation in addition to improved thermal stability (Bezsudnova et al. 2012). These features include a high frequency of salt bridges, a large number of surface ion pairs, a small hydrophobic core, more hydrogen bonding, the presence of the highest arginine, aspartate, and glutamate amino acid, loop shortening and additional alpha-helices in the structure of an enzyme (Arbab et al. 2021). A few of the extremozymes that contribute to the survival of thermophilic microbes in high-temperature conditions are lipases, proteases, amylases, pullulanase, Taq and Pfu DNA polymerases, glutamate synthetase, and aminotransferases (Rabbani et al. 2023). These enzymes remain active at temperatures above 100 °C.

15.2.4 Adaptations at Proteomic Level

Thermophiles have higher 23S-rRNA gene affinities in both bacterial and archaeal ribosomal protein complexes than mesophiles (Running and Reilly 2009; Shcherbakov et al. 2006). Proteomics studies revealed changes in the ribosomal proteins of thermophiles (van den Elzen et al. 2023). According to a proteomics investigation of *T. tengcongensis*, the size of ribosomal protein S1 was shown to be much larger under higher temperatures in comparison to original growth temperatures (Chen et al. 2013). When the thermophilic *Bacillus methanolicus* MGA3 grows at a higher temperature, several ribosomal protein members, e.g., L17, 50S, L14, 30S, S18, and ribosome-binding factor A, are found to be vastly increased (Müller et al. 2014). Trauger et al. (2008) with combining of the proteome and transcriptome study in *Pyrococcus furiosus* discovered that ribosomal proteins, like the larger subunit (LSU) ribosomal proteins L10E, L12A, and L7AE, exhibited larger abundances at 90 °C than at 70 °C. These results strongly suggest that thermophiles have stable and effective translational machinery that allows for the adequate synthesis of functional proteins. There is mounting evidence that certain gene expression products are essential to thermophiles' ability to adapt to heat. These genes are typically thought to have particular roles at higher temperatures and are elevated in response to temperature changes at either the transcriptome or proteomic level. The expression of gene clusters responding to changes in environmental temperature is under the global control of several regulators. Some global regulators have been identified in *Escherichia Coli*, which includes cAMP Receptor Protein (*CRP*), Integration Host Factor (IHF), Fumarate Nitrate Reductase (FNR), Factor for Inversion Stimulation (FIS), Aerobic Respiration Control (ArcA), Leucine Responsive Regulatory Protein (Lrp), and Histone-like Nucleoid Structuring protein (H-NS) (Browning and Busby 2004). The global regulatory systems occasionally overlap and can interact with a heat-shock transcription factor, which controls both the production of chaperon proteins and the heat-shock response proteins. The protein synthesis mechanism in thermophiles is anticipated to be stable and effective under high temperatures since retaining a sufficient number of proteins is important for functional performance.

15.2.5 Adaptations Through Stabilization of Nucleic Acids (DNA and RNA)

At molecular levels, the DNA repair system should be stricter in thermophiles to ensure genomic stability because DNA is more fragile at higher temperatures (Wani et al. 2022). According to several studies (Boto 2014; Feng et al. 2014), one of the primary mechanisms underlying bacterial adaptation to high temperatures is the horizontal transfer of DNA segments among various species. The homologous gene sequences are altered through horizontal gene transfer (HGT) either by replacing them with the new sequences or with incorporation through transformation, conjugation, or transduction. The majority of the genes acquired through HGT exhibited thermophilic characteristics necessary for survival at high temperatures. Reverse gyrase, for instance, is a thermophilic adaptation enzyme that is thought to have been passed from thermophilic archaea to bacteria. It is a particular variety of DNA topoisomerase that helps dsDNA to supercoil. In general, compared to linear DNA, circular DNA is more heat resistant. Besides incorporating DNA sequences from other species, mutation also occurs in thermophilic microbes under heat stress conditions viz., *T. thermophilus* and *Sulfolobus acidocaldarius* (Drake 2009). It has been reported through several studies that base substitutions are less common in thermophiles than in mesophiles. This raises the possibility that a thermophile will experience more negative effects in the amino acid substitutions from gene mutations than a non-thermophile. It's interesting to note that thermophile genome sizes are also smaller than non-thermophile genome sizes (Basak et al. 2004). Furthermore, high GC content contributes to genome thermostability (Musto et al. 2005; Musto et al. 2006). However, thermophiles have a selective response for survival that has a significantly high AG content in their mRNAs. Thermophilic mRNAs have a higher proportion of purines and purine clusters with notably high purine/pyrimidine ratios when compared to mesophilic species (Basak et al. 2004; Paz et al. 2004). The frequency of charged residues (Glu, Arg, and Lys) is on the rise, while the frequency of polar uncharged residues (Asn, Gln, Ser, and Thr) is on the decline, the frequency of thermo-labile amino acids (His, Gln, and Thr) is on the rise, and the ratio of (Glu and Lys) to (Gln and His) is on the rise in thermophiles (Kreil 2001; Tekaia et al. 2002; Singer and Hickey 2003). Therefore, the particular amino acid consumption is justified as a thermo-adaptation approach. A few investigations have also demonstrated that polyamines modify DNA in thermophiles. Thermophiles have polyamines that are better at stabilizing dsDNA, while branching polyamines are better at stabilizing ssDNA and tRNA, according to Terui et al. (2005). Salts that are known to stabilize proteins and dsDNA are frequently found in high concentrations in thermophilic microorganisms (Hensel and König 1988). The negative charges of KCl, $MgCl_2$, and phosphate groups are screened out by divalent and monovalent salts, which enhance nucleic acid stability by shielding DNA from hydrolysis and depurination. In the case of rRNA and tRNA, again salt concentration is essential. As opposed to genomic DNA, the amount of GC in rRNA is correlated with the organism's growth. Only dsRNA is affected by this association. Another element influencing thermal stability is the increased affinity of ribosomal proteins for rRNA. Post-translational RNA modifications are frequently observed in *S. solfataricus* (a hyperthermophile) rRNA, particularly for ribose methylation at the 20-hydroxyl position (Charlier and Droogmans 2005). Since tRNAs are vulnerable to extreme temperatures, some thermophiles also possess non-canonical enzymes or pathways to get beyond the translational barrier (Ibba and Soll 2004). According to Stepanov and Nyborg (2002), a rapid turnover in conjunction with a high level of aminoacyl tRNA synthesis may be able to counteract the aminoacyl tRNA degradation that occurs at high temperatures.

15.3 Significant Role of Thermophilic Microbes in Plant Growth Promotion

The global population is increasing gradually, and it is estimated that in 2050, there will be 10 billion people on the planet (Gu and Andreev 2021; Sadigov 2022). The biggest challenge facing agriculture is fulfilling the food demands of an increasing population. In this context, there is an urgent need for sustainable agricultural practices that easily meet all demands without causing any undesirable effects. One of the most crucial aspects of achieving sustainable agriculture is using plant growth-promoting microorganisms (PGPMs). Among PGPMs, the application of thermophilic microbes in agriculture has increased in recent years because they have various mechanisms that help in plant growth and development (Meng et al. 2020; Thakur et al. 2021). The different mechanisms through which microbes promote plant growth are nitrogen fixing, potassium solubilization, phosphate solubilization, siderophore production, heavy metal resistance, indole compound, hydrogen cyanide production, and ACC deaminase production, providing thermo-tolerance and halotolerance (Verma et al. 2018; Mukherjee et al. 2020). Moreover, by secreting certain compounds that directly contribute to plant development and growth (growth hormones) or by changing the form of specific inefficient soil components to make them effective for uptake (such as nitrogen fixation and phosphorous solubilization), they actively boost plant growth. Also, these microbes indirectly reduce the negative effects of specific plant diseases that hamper the yield and growth of plants (Rahman et al. 2023).

Owing to their habitat and surroundings, microbes acquire notable characteristics for their survival. The propensity of thermophilic microbes to thrive in extremely high-temperature environments has marked their importance in exploiting them and their biomolecules. Natural hot spring water is a unique source of these microbes and other microbes (Abussaudet al. 2013; Patel et al. 2017). Another source is the rhizosphere of heat-tolerant crop plants (Dastogeer et al. 2022). These microbes from the natural source have been exclusively studied for their plant growth-promoting capabilities (Mukherjee et al. 2020). Recent technological developments have also led to the identification of thermophilic microorganisms as potential producers of novel pigments (food additives) and enzymes that can be useful in agriculture as bio-inoculants for promoting plant growth and suppressing plant diseases as biocontrol agents (Srinivas et al. 2009).

Essential nutrients, including phosphorus, calcium, magnesium, and molybdenum, are less accessible in extreme temperatures and disturbed pH soils. In arid deserts, where the ecosystem often consists of poor soil quality with little organic content and accessible inorganic nutrient availability, this is a big challenge for crop cultivation. Microbial communities present in such harsh conditions play a vital role in elevating soil productivity and balancing soil carbon storage as a result of climate change, which favors crop production in hot conditions in arid regions. As a result, in these harsh settings, a diversity of stress-resistant microorganisms can be found and studied. The thermophilic microbial consortium seems to be a beneficial and innovative component for sustainable agriculture production under these conditions. Ravikiran and co-workers (2020) found that field crops are more susceptible to heat damage during the reproductive phase than their vegetative phase. In the summer, heat is the main factor that restricts their growth and yield. In these situations, applying thermophilic microbial cultures can help reduce stress. A *Bacillus subtilis* strain isolated from the rhizospheric soil of heat-stressed spinach was found to effectively solubilize phosphorus; it also generates indole acetic acid (IAA), amylase, and protease. Further investigations revealed that the strain of *B. subtilis* had higher levels of soluble protein, soluble sugar, and proline, which altered their osmotic potential. Moreover, five thermophilic bacteria with significant phosphate solubilizing ability were identified (Li et al. 2023). Thermotolerant actinomycetes, fungi, and bacteria could dissolve phosphate in all of its diverse forms, including calcium phosphate, rock phosphate, aluminum phosphate, iron phosphate, and hydroxyapatite (Xiao et al. 2011; Chang and Yang 2009). Singh and Satyanarayana (2009) investigated the role of extracellular phytase from *Sporotrichum thermophile* in the hydrolysis of insoluble phytates and plant development. *Aspergillus fumigatus* is a thermophilic fungus with three potential mechanisms by which the fungus significantly releases potassium: (1) the complexation of soluble organic ligands; (2) responding to immobile biopolymers; and (3) mechanical forces linked to direct physical contact between cells and mineral particles (Lian et al. 2008). Iron transport systems in thermophilic bacteria are uncommon and poorly characterized (Guerry et al. 1997). Iron and other metals from volcanic rocks were reported to be chelated from the rock matrix to the aqueous phase of glacial melt water, enhancing iron bioavailability (Bau et al. 2013).

Thermophilic microbes show various growth-inducing activities like seed germination and vigor index, plant height, crop yield, and nutritional quality of crops. Tomato seeds treated with bacterial isolates significantly increased shoot height, fresh shoot weight, root length, and fresh root weight of tomato seedlings over the control, with corresponding values ranging from 3.12% to 74.37%, 33.33% to 350.0%, 16.06% to 130.41%, and 36.36% to 318.18% (Patel et al. 2017). The thermophilic bacterial species *Klebsiella* sp. PMnew was isolated by Mukherjee et al. (2020) from Paniphala hot springs. He evaluated this strain's qualities that promote plant growth. To investigate its impact on the vegetative and reproductive growth of rice plants, he conducted a greenhouse experiment using rice seeds (var. Swarna). The PMnew strain produces indole compounds, siderophores, and ACC deaminase, as well as organic acids that help solubilize phosphate and fix nitrogen. After receiving the PMnew strain treatment, the infected rice plants considerably outgrew the control in terms of growth parameters and yield of the plant. Three thermophilic bacterial strains—*B. subtilis* BHUJP-H1, *Bacillus* sp. BHUJP-H2, and *B. licheniformis* BHUJP-H3—isolated from hot springs in the Leh and Ladakh region of

India were found to work extremely well together as a treatment to improve mung-bean plant growth characteristics (Verma et al. 2018). The mungbean plants treated with the consortium of *B. subtilis* BHUJP-H1+ *Bacillus* sp. BHUJP-H2+ *B. licheniformis* BHUJP-H3 significantly improved plant growth and production of mungbean in field conditions. As a result, different *Bacillus* spp.. that can produce thermostable enzymes have been isolated from hot springs in this context. In order to produce thermostatic enzymes, sugars, suitable solutes, and antibiotics, hot springs have recently attracted increased interest (Satyanarayana et al. 2013). Rajput et al. (2020) found that tomato seeds treated with a combination of thermophilic *Trichoderma pseudokoningii* BHUR2 and vermiwash resulted in an increase in the nutritional profile of tomato fruit, including its concentration of lycopene, total soluble sugar, and total protein. Furthermore, he discovered that, as compared to untreated plants in heat-stressed and greenhouse circumstances, this consortium boosted both fresh and dry shoot and root weight and the yield of fruits per plant. However, microbial interactions with plants promote ion homeostasis by controlling osmoprotectant levels and activating the antioxidant defense system. According to numerous transcriptome studies, plants adapted to heat stress in a variety of ways, including through signal transduction, translation, carbohydrate metabolism, and environmental adaptation. Folding, sorting, and degradation pathways were improved after plant growth-promoting rhizobacteria (PGPR)) inoculation and under heat stress, along with other fundamental mechanisms. Based on these, it can be convincing saying that research can be improved to comprehend the mechanisms by which plants resist and tolerate heat stress as well as those that promote plant growth, opening the door for its widespread use and future research.

15.4 Role of Thermophilic Microbes in Mitigation of Biotic Stress in Crop Plants

The utilization of thermophilic microorganisms in mitigating biotic stress in crop plants holds notable significance owing to their capacity to thrive in high-temperature environments and produce enzymes characterized by exceptional heat stability. Biotic stress denotes adverse circumstances wherein plants struggle to sustain their normal growth due to interactions with detrimental microorganisms, encompassing fungi, bacteria, viruses, viroids, phytoplasma, and nematodes which results in disturbances in the intracellular K^+ homeostasis (Shabala and Pottosin 2014).

Raza et al. (2019) revealed that the progressive rise in temperature has negative impacts on morphological, physiological, genetic, and biochemical attributes, which have a significant effect on crop productivity. Reactive oxygen species (ROS) are produced in high quantities, membrane permeability is altered, the mitochondrial and chlorophyll functions are disrupted, cellular structure is compromised, and even cell death occurs in plants that are exposed to long-term high-temperature stress (Liu et al. 2019; Kim et al. 2021). To adapt to shifting environmental conditions, microorganisms frequently modify their genetic makeup and metabolic pathways. These modifications may cause the form of novel disease (Jones and Naidu 2019). For instance, *Sclerotium rolfsii*, a corticioid species (Family: Atheliaceae), is a highly destructive necrotrophic pathogen with a wide host range, infecting over 400 plant species worldwide, established particularly in hot and humid regions (Paparu et al. 2020; Sun et al. 2020). This pathogen exhibits aggressive proliferation in warm and humid conditions, affecting various parts of the plant and resulting in stem canker, crown blight, and the prevalent collar rot (Kator et al. 2015; Sahu et al. 2019; Sun et al. 2020). Rajput et al. (2020) suggested that thermophilic microorganisms such as *T. pseudokoningii* BHUR2 and vermiwash enhance the resilience of tomato crops against *S. rolfsii* under heat stress conditions. This effect was attributed to heightened antioxidant activity and activation of the phenylpropanoid pathway, which in turn mitigated oxidative reaction, curbing pathogenic infiltration. *Trichoderma* spp. possesses substantial sustainable potential in antagonizing various phytopathogens. Through rapid colonization of soil and host plants, competition for nutrients and space, direct or indirect stimulation of the plant's defense mechanisms, and production of cell wall-degrading enzymes, metabolites, and antibiotics for cell lysis that ultimately inhibit the growth of pathogens, *Trichoderma* species uphold a sustainable potential to combat several phytopathogens (Beneduzi et al. 2012; Saravanakumar and Wang 2020; Kaur et al. 2020). Because they successfully reduce soil- and seed-borne phytopathogens and have long-lasting effects from their introduction and proliferation in soil, *Trichoderma* spp. is now acknowledged as an environmentally friendly alternative in organic farming (Ram et al. 2018; Vinci et al. 2018). In studies conducted by Yildirim et al. (2006) and Bisen et al. (2014), it was found that specific kinds of *Trichoderma* spp. were mutualistic fungi that improved the root expansion, water holding capacity, fertilizer use efficiency, and overall plant tolerance to stress conditions. *Trichoderma harzianum* strain BHU P4 (Singh et al. 2021) is a thermophilic strain and has antagonistic properties against *S. rolfsii*, which is the cause of collar rot disease. Additionally, compared to control and pathogen-challenged plants, the strain was successful in enhancing the growth parameters of plants in okra plants. It has been demonstrated that applying *Trichoderma* spp. is an efficient tactic for controlling phytopathogens that produce sclerotia (Poosapati et al. 2014; Bisen et al. 2014). *Trichoderma* species also use mechanisms, e.g., competition for nutrients and space, antibiosis, and mycoparasitism as biocontrol mechanisms for plant diseases. Additionally, they directly affect plant development and growth by promoting the generation of phytohormones, improving stress tolerance, and triggering plant defense mechanisms (Rajput et al. 2019). Thermophilic plant growth-promoting rhizobacteria (PGPR) play a pivotal role in producing thermopile enzymes that could be advantageous for plant growth promotion, offering an eco-friendly alternative to chemical solutions. Patel et al. (2017) shed light on the multifaceted plant growth-promoting (PGP) capabilities of bacteria thriving in hot springs, highlighting their antagonistic effects on diverse phytopathogens. Through 16S-rRNA gene sequencing, they identified 12 versatile bacteria, primarily belonging to *Aneurinibacillus aneurinilyticus* and *Bacillus* spp. Inoculating tomato seeds with these thermophilic bacteria resulted in notable improvements in shoot height, fresh shoot weight, root length, and fresh root weight of tomato seedlings under biotic stress. Soil-borne crop diseases, primarily caused by phytopathogenic genera like *Pythium, Fusarium, Phytophthora, Rhizoctonia*, and *Sclerotium* (Rao 2007; Wightwick et al. 2009), pose significant challenges. Employing antagonistic bacteria such as *B. subtilis* GB-03 against *F. oxysporum* f. sp. vasinfectum and *Rhizoctania solani* in cotton (Brannen and Kenney 1997), or *Bacillus cereus* against *Phytophthora sojae* in soybean has proven effective in managing biotic stresses. The endospore-forming nature of antagonistic thermophilic bacteria grants them resilience to high temperatures and desiccation, making them attractive candidates for biocontrol of plant diseases (Wightwick et al. 2009). According to Pawar and Borkar (2018), three thermophilic bacterial isolates demonstrated robust antagonistic effects against various soil-borne fungal plant pathogens. Specifically, *B. licheniformis* exhibited the highest inhibition zone of 67.67 mm against *Rhizoctonia bataticola*, followed by *Bacillus stearothermophilus* from compost manure (51.67 mm against *R. solani*) and *B. stearothermophilus* from tomato rhizospheric soil (38.33 mm against *Pythium aphanidermatum*). Antagonistic thermophilic *Bacillus* isolates showcase an ability to thrive in high temperatures (>55 °C), a pH range of 6–8 and tolerate salt concentrations up to 8%. Additionally, their resistance to antibiotics positions them as promising biocontrol candidates, particularly in stressed rhizosphere environments where other biocontrol agents may prove ineffective. For instance, the strain *Bacillus licheniformis* SB3086 has been identified to secrete novozymes through its spores, playing a pivotal role as a phosphate solubilizer and offering protection against biotic stress, such as Dollar spot disease in plants (Saharan and Verma 2014). As reported by Kayasth et al. (2013), *B. licheniformis* has been explored as a potential plant growth-promoting rhizobacteria (PGPR)) strain, developed into a multifunctional biofertilizer for diverse crop production. Additionally, *B. licheniformis*, along with *B. cereus, B. circulans, B. subtilis*, and *B. thuringiensis*, have been identified as potential biocontrol agents with chitinolytic activities (Sadfi et al. 2002; Verma et al. 2018).

15.5   Conclusion

In conclusion, the imposition of biotic stress significantly results in yield losses in major crops, with viral, fungal, and bacterial infections adversely impacting leaf health and leading to reduced levels of photosynthesis. Given these challenges, there is an urgent need for sustainable alternatives to whole-organism formulations. These alternatives must not only withstand both biotic and abiotic stresses but also possess biocontrol and plant growth-promoting capabilities. Consequently, in response to the adverse consequences associated with chemical methods, biological interventions have gained prominence as supplementary treatments. Whether used individually or in combination, these interventions represent fewer environmental repercussions. Considering these advancements, it is evident that thermophilic microorganisms play a pivotal role in promoting plant growth and mitigating biotic stress. These resilient microorganisms offer a promising avenue for sustainable crop management, providing a viable alternative to traditional chemical approaches. By harnessing the biocontrol and growth-enhancing properties of thermophilic microbes, we have the potential to revolutionize agricultural practices, contributing to both food security and environmental sustainability.

Acknowledgments The research was financially supported by the Ministry of Science and Higher Education of the Russian Federation (no. FENW-2023-0008)

References

Abbass K, Qasim MZ, Song H et al (2022) A review of the global climate change impacts, adaptation, and sustainable mitigation measures. Environ Sci Pollut Res 29:42539–42559. https://doi.org/10.1007/s11356-022-19718-6

Adiguzel A, Ozkan H, Baris O et al (2009) Identification and characterization of thermophilic bacteria isolated from hot springs in Turkey. J Microbiol Methods 79:321–328. https://doi.org/10.1016/j.mimet.2009.09.026

Alsajri FA, Wijewardana C, Bheemanahalli R et al (2022) Morpho-physiological, yield, and transgenerational seed germination responses of soybean to temperature. Front Plant Sci 13. https://doi.org/10.3389/fpls.2022.839270

Arbab S, Ullah H, Khan MIU et al (2021) Diversity and distribution of thermophilic microorganisms and their applications in biotechnology. J Basic Microbiol 62:95–108. https://doi.org/10.1002/jobm.202100529

Arias PM, Butler J, Randhawa GS et al (2023) Environment and taxonomy shape the genomic signature of prokaryotic extremophiles. Sci Rep 13:16105. https://doi.org/10.1038/s41598-023-42518-y

Arora NK, Panosyan H (2019) Extremophiles: applications and roles in environmental sustainability. Environ Sustain 2:217–218. https://doi.org/10.1007/s42398-019-00082-0

Basak S, Banerjee T, Gupta SK, Ghosh TC (2004) Investigation on the causes of codon and amino acid usages variation between ThermophilicAquifex aeolicusand MesophilicBacillus subtilis. J Biomol Struct Dyn 22:205–214. https://doi.org/10.1080/07391102.2004.10506996

Bau M, Tepe N, Mohwinkel D (2013) Siderophore-promoted transfer of rare earth elements and iron from volcanic ash into glacial meltwater, river and ocean water. Earth Planet Sci Lett 364:30–36. https://doi.org/10.1016/j.epsl.2013.01.002

Becerra A, Delaye L, Lazcano A, Orgel LE (2007) Protein disulfide oxidoreductases and the evolution of thermophily: was the last common ancestor a heat-loving microbe? J Mol Evol 65:296–303. https://doi.org/10.1007/s00239-007-9005-0

Beneduzi A, Ambrosini A, Passaglia LMP (2012) Plant growth-promoting rhizobacteria (PGPR): their potential as antagonists and biocontrol agents. Genet Mol Biol 35:1044–1051. https://doi.org/10.1590/s1415-47572012000600020

Berezovsky IN, Shakhnovich EI (2005) Physics and evolution of thermophilic adaptation. Proc Natl Acad Sci 102:12742–12747. https://doi.org/10.1073/pnas.0503890102

Bezsudnova EY, Boyko KM, Polyakov KM et al (2012) Structural insight into the molecular basis of polyextremophilicity of short-chain alcohol dehydrogenase from the hyperthermophilic archaeon Thermococcus sibiricus. Biochimie 94:2628–2638. https://doi.org/10.1016/j.biochi.2012.07.024

Bisen K, Keswani C, Mishra S et al (2014) Unrealized potential of seed biopriming for versatile agriculture. In: Nutrient use efficiency: from basics to advances, pp 193–206. https://doi.org/10.1007/978-81-322-2169-2_13

Boto L (2014) Horizontal gene transfer in the acquisition of novel traits by metazoans. Proc R Soc B Biol Sci 281:20132450. https://doi.org/10.1098/rspb.2013.2450

Brannen PM, Kenney DS (1997) Kodiak®—a successful biological-control product for suppression of soil-borne plant pathogens of cotton. J Ind Microbiol Biotechnol 19:169–171. https://doi.org/10.1038/sj.jim.2900439

Brock TD (2012) Thermophilic microorganisms and life at high temperatures. Springer Science & Business Media

Browning DF, Busby SJW (2004) The regulation of bacterial transcription initiation. Nat Rev Microbiol 2:57–65. https://doi.org/10.1038/nrmicro787

Buchner J (1996) Supervising the fold: functional principles of molecular chaperones. FASEB J 10:10–19. https://doi.org/10.1096/fasebj.10.1.8566529

Camargo FP, Sakamoto IK, Delforno TP et al (2022) Microbial and functional characterization of granulated sludge from full-scale UASB thermophilic reactor applied to sugarcane vinasse treatment. Environ Technol 44:3141–3160. https://doi.org/10.1080/09593330.2022.2052361

Cavicchioli R, Amils R, Wagner D, McGenity T (2011) Life and applications of extremophiles. Environ Microbiol 13:1903–1907. https://doi.org/10.1111/j.1462-2920.2011.02512.x

Chang C-H, Yang S-S (2009) Thermo-tolerant phosphate-solubilizing microbes for multi-functional biofertilizer preparation. Bioresour Technol 100:1648–1658. https://doi.org/10.1016/j.biortech.2008.09.009

Charlier D, Droogmans L (2005) Microbial life at high temperature, the challenges, the strategies. Cell Mol Life Sci CMLS 62:2974–2984. https://doi.org/10.1007/s00018-005-5251-8

Chen Z, Wang Q, Lin L et al (2012) Comparative evaluation of two isobaric labeling tags, DiART and iTRAQ. Anal Chem 84:2908–2915. https://doi.org/10.1021/ac203467q

Chen Z, Wen B, Wang Q et al (2013) Quantitative proteomics reveals the temperature-dependent proteins encoded by a series of cluster genes in Thermoanaerobacter Tengcongensis. Mol Cell Proteomics 12:2266–2277. https://doi.org/10.1074/mcp.m112.025817

Chen X, Shi X, Cai X et al (2020) Ice-binding proteins: a remarkable ice crystal regulator for frozen foods. Crit Rev Food Sci Nutr 61:3436–3449. https://doi.org/10.1080/10408398.2020.1798354

Colman DR, Poudel S, Hamilton TL et al (2017) Geobiological feedbacks and the evolution of thermoacidophiles. ISME J 12:225–236. https://doi.org/10.1038/ismej.2017.162

Cooper CR, Daugherty AJ, Tachdjian S et al (2009) Role of vapBC toxin–antitoxin loci in the thermal stress response of Sulfolobus solfataricus. Biochem Soc Trans 37:123–126. https://doi.org/10.1042/bst0370123

Counts JA, Zeldes BM, Lee LL et al (2017) Physiological, metabolic and biotechnological features of extremely thermophilic microorganisms. WIREs Syst Biol Med 9. https://doi.org/10.1002/wsbm.1377

Dastogeer KMG, Zahan MI, Rhaman MS et al (2022) Microbe-mediated thermotolerance in plants and pertinent mechanisms-a meta-analysis and review. Front Microbiol 13:833566

de la Haba RR, Antunes A, Hedlund BP (2022) Editorial: extremophiles: microbial genomics and taxogenomics. Front Microbiol 13. https://doi.org/10.3389/fmicb.2022.984632

Drake JW (2009) Avoiding dangerous missense: thermophiles display especially low mutation rates. PLoS Genet 5:e1000520. https://doi.org/10.1371/journal.pgen.1000520

Feder ME, Hofmann GE (1999) Heat-shock proteins, molecular chaperones, and the stress response: evolutionary and ecological physiology. Annu Rev Physiol 61:43

Feng S, Powell SM, Wilson R, Bowman JP (2014) Extensive Gene Acquisition in the extremely psychrophilic bacterial species Psychroflexus torquis and the link to sea-ice ecosystem specialism. Genome Biol Evol 6:133–148. https://doi.org/10.1093/gbe/evt209

Gao H, Wang Y, Liu X et al (2004) Global transcriptome analysis of the heat shock response of Shewanella oneidensis. J Bacteriol 186:7796–7803. https://doi.org/10.1128/jb.186.22.7796-7803.2004

Ghati A, Sarkar K, Paul G (2013) Production and characterization of an alkalothermostable, organic solvent tolerant and surfactant tolerant esterase produced by a thermophilic bacterium Geobacillus sp. AGP-04, isolated from Bakreshwar Hot Spring, India. J Microbiol Biotechnol food Sci 3:155–162

Gong P, Lei P, Wang S et al (2020) Post-translational modifications aid archaeal survival. Biomolecules 10:584. https://doi.org/10.3390/biom10040584

Gu D, Andreev KE, Dupre M (2021) Major trends in population growth around the world. China CDC Weekly 3:604–613. https://doi.org/10.46234/ccdcw2021.160

Guerry P, Perez-Casal J, Yao R et al (1997) A genetic locus involved in iron utilization unique to some Campylobacter strains. J Bacteriol 179:3997–4002. https://doi.org/10.1128/jb.179.12.3997-4002.1997

Gull A, Wani NUI, Lone AAA (2018) Biotic and abiotic stresses in plants

Gupta GN, Srivastava S, Khare SK, Prakash V (2014) Extremophiles: an overview of microorganism from extreme environment. Int J Agric Environ Biotechnol 7:371. https://doi.org/10.5958/2230-732x.2014.00258.7

Harirchi S, Etemadifar Z, Mahboubi A et al (2020) The effect of calcium/magnesium ratio on the biomass production of a novel thermoalkaliphilic Aeribacillus pallidus strain with highly heat-resistant spores. Curr Microbiol 77:2565–2574. https://doi.org/10.1007/s00284-020-02010-6

Helmann JD, Wu MFW, Kobel PA et al (2001) Global transcriptional response of Bacillus subtilis to heat shock. J Bacteriol 183:7318–7328. https://doi.org/10.1128/jb.183.24.7318-7328.2001

Hendrick JP, Hartl F-U (1993) Molecular chaperone functions of heat-shock proteins. Annu Rev Biochem 62:349–384. https://doi.org/10.1146/annurev.bi.62.070193.002025

Hensel R, König H (1988) Thermoadaptation of methanogenic bacteria by intracellular ion concentration. FEMS Microbiol Lett 49:75–79. https://doi.org/10.1111/j.1574-6968.1988.tb02685.x

Holden JF, Summit M, Baross JA (1998) Thermophilic and hyperthermophilic microorganisms in 3–30°C hydrothermal fluids following a deep-sea volcanic eruption. FEMS Microbiol Ecol 25:33–41. https://doi.org/10.1111/j.1574-6941.1998.tb00458.x

Huber R, Wilharm T, Huber D et al (1992) Aquifex pyrophilus gen. Nov. sp. nov., represents a novel group of marine hyperthermophilic hydrogen-oxidizing bacteria. Syst Appl Microbiol 15:340–351. https://doi.org/10.1016/s0723-2020(11)80206-7

Ibba M, Söll D (2004) Aminoacyl-tRNAs: setting the limits of the genetic code. Genes Dev 18:731–738

Jiang X, Van Horn DJ, Okie JG et al (2022) Limits to the three domains of life: lessons from community assembly along an Antarctic salinity gradient. Extremophiles 26:15. https://doi.org/10.1007/s00792-022-01262-3

Jones RAC, Naidu RA (2019) Global dimensions of plant virus diseases: current status and future perspectives. Annual Rev Virol 6:387–409. https://doi.org/10.1146/annurev-virology-092818-015606

Kator L, Hosea ZY, Oche OD (2015) Sclerotium rolfsii: Causative organism of southern blight, stem rot, white mold and sclerotia rot disease. Ann Biol Res 6:78–89

Kaur R, Kalia A, Lore JS, Sandhu JS (2020) Antifungal effect of Trichoderma spp. β-1,3-glucanase on Phytophthora parasitica: hyphal morphological distortions. J Phytopathol 168:700–706. https://doi.org/10.1111/jph.12950

Kayasth M, Kumar V, Gera R (2013) Exploring the potential of PGPR strain Bacillus licheniformis to be developed as multifunctional biofertilizer. Cent Eur J Biol 2:12–17

Kecha M, Benallaoua S, Touzel JP et al (2006) Biochemical and phylogenetic characterization of a novel terrestrial hyperthermophilic archaeon pertaining to the genus Pyrococcus from an Algerian hydrothermal hot spring. Extremophiles 11:65–73. https://doi.org/10.1007/s00792-006-0010-9

Khan MIR, Asgher M, Khan NA (2013) Rising temperature in the changing environment: a serious threat to plants. Climate Change Environ Sustain 1:25. https://doi.org/10.5958/j.2320-6411.1.1.004

Kim S-E, Lee C-J, Park S-U et al (2021) Overexpression of the Golden SNP-carrying Orange gene enhances carotenoid accumulation and heat stress tolerance in Sweetpotato plants. Antioxidants 10:51. https://doi.org/10.3390/antiox10010051

Kreil DP (2001) Identification of thermophilic species by the amino acid compositions deduced from their genomes. Nucleic Acids Res 29:1608–1615. https://doi.org/10.1093/nar/29.7.1608

Kumar V, Nautiyal CS (2022) Plant abiotic and biotic stress alleviation: from an endophytic microbial perspective. Curr Microbiol 79:311. https://doi.org/10.1007/s00284-022-03012-2

Kumar S, Sravani B, Korra T et al (2022) Psychrophilic microbes: biodiversity, beneficial role and improvement of cold stress in crop plants. In: New and future developments in microbial biotechnology and bioengineering, pp 177–198. https://doi.org/10.1016/b978-0-323-85163-3.00002-8

Labes A (2018) Secondary metabolites from microalgal extremophiles and their "Extreme-Loving" neighbors. In: Extremophiles. CRC Press, pp 305–314

Lal MK, Tiwari RK, Altaf MA et al (2023) Editorial: abiotic and biotic stress in horticultural crops: insight into recent advances in the underlying tolerance mechanism. Front Plant Sci 14. https://doi.org/10.3389/fpls.2023.1212982

Lee Y-J, Ganbat D, Oh D et al (2022) Isolation and characterization of thermophilic bacteria from Hot Springs in Republic of Korea. Microorganisms 10:2375. https://doi.org/10.3390/microorganisms10122375

Lhoste E, Comte F, Brown K et al (2023) Bacterial, archaeal, and eukaryote diversity in planktonic and sessile communities inside an abandoned and flooded iron mine (Quebec, Canada). Appl Microbiol 3:45–63. https://doi.org/10.3390/applmicrobiol3010004

Li H, Ji X, Zhou Z et al (2010) Thermus thermophilus proteins that are differentially expressed in response to growth temperature and their implication in thermoadaptation. J Proteome Res 9:855–864. https://doi.org/10.1021/pr900754y

Li SS, Yang ZC, Wang D et al (2023) Role of Bacillus subtilis BE-L21 in enhancing the heat tolerance of spinach seedlings. Biol Plant 67:36–44

Lian B, Wang B, Pan M et al (2008) Microbial release of potassium from K-bearing minerals by thermophilic fungus Aspergillus fumigatus. Geochim Cosmochim Acta 72:87–98. https://doi.org/10.1016/j.gca.2007.10.005

Liu C, LiCata VJ (2013) The stability of Taq DNA polymerase results from a reduced entropic folding penalty; identification of other thermophilic proteins with similar folding thermodynamics. Proteins 82:785–793. https://doi.org/10.1002/prot.24458

Liu B, Gumpertz ML, Hu S, Ristaino JB (2007) Long-term effects of organic and synthetic soil fertility amendments on soil microbial communities and the development of southern blight. Soil Biol Biochem 39:2302–2316. https://doi.org/10.1016/j.soilbio.2007.04.001

Liu J, Hasanuzzaman M, Wen H et al (2019) High temperature and drought stress cause abscisic acid and reactive oxygen species accumulation and suppress seed germination growth in rice. Protoplasma 256:1217–1227. https://doi.org/10.1007/s00709-019-01354-6

Mehta D, Satyanarayana T (2013) Diversity of hot environments and thermophilic microbes. In: Thermophilic microbes in environmental and industrial biotechnology. Springer, Dordrecht, pp 3–60

Meng L, Xie L, Suenaga T et al (2020) Eco-compatible biochar mitigates volatile fatty acids stress in high load thermophilic solid-state anaerobic reactors treating agricultural waste. Bioresour Technol 309:123366. https://doi.org/10.1016/j.biortech.2020.123366

Merino N, Aronson HS, Bojanova DP et al (2019) Living at the extremes: extremophiles and the limits of life in a planetary context. Front Microbiol 10. https://doi.org/10.3389/fmicb.2019.00780

Milojevic T, Cramm MA, Hubert CRJ, Westall F (2022) "Freezing" thermophiles: from one temperature extreme to another. Microorganisms 10:2417. https://doi.org/10.3390/microorganisms10122417

Miryeganeh M (2021) Plants' epigenetic mechanisms and abiotic stress. Genes 12:1106. https://doi.org/10.3390/genes12081106

Mukherjee T, Banik A, Mukhopadhyay SK (2020) Plant growth-promoting traits of a thermophilic strain of the Klebsiella group with its effect on rice plant growth. Curr Microbiol 77:2613–2622. https://doi.org/10.1007/s00284-020-02032-0

Müller JEN, Litsanov B, Bortfeld-Miller M et al (2014) Proteomic analysis of the thermophilic methylotroph bacillus methanolicusMGA3. Proteomics 14:725–737. https://doi.org/10.1002/pmic.201300515

Musto H, Naya H, Zavala A et al (2005) The correlation between genomic G + C and optimal growth temperature of prokaryotes is robust: a reply to Marashi and Ghalanbor. Biochem Biophys Res Commun 330:357–360. https://doi.org/10.1016/j.bbrc.2005.02.133

Musto H, Naya H, Zavala A et al (2006) Genomic GC level, optimal growth temperature, and genome size in prokaryotes. Biochem Biophys Res Commun 347:1–3. https://doi.org/10.1016/j.bbrc.2006.06.054

Neifar M, Maktouf S, Ghorbel RE et al (2015) Extremophiles as source of novel bioactive compounds with industrial potential. In: Biotechnology of bioactive compounds, pp 245–267. https://doi.org/10.1002/9781118733103.ch10

Nemoto N, Shida Y, Shimada H et al (2003) Characterization of the precursor of tetraether lipid biosynthesis in the thermoacidophilic archaeon Thermoplasma acidophilum. Extremophiles 7:235–243. https://doi.org/10.1007/s00792-003-0315-x

Paparu P, Acur A, Kato F et al (2020) Morphological and pathogenic characterization of Sclerotium rolfsii, the causal agent of southern blight disease on common bean in Uganda. Plant Dis 104:2130–2137. https://doi.org/10.1094/pdis-10-19-2144-re

Patel KS, Naik JH, Chaudhari S, Amaresan N (2017) Characterization of culturable bacteria isolated from hot springs for plant growth promoting traits and effect on tomato (Lycopersicon esculentum) seedling. C R Biol 340:244–249. https://doi.org/10.1016/j.crvi.2017.02.005

Pati S, Banerjee S, Sengupta A et al (2023) Adaptation strategies of thermophilic microbes. In: Bacterial survival in the hostile environment, pp 231–249. https://doi.org/10.1016/b978-0-323-91806-0.00012-6

Pawar RR, Borkar SG (2018) Isolation and characterization of thermophilic bacteria from different habitats and their assessment for antagonism against soil-borne fungal plant pathogens. African J Microbiol Res 12:556–566

Paz A, Mester D, Baca I et al (2004) Adaptive role of increased frequency of polypurine tracts in mRNA sequences of thermophilic prokaryotes. Proc Natl Acad Sci 101:2951–2956. https://doi.org/10.1073/pnas.0308594100

Peng C, Shi Y, Wang S et al (2023) Genetic and functional characterization of multiple thermophilic organosulfur-removal systems reveals desulfurization potentials for waste residue oil cleaning. J Hazard Mater 446:130706. https://doi.org/10.1016/j.jhazmat.2022.130706

Phour M, Sindhu SS (2022) Mitigating abiotic stress: microbiome engineering for improving agricultural production and environmental sustainability. Planta 256:85. https://doi.org/10.1007/s00425-022-03997-x

Poosapati S, Ravulapalli PD, Tippirishetty N et al (2014) Selection of high temperature and salinity tolerant Trichoderma isolates with antagonistic activity against Sclerotium rolfsii. Springerplus 3. https://doi.org/10.1186/2193-1801-3-641

Priyadharshini R, Brindha T, Uthandi S (2023) Thermophilic microbes producing industrially important enzymes from the Manikaran geothermal springs of Himachal Pradesh (India) and their application in biomass saccharification. Biomass Convers Biorefinery 13:15161–15172. https://doi.org/10.1007/s13399-023-04479-0

Rabbani G, Ahmad E, Ahmad A, Khan RH (2023) Structural features, temperature adaptation and industrial applications of microbial lipases from psychrophilic, mesophilic and thermophilic origins. Int J Biol Macromol 225:822–839. https://doi.org/10.1016/j.ijbiomac.2022.11.146

Rahman M, Borah SM, Borah PKR et al (2023) Deciphering the antimicrobial activity of multifaceted rhizospheric biocontrol agents of solanaceous crops viz., Trichoderma harzianum MC2, and Trichoderma harzianum NBG. Front Plant Sci 14. https://doi.org/10.3389/fpls.2023.1141506

Rajashree RP, Borkar SG (2018) Isolation and characterization of thermophilic bacteria from different habitats and their assessment for antagonism against soil-borne fungal plant pathogens. Afr J Microbiol Res 12:556–566. https://doi.org/10.5897/ajmr2015.7511

Rajput RS, Singh P, Singh J et al (2019) Seed biopriming through beneficial Rhizobacteria for mitigating soil-borne and seed-borne diseases. In: Plant growth promoting rhizobacteria for sustainable stress management, pp 201–215. https://doi.org/10.1007/978-981-13-6986-5_7

Rajput RS, Singh J, Singh P et al (2020) Influence of seed biopriming and Vermiwash treatment on tomato Plant's immunity and nutritional quality upon Sclerotium rolfsii challenge inoculation. J Plant Growth Regul 40:1493–1509. https://doi.org/10.1007/s00344-020-10205-1

Ram RM, Keswani C, Bisen K et al (2018) Biocontrol technology. Omics technologies and bioengineering, pp 177–190. https://doi.org/10.1016/b978-0-12-815870-8.00010-3

Rao DLN (2007) Microbial diversity, soil health and sustainability. J Indian Soc Soil Sci 55:392–403

Rasuk MC, Ferrer GC, Moreno JR et al (2016) The diversity of microbial extremophiles. In: Rodriguez TB, Silva AE (eds) Molecular diversity of environmental prokaryotes1st edn, pp 87–126

Raza MA, Feng LY, van der Werf W et al (2019) Narrow-wide-row planting pattern increases the radiation use efficiency and seed yield of intercrop species in relay-intercropping system. Food Energy Secur 8. https://doi.org/10.1002/fes3.170

Rothschild LJ, Mancinelli RL (2001) Life in extreme environments. Nature 409:1092–1101. https://doi.org/10.1038/35059215

Running WE, Reilly JP (2009) Ribosomal proteins of Deinococcus radiodurans: their solvent accessibility and reactivity. J Proteome Res 8:1228–1246. https://doi.org/10.1021/pr800544y

Sadfi N, Chérif M, Hajlaoui MR, Boudabbous A (2002) Biological control of the potato tubers dry rot caused by Fusarium roseum var. sambucinum under greenhouse, field and storage conditions using Bacillus spp. Isolates Journal of Phytopathology 150:640–648. https://doi.org/10.1046/j.1439-0434.2002.00811.x

Sadigov R (2022) Rapid growth of the world population and its socioeconomic results. Sci World J 2022:1–8. https://doi.org/10.1155/2022/8110229

Saharan BS, Verma S (2014) Potential plant growth promoting activity of Bacillus licheniformis UHI (II) 7. Int J Microb Res Technol 2:22–27

Sahu PK, Singh S, Gupta A et al (2019) Antagonistic potential of bacterial endophytes and induction of systemic resistance against collar rot pathogen Sclerotium rolfsii in tomato. Biol Control 137:104014. https://doi.org/10.1016/j.biocontrol.2019.104014

Salazar-Badillo FB, Salas-Muñoz S, Mauricio-Castillo JA et al (2017) The rhizospheres of arid and semi-arid ecosystems are a source of microorganisms with growth-promoting potential. Adv PGPR Res:187–196. https://doi.org/10.1079/9781786390325.0187

Saravanakumar K, Wang M-H (2020) Isolation and molecular identification of Trichoderma species from wetland soil and their antagonistic activity against phytopathogens. Physiol Mol Plant Pathol 109:101458. https://doi.org/10.1016/j.pmpp.2020.101458

Sarma J, Sengupta A, Laskar MK et al (2023) Microbial adaptations in extreme environmental conditions. In: Bacterial survival in the hostile environment, pp 193–206. https://doi.org/10.1016/b978-0-323-91806-0.00007-2

Satyanarayana T, Littlechild J, Kawarabayasi Y (2013) Thermophilic microbes in environmental and industrial biotechnology. Springer Science & Business Media

Shabala S, Pottosin I (2014) Regulation of potassium transport in plants under hostile conditions: implications for abiotic and biotic stress tolerance. Physiol Plant 151:257–279. https://doi.org/10.1111/ppl.12165

Shah A, Ishaq K, Fariq A et al (2023) Diversity of culturable thermophilic bacteria in hotspring of Kotli Azad Jammu and Kashmir. ARPHA Prepr 4:e100831

Shcherbakov D, Dontsova M, Tribus M et al (2006) Stability of the 'L12 stalk' in ribosomes from mesophilic and (hyper)thermophilic Archaea and Bacteria. Nucleic Acids Res 34:5800–5814. https://doi.org/10.1093/nar/gkl751

Shih T-W, Pan T-M (2011) Stress responses of thermophilic Geobacillus sp. NTU 03 caused by heat and heat-induced stress. Microbiol Res 166:346–359. https://doi.org/10.1016/j. micres.2010.08.001

Singer GAC, Hickey DA (2003) Thermophilic prokaryotes have characteristic patterns of codon usage, amino acid composition and nucleotide content. Gene 317:39–47. https://doi.org/10.1016/s0378-1119(03)00660-7

Singh B, Satyanarayana T (2009) Plant growth promotion by an extracellular HAP-Phytase of a thermophilic mold sporotrichum thermophile. Appl Biochem Biotechnol 160:1267–1276. https://doi.org/10.1007/s12010-009-8593-0

Singh J, Singh Rajput R, Singh P et al (2021) Screening, isolation and characterization of heat stress tolerant Trichoderma isolates: sustainable alternative to climate change. Plant Arch 21. https://doi.org/10.51470/plantarchives.2021.v21.no1.235

Sinha T, Nandi K, Das R et al (2022) Microbe-mediated biotic and abiotic stress tolerance in crop plants. In: Microbes and microbial biotechnology for green remediation, pp 93–116. https://doi.org/10.1016/b978-0-323-90452-0.00015-3

Slobodkin AI, Reysenbach A-L, Slobodkina GB et al (2013) Dissulfuribacter thermophilus gen. nov., sp. nov., a thermophilic, autotrophic, sulfur-disproportionating, deeply branching delta-proteobacterium from a deep-sea hydrothermal vent. Int J Syst Evol Microbiol 63:1967–1971. https://doi.org/10.1099/ijs.0.046938-0

Somayaji A, Dhanjal CR, Lingamsetty R et al (2022) An insight into the mechanisms of homeostasis in extremophiles. Microbiol Res 263:127115. https://doi.org/10.1016/j.micres.2022.127115

Srinivas TNR, Nageswara Rao SSS, Vishnu Vardhan Reddy P et al (2009) Bacterial diversity and bioprospecting for cold-active lipases, amylases and proteases, from culturable bacteria of Kongsfjorden and Ny-Ålesund, Svalbard, Arctic. Curr Microbiol 59:537–547. https://doi.org/10.1007/s00284-009-9473-0

Stepanov V, Nyborg J (2002) Thermal stability of aminoacyl-tRNAs in aqueous solutions. Extremophiles 6:485–490. https://doi.org/10.1007/s00792-002-0285-4

Stillman JH (2019) Heat waves, the new Normal: summertime temperature extremes will impact animals, ecosystems, and human communities. Physiology 34:86–100. https://doi.org/10.1152/physiol.00040.2018

Suddin S, Mokosuli YS, Marcelina W et al (2019) Molecular barcoding based 16S rRNA gene of thermophilic bacteria from Vulcanic Sites, Linow Lake, Tomohon. Mater Sci Forum 967:83–92. https://doi.org/10.4028/www.scientific.net/msf.967.83

Sun S, Sun F, Deng D et al (2020) First report of southern blight of mung bean caused by Sclerotium rolfsii in China. Crop Prot 130:105055. https://doi.org/10.1016/j.cropro.2019.105055

Tam LT, Antelmann H, Eymann C et al (2006) Proteome signatures for stress and starvation inBacillus subtilis as revealed by a 2-D gel image color coding approach. Proteomics 6:4565–4585. https://doi.org/10.1002/pmic.200600100

Tekaia F, Yeramian E, Dujon B (2002) Amino acid composition of genomes, lifestyles of organisms, and evolutionary trends: a global picture with correspondence analysis. Gene 297:51–60. https://doi.org/10.1016/s0378-1119(02)00871-5

Terui Y, Ohnuma M, Hiraga K et al (2005) Stabilization of nucleic acids by unusual polyamines produced by an extreme thermophile, Thermus thermophilus. Biochem J 388:427–433. https://doi.org/10.1042/bj20041778

Thakur V, Kumar V, Kumar V, Singh D (2021) Genomic insights driven statistical optimization for production of efficient cellulase by Himalayan thermophilic bacillus sp. PCH94 using agricultural waste. Waste Biomass Valoriz 12:6917–6929. https://doi.org/10.1007/s12649-021-01491-1

Tiwari YK, Yadav SK (2019) High temperature stress tolerance in maize (Zea mays L.): physiological and molecular mechanisms. J Plant Biol 62:93–102. https://doi.org/10.1007/s12374-018-0350-x

Trauger SA, Kalisak E, Kalisiak J et al (2008) Correlating the transcriptome, proteome, and metabolome in the environmental adaptation of a hyperthermophile. J Proteome Res 7:1027–1035. https://doi.org/10.1021/pr700609j

Trent JD (1996) A review of acquired thermotolerance, heat-shock proteins, and molecular chaperones in archaea. FEMS Microbiol Rev 18:249–258. https://doi.org/10.1111/j.1574-6976.1996.tb00241.x

Tse C, Ma K (2016) Growth and metabolism of Extremophilic microorganisms. Biotechnol Extremophil:1–46. https://doi.org/10.1007/978-3-319-13521-2_1

Tuteja N, Gill SS, Tuteja R (2011) Omics and plant abiotic stress tolerance. Bentham Science Publishers

van den Elzen A, Helena-Bueno K, Brown CR et al (2023) Ribosomal proteins can hold a more accurate record of bacterial thermal adaptation compared to rRNA. Nucleic Acids Res 51:8048–8059. https://doi.org/10.1093/nar/gkad560

Vavitsas K, Glekas PD, Hatzinikolaou DG (2022) Synthetic biology of thermophiles: taking bioengineering to the extremes? Appl Microbiol 2:165–174. https://doi.org/10.3390/applmicrobiol2010011

Verma JP, Jaiswal DK, Krishna R et al (2018) Characterization and screening of thermophilic bacillus strains for developing plant growth promoting consortium from hot spring of Leh and Ladakh Region of India. Front Microbiol 9. https://doi.org/10.3389/fmicb.2018.01293

Vinci G, Cozzolino V, Mazzei P et al (2018) An alternative to mineral phosphorus fertilizers: the combined effects of Trichoderma harzianum and compost on Zea mays, as revealed by 1H NMR and GC-MS metabolomics. PLoS One 13:e0209664. https://doi.org/10.1371/journal.pone.0209664

Wani AK, Akhtar N, Sher F et al (2022) Microbial adaptation to different environmental conditions: molecular perspective of evolved genetic and cellular systems. Arch Microbiol 204:144. https://doi.org/10.1007/s00203-022-02757-5

Wightwick AM, Salzman SA, Reichman SM et al (2009) Inter-regional variability in environmental availability of fungicide derived copper in vineyard soils: an Australian case study. J Agric Food Chem 58:449–457. https://doi.org/10.1021/jf9030647

Wongwilaiwalin S, Rattanachomsri U, Laothanachareon T et al (2010) Analysis of a thermophilic lignocellulose degrading microbial consortium and multi-species lignocellulolytic enzyme system. Enzym Microb Technol 47:283–290. https://doi.org/10.1016/j.enzmictec.2010.07.013

Xia Y, Wang Y, Fang HHP et al (2014) Thermophilic microbial cellulose decomposition and methanogenesis pathways recharacterized by metatranscriptomic and metagenomic analysis. Sci Rep 4. https://doi.org/10.1038/srep06708

Xiao CQ, Chi RA, Li WS, Zheng Y (2011) Biosolubilization of phosphorus from rock phosphate by moderately thermophilic and mesophilic bacteria. Miner Eng 24:956–958. https://doi.org/10.1016/j.mineng.2011.01.008

Yadav AN, Kaur T, Devi R et al (2021) Biodiversity and biotechnological applications of extremophilic microbiomes: current research and future challenges. In: Microbiomes of extrem environments. CRC Press, pp 278–290

Yildirim E, Taylor AG, Spittler TD (2006) Ameliorative effects of biological treatments on growth of squash plants under salt stress. Sci Hortic 111:1–6. https://doi.org/10.1016/j.scienta.2006.08.003

Chapter 16
Biotechnology of Promising Genes from Extremophiles to Produce Stress-Resilient Plants and Microbes for Sustainable Agriculture

Manmeet Kaur, Diksha Singla, Kamal Kapoor, Gautam Chhabra, Sezai Ercisli, Mehmet Ramazan Bozhuyuk, Shiv K. Yadav, and Ravish Choudhary

16.1 Introduction

Extremozymes are enzymes produced by extremophiles. These are microorganisms that need harsh environments to grow. Some of them have a wide range of serious environmental influences. These conditions can be physically harsh in terms of heat, ionizing radiation and pressure, moisture loss, salinity, optimal pH, and oxygen redox potential (Rothschild and Manicinelli 2001). Extremotolerant or extremely

---

The original version of the chapter has been revised. A correction to this chapter can be found at
https://doi.org/10.1007/978-3-031-70203-7_22

M. Kaur
Department of Microbiology, Punjab Agricultural University, Ludhiana, India

D. Singla
Department of Biochemistry, Punjab Agricultural University, Ludhiana, India

K. Kapoor
Apex Institute of Management & Science, Rajasthan Technical University, Jaipur, India

G. Chhabra
School of Agricultural Biotechnology, Punjab Agricultural University, Ludhiana, India

S. Ercisli
Faculty of Agriculture, Ataturk University, HGF Agro, Ata Teknokent, Erzurum, Turkey

M. R. Bozhuyuk
Department of Horticulture, Faculty of Agriculture, Igdir University, Igdir, Turkey

S. K. Yadav · R. Choudhary (✉)
Division of Seed Science and Technology, ICAR-Indian Agricultural Research Institute, New Delhi, India

resistant microorganisms can survive or withstand extreme conditions, such as low water activity, high radiation, and heavy metals, but are unable to reproduce well in these environments (Sghaier et al. 2008). Extremophiles are living things that have unique metabolic processes or physical properties that allow them to survive in different environmental conditions. Extremotolerant and extremo-resistant species exhibit contrasting morphological characteristics and/or metabolic resilience. Two of the most important nutrients for all species are sources of carbon and nitrogen, which are present in varying amounts and forms in their environment. It is incredibly fascinating to understand how these microorganisms use different substrates and metabolic pathways for their survival and proliferation. Microorganisms are the least understood source of genetic and biochemical diversity on the planet. They have a wide variety of physiological variations and adaptations that allow them to survive and thrive in almost any environment (Singh 2021). Improved nutrition of crops and continuous improvement of germplasm are prerequisites for sustained growth in agricultural productivity. The ability to respond to high-stress conditions increase yield potential and stability through stress tolerance. The focus is now on genetically engineering crops to alleviate stress by studying microbiota for their precise role and function. The current focus is on genetically engineering crops to reduce stress by determining the specific role and function of the microbiota. Gene prospecting is a technique of researching and locating desired genes responsible for the above traits, which can then be used to improve crops (Vaishnav et al. 2020). The use of biotechnological technologies to search for candidate genes with increased resistance to combinatorial stress and to promote their use in the development of stress-tolerant food crops would therefore be of enormous benefit to agriculture.

16.2 Growth, Substrates, and Transporters

According to Kazak et al. (2010) monosaccharides, cellulose, disaccharides, peptides, amino acids, hemicelluloses, and polysaccharides are just some of the many substrates on which extremophiles can survive. Various substrates are studied in pure culture to find out which ones can be used to support their growth (Table 16.1). Different solutes are transported into the cell by different transporters for use (Albers and Driessen 2007). The ATP-binding cassette transporter (ABC transporter) is a major class of important transporters found in both bacteria and archaea. Consensus sequences for the ATPase subunit of transporters in both domains include a Q loop, a Walker sequence, and an H region (Higgins 1995). The cytoplasmic ATPase domains of the system are terminated by two permeases and an external binding protein (Albers et al. 2004). According to Schneider (2001), the ABC transporters of bacteria can be divided into two groups: those that transport di- or oligopeptides and those that absorb carbohydrates. Phosphatidyl pyruvate (PEP) is another enzyme that aids in glucose translocation in the sugar phosphotransferase system (PTS), which is absent in archaea (Siebold et al. 2001). Numerous proteins transfer the phosphoryl group of the PEP to the supplied sugar, while PTS burns PEP for fuel. Enzyme I, the histidine protein, and a specific form of enzyme II, also

Table 16.1 Growth of microorganisms under harsh environment

Extremophiles	Substrates	Growth conditions			References
		Temperature (°C)	pH	Generation time (h)	
Hyperthermophiles	Xylose, glucose, yeast extract, glycogen	80–85	6.5–7.5	1.5–2.0	Huber et al. (1986)
Thermophiles	Arabinose, cellobiose, lactose, starch, tryptone	48–75	4.0–9.5	2–5	Bouchotroch et al. (2001)
	Glucose, galactose, dextrose, casein	50–75	3.0–8	1.5–2.0	Huber et al. (2000)
Halophiles	Lactose, trehalose, proline, glutamic acid, gelatin	30–50	6.5–8.0	3–18	Bouchotroch et al. (2001)
	Arabinose, starch, arginine, tryptophan, aspartate	30–45	6.5–9.5	1.5–3	Burns et al. (2007)
Xerophiles	Fructose, sorbitol, gluconate, glucose	20–40	5.5–6.8	2.0–4.0	Marco et al. (2009)
Osmophiles	Fructose, sucrose, glucose	28–35	3.5–5.5	2–12	Restaino et al. (1983)
Acidophiles	Starch, tryptone, yeast extract, peptone	53–95	1–2.2	4–5	Dopson et al. (2004)
	Agar, lactose, starch, glycogen, cellobiose	30–52	1.3–3	2–3	Hallberg et al. (2010)
Psychrophiles	Ribose, glycerol, alanine, glucosamine	5–15	5–10	1–6	Yumoto et al. (2003)
Alkaliphiles	Xylan, formate, trehalose, glycogen	45–60	9–13	3–4	Takai et al. (2001)

known as sugar-specific transport complexes, are the two cytoplasmic proteins that makeup PTS. The histidine protein receives phosphoryl groups from enzyme I, which subsequently carry them to other transport complexes. For example, as glucose travels through the plasma membrane, it is phosphorylated by an enzyme II protein to produce glucose-6-phosphate, which then enters the Embden-Meyerhof (EM) pathway (Bettenbrock et al. 2007). In addition to ABC transporters, secondary transporters are essential for solute transport in archaea (Paulsen et al. 2000). Secondary transporters typically rely on the electrochemical gradient of sodium ions or protons across the cytoplasmic membrane. However, the ABC transporter is not restricted by this and can accumulate substrates in cells in much larger concentrations. The main difference between bacteria and archaea ABC transporters is formed by the substrate-binding proteins (Albers et al. 2004). Quiocho and Ledvina (1996) claim that the bacterial extracellular protein captures the substrate, which is then released and transported across the membrane by permeases. Archaeal-binding proteins, on the other hand, are glycosylated and may potentially protect extracellular proteins from degradation by proteases (Erra-Pujada et al. 2001). These binding proteins have been studied extensively by hyperthermophilic archaea. The binding proteins of both groups have different lengths and compositions and bind to different substrates (Koning et al. 2002). Albers et al. (2004) claim that the ABC transporters in archaea for monosaccharides and most disaccharides resemble bacterial carbohydrate uptake transporters. However, hyperthermophile transporters for various disaccharides and oligosaccharides, including cellobiose, β-glucoside, and cello-oligomer, show sequence similarities to those of the di−/oligopeptide class of transporters in bacteria (Elferink et al. 2001). In addition, some transporter operons known as di−/oligopeptide transporters have been shown to correspond to those for enzymes that degrade sugars, suggesting that they may be involved in catalysing the uptake of di−/oligosaccharides (Nelson et al. 1999).

16.3   Metabolic Pathways for Utilizing Carbohydrates and Adaptation to Extreme Environments

Extremophiles can flourish on many different surfaces. Before polysaccharides may enter the cells, they must first be degraded by hydrolases released by the bacteria. Table 16.2 lists their primary metabolic pathways for metabolizing carbohydrates.

Table 16.2 Enzyme activities and metabolic pathways of extremophiles under stressful conditions

Microorganisms	Enzyme activities	Metabolic pathways	References
Thermophiles	Characterized protease and peptidase, genes	TCA cycle, EMP pathway	Feng et al. (2009)
Alkaliphile	Characterized proteases genes	EMP, PP pathway	Detkova and Kevbrin (2009)
Halophiles	Characterized serine peptidase	TCA cycle	Fine et al. (2006)
Osmophiles	Characterized alkaline protease	TCA cycle	Seo et al. (2005)
Acidophiles	Characterized proteases and peptidase	EMP, TCA cycle	Valdés et al. (2008)
Xerophiles	–	ED and PP pathway	Leong et al. (2014)
Piezophile	Proteases and peptidases up to 46 are detected in the genome	EMP pathway	Lucas et al. (2012)
Hyperthermophile	Characterized proteases	TCA cycle, ED pathway	Kaushik et al. (2002)

16.3.1 Acidophiles and Alkaliphiles

Bacteria that can live at low and very high pH, so-called acidophiles and alkaliphiles, are special because pH affects the acidic or ionization state of amino acids and enzymes. It has been shown that the cell membranes and intracellular environment of extremophiles have homeostatic mechanisms that maintain a constant pH (approximately neutral), thereby creating a stable environment in the cell in which enzymes can act (Jolivet et al. 2004). There are many potential strategies that acidophiles can use to maintain pH balance. One possible way to create a chemiosmotic barrier against proton penetration is to form a barrier on the positive core of the membrane via the reverse reaction of potassium ions with ATPase (Baker-Austin and Dopson 2007). Alkaliphiles rely mainly on four mechanisms to maintain pH homeostasis (Padan et al. 2005). Expression and activity of monovalent cation/proton antiporters, often in cooperation with Na^+/H^+ antiporters, is a pathway still used by bacteria to maintain pH homeostasis. The metabolism of sucrose and amino acids also accelerates and the pH of the cell rises (Richard and Foster 2004). In addition, the ability of ATP synthase to bind H^+ uptake has been improved (Rozen and Belkin 2001). They also alter the availability of H^+ and Na^+ for pH homeostasis and increase cell surface acidity to facilitate cation binding (Wang et al. 2004). For the proper functioning of metabolic enzymes and metabolic pathways, pH homeostasis is controlled by the system of acidophilic and alkalophilic bacteria. The Embden-Meyerhof (EM), Entner-Doudoroff (ED), and pentose phosphate (PP) pathways are present in both acidophilic and alkalophilic bacteria; however, archaea exhibit altered EM and ED pathways. The complete tricarboxylic acid (TCA) cycle and all essential enzymes are present in most bacteria, including acidophilic and alkalophilic bacteria.

16.3.2 Piezophiles

Extremophiles with more areas and carbon areas are barophiles. According to Simonato et al. (2006), high pressure, unlike temperature, changes the membrane fluidity. To avoid this, piezophiles increase the proportion of monounsaturated fatty acids in the lipid bilayer (Valentine and Valentine 2004). Also, this changes the structure of the membrane to withstand the flexibility of high pressure (Attard et al. 2000) and helps regulate ion permeability for bioenergetic purposes (Vossenberg et al. 1995). Reducing transport is a second concern (Simonato et al. 2006). A well-studied example is the tat 2 transporter, an efficient tryptophan permease in the liver that also facilitates tryptophan uptake (Abe 2003). Because piezophiles can survive in hot and cold environments, they produce many heat shock and cold shock proteins. In addition, they produce hydroxybutyrate and its oligomers as intracellular solutes to protect against hydrostatic pressure (Martin et al. 2002). There are many complex polymers in the deep ocean, including several piezoelectricophiles. Piezophiles can degrade polymers using a variety of enzymes that also have effective applications (Simonato et al. 2006).

16.3.3 Psychrophiles

There are two main physical problems at high temperatures: low thermal energy and excessive viscosity. Because the protein is involved in many cellular survival processes, it is a target for modification (D'Amico et al. 2006). Exopolysaccharides, chaperones, antifreeze proteins, trehalose, and cold acclimation/cold shock proteins in psychrophiles contribute to protein folding. They are also important for cryoprotection. Many enzymes involved in important processes, such as translation and transcription, have evolved to work better at low temperatures, making their structures more flexible and reducing the enthalpy interactions that must be affected during catalysis (Violot et al. 2005). The membranes of psychrophiles contain more unsaturated, polyunsaturated, and methyl-branched fatty acids and have shorter acyl chain lengths, which increases the temperature resistance of the membrane (Chintalapati et al. 2004). The resource is used to regulate the activity of metabolic enzymes, a process that allows psychrophiles to reach their potential at high temperatures. In psychrophilic bacteria and archaea, glycolysis and gluconeogenesis processes are weak and most of these organisms use carbon monoxide via EM (Medigue et al. 2005). A challenge in the cold environment is the reduction of enzymes involved in the function of the body, such as the synthesis of ATP and cofactors in important metabolic processes. Studies based on proteomic, transcriptomic, and enzymatic analyses have shown that psychrophiles can overcome this problem by promoting or increasing the activity of important enzymes in important metabolic processes (Amato 2013). To cope with cold stress, *Pseudomonas salinae* also increases the activity of TCA cycle enzymes (Piette et al. 2010). Increased TCA enzyme activity requires the provision of substrates from the TCA cycle intermediate to enhance ATP synthesis and drive catabolic processes. A psychrophilic archaea named *Methanococcoides burtonii* has a short directional TCA oxidation cycle and does not have many genes for conversion of oxaloacetate to 2-oxoglutarate (Goodchild et al. 2004).

16.3.4 Halophiles

Halophiles tolerate the pressure by producing osmolytes in the cells and actively removing salt from the cytoplasm (Roberts 2004). High osmotic pressure results from high-salt concentrations. All halophiles possess transport mechanisms comparable to the Na^+/H^+ antiporters for removing intracellular Na^+ ions. They use various techniques to increase the osmotic pressure in the cytoplasm. The first approach is only used by a few bacteria and archaea due to its disadvantages. It increases the cytoplasmic KCl concentration to a level at least equal to the ambient NaCl concentration to deal with the stress (Pfluger and Muller 2004). The process of enriching organic solutes such as glycerol, sucrose, betaine, ectoine, glycine, and trehalose are among the most commonly used (Imhoff and Rodriguez-Valera 1984). Unlike the first method, this method does not require the internal enzyme system to adapt to the high KCl concentration, hence more energy is required to build up the solute. In addition, halophiles have engineered their enzymes to function efficiently in high-salt conditions. In the main metabolic pathway of halophiles, enzymes are altered to adapt to the environment. For example, the halophilic bacterium *Salinibacter ruber* uses a salt-dependent NADP-linked glucose-6-phosphate dehydrogenase and a salt-inhibited constitutive hexokinase for its glucose metabolism. Fructose-1, 6-bisphosphate aldolase and glucose dehydrogenase activity cannot be identified, therefore the standard ED pathway rather than the EM pathway is used to metabolize glucose (Oren and Mana 2003).

16.3.5 Hyperthermophiles

Hyperthermophilic archaea, including *Pyrococcus*, *Desulfococcus*, and *Archaeococcales*, use modified EM and specific enzymes such as 'glyceraldehyde-3 phosphate dehydrogenase', 'hexose kinase,' 'fructose' phosphate, 'glycerate kinase', 'pyruvate kinase', and 'glyceraldehyde-3-phosphate oxidoreductase' (Bertoldo and Antranikian 2006). Several acidophilic hyperthermophiles, including marine *Thermotoga*, have adapted adaptations of ED and EM (Siebers and Schönheit 2005). Specific enzymes that modulate the EM pathway include 'ADP-dependent kinases' and 'ADP-dependent phosphofructokinases', as well as 'ADP-dependent kinases' that are ATP-dependent on the mesophilic pathway. The free energy change in the hydrolysis of ADP is similar to that in the hydrolysis of ATP. Also, ADP is thought to be more stable than ATP at higher temperatures, especially in the presence of different specific metals (Hongo et al. 2006).

16.3.6 Osmophiles and Xerophiles

According to Pfluger and Muller (2004), osmoregulation systems are present in microorganisms that can survive and function properly when exposed to solutions with high solute concentrations. Osmophilic yeasts are abundant in the environment and follow similar central metabolic pathways as yeasts, progressing via the EM, ED, and PP pathways to produce pyruvate, which enters the TCA cycle. Nevoigt and Stahl (1997) claim that glycerol is essential for osmoregulation. The rate of glycerol synthesis is significantly enhanced by an increase in cytoplasmic glycerol-3-phosphate dehydrogenase activity in response to a decrease in extracellular water activity (Pahlman et al. 2001). Fps1p channel is also turned off to maintain osmotic balance with the environment (Tamas et al. 1999). Trehalose is another osmoregulation favourable solute (De Smet et al. 2000). The EM pathway is used by xerophilic fungi to metabolize carbohydrates. In a serious example, all genes for the enzymes that catalyse the synthesis of secondary metabolites are lost in *Xeromyces bisporus* (Leong et al. 2014).

16.3.7 Thermophiles

Thermophiles and hyperthermophiles must be able to withstand the high temperatures required to produce protein (Jaenicke and Bohm 1998). In addition, the method
is dependent on the history of extremophiles. Two different methods can be used to
stabilize proteins and prevent their denaturation at high temperatures. The first process changes the protein's structure to be more adaptive than its mesophilic counterparts. It is found in organisms that thrive in very hot environments, including
thermophiles and hyperthermophiles. The second mechanism is a sequence. These
structures do not differ from their mesophilic homologues due to changes that form
some strong bonds for high thermal stability. Chaperones or heat shock proteins are
another class of proteins that cannot withstand cold environments (Feder and
Hofmann 1999). Chaperones also prevent the aggregation of temperature-insensitive
proteins and redirect misfolded and denatured proteins to the cellular protein degradation process. The most common companion is the Hsp60 complex, also known as
the thermosome (Sterner and Liebl 2001). At low temperatures, most thermophilic
bacteria use the EM pathway, which includes all enzymes of this pathway (Selig
et al. 1997). Important enzymes 'phosphogluconate dehydratase' and '2-keto-3-
deoxy-6-phosphogluconate aldolase' have been found in many thermophilic bacteria, including *Geobacillus thermoglucosidasius* (Tang et al. 2009). In addition, it
has been shown that *G. thermoglucosidasius* uses the PP pathway differently
depending on the oxygen available. During glucose fermentation, the oxidative
componentofthePPsignallingpathwayislessactiveasthe6-phosphogluconolactonase
gene is absent (Tang et al. 2009).

16.4 Quorum-Sensing System Prevalent in Extremophiles

The quorum-sensing (QS) system in extremophiles and their role in extreme environments has been described below:

16.4.1 Bacteria Thriving at Extreme Temperature

(a) *Thermophile*:
  Thermophilic *Caminibacter media atlanticus* and mesophilic *Sulfurovum litho-
  tropicum* from deep-sea hydrothermal vents both produced bioluminescence in
  the *Vibrio harveyi* AI 2 assay, providing evidence that the *Proteobacterial* epsi-
  lon line of the LuxS enzyme originates from high-temperature geothermal.
  LuxS has been found to be involved in the formation of deep ocean hydrothermal vent biofilms. One hypothesis is that the reduction of sulfur at low temperatures is not limiting, making de novo methionine synthesis an endeavour.

Therefore, the contribution of LuxS to the active methyl cycle is not large compared to other sources. Therefore, the presence of Lux S in thermophilic organisms from deep-sea hydrothermal vents may lead to AI-2 formation rather than methionine recycling. Based on non-redundant data analysis, the LuxS gene was found in members of the *Deinococcus thermus* phylum with LuxS protein, including *Thermus thermophilus* HB27, *Truepera radiovictrix* DSM 17.093, *Oceanithermus profundus* DSM 14977 and *Meiothermus silvanus* DSM 46 DSM. The research model of *Truepera radiovictrix* representing this phylum shows that the N-terminal region of the protein appears different from other LuxS sequences of mesophilic representatives (Rao et al. 2016).

(b) *Psychrophiles*

Bacteria can live in environments that can withstand temperatures of up to 15 °C. The cyclic diguanylate system is known to regulate genes involved in adhesion, biofilm formation, and motility. Regulators of the DiGMP cyclic signalling system have been identified in *Chromophila ingrahami*, which may play a role in the production of exopolysaccharides that reduce the freezing of cell surfaces (Riley et al. 2008). In addition, the biofilm control regulator HapR has an ortholog in *Vibrio cholerae*. The large cell rapid regulator HapR controls virulence and nucleo-sensing-dependent biofilm formation in *V. cholerae* and is a homologue of LuxR in the body. Papenfort et al. (2017) examined liquid cultures of *V. cholerae* hapR mutants lacking CAI-1 and AI-2 synthesis to demonstrate the importance of hapR in QS. The QS mechanism was determined to be unclear because only hapR was detected in *Chromomonas* and no autoinducible products were detected. The bioluminescence of the psychrophilic *Aliivibrio logei* is controlled by the AI-1 type QS with two copies of the LuxR gene (luxR1 and luxR2). Autoinducer should activate LuxR1 100 times more than LuxR2. Lux R2 is a substrate for Lon protease and is required for folding, but LuxR1 does not require GroEL/ES chaperones or is degraded by Lon. In addition, GroEL/ES and Lon of luxR1 and luxR2 products do not activate the PR promoter of the *A. logei* lux operon in *E. coli*. This combination activates LuxR2 at high cellular levels even in the absence of GroEL/ES (Khrulnova et al. 2016).

16.4.2 Bacteria Thriving at Extreme pH

(a) *Acidophiles*

Eleno reductase (ER) is a flavin-dependent oxidoreductase that has been reported to be present in many acidophilic bacteria and may be involved in QS-mediated oxidative stress in these bacteria (Toogood et al. 2010). The role of QS in *Acidithrix* str. and the strain of *Acidiphilum*, two organisms coexisting from iron-rich aggregates. Counting the cell-free supernatant from *Acidiphilum* increases the rate at which *Acidithrix* oxidizes nucleic acids and Fe(II). The polar flagella of *Acidiphilum* inhibits cell aggregation. However, when the cell-free supernatant of *Acidithrix* was added to *Acidiphilum*, macroscopic aggregates were formed (Mori et al. 2017). QS molecules need further study because these organisms can produce them through chemical communication. Another bacterium thought to be able to live in the acidic environment of the stomach is *H. pylori*. Iron oxide indicates that QS regulates the *N*-acyl-L-homoserine lactone (AHL) synthase afeI but not the transcriptional regulator afeR. In addition, biofilm formation was found to affect 42.5% of the QS network, which accounts for 4.5% of the *Ferrooxidans* ATCC 23270 T genome (Mamani et al. 2016). Therefore, it can be assumed that QS regulates the biofilm process of these bacteria. Disease pathogenesis also relies on the QS system in complex conditions such as gastric acidification. *V. cholerae* survive in an acidic environment by forming a thick, sticky membrane. This is done by the QS regulator 'Hap R', which inhibits the expression of the *Vibrio* polysaccharide operon. Once the bacteria leave the acidic environment of the stomach, biofilm protection is no longer required, causing hapR synthesis to resume and the biofilm to change shape (March and Bentley 2004).

(b) *Alkaliphiles*

Haloalkaliophilic archaeal occult *Nadococcus* has been reported to produce cellular proteases during late exponential and steady-state growth phases. When low cell culture is subtracted from late exponential culture, the production of cellular proteases changes to early exponential, suggesting that the increase in cell number during the exponential phase may be due to QS. The presence of AI-1 molecules (N-3-oxoctanoyl and N-3-oxoctanoyl homoserine lactones) in this archaea was confirmed using *Agrobacterium* biosensor strains. Ramanathan and Ting (2015) demonstrated the presence of QS in pure cultures of four alkaliphilic bacteria (i.e. '*Alcalibacterium*' sp., '*Agrobacterium aureus*', '*Bacillus foraminis*', and '*Alkalibacter pelagium*') for the copper bioleach. They speculated that these basophilic bacteria could use QS for coordinated biological leaching.

16.4.3   Bacteria Resistant to DNA-Damaging Agents

(a) *Radiation/ oxidative stress-resistant bacteria*

The QS mechanism of *Deinococcus radiodurans* is well understood. This resistance is due to its incredible ability to repair DNA damage and have four to ten copies of the genome. *D. radiodurans* is one of the organisms that communicate through the AI-1 and AI-2 systems. AI-1-mediated QS is essential for oxidative stress mediated by DqsI/DqsR regulatory mechanisms. In the absence of stress, the quencher group (QqaR and QqlR) turned off the DqsI/DqsR regulation process. The study also found that AHL levels were lower in non-stressful conditions, suggesting that oxidative stress is the only factor they affect. Unlike *E. coli*, where extracellular death factor (EDF) or QS factor is not required to activate MazF-mediated cell death, AHL is activated in *D. radiodurans*, thereby producing the death phase of oxidative stress (Lin et al. 2016).

16.4.4 Other Extremophiles

(a) *Halophiles*

Most of the halophiles can survive in saline areas with salinity between 3.4 and 5.1 M. A chlorine-dependent homologue of Lux S was found in the small halophilic bacterium *Halobacillus halophilus* (Sewald et al. 2007). Transcription of Lux S was salt and chloride-dependent when the highest mRNA concentration in the growth medium was 2.0 M NaCl. However, neither the occurrence of AI-2 nor the activity of QS was confirmed. Members of the genus *Halanaerobium* commonly found in oil reservoirs have been found to have a QS mechanism. It can also be detected in water used for hydraulic fracturing (Monzon et al. 2016). To determine the presence of AI-1 type QS in 43 members of the Halomonadaceae family, Tahrioui et al. (2013) used an AHL detection method based on the *Chromobacterium violaceum* strain CV026 and *Agrobacterium tumefaciens* NTL4 (pZLR4). TLC analysis showed the presence of '*N*-octanoyl-homoserine lactone' and '*N*-hexanoyl-L-homoserine lactone'. In addition, the AI-1 synthase gene was analysed using PCR primers targeting the Lux I active site. Lux I-like sequences were identified in 29 species, but not in 14 other species.

(b) *Piezophiles*

Piezophiles or barophiles are microorganisms that can survive in high-pressure conditions. The deep sea is characterized by high hydrostatic pressure, ice, darkness, and a lack of biological life. *Photobacterium profundum* belongs to the Vibrionaceae family together with the bacteria *Vibrio harveyi* and *Aliivibrio fischeri*. Despite belonging to the same Vibrionaceae family, the genomes of *P. profundum* and *V. harveyi* have different quorum-sensing genes. Rezzonico and Duffy (2008) attempted comparative genomic studies to search for the AI-2 signalling system in *P. profundum* and found that while the LuxS homologue is present, the Lux-P homologue (AI-2 receptor) is rare in Vibrionales and has the QS type AI-2, and no alternative receptor for AI-2 has been described in Vibrionales bacteria. Therefore, it appears that Lux S exerts only a metabolic function in *P. profundum*, suggesting that *P. profundum* may utilize some as-yet-to-be-discovered QS processes (Montgomery et al. 2013).

16.5 Protein Adaptations in Extremophiles

According to Basak et al. (2020), extremophiles require certain survival mechanisms, such as genetic changes, which lead to further alterations in protein sequence and structure.

16.5.1   Halophiles

Salt is an important factor influencing protein solubility, stability, and structure, and halophiles are highly adapted to survive in low salinity (Mokashe et al. 2018). By inhibiting the entry of inorganic salts and the synthesis of organic osmotic agents, the osmotic pressure in the cell remains constant (Raval et al. 2018). When salt and water have a certain salinity (> 0.1 M), the amount of water contained in the protein decreases (Basak et al. 2020). Aggregates result from enhanced interactions between hydrophobic amino acids due to dehydration. Halotropic proteins such as the P45 protein (which inhibits denaturation and resists malate dehydrogenase inactivation), transcription-binding protein (TBP), and TATA box protein are specific proteins found under normal conditions and with relatively large salt bridges. However, binding proteins enhance DNA interactions (Kumar et al. 2018).

16.5.2   Psychrophiles

Cold shock protein (CSP) and cold adaptation protein (CAP) prove that psychrophiles have the necessary adaptive capacity to withstand the cold environment. CAP is overexpressed after a severe shock of up to 4 °C, while CSP is only expressed when the environment is appropriate. When valine is converted to alanine, lysine to arginine, and alanine to glutamate, the molecular structure of the protein changes (Basak et al. 2020). Enzymes are stable and selective at low temperatures because their nuclei do not contain much protein (Kochhar et al. 2022).

16.5.3   Acidophiles

Protonation reduces membrane permeability while maintaining the proton gradient across it by altering the charge of polar residues and proteins in acidic environments. Acidophils have cytoplasmic buffering, which keeps the intracellular pH at a neutral level. By shrinking the membrane hole, acidophiles lessen membrane permeability (Basak et al. 2020). *Thiobacillus ferrooxidans* has an unusually wide outer loop that decreases pore size and ion selectivity (Kumar et al. 2018).

16.5.4   Thermophiles

Protein folding irreversibly caused by extremely high temperatures exposes the hydrophobic centres and causes aggregation. To stabilize these, huge hydrophobic cores, a larger number of disulfide bonds, surface charges, and salt bridges are found in thermophilic and hyperthermophilic proteins (Basak et al. 2020).

16.5.5 Piezophiles

Piezophilic proteins form protein multimers via hydrogen bonding between their protein subunits, have hydrophobic cores, and contain fewer amino acids (Basak et al. 2020). Additionally, they have fewer proline and glycine residues, which weaken helixes, constrict conformational space, and lessen protein flexibility. By building up trace levels of the organic osmolyte mannosylglycerate at room temperature, *Thermococcus barophilus* lowers the hydration layer surrounding the protein (Brininger et al. 2018). In reaction to temperature and pressure, bacteria at deep-sea hydrothermal vents have a pressure-sensitive operon system that controls their growth (Basak et al. 2020).

16.5.6 Alkaliphiles

Phosphoserine aminotransferase is an enzyme that can form a homodimer is dependent on vitamin B6 and is found in alkaliphiles (Kumar et al. 2018). Their negatively charged amino acid residues, stronger hydrogen bonds, and improved hydrophobic contacts at the dimer interface facilitate their stability and activity under extremely alkaline circumstances (Basak et al. 2020).

16.6 Functional Characterization of Genes

As a result, several approaches to explaining gene function have been created. These techniques, when combined with conventional biochemical and physiological experiments, have been effective in identifying the relevance of various signalling pathways and mapping stress signalling networks, providing researchers with a thorough systems biology perspective on plant stress response mechanisms. Below is a description of how some of these methods can be used to investigate the role of stress-related genes or Quantitative Trait Locus (QTLs), and Table 16.3 shows a large number of genes involved in the stress response.

16.6.1 RNA-mediated Gene Silencing

RNA-mediated gene silencing is one of the most widely used techniques for figuring out how genes function. A gene's low expression occurs when it is silenced. Short RNAs (sRNAs), which are split into two classes: small interfering RNAs (siRNAs) and microRNAs (miRNAs), are frequently used to do this through the targeted inhibition of transcript accumulation (Eamens et al. 2008). To characterize the

Table 16.3 Genes involved in stress response

Genes	Function	Reference
SOS3	Calcium sensing	Gong et al. (2004)
NAC transcription factors	ABA-dependent and independent signalling pathways	Nakashima et al. (2012)
AtOSCA 1	Increase upon osmotic stress imposition while utilizing calcium	Yuan et al. (2014)
Calcium-dependent protein kinases	Mediate salinity through protein phosphorylation	Schulz et al. (2013)
The NHX-type cation	Generate pH gradients	Reguera et al. (2014)
SOS1	Na^+/H^+ antiporter for NA^+ efflux	Brini and Masmoudi (2012)
Leucine basic zipper transcription factors	ABA-dependent signalling in responses to stress conditions	Uno et al. (2000)
Histidine kinases	Function as receptors for cytokinin and ethylene	Tran et al. (2007)

function of genes, silence based on the transgenic production of sRNA has been employed extensively. RNA-mediated gene silencing is one of the most popular methods for studying how genes function. When a gene is silenced, its expression is minimal. Short RNAs (sRNAs), which can be separated into two types by targeting the accumulation of transcripts, are frequently utilized for this purpose. These classes include microRNAs (miRNAs) and small interfering RNAs (siRNAs) (Eamens et al. 2008). Gene function has been extensively studied using silence based on the transgenic generation of sRNA. Researchers demonstrated that 'OsNAC5' positively modulates the response to abiotic stress in rice through the degradation of 'OsNAC5' by RNA interference (RNAi). The study demonstrated that 'RNAi'' silencing of the 'SOS1' gene from *Thellungiella salsuginea* improved salt sensitivity and eliminated halophytes in salt cress transgenes, supporting the significance of 'ThSOS1' as a critical regulator of halophyte salt tolerance. 'RNAi' is thus a practical method for learning how genes function in plants (Song et al. 2011).

16.6.2 Mutagenesis

The easiest way to determine the function of a gene is to examine the phenotypic changes that result from the inactivation of the gene in the plant (Bouche and Bouchez 2001). Therefore, knockdown and loss-of-function mutants are useful tools for characterizing how genes function. There are three categories of mutagens: chemical, physical, and biological. While minor insertions/deletions and point mutations can be induced by chemical and physical mutagens, disruption of gene function can be achieved by the insertion of large pieces of DNA such as T-DNA and transposable elements (Bolle et al. 2011). Chemical and physical mutagenesis has the disadvantage of being random events that can occur anywhere in the genome and often. There is no mechanism to direct the mutational event to target our target gene. However, the discovery of Targeting Induced Local Lesions IN Genome (TILLING) technology has led to a resurgence in the use of chemically induced mutagenesis to study gene function. The detection of SNPs and mutations in the target gene in chemically engineered mutants can be performed with high throughput and low cost using TILLING (Colbert et al. 2001). A faster method to find the desired mutant than chemical or physical mutagenesis is insertional mutagenesis using T-DNA and/or transposable elements. Since the inserted sequences are known, methods based on PCR and cloning can be used to identify the region around the inserts. Wang (2008) studied the efficiency of insertional mutagenesis for gene knockout and his results show that there is a 90% knockout when the insertion is in the protein-coding region and a 25% knockout when the insertion is in the protein-coding region occurs before the start codon. Hundreds of thousands of insertion mutations have been generated in *Arabidopsis* to fully saturate the genome. The *Arabidopsis* Biological Resource Center (ABRC) has insertion mutants of most of the known *Arabidopsis* genes. The Tos17 retrotransposon has also been used to perform large-scale insertional mutagenesis in rice, and the mutant lines can be purchased online through the Rice Tos17 Insertion Mutant Database (Nongpiur et al. 2016).

16.6.3 Gain of Functional Lines

Gene overexpression to produce a gain-of-function phenotype is one of the most widely used methods to determine gene activity in plants. The first method uses randomly inserted transcriptional enhancers to activate the endogenous gene, while the second method uses mutations to stimulate the expression of the transgene. The effect of overexpression becomes clear when measuring the activity of some genes that are members of poorly functioning gene families. This technique is widely used to characterize the function of genes, and genes can also be produced heterologously in yeast. The activity of many genes, particularly those involved in the response to abiotic stress, is characterized by ectopic overexpression of cDNA under the control of the 'CaMV 35S' promoter (Orellana et al. 2010). In addition, transgenic methods in model organisms can be used to measure gene function in plants. Therefore, the gain-of-function line is a crucial tool for identifying genes in species that are well-developed and lack scientific tools.

16.7 Genome Editing

The targeted mutation of genomes is called genome editing. The approach uses cellular DNA repair pathways and specially designed DNA cleavage reagents (Orellana et al. 2010). The reagents are often designed nucleases that cleave target DNA at user-specified sites and then repair the double-strand breaks by either 'non-homologous end-joining (NHEJ)' or 'homologous recombination (HR)'. Manipulation of these DNA repair processes produces specific DNA changes, including 'deletions,' 'insertions,' and 'donor cassette insertions'. According to Curtin et al. (2012), there are four different types of engineered nucleases used to alter the plant genome.

16.7.1 The Clustered Regularly Short Interspaced Palindromic Repeats/CRISPR-Associated Sequence (CRISPR/Cas) RNA-Guided Nucleases

A recent addition to the nuclease repertoire for genome engineering purposes is the 'CRISPR/Cas' RNA-guided nuclease obtained by modifying immune system proteins found in bacteria and archaea (Sorek et al. 2013). All classes (I-III) of CRISPR/Cas contain the 'Cas nuclease group'; a 'non-coding RNA'; and a set of 'direct repeats' (Ran et al. 2013). Cleavage of complex invasive nucleic acids such as viral DNA by RNA-guided sequence-specific cleavage (Makarova et al. 2011). Take the foreign DNA entering the body, called the spacer, between the two regions near the end of the CRISPR, generating CRISPR RNA (crRNA) from the spacer containing the CRISPR template. It induces interference on the target in the crRNA direction by recognizing and separating protospacer sequences in foreign DNA (Sorek et al. 2013). Target DNA also requires a 'protospacer adjacent motif (PAM)' for cleavage to occur. Genome engineering currently uses *Streptococcus pyogenes* type II CRISPR/Cas9 technology. There are three different versions of the CRISPR/Cas system. Two short RNAs, transacting crRNA (tracrRNA), and complementary crRNA have been shown to direct Cas9 nuclease to its target sequence. They also showed that Cas9 DNA cleavage can be performed efficiently by a chimeric RNA (sgRNA) containing tracrRNA and crRNA, and target selectivity can be altered by changing the crRNA as little as 20 nucleotides (Jinek et al. 2012).

16.7.2 Zinc Finger Nucleases (ZFNs)

ZFNs are chimeric proteins formed by linking the cleavage domain of the 'Fok I' restriction enzyme with a specially designed 'Cys2-His2' zinc finger domain (DNA binding) (Bortesi and Fischer 2014). Individual ZFN monomers are engineered to surround a 56 bp spacer region to achieve site-specific cleavage at this spacer sequence. The only way to achieve this is through ZFN heterodimerization. Each of the three to six fingers that make up a zinc finger domain can recognize DNA sequences up to three base pairs (bp) in length. Multiple fingers can be arranged in groups of three to six fingers to recognize any desired sequence. As a result, a ZFN heterodimer, which consists of two DNA-binding domains, each with four fingers, recognizes a sequence of approximately 24 bp that can only occur once in the genome of an organism. ZFNs are therefore very strong and specific for targeted genome modification (Urnov et al. 2010).

16.7.3 The Transcription Activator-like Effector Nucleases (TALENs)

The 'Fok I' endonuclease is part of the chimeric proteins known as 'TALENs', which also contain a variety of DNA-binding elements. The DNA-binding elements of highly conserved amino acid repeats in this case are variants of 'transcription activator-like effectors (TALEs)' produced by the bacterial plant pathogen *Xanthomonas* spp. to be produced. After infection, TALEs are released into host cells via a type III secretory pathway, where they bind to host DNA and regulate the expression of host genes (Boch and Bonas 2010). The fact that 'TALENs' are specifically designed to recognize the target DNA sequence is the only difference between 'TALENs' and 'ZFNs' technologies. TALENs typically have 1620 tandem repeats of approximately 34 amino acids in their DNA-binding domain (Curtin et al. 2012). TALENs have only recently been used in plant studies. Some successful applications of TALENs for gene replacement and gene regulation in plants include the development of disease-resistant rice by disrupting the 'Os11N3' promoter or the targeted in-frame insertion of the 'YFP' gene into the 'Sur A' and 'Sur B' genes in tobacco protoplasts (Zhang et al. 2013). Therefore, TALENs could be an extremely useful tool to study gene function in plants.

16.7.4 Meganucleases

Meganucleases, often referred to as homing endonucleases, are the third class of DNA-cleaving restriction enzymes used for genome modification. According to Puchta and Fauser (2014), meganucleases were the first double-strand break-producing nucleases used in genetic engineering. According to their structural and sequence motifs, meganucleases can be divided into five groups: 'LAGLIDADG', 'GIY-YIG', 'HNH', 'His-Cys Box', and 'PD-(D/E) XK'. The class of meganucleases that has attracted the most attention is the 'LAGLIDADG' homing endonucleases (LHEs). LHEs can exist in two different functional states: homodimeric forms and single peptides, which consist of two repeats of a monomer joined together by a linker. LHEs are highly sequence-specific and bind to 20–30 bp DNA sequences. However, their use is limited because modifications to their binding domain intended to make them more specific often undermine their ability to function as a nuclease. This makes the development of tailored meganucleases more challenging and time-consuming than the development of ZFNs and TALENs. Meganucleases have therefore not been widely used to modify the plant genome (Curtin et al. 2012).

16.8 Extremophile Alleles for Stress Tolerance

16.8.1 Sodium Proton Antiporters

Transgenic tobacco plants that overexpressed the 'SbSOS1' gene of the extreme halophyte *Salicornia brachiata* exhibited better salt tolerance than wild-type plants (Yadav et al. 2012). Plants overexpressing the halophyte SsNHX1 gene continued to grow normally after being exposed to 200 mM NaCl, despite the fact that identical findings have previously been seen in Arabidopsis plants using 'AtNHX1' or 'AtSOS1' overexpressing plants (Shi et al. 2003). However, 'AtNHX1' or 'AtSOS1' overexpressors could only endure NaCl concentrations up to 200 mM, demonstrating weak salt resistance capabilities (Pehlivan et al. 2016). Therefore, it is anticipated that these halophyte genes will function under stress better than their glycophyte equivalents.

16.8.2 MBF1c

'*Polytrichastrum alpinum*' is a species of polar moss that can withstand the harsh Antarctic climate. The multiprotein bridging factor '1c (PaMBF1c)', an extremophile with a stress-responsive transcription coactivator gene, to *Arabidopsis* 'MBF1c (AtMBF1c)'. The overexpressing 'PaMBF1c' lines had better resistance to salt (NaCl) and ionic (LiCl) stressors compared to the overexpressing WT and 'AtMBF1c' lines at both the germination and seedling stages. In other words, overexpression of 'PaMBF1c' conferred tolerance to a variety of stressors such as heat, salt, and ionic stress, but overexpression of 'AtMBF1c' did not. By maximizing amino acid residues for multifunctional activity, 'PaMBF1c' likely evolved to help *P. alpinum* adapt to the harsh salt conditions of Antarctica.

16.8.3 Galactinol Synthase 1

According to Ge et al. (2005), the arid shrub *Ammopiptanthus nanus* can survive extreme heat. The galactinol biosynthetic gene 'galactinol synthase 1' (GolS1), which is linked to stress tolerance, has been demonstrated to be cold-inducible in *A. nanus* (Selvaraj et al. 2017). The overexpression of *A. nanus* 'GolS1' in tomatoes promoted the development of cold tolerance by increasing the amount of galactinol in early leaves under cold stress, activating ethylene signalling, and producing ethylene response factors (ERFs). Only 'SlGolS2', though at low levels, was cold-inducible of the four tomato GolS molecules ('SlGolS1', 'SlGolS2', 'SlGolS3', and 'SlGolS4') that were functionally identical to AnGolS1. 'SlGolS2' has less catalytic activity than 'AnGolS1'. 'AnGolS1' probably has stronger resistance to cold stress than 'SlGolS2' due to its characteristics, such as its higher inducibility and catalytic activity (Liu et al. 2020).

16.8.4 HKT1

The high-affinity K^+ transporter is crucial for regulating Na^+ levels in the shoot and maintaining the equilibrium between Na^+ and K^+ ions during salt stress. The ice plant is an extremophile that is indigenous to the Namibian desert in southern Africa and is frequently referred to as *Mesembryanthemum crystallinum*. It can resist concentrations of up to 0.5 M NaCl due to its excellent salt stress tolerance (Bohnert and Cushman 2000). Salt sensitivity is exhibited by Arabidopsis when 'AtHKT1' is overexpressed (Ali et al. 2018). However, excessive 'McMKT2' expression in *Arabidopsis* led to a startlingly high-salt tolerance. Therefore, it is believed that these functional differences brought on by the halophyte M. crystallinum's overexpression of 'McMKT2' are due to the protein's intrinsic variants. In fact, 'HKT1' orthologs with spontaneous mutations from maize and soybeans boosted salt tolerance in tobacco plants when overexpressed (Ren et al. 2015).

16.9 Applications of Extremophiles in Agriculture

Extremophiles act as bio-fertilizers, bio-vaccines, and biocontrol agents, protecting crop development and agricultural yield in regions with harsh climatic conditions such as high salinity, low temperatures, and arid conditions (Yadav and Saxena 2018).

16.9.1 Biocontrol Strategies

Erwinia carotovora must have acylated homoserine lactones (AHLs) to be harmful to tobacco plants. The ability of transgenic tobacco plants to produce AHLs allows them to trap bacteria with low cell densities into virulent forms, making the bacteria vulnerable to the tobacco plant's defence mechanisms. This type of biocontrol technique can be applied when the AHLs produced by aggressive bacteria are detected (Mae et al. 2001). By employing their autoinducers in transgenic plants, which can also confer resistance to plants, extremophiles can use autoinducer molecules in the development of biocontrol techniques.

16.9.2   Biosurfactants

Insecticides that support soil bioremediation can be made by synthesizing biosurfactants in place of chemical surfactants. They boost plant defences and have an antibacterial impact. Rhamnolipids can be utilized to suppress *Phytophthora* zoospores. They allow for the expansion of sustainable agriculture in arid locations by decreasing water infiltration into soils that are already hydrophilized (Markande et al. 2021).

16.9.3   Biocontrol Agents

Extremophiles have particular gene expression that is being investigated for a range of biotechnological and industrial uses that enable them to endure in challenging settings. These microbes can be employed to treat biological diseases (Mehetre et al. 2021). By creating siderophores (compounds that chelate iron), chitinases, ammonia, hydrocyanic acid, and a variety of secondary metabolites, rhizobacteria shield plants from diseases (Pandey et al. 2021). By interfering with the nematodes' reproductive cycles and making them feed, these biocontrol drugs prevent the spread of diseases and a variety of worms. Bacteria that shield plants against disease include '*Bacillus*', '*Microbacterium*', '*Pseudomonas*', and '*Clavibacter*' (Verma et al. 2017).

16.9.4   Bio-fertilizers and Bioinoculants

Microorganisms enable nutrient cycling, nutrient fixation, mineralization, and solubilization in addition to acting as bio-fertilizers, bio-vaccines, and alternatives to conventional agricultural practices. They can be utilized as biocontrol agents and can promote resistance (Tiwari et al. 2019). Their genetic variety can be leveraged to support affordable, sustainable agriculture, and chemical alternatives in the agro-industrial sector (Chakraborty and Akhtar 2021). Due to their ability to solubilize nutrients, fix nitrogen, produce phytohormones, and make siderophores, psychrophilic extremophiles are employed as bio-vaccines to improve low-temperature plant development and infection resistance (Yadav et al. 2017). Due to their abilities to promote plant development, acidophilic extremophiles like '*Azotobacter*', '*Bacillus*', '*Flavobacterium*', and '*Pseudomonas*' are used as bioinoculants and biocontrol agents in acidic soils. Using phosphorus- and drought-tolerant extremophiles for dryland agriculture could help provide food security for the rapidly expanding global population (Verma et al. 2017). Under salt stress, halophilic extremophiles enhance seedling growth, root and shoot length, biomass, yield, and chlorophyll content. According to Yadav and Saxena (2018), *Haloferax alexandrinus* enhances phosphorus content in hypersaline soils and exhibits phosphorus solubilization.

16.10 Conclusion

Extremophiles are inherently unique in that they live in hostile conditions and can utilize one or more of the EM, ED, PP, and TCA cycles. Most physiological characterization studies have traditionally relied on genes from model plants or modern cultivars, either through mutant or overexpression-based approaches, to elucidate gene function. Recent findings from various studies on the utilization of genetic resources from progenitors and extremophiles of wild plants showed that many of them increased plants' stress resistance more effectively than alleles from model plants. To better understand the genetic variations that occur naturally in wild ancestors who have adapted to harsh environments, we would need to accelerate our research. Understanding how they resist stress can help us plan and design breeding and biotechnology programmes to produce plants with a range of stress tolerances.

References

Abe F (2003) The role of tryptophan permease Tat2 in cell growth of yeast under high-pressure condition. In: Winter R (ed) Advances in high pressure bioscience and biotechnology II. Springer, Berlin, pp 271–274

Albers SV, Driessen AJ (2007) Membranes and transport proteins of thermophilic microorganisms. In: Robb F, Antranikian G (eds) Thermophiles: biology and technology at high temperatures. CRC Press, New York, pp 39–54

Albers SV, Koning SM, Konings WN, Driessen AJ (2004) Insights into ABC transport in archaea. J Bioenerg Biomembr 36:5–15

Ali A, Khan IU, Jan M, Khan HA, Hussain S, Nisar M, Chung WS, Yun DJ (2018) The high-affinity potassium transporter EpHKT1; 2 from the extremophile *Eutrema parvula* mediates salt tolerance. Front Plant Sci 9:1108

Amato P (2013) Energy metabolism in low-temperature and frozen conditions in cold-adapted microorganism. In: Yumoto I (ed) Cold-adapted microorganisms. Caister Academic Press, Norfolk, pp 71–96

Attard GS, Templer RH, Smith WS, Hunt AN, Jackowski S (2000) Modulation of CTP: phosphocholine cytidylyltransferase by membrane curvature elastic stress. Proc Natl Acad Sci 97:32–36

Baker-Austin C, Dopson M (2007) Life in acid: pH homeostasis in acidophiles. Trends Microbiol 15(4):165–171

Basak P, Biswas A, Bhattacharyya M (2020) Exploration of extremophiles genomes through gene study for hidden biotechnological and future potential. In: Physiological and biotechnological aspects of extremophiles. Academic Press, pp 315–325

Bertoldo C, Antranikian G (2006) The order Thermococcales. In: Dworkin M, Falkow S, Rosenberg E, Schleifer K, Stackebrandt E (eds) The Prokaryotes. Springer, New York, pp 69–81

Bettenbrock K, Sauter T, Jahreis K, Kremling A, Lengeler JW, Gilles ED (2007) Correlation between growth rates, EIIACrr phosphorylation, and intracellular cyclic AMP levels in *Escherichia coli* K-12. J Bacteriol 189:6891–6900

Boch J, Bonas U (2010) *Xanthomonas* AvrBs3 family-type III effectors: discovery and function. Annu Rev Phytopathol 48:419–436

Bohnert HJ, Cushman JC (2000) The ice plant cometh: lessons in abiotic stress tolerance. J Plant Growth Regul 19:334–346

Bolle C, Schneider A, Leister D (2011) Perspectives on systematic analyses of gene function in *Arabidopsis thaliana*: New Tools, Topics and Trends. Curr Genomics 12:1–14

Bortesi L, Fischer R (2014) The CRISPR/Cas9 system for plant genome editing and beyond. Biotechnol Adv 33:41–52

Bouche N, Bouchez D (2001) *Arabidopsis* gene knockout: phenotypes wanted. Curr Opin Plant Biol 4:111–117

Bouchotroch S, Quesada E, del Moral A, Llamas I, Bejar V (2001) *Halomonas maura* sp. nov., a novel moderately halophilic, exopolysaccharide-producing bacterium. Int J Syst Evol Microbiol 51:25–32

Brini F, Masmoudi K (2012) Ion transporters and abiotic stress tolerance in plants. ISRN Mol Biol 20:927436

Brininger C, Spradlin S, Cobani L, Evilia C (2018) The more adaptive to change, the more likely you are to survive: protein adaptation in extremophiles. In: Seminars in cell & developmental biology. Academic Press. Bruins, pp 158–169

Burns DG, Janssen PH, Itoh T, Kamekura M, Li Z, Jensen G, Rodríguez-Valera F, Bolhuis H, Dyall-Smith ML (2007) Haloquadratum walsbyi gen. nov., sp. nov., the square haloarchaeon of Walsby, isolated from saltern crystallizers in Australia and Spain. Int J Syst Evol Microbiol 57:87–92

Chakraborty T, Akhtar N (2021) Biofertilizers: prospects and challenges for future. In: Biofertilizers: study and impact, pp 575–590

Chintalapati S, Kiran M, Shivaji S (2004) Role of membrane lipid fatty acids in cold adaptation. Cell Mol Biol 50:31–42

Colbert T, Till BJ, Tompa R, Reynolds S, Steine MN, Yeung AT, Mccallum CM, Comai L, Henikoff S, Division BS, Hutchinson F, Washington TC (2001) High-throughput screening for induced point mutations. Plant Physiol 98109:80–84

Curtin SJ, Voytas DF, Stupar RM (2012) Genome engineering of crops with designer nucleases. Plant Genome J 5:42

D'Amico S, Collins T, Marx JC, Feller G, Gerday C (2006) Psychrophilic microorganisms: chal lenges for life. EMBO Rep 7:85–89

De Smet KA, Weston A, Brown IN, Young DB, Robertson BD (2000) Three pathways for trehalose biosynthesis in mycobacteria. Microbiology 146:199–208

Detkova E, Kevbrin V (2009) Cellobiose catabolism in the haloalkaliphilic hydrolytic bacterium *Alkaliflexus imshenetskii*. Microbiology 78:67–72

Dopson M, Baker-Austin C, Hind A, Bowman JP, Bond PL (2004) Characterization of *Ferroplasma* isolates and *Ferroplasma acidarmanus* sp. nov., extreme acidophiles from acid mine drainage and industrial bioleaching environments. Appl Environ Microbiol 70:79–88

Eamens A, Wang MB, Smith NA, Waterhouse PM (2008) RNA silencing in plants: yesterday, today, and tomorrow. Plant Physiol 147:456–468

Elferink MG, Albers SV, Konings WN, Driessen AJ (2001) Sugar transport in *Sulfolobus solfatari-cus* is mediated by two families of binding protein-dependent ABC transporters. Mol Microbiol 39(6):1494–1503

Erra-Pujada M, Chang-Pi-Hin F, Debeire P, Duchiron F, O'Donohue MJ (2001) Purification and properties of the catalytic domain of the thermostable pullulanase type II from *Thermococcus* hydrothermalis. Biotechnol Lett 23:73–77

Feder ME, Hofmann GE (1999) Heat-shock proteins, molecular chaperones, and the stress response: evolutionary and ecological physiology. Annu Rev Physiol 61:43–82

Feng X, Mouttaki H, Lin L, Huang R, Wu B, Hemme CL, He Z, Zhang B, Hicks LM, Xu J (2009) Characterization of the central metabolic pathways in *Thermoanaerobacter* sp. strain X514 via isotopomer-assisted metabolite analysis. Appl Environ Microbiol 75:1–8

Fine A, Irihimovitch V, Dahan I, Konrad Z, Eichler J (2006) Cloning, expression, and purification of functional Sec11a and Sec11b, type I signal peptidases of the archaeon *Haloferax volcanii*. J Bacteriol 188:11–19

Ge XJ, Yu Y, Yuan YM, Huang HW, Yan C (2005) Genetic diversity and geographic differentiation in endangered *Ammopiptanthus* (*Leguminosae*) populations in desert regions of northwest China as revealed by ISSR analysis. Ann Bot 95:43–51

Gong D, Guo Y, Schumaker KS, Zhu JK (2004) The SOS 3 family of calcium sensors and SOS 2 family of protein kinases in *Arabidopsis*. Plant Physiol 134:19–26

Goodchild A, Saunders NF, Ertan H, Raftery M, Guilhaus M, Curmi PM, Cavicchioli R (2004) A proteomic determination of cold adaptation in the antarctic archaeon, *Methanococcoides burtonii*. Mol Microbiol 53:9–21

Hallberg KB, Gonzalez-Toril E, Johnson DB (2010) *Acidithiobacillus ferrivorans*, sp. nov.; facultatively anaerobic, psychrotolerant iron-, and sulfur-oxidizing acidophiles isolated from metal mine-impacted environments. Extremophiles 14:9–19

Higgins CF (1995) The ABC of channel regulation. Cell 82:93–96

Hongo K, Hirai H, Uemura C, Ono S, Tsunemi J, Higurashi T, Mizobata T, Kawata Y (2006) A novel ATP/ADP hydrolysis activity of hyperthermostable group II chaperonin in the presence of cobalt or manganese ion. FEBS Lett 580:34–40

Huber R, Langworthy TA, Konig H, Thomm M, Woese CR, Sleytr UB, Stetter KO (1986) *Thermotoga maritima* sp. nov. represents a new genus of unique extremely thermophilic eubacteria growing up to 90 °C. Arch Microbiol 144:24–33

Huber H, Burggraf S, Mayer T, Wyschkony I, Rachel R, Stetter KO (2000) Ignicoccus gen. nov., a novel genus of hyperthermophilic, chemolithoautotrophic Archaea, represented by two new species, *Ignicoccus islandicus* sp nov and *Ignicoccus pacificus* sp nov. and *Ignicoccus pacificus* sp. nov. Int J Syst Evol Microbiol 50:2093–2100

Imhoff JF, Rodriguez-Valera F (1984) Betaine is the main compatible solute of halophilic eubac teria. J Bacteriol 160:78–79

Jaenicke R, Bohm G (1998) The stability of proteins in extreme environments. Curr Opin Struct Biol 8:38–48

Jinek M, Chylinski K, Fonfara I, Hauer M, Doudna JA, Charpentier E (2012) A programmable dual-RNA–guided DNA endonuclease in adaptive bacterial immunity. Science 337:16–22

Jolivet E, Corre E, L'Haridon S, Forterre P, Prieur D (2004) *Thermococcus marinus* sp. nov. and *Thermococcus radiotolerans* sp. nov., two hyperthermophilic archaea from deep-sea hydrothermal vents that resist ionizing radiation. Extremophiles 8:19–27

Kaushik JK, Ogasahara K, Yutani K (2002) The unusually slow relaxation kinetics of the folding unfolding of pyrrolidone carboxyl peptidase from a hyperthermophile, *Pyrococcus furiosus*. J Mol Biol 316:991–1003

Kazak H, Oner E, Dekker RF (2010) Extremophiles as sources of exopolysaccharides. In: Matsuo Y (ed) Ito R. Handbook of carbohydrate polymers development and properties, Nova

Khrulnova SA, Baranova A, Bazhenov SV, Goryanin II, Konopleva MN, Maryshev IV et al (2016) Lux-operon of the marine psychrophilic bacterium *Aliivibrio logei*: a comparative analysis of the LuxR1/LuxR2 regulatory activity in *Escherichia coli* cells. Microbiology 162:717. https://doi.org/10.1099/mic.0.000253

Kochhar N, Kavya KI, Shrivastva S, Ghosh A, Rawat VS, Sodhi KK, Kumar M (2022) Perspectives on the microorganism of extreme environments and their applications. Curr Res Mic Sci 3:100134

Koning SM, Albers SV, Konings WN, Driessen AJ (2002) Sugar transport in (hyper) thermophilic archaea. Res Microbiol 153(2):61–67

Kumar A, Alam A, Tripathi D, Rani M, Khatoon H, Pandey S, Ehtesham NZ, Hasnain SE (2018) Protein adaptations in extremophiles: an insight into extremophilic connection of mycobacterial proteome. Cell Dev Biol 84:47–57

Leong SL, Lantz H, Pettersson OV, Frisvad JC, Thrane U, Heipieper HJ, Dijksterhuis J, Grabherr M, Pettersson M, Tellgren-Roth C (2014) Genome and physiology of the ascomycete filamentous fungus *Xeromyces bisporus*, the most xerophilic organism isolated to date. Environ Microbiol 17:496–513

Lin L, Dai S, Tian B, Li T, Yu J, Liu C (2016) DqsIR quorum sensing-mediated gene regulation of the extremophilic bacterium *Deinococcus radiodurans* in response to oxidative stress. Mol Microbiol 100:527–541

Liu Y, Zhang L, Meng S, Liu Y, Zhao X, Pang C, Zhang H, Xu T, He Y, Qi M (2020) Expression of galactinol synthase from *Ammopiptanthus nanus* in tomato improves tolerance to cold stress. J Exp Bot 71:35–49

Lucas S, Han J, Lapidus A, Cheng JF, Goodwin LA, Pitluck S, Peters L, Mikhailova N, Teshima H, Detter JC (2012) Complete genome sequence of the thermophilic, piezophilic, heterotrophic bacterium *Marinitoga piezophila* KA3. J Bacteriol 194:74–75

Mae A, Montesano M, Koiv V, Tapio Palva E (2001) Transgenic plants producing the bacterial pheromone N-Acyl-homoserine lactone exhibit enhanced resistance to the bacterial phytopathogen *Erwinia carotovora*. Mol Plant-Microbe Interact 14:35–42

Makarova KS, Haft DH, Barrangou R, Brouns SJJ, Charpentier E, Horvath P, Moineau S, Mojica FJM, Wolf YI, Yakunin AF, van der Oost J, Koonin EV (2011) Evolution and classification of the CRISPR-Cas systems. Nat Rev Microbiol 9:467–477

Mamani S, Moinier D, Denis Y, Soulere L, Queneau Y, Talla E (2016) Insights into the quorum sensing regulon of the acidophilic *Acidithiobacillus ferrooxidans* revealed by transcriptomic in the presence of an acyl homoserine lactone superagonist analog. Front Microbiol 7:1365

March JC, Bentley WE (2004) Quorum sensing and bacterial cross-talk in bio technology. Curr Opin Biotechnol 15:495–502

Marco MG, Rodríguez LV, Ramos EL, Renovato J, Cruz-Hernández MA, Rodríguez R, Contreras J, Aguilar CN (2009) A novel tannase from the xerophilic fungus *Aspergillus niger* GH1. J Microbiol Biotechnol 1:1–10

Markande AR, Patel D, Varjani S (2021) A review on biosurfactants: properties, applications and current developments. Bioresour Technol 24:124963

Martin D, Bartlett DH, Roberts MF (2002) Solute accumulation in the deep-sea bacterium *Photobacterium profundum*. Extremophiles 6(6):507–514

Medigue C, Krin E, Pascal G, Barbe V, Bernsel A, Bertin PN, Cheung F, Cruveiller S, D'Amico S, Duilio A (2005) Coping with cold: the genome of the versatile marine Antarctica bacterium *Pseudoalteromonas haloplanktis* TAC125. Genome Res 15(10):1325–1335

Mehetre G, Leo VV, Singh G, Dhawre P, Maksimov I, Yadav M, Upadhyaya K, Singh BP (2021) Biocontrol potential and applications of extremophiles for sustainable agriculture. In: Microbiomes of extreme environments. CRC Press, pp 230–242

Mokashe N, Chaudhari B, Patil U (2018) Operative utility of salt-stable proteases of halophilic and halotolerant bacteria in the biotechnology sector. Int J Biol Macromol 117:493–522

Montgomery K, Charlesworth JC, Lebard R, Visscher PT, Burns BP (2013) Quorum sensing in extreme environments. Life 3:31–48

Monzon O, Yang Y, Li Q, Alvarez PJJ (2016) Quorum sensing autoinducers enhance biofilm formation and power production in a hypersaline microbial fuel cell. Biochem Eng J 109:222–227

Mori JF, Ueberschaar N, Lu S, Cooper RE, Pohnert G, Kusel K (2017) Sticking together: interspecies aggregation of bacteria isolated from iron snow is controlled by chemical signaling. ISME J 11:75–86

Nakashima K, Takasaki H, Mizoi J, Shinozaki K, YamaguchiShinozaki K (2012) NAC transcription factors in plant abiotic stress responses. Biochim Biophys Acta 1819:97–103

Nelson KE, Clayton RA, Gill SR, Gwinn ML, Dodson RJ, Haft DH, Hickey EK, Peterson JD, Nelson WC, Ketchum KA (1999) Evidence for lateral gene transfer between Archaea and bacteria from genome sequence of *Thermotoga maritima*. Nature 399:23–29

Nevoigt E, Stahl U (1997) Osmoregulation and glycerol metabolism in the yeast *Saccharomyces cerevisiae*. FEMS Microbiol Rev 21:31–41

Nongpiur RC, Pareek SLS, Pareek A (2016) Genomics approaches for improving salinity stress tolerance in crop plants. Curr Genomics 17:43–57

Orellana S, Yanez M, Espinoza A, Verdugo I, González E, Ruiz-Lara S, Casaretto J (2010) The transcription factor SlAREB1 confers drought, salt stress tolerance and regulates biotic and abiotic stress-related genes in tomato. Plant Cell Environ 33:191–208

Oren A, Mana L (2003) Sugar metabolism in the extremely halophilic bacterium *Salinibacter ruber*. FEMS Microbiol Lett 223:83–87

Padan E, Bibi E, Ito M, Krulwich TA (2005) Alkaline pH homeostasis in bacteria: new insights. BBA-Biomembranes 1717:67–88

Pahlman AK, Granath K, Ansell R, Hohmann S, Adler L (2001) The yeast glycerol 3-phosphatases Gpp1p and Gpp2p are required for glycerol biosynthesis and differentially involved in the cellular responses to osmotic, anaerobic, and oxidative stress. J Biol Chem 276:55–63

Pandey KD, Patel AK, Singh M, Kumari A (2021) Secondary metabolites from bacteria and viruses. Na Bioact Compd 25:19–40

Papenfort K, Silpe JE, Schramma KR, Cong JP, Seyedsayamdost MR, Bassler BL (2017) A *Vibrio cholerae* autoinducer-receptor pair that controls biofilm formation. Nat Chem Biol 13:551–557

Paulsen IT, Nguyen L, Sliwinski MK, Rabus R, Saier MH (2000) Microbial genome analyses: comparative transport capabilities in eighteen prokaryotes. J Mol Biol 301(1):75–100

Pehlivan N, Sun L, Jarrett P, Yang X, Mishra N, Chen L, Kadioglu A, Shen G, Zhang H (2016) Co-overexpressing a plasma membrane and a vacuolar membrane sodium/proton antiporter significantly improves salt tolerance in transgenic *Arabidopsis* plants. Plant Cell Physiol 57:69–84

Pfluger K, Muller V (2004) Transport of compatible solutes in extremophiles. J Bioenerg Biomembr 36:17–24

Piette F, D'Amico S, Struvay C, Mazzucchelli G, Renaut J, Tutino ML, Danchin A, Leprince P, Feller G (2010) Proteomics of life at low temperatures: trigger factor is the primary chaperone in the antarctic bacterium *Pseudoalteromonas haloplanktis* TAC125. Mol Microbiol 76:120–132

Puchta H, Fauser F (2014) Synthetic nucleases for genome engineering in plants: prospects for a bright future. Plant J 78:727–741

Quiocho FA, Ledvina PS (1996) Atomic structure and specificity of bacterial periplasmic receptors for active transport and chemotaxis: variation of common themes. Mol Microbiol 20:17–25

Ramanathan T, Ting YP (2015) Selective copper bioleaching by pure and mixed cultures of alkaliphilic bacteria isolated from a fly ash landfill site. Water Air Soil Pollut 226. https://doi.org/10.1007/s11270-015-2641-x

Ran FA, Hsu PD, Wright J, Agarwala V, Scott D, Zhang F (2013) Genome engineering using the CRISPR-Cas9 system. Nat Protoc 8:2281–2308

Rao RM, Pasha SN, Sowdhamini R (2016) Genome-wide survey and phylogeny of S Ribosylhomocysteinase (LuxS) enzyme in bacterial genomes. BMC Genomics 17:742

Raval VH, Bhatt HB, Singh SP (2018) Adaptation strategies in halophilic bacteria. In: Extremophiles. CRC Press, pp 137–164

Reguera M, Bassil E, Blumwald E (2014) Intracellular NHX-Type cation/H+ antiporters in plants. Mol Plant 7:61–63

Ren Z, Liu Y, Kang D, Fan K, Wang C, Wang G, Liu Y (2015) Two alternative splicing variants of maize HKT1;1 confer salt tolerance in transgenic tobacco plants. Plant Cell Tissue Organ Cult 123:569–578

Restaino L, Bills S, Tscherneff K, Lenovich LM (1983) Growth characteristics of *Saccharomyces rouxii* isolated from chocolate syrup. Appl Environ Microbiol 45:14–21

Rezzonico F, Duffy B (2008) Lack of genomic evidence of AI-2 receptors suggests a non-quorum sensing role for luxS in most bacteria. BMC Microbiol 8:154

Richard H, Foster JW (2004) *Escherichia coli* glutamate-and arginine-dependent acid resistance systems increase internal pH and reverse transmembrane potential. J Bacteriol 186:6032–6604

Riley M, Staley JT, Danchin A, Zhang TZ, Brettin TS, Hauser LJ (2008) Genomics of an extreme psychrophile, *Psychromonas ingrahamii*. BMC Genomics 9:210

Roberts MF (2004) Osmoadaptation and osmoregulation in archaea: update 2004. Front Biosci 9:1999–2019

Rothschild LJ, Manicinelli RL (2001) Life in extreme environments. Nature 409:1092–1101

Rozen Y, Belkin S (2001) Survival of enteric bacteria in seawater. FEMS Microbiol Rev 25:13–29

Schneider E (2001) ABC transporters catalyzing carbohydrate uptake. Res Microbiol 152:303–310

Schulz P, Herde M, Romeis T (2013) Calcium-dependent protein kinases: hubs in plant stress signaling and development. Plant Physiol 163:23–30

Selig M, Xavier KB, Santos H, Schonheit P (1997) Comparative analysis of Embden-Meyerhof and Entner-Doudoroff glycolytic pathways in hyperthermophilic archaea and the bacterium *Thermotoga*. Arch Microbiol 167:17–32

Selvaraj MG, Ishizaki T, Valencia M, Ogawa S, Dedicova B, Ogata T, Yoshiwara K, Maruyama K, Kusano M, Saito K (2017) Overexpression of an *Arabidopsis thaliana* galactinol synthase gene improves drought tolerance in transgenic rice and increased grain yield in the field. Plant Biotechnol J.15:65–77

Seo JS, Chong H, Park HS, Yoon KO, Jung C, Kim JJ, Hong JH, Kim H, Kim JH, Kil JI (2005) The genome sequence of the ethanologenic bacterium *Zymomonas mobilis* ZM4. Nat Biotechnol 23:63–68

Sewald X, Saum SH, Palm P, Pfeiffer F, Oesterhelt D, Muller V (2007) Autoinducer-2-producing protein LuxS, a novel salt- and chloride-induced protein in the moderately halophilic bacterium *Halobacillus halophilus*. Appl Environ Microbiol 73:71–79

Sghaier H, Ghedira K, Benkahla A, Barkallah I (2008) Basal DNA repair machinery is subject to positive selection in ionizing-radiation-resistant bacteria. BMC Genomics 9:297

Shi H, Lee BH, Wu SJ, Zhu JK (2003) Overexpression of a plasma membrane Na$^+$/H$^+$ antiporter gene improves salt tolerance in *Arabidopsis thaliana*. Nat Biotechnol 21:81–85

Siebers B, Schönheit P (2005) Unusual pathways and enzymes of central carbohydrate metabolism in Archaea. Curr Opin Microbiol 8:695–705

Siebold C, Flükiger K, Beutler R, Erni B (2001) Carbohydrate transporters of the bacterial phosphoenolpyruvate: sugar phosphotransferase system (PTS). FEBS Lett 504:4–11

Simonato F, Campanaro S, Lauro FM, Vezzi A, D'Angelo M, Vitulo N, Valle G, Bartlett DH (2006) Piezophilic adaptation: a genomic point of view. J Biotechnol 126:11–25

Singh BP (2021) Biocontrol potential and applications of extremophiles for sustainable agriculture. In: Microbiomes of extreme environments. CRC Press, pp 230–242

Song SY, Chen Y, Chen J, Dai XY, Zhang WH (2011) Physiological mechanisms underlying OsNAC5-dependent tolerance of rice plants to abiotic stress. Planta 234:31–45

Sorek R, Lawrence CM, Wiedenheft B (2013) CRISPR-mediated adaptive immune systems in bacteria and archaea. Annu Rev Biochem 82:37–66

Sterner R, Liebl W (2001) Thermophilic adaptation of proteins. Crit Rev Biochem Mol Biol 36:39–106

Tahrioui A, Schwab M, Quesada E, Llamas I (2013) Quorum sensing in some representative species of Halomonadaceae. Life 3:260–275

Takai K, Moser DP, Onstott TC, Spoelstra N, Pfiffner SM, Dohnalkova A, Fredrickson JK (2001) Alkaliphilus transvaalensis gen. nov., sp. nov., an extremely alkaliphilic bacterium isolated from a deep South African gold mine. Int J Syst Evol Microbiol 51:45–56

Tamas MJ, Luyten K, Sutherland FCW, Hernandez A, Albertyn J, Valadi H, Li H, Prior BA, Kilian SG, Ramos J (1999) Fps1p controls the accumulation and release of the compatible solute glycerol in yeast osmoregulation. Mol Microbiol 31:87–104

Tang YJ, Sapra R, Joyner D, Hazen TC, Myers S, Reichmuth D, Blanch H, Keasling JD (2009) Analysis of metabolic pathways and fluxes in a newly discovered thermophilic and ethanol-tolerant *Geobacillus* strain. Biotechnol Bioeng 102:77–86

Tiwari S, Prasad V, Lata C (2019) *Bacillus*: plant growth promoting bacteria for sustainable agriculture and environment. In: New and future developments in microbial biotechnology and bioengineering. Elsevier, pp 43–55

Toogood HS, Gardiner JM, Scrutton NS (2010) Biocatalytic reductions and chemical versatility of the old yellow enzyme family of flavoprotein oxidoreductases. Chem Cat Chem 2:892–914

Tran LP, Urao T, Qin F, Maruyama K, Kakimoto T, Shinozaki K, Yamaguchi-shinozaki K (2007) Functional analysis of AHK1 / ATHK1 and cytokinin receptor histidine kinases in response to abscisic acid, drought, and salt stress in Arabidopsis. Proc Natl Acad Sci 14:3–8

Uno Y, Furihata T, Abe H, Yoshida R, Shinozaki K, Yamaguchi-Shinozaki K (2000) *Arabidopsis* basic leucine zipper transcription factors involved in an abscisic acid-dependent signal transduction pathway under drought and high-salinity conditions. Proc Natl Acad Sci 97:32–37

Urnov FD, Rebar EJ, Holmes MC, Zhang HS, Gregory PD (2010) Genome editing with engineered zinc finger nucleases. Nat Rev Genet 11:36–46

Vaishnav A, Sahu J, Singh HB (2020) Genomics of extremophiles for sustainable agriculture and biotechnological applications. Curr Genomics 21:78–79

Valdés J, Pedroso I, Quatrini R, Dodson RJ, Tettelin H, Blake R, Eisen JA, Holmes DS (2008) Acidithiobacillus ferrooxidans metabolism: from genome sequence to industrial applications. BMC Genomics 9(1):597

Valentine RC, Valentine DL (2004) Omega-3 fatty acids in cellular membranes: a unified concept. Prog Lipid Res 43(5):383–402

Verma P, Yadav AN, Kumar V, Singh DP, Saxena AK (2017) Beneficial plantmicrobes interactions: biodiversity of microbes from diverse extreme environments and its impact for crop improvement. In: Plant-microbe interactions in agroecological perspectives. Springer, Singapore, pp 543–580

Violot S, Aghajari N, Czjzek M, Feller G, Sonan GK, Gouet P, Gerday C, Haser R, Receveur B, V. (2005) Structure of a full length psychrophilic cellulase from *Pseudoalteromonas haloplanktis* revealed by X-ray diffraction and small angle X-ray scattering. J Mol Biol 348:11–24

Vossenberg JL, Ubbink-Kok T, Elferink MG, Driessen AJ, Konings WN (1995) Ion permeability of the cytoplasmic membrane limits the maximum growth temperature of bacteria and archaea. Mol Microbiol 18:25–32

Wang YH (2008) How effective is T-DNA insertional mutagenesis in *Arabidopsis*? J Biochem Technol 1:11–20

Wang Z, Hicks DB, Guffanti AA, Baldwin K, Krulwich TA (2004) Replacement of amino acid sequence features of a-and c-subunits of ATP synthases of alkaliphilic Bacillus with the *Bacillus* consensus sequence results in defective oxidative phosphorylation and non-fermentative growth at pH 10.5. J Biol Chem 279:46–54

Yadav AN, Saxena AK (2018) Biodiversity and biotechnological applications of halophilic microbes for sustainable agriculture. J Appl Biol Biotechnol 6(1):48–55

Yadav NS, Shukla PS, Jha A, Agarwal PK, Jha B (2012) The SbSOS1 gene from the extreme halophyte Salicornia brachiata enhances Na+ loading in xylem and confers salt tolerance in transgenic tobacco. BMC Plant Biol 12

Yadav AN, Verma P, Kumar V, Sachan SG, Saxena AK (2017) Extreme cold environments: a suitable niche for selection of novel psychrotrophic microbes for biotechnological applications. Adv Biotechnol Microbiol 2(2):1–4

Yuan F, Yang H, Xue Y, Kong D, Ye R, Li C, Zhang J, Theprungsirikul L (2014) (2014) OSCA1 mediates osmotic-stress-evoked Ca^{2+} increases vital for osmosensing in Arabidopsis. Nature 514:367–371

Yumoto I, Hirota K, Sogabe Y, Nodasaka Y, Yokota Y, Hoshino T (2003) Psychrobacter okhotskensis sp. nov., a lipase-producing facultative psychrophile isolated from the coast of the Okhotsk Sea. Int J Syst Evol Microbiol 53:85–89

Zhang Y, Zhang F, Li X, Baller JA, Qi Y, Starker CG, Bogdanove AJ, Voytas DF (2013) Transcription activator-like effector nucleases enable efficient plant genome engineering. Plant Physiol 161:20–27

Chapter 17
Insight into Soil Nutrient Management in Agriculture by Acidophilus Microbes

Vaibhav Mishra, Neeraj Shrivastava, Smriti Shukla, and Rupesh Kumar Basniwal

17.1 Introduction

Soil serves as the foundation for agricultural productivity, providing essential nutrients to plants. Ensuring an optimal nutrient supply to crops is paramount for achieving high yields and food security. A total of 14 mineral elements are responsible for adequate plant nutrition. The elements that are required in larger amounts, i.e., macronutrients include nitrogen (N), phosphorus (P), potassium (K), calcium (Ca), magnesium (Mg), and sulfur (S). In addition, these plants require some trace elements that are required in smaller amounts. These micronutrients include iron (Fe), zinc (Zn), boron (B), copper (Cu), molybdenum (Mo), manganese (Mn), nickel (Ni), and chlorine (Cl). Both macro and micronutrients play vital roles in plant growth, development, and food grain production. These also play vital roles in enhancing tolerance of abiotic/biotic stress (Saleem et al. 2023).

Various available and nonavailable soil nutrients are important for the health of plants and are required by plants in low and high quantities. The macronutrients are Lack of these mineral nutrients from minimum required quantity results in abnormal growth and stress that reduces the plant's immunity and fitness. Moreover, if nutrients are present in excess quantity, they may be toxic to the plant. Crops, vegetables, and plants with adequate quantities of nutrients provide healthy food to the world's population. So, the management of soil nutrients is essential for food production and at the same time retaining the fertility of soil. The management of soil fertility is a very essential component to regulate the various metabolic functions of plant tissue. At the same time, soil nutrient balance is directly and indirectly regulated by various other conditions like fertilizer, climatic conditions, soil

V. Mishra · N. Shrivastava
Amity Institute of Microbial Technology, Amity University, Noida, UP, India

S. Shukla
Amity Institute of Environmental Toxicology Safety and Management, Amity University, Noida, UP, India

R. K. Basniwal (✉)
Amity Institute of Advanced Research and Studies (M&D), Amity University, Noida, UP, India
e-mail: rkbasniwal@amity.edu

water, and plant capability. Apart from physiochemical properties, biological properties also play a major role in regulation of soil properties (Kanekar and Kanekar 2022; Tamreihao et al. 2018; Vaishnav et al. 2020).

Moreover, to feed the large population, the modern agriculture system is majorly dependent upon chemical fertilizers, which has raised a prime concern for agriculture. We have targeted sustainable agriculture along with food safety so we cannot ignore the harmful effect of chemical fertilizers which plays a major role in modern agriculture systems. Traditional approaches to soil nutrient management often involve chemical fertilizers, which have raised concerns about environmental degradation and sustainability. In response to these challenges, acidophilus microbes have emerged as promising agents for sustainable soil nutrient management in agriculture (Poomthongdee et al. 2015; Sharma et al. 2012). This chapter explores the role of acidophilus microbes in agriculture, shedding light on their ability to enhance nutrient availability, improve soil structure, and promote plant growth. By harnessing the potential of acidophilus microbes, farmers can reduce their dependence on chemical fertilizers, mitigate environmental impacts, and achieve sustainable agricultural practices. Soil pH can have a positive impact on soil nutrient management by enhancing the availability of essential nutrients for plant growth. In addition, we are focusing on acidophilus microbes and their contributions that can improve crop productivity and overall soil health.

17.2   Definition and Classification

Acidophilus microbes, often referred to as acidophiles, are a group of microorganisms characterized by their unique ability to thrive in acidic environments, typically with a pH level of 3.0 or lower (Johnson et al. 2020). These microbes have adapted to acidic conditions, and many of them play crucial roles in various natural ecosystems, including acidic soils. Acidophilus microbes encompass a diverse array of microorganisms, representing multiple domains of life, including bacteria, archaea, and fungi.

Bacteria
Acidophilic Bacteria: These are bacteria that can grow and reproduce in acidic conditions. They are commonly found in environments such as acid mine drainage (AMD), peat bogs, and acidic soils. Prominent examples include *Acidithiobacillus ferrooxidans* (Mahmoud et al. 2005) and *Acidobacterium* spp.

Archaea
Acidophilic Archaea: Similar to acidophilic bacteria, acidophilic archaea thrive in acidic environments. They are often found in extreme environments like hot springs, volcanic vents, and acidic soils. Notable acidophilic archaea include *Ferroplasma* spp. and *Sulfolobus* spp. (Jung et al. 2020).

Fungi
Acidophilic Fungi: Some fungi are also acidophilic and can tolerate low pH conditions. These fungi are essential for decomposing organic matter in acidic soils and contributing to nutrient cycling. Well-known acidophilic fungi include *Penicillium* spp. and *Aspergillus* spp. Acidophilic fungi are proficient decomposers of organic matter, breaking down complex organic compounds and releasing nutrients into the soil, enriching its fertility (Rehman Javaid et al. 2019).

17.3   Acidophiles in Different Habitants

Acidophilus microbes, specifically those belonging to the group of acidophilic microorganisms, can play a significant role in soil nutrient management through various mechanisms. Acidophilic microbes are microorganisms that thrive in acidic environments (Irwin 2020), and they can contribute to nutrient cycling and availability in soils. Here are some of the mechanisms through which acidophilus microbes influence soil nutrient management (Mehmet et al. 2023). Moreover, acidophilic microbes can indirectly influence nutrient management by suppressing soil-borne pathogens and diseases. By reducing the impact of diseases on plants, these microbes can enhance plant health and nutrient uptake, leading to improved nutrient management. Acidophilus microbes can antagonize pathogenic microorganisms, helping to suppress soil-borne diseases and promoting healthier plant growth. Understanding their presence in nature is essential for recognizing their potential applications in agriculture and soil nutrient management. In this chapter, we are trying to explore some of the key natural environments where acidophilus microbes are prevalent.

Acidophiles in Acidic Soils
Acidic soils, with pH levels below 5.5, are prime habitats for acidophilus microbes. They are commonly found in regions with high precipitation and leaching, where acidification occurs due to the loss of basic cations. Acidophilic bacteria, such as *Acidobacterium* spp., and fungi, including *Penicillium* spp., play critical roles in nutrient cycling and organic matter decomposition in these soils (Mahmoud et al. 2005).

Acidophilus Microbes in Acid Mine Drainage (AMD)
One of the most extreme acidic environments on Earth, AMD results from the exposure of sulfide minerals in mines to air and water, leading to the generation of highly acidic, metal-laden runoff. Acidophilus microbes, particularly acidophilic bacteria like *A. ferrooxidans*, thrive in AMD and contribute to the oxidation of sulfide minerals, which can be both a natural phenomenon and a significant environmental concern (Mahmoud et al. 2005).

Acidophiles in Hot Springs and Volcanic Vents
Certain acidophilic archaea, such as *Sulfolobus* spp., inhabit hot springs and volcanic vents characterized by low pH levels and high temperatures. These extreme environments are rich in sulfur compounds and heavy metals, and acidophilus microbes are key players in the geochemical processes occurring in these locations (Bohlool Ben 1975).

Acidophiles in Peat Bogs
Peat bogs are acidic wetlands with pH levels ranging from 3.5 to 4.5. Acidophilic microbes, including fungi like *Aspergillus* spp. (Dedyesh et al. 1998), thrive in the anaerobic conditions of peat bogs. They contribute to the decomposition of organic

matter and the formation of peat, which can be used as a soil amendment to improve soil structure and fertility (Albert et al. 2005).

Acidophiles in Geothermal Environment

Geothermal areas, such as geysers and fumaroles, are characterized by acidic and high-temperature conditions. Acidophilic microorganisms in these extreme environments are crucial for sulfur oxidation and other geochemical processes (Romoli 2011).

Acidophiles in Acidic Lakes and Ponds

Some freshwater bodies, especially in volcanic regions, have low pH levels due to volcanic activity and the release of acidic gases. Acidophilus microbes are known to inhabit these environments and are involved in nutrient cycling and microbial interactions (X chen et al. 2021).

Understanding where acidophilus microbes naturally thrive underscores their adaptability to challenging conditions and provides insights into how they can be harnessed for agricultural applications.

17.4 Traditional Approaches to Soil Nutrient Management

Traditional approaches to soil nutrient management have been in practice for centuries and have evolved based on farmers' experiences and available knowledge. These approaches typically involve practices that aim to maintain soil fertility and support crop growth. Some traditional soil nutrient management techniques include:

Crop Rotation Farmers traditionally practiced crop rotation, which involves growing different crops in sequential seasons on the same land. This helps prevent nutrient depletion and the buildup of pests and diseases associated with specific crops (Albert et al. 2005). Crop rotation also provides a special window to farmers to grow crops without chemical fertilizers. It also improves soil health because of versatility use of nutrients from the soil.

Manure and Organic Matter The use of animal manure and organic matter (such as compost) is a traditional method to improve soil fertility. These organic materials provide essential nutrients and enhance soil structure and water-holding capacity. It is mostly obtained from biomass products, which are full of carbon and micronutrients. After addition of manure, it increases the crop yield. We all know that in nature three different types of manure are available animals, plants, and compost (Gross and Glaser 2021).

Ash and Residue Recycling In some regions, burning crop residues and using ash as a source of nutrients for subsequent crops is a traditional practice. This can provide potassium and other trace elements to the soil.

Intercropping Planting multiple crops in the same field simultaneously or in close proximity is a traditional approach to diversify nutrient sources and increase overall productivity. It is a common practice used by developing countries to enhance their production. Many of the studies were planned and executed to endow more crops (Cheng et al. 2019).

Fallows Allowing fields to lie fallow (uncultivated) for a period allows the soil to naturally replenish nutrients and regain fertility.

Livestock Integration Combining crop farming with livestock rearing allows for nutrient recycling. Livestock manure can be used to fertilize fields, closing nutrient cycles.

17.5 Exploring the Role of Acidophiles for Soil Nutrient Management in Agriculture

In the subsequent sections of this chapter, we will explore how acidophilus microbes can be applied to improve soil nutrient management in agriculture, drawing from their natural abilities and adaptations.

17.5.1 Nutrient Mineralization by Acidophilic Microbes

Increase in the human population has put immense pressure on agriculture, which has led to the use of nitrogen and phosphorus-based fertilizers. Prolonged use of chemical fertilizers and increasing pollution (acid rain and mining of coal and ores) have become a cause of declining soil pH (below 5), which affects the physiology of neutrophilic bacteria and promotes the growth of pathogenic fungi (Poomthongdee et al. 2015; Sharma et al. 2012). Acidophiles or acid-tolerant bacteria would serve as a desirable alternative for agricultural practices as they can sustain under acidic soil environments (solfataric fields and sulfuric pools). One such group is Actinobacteria with anti-fungal and growth-promoting abilities by solubilizing nutrients and has tremendous potential to serve in the development of novel biocontrol or biofertilizer products. Although there has been research on extremophiles as a potential source of biofertilizer preparation (thermophiles, halophiles, and psychrophiles), acidophilic or acid-tolerant microbes have also pivotal role in sustainable agriculture (Tamreihao et al. 2018; Vaishnav et al. 2020). Acidophiles are metabolically and physiologically heterogenous, and they can use different electron acceptors such as ferric ions, sulfur ions, and oxygen. Reduced sulfur compounds and ferrous ions which form from mineral weathering and geothermal activities serve as an energy source for acidophiles. Different organic compounds, including short-chain carbon or organic compounds, including glucose as carbon source for their growth. They can serve as a potential candidate to produce next-generation bioformulation/biofertilizer, especially for the cultivable lands to enhance the crop production of the major crops, including rice and wheat which are prone to such environmental conditions. They have been shown to possess Plant growth-promoting (PGP) traits such as IAA production, phosphate solubilization, siderophore production, and nitrogen fixation at pH below 5 indicating that they can thrive in acidic environments retaining PGP traits. Examples of acidophilic bacteria include Actinobacteria, *Streptomyces* sp., *Saccharothrix* sp., and *Amycolatopsis* sp. These microorganisms have evolved cellular adaptations to regulate lower pH inside the cell. Enzymes such as proteases, amylases, cellulases, α-glucosidases, esterases, ligases, xylanases, and endoglucanases have been found to be stable at lower pH. Apart from conventional techniques, true potential of these microbes can be investigated via high-throughput integrated omics techniques such as proteomics, metabolomics, transcriptomics, and WGS, which could help in strain selection. Further research is being done in improving the strains by genetic engineering to boost their industrial applications. Due to their acidophilic and/or acid-tolerant nature, Actinobacteria, particularly *Streptomyces*, can thrive and reproduce in acidic environments. As a result, they could improve soil productivity and health, avoid plant diseases, and eventually boost agricultural crop yields. The process of deacetylation and deamination results in the liberation of ammonia. The production of chitinase by *Streptomyces* bacteria may increase the pH of the soil by breaking down *N*-acetylglucosamine residues. This creates favorable conditions for other neutrophilic plant growth-promoting bacteria (PGPB) to establish themselves and outcompete pathogens. Hence, application of acidophiles can enhance soil fertility, improve plant nutrient uptake, and mitigate nutrient deficiencies in diverse cropping systems (Kanekar and Kanekar 2022; Tamreihao et al. 2018; Vaishnav et al. 2020; Poomthongdee et al. 2015; Sharma et al. 2012).

Acidophilic microbes are involved in the process of nutrient mineralization. They break down organic matter in the soil, such as dead plant material or organic residues, into simpler compounds. This decomposition releases nutrients like nitrogen, phosphorus, and sulfur in forms that plants can readily absorb. By facilitating nutrient mineralization, acidophilus microbes contribute to the availability of essential nutrients for plant growth.

17.5.2 *Phosphorus Solubilization*

Phosphorus is often present in soil in forms that are not easily accessible to plants. Acidophilic microbes, especially certain types of bacteria and fungi, can produce organic acids and enzymes that solubilize phosphorus compounds, making this vital nutrient more available to plants (Souza et al. 2022).

17.5.3 Enhancing Availability of Metal Ions in Soil

Acidophilic bacteria and fungi are proficient at solubilizing essential nutrients, such as iron (Fe) and phosphorus (P), from insoluble forms in the soil. Some acidophilic microbes produce organic acids as metabolic byproducts. These organic acids can lower the pH of the soil, making certain micronutrients like iron (Fe) and manganese (Mn) more soluble and available for plant uptake. Improved micronutrient availability contributes to healthier plants and better overall soil fertility (Yang et al. 2018).

17.5.4 Enhancing Availability of Sulfur Compounds in Soil

Acidophilic microbes are known for their ability to oxidize sulfur compounds in soils. This oxidation process can release sulfate ions (SO_4^-) from sulfur-containing minerals, making sulfur available for plant uptake. Sulfur is an essential nutrient for plant growth, and acidophilus microbes contribute to its cycling in the soil (Leandro et al. 2023). Furthermore, acidophilus microbes generate organic acids, such as citric acid or oxalic acid, as metabolic byproducts. These organic acids can lower the pH of the soil, making certain nutrients, like iron (Fe) and manganese (Mn), more soluble and available for plants. This increased solubility enhances nutrient uptake by plants.

17.5.5 Biofertilizer Production

Some acidophilic microbes have been explored for their potential as biofertilizers. These microbial inoculants can be applied to soil to improve nutrient availability and nutrient use efficiency by plants. They may also possess the ability to fix atmospheric nitrogen (N) into a plant-available form. Acidophilic microbes can contribute to the formation of stable soil aggregates. Well-aggregated soils have improved water infiltration and retention properties, which can enhance nutrient transport and availability to plant roots (Chaudhary et al. 2022).

17.5.6 Soil pH Regulation

Acidophilus microbes can influence soil pH, particularly in acidic soils. By moderating soil pH, these microbes can create conditions that are more favorable for nutrient availability. In some cases, acidophilus microbes can help mitigate soil acidity problems.

17.5.7 Improving Soil Fertility and Their Texture

Acidophilic microbes can contribute to the formation of stable soil aggregates. Well-aggregated soils have improved water infiltration and retention properties, which enhance nutrient transport and availability to plant roots. It's important to note that the specific mechanisms and effects of acidophilus microbes can vary depending on the microbial species present and the local soil conditions, including pH, organic matter content, and nutrient composition. Therefore, the role of acidophilus microbes in soil nutrient management should be considered within the context of a particular soil ecosystem (Leandro et al. 2023).

17.6 Role of Microbes in Sustainable Soil Ecology, Productivity, and Nutrient Cycling

Acidophilic microbes are involved in solubilizing essential nutrients, such as iron and phosphorus, making them more available to plants. This ability aids in nutrient uptake by crops. The importance of soil nutrient management in modern agriculture cannot be overstated. It plays a critical role in ensuring sustainable and productive farming practices. In this chapter, we are also trying to highlight the utmost importance of soil nutrient management. Appropriate nutrient management supports soil health by maintaining soil fertility. Healthy soils have improved structure, better water-holding capacity, and increased microbial activity, all of which promote crop growth and resilience to environmental stressors.

Adequate soil nutrient levels are essential for crop growth and development. Nutrient management practices ensure that crops receive the necessary nutrients, leading to healthier plants, increased yields, and better-quality produce (Albert et al. 2005). Appropriate nutrient management ensures a balanced supply of essential nutrients such as nitrogen (N), phosphorus (P), and potassium (K). An imbalance in nutrient levels can lead to nutrient deficiencies or toxicities, which can hinder crop growth and reduce yields (Javaid et al. 2020). Effective nutrient management helps farmers use fertilizers and other inputs more efficiently. This not only reduces production costs but also minimizes the environmental impact associated with excessive fertilizer use, including water pollution and greenhouse gas emissions.

17.6.1 Mechanisms of Acidophilus Microbes in Soil

Acidophilus microbes exhibit several essential mechanisms in soil that contribute to nutrient cycling, soil improvement, and overall plant health (Table 17.1.). These mechanisms are crucial for understanding how these microbes can be harnessed for soil nutrient management in agriculture. Below, we delve into the key mechanisms employed by acidophilus microbes in soil environments.

17.6.2 Plant Growth Promotion

Production of Growth-Promoting Compounds: Some acidophilic microbes synthesize plant growth-promoting compounds such as auxins, gibberellins, and cytokinins. These compounds stimulate root development and overall plant growth. Acidophilus microbes enhance nutrient uptake by plants through various mechanisms, including nutrient solubilization, root exudate modification, and mycorrhizal associations. Understanding these mechanisms is critical for harnessing the potential of acidophilus microbes in agriculture. In the subsequent sections of this chapter, we will explore practical applications of acidophilus microbes in soil nutrient management, including their use as soil inoculants, biofertilizers, and tools for improving soil health and crop productivity (Javaid et al. 2020).

17.6.3 Soil Nutrient Management

Soil nutrient management is a critical aspect of modern agriculture, as it plays a vital role in ensuring sustainable crop production and food security. The importance of soil nutrient management can be highlighted in several key areas:

Optimizing Crop Growth Nutrients such as nitrogen (N), phosphorus (P), and potassium (K) are essential for plant growth and development. Proper management of these nutrients ensures that crops have an adequate supply, leading to healthier plants and higher yields.

Balanced Nutrition Soil nutrient management helps maintain a balanced supply of essential nutrients for crops. This balance is crucial because an imbalance can lead to nutrient deficiencies or toxicities, which can negatively impact crop health and productivity.

17.7 The Need for Sustainable Solutions

While traditional approaches to soil nutrient management have their merits, modern agriculture faces new challenges that require more sustainable and science-based solutions. Here's why sustainable approaches are needed:

- *Growing Population and Food Demand:* The global population is increasing, and there is a growing demand for food. Sustainable nutrient management practices

Table 17.1 Important applications of Acidobacteria in agriculture

Si No.	Importance	Microorganisms	References
1	Biofilms	*Acidobacteria*	Belova et al. (2018)
2	Carbon cycle	*Terriglobus*	Eichorst et al. (2018)
3	Exopolysaccharides	*Proteobacteria*	Ward et al. (2009)
3	Plant growth promotion	*Acidobacteria*	Kielak et al. (2016)
4	Nitrogen cycle	*Desulfovibrio vulgaris*	Rajeev et al. (2015)
5	Sulfur cycle	*Acidobacteria*	Hausmann et al. (2018)
6	Secondary metabolites	*Verrucomicobia and Gemmatimonadetes*	Crits-Christoph et al. (2018)
7	Stress	*Acidobacteria*	Pinto et al. (2020)

are essential to increase crop yields and meet these demands without depleting soils.

- *Regulatory Compliance*: Many regions have regulations governing nutrient use in agriculture to protect water quality and ecosystems. Sustainable practices help farmers meet these regulatory requirements.
- *Economic Viability*: Sustainable nutrient management can lead to cost savings for farmers in the long run, making their operations more economically viable.

17.7.1 Advantages of Sustainable Nutrient Management Practices

Environmental Protection Sustainable nutrient management practices help prevent nutrient runoff into water bodies, which can lead to water pollution and harm aquatic ecosystems. By reducing nutrient losses, soil nutrient management contributes to environmental protection (Arora et al. 2018).

Climate Change Mitigation Nutrient management practices can help mitigate climate change. For instance, precision application techniques can reduce the release of nitrous oxide (a potent greenhouse gas) from fertilizers. Additionally, healthy soils act as carbon sinks, sequestering carbon dioxide from the atmosphere (Arora et al. 2018). Climate change is leading to more unpredictable weather patterns, which can affect nutrient availability in the soil. Proper nutrient management practices help farmers adapt to changing conditions and ensure stable crop production.

Food Security As the global population continues to grow, there is increasing pressure on agriculture to produce more food. Effective soil nutrient management is crucial for meeting this demand and ensuring food security for billions of people worldwide (Chaudhary et al. 2022).

Long-Term Sustainability Sustainable nutrient management practices aim to maintain soil fertility over the long term. Continuous nutrient replenishment, crop rotation, and other practices ensure that the soil remains productive for future generations (Arora et al. 2018).

Reducing Food Waste

Well-nourished crops are more likely to have longer shelf lives and be less susceptible to spoilage and waste. Nutrient management can indirectly contribute to reducing post-harvest losses.

Role in Soil Environments

In acidic soils, acidophilus microbes play vital roles in nutrient cycling, soil development, and plant-microbe interactions. Some acidophilic bacteria produce extracellular polymeric substances (EPS), which contribute to soil aggregation and stability. This enhances soil structure and water retention capacity. Understanding the diverse classification and ecological roles of acidophilus microbes is essential for harnessing their potential in soil nutrient management strategies. In the subsequent sections of this chapter, we will delve deeper into the mechanisms through which acidophilus microbes influence soil nutrient dynamics and their practical applications in agriculture.

Economic Efficiency Effective nutrient management can lead to cost savings for farmers. By supplying only the nutrients that crops need, farmers can reduce the wastage of fertilizers and other inputs, making their operations more economically sustainable. Proper soil nutrient management helps prevent the overuse and runoff of fertilizers, which can lead to water pollution and harm ecosystems. Sustainable nutrient management practices protect water quality and reduce environmental degradation (Arora et al. 2018).

17.8   Conclusion and Future Prospect

The chapter presents collective information on acidophilus microorganisms and highlights their potential application in the agricultural sector. Adaptation of sustainable nutrient management practices results in environmental protection, climate change mitigation, food security, and long-term sustainability. This article will generate perception among scientists, researchers, and industrialists and will provoke them to use acidophilic microorganisms for soil nutrient management and for target-specific research for improving agricultural crop yield.

Screening, identification, and application of acidophilic microorganisms for sustainable agricultural crop production is the future demand for good soil nutrient management practices. Due to the vast applications of acidophiles, demand will further increase in the agriculture, food, medicine, and other enzyme industries (Akanksha and Kesari 2020). Extensive research will give warranty toward the exploration of acidophiles, their genes, and metabolic pathways to harness their potential for industrial processes. The biotechnological potential of acidophiles in agriculture has become more relevant in recent years for the development of new microbial inoculants. Hence, the manipulation of acidophilic genes/enzymes through genetic engineering could generate probable transgenics having superior phenotypes for various biotechnological applications in the foreseeable future (Garg et al. 2024).

References

Akanksha MV, Kesari KK (2020) Microbial cholesterol oxidase enzyme role in cardiovascular disease. In: Arora NK, Mishra V, Mishra J (eds) Microbial enzymes; role and application in Industries. Springer-Nature Publisher Germany, pp 303–317

Albert RA, Archambault J, Rosselló-Mora R et al (2005) Bacillus acidicola sp. nov., a novel mesophilic, acidophilic species isolated from acidic Sphagnum peat bogs in Wisconsin. Int J Syst Evol Microbiol 55:2125–2130. https://doi.org/10.1099/ijs.0.02337-0

Arora NK, Fatima T, Mishra I et al (2018) Environmental sustainability: challenges and viable solutions. Environ Sustain 1:309–340. https://doi.org/10.1007/s42398-018-00038-w

Belova SE, Ravin NV, Pankratov TA et al (2018) Hydrolytic capabilities as a key to environmental success: chitinolytic and cellulolytic acidobacteria from acidic sub-arctic soils and boreal peatlands. Front Microbiol 9. https://doi.org/10.3389/fmicb.2018.02775

Ben B (1975) Occurance of Sulfolobus acidophilus an extremely thermophilic acidophile bacterium, in New Zealand hot spring. Arch Microbiol 106:171–174

Chaudhary P, Singh S, Chaudhary A et al (2022) Overview of biofertilizers in crop production and stress management for sustainable agriculture. Front Plant Sci 13. https://doi.org/10.3389/fpls.2022.930340

Cheng DH, Liu QG, Xioa, Chaoochun (2019) Effect of intercropping on maize grain yield and yield components. Journal of integrative. Agriculture 18(8):1690–1700

Crits-Christoph A, Diamond S, Butterfield CN et al (2018) Novel soil bacteria possess diverse genes for secondary metabolite biosynthesis. Nature 558:440–444. https://doi.org/10.1038/s41586-018-0207-y

de Souza TSP, de Andrade CJ, Koblitz MGB, Fai AEC (2022) Microbial peptidase in food processing: current state of the art and future trends. Catal Lett 153:114–137. https://doi.org/10.1007/s10562-022-03965-w

Dedysh SN, Panikov NS, Tiedje JM (1998) Acidophilic Methanotrophic Communities from Sphagnum Peat Bogs. Appl Environ Microbiol. 64(3):922–929. https://doi.org/10.1128/aem.64.3.922-929.1998

Garg T, Dwivedi P, Mishra M et al (2024) Artificial intelligence in plant disease identification: empowering agriculture. In: Methods in microbiology. ISSN-0580-9517. https://doi.org/10.1016/bs.mim.2024.05.007

Gross & Glaser (2021) Meta analysis on manure application changes soil organic carbon storage. Sci Rep 11:5516

Hausmann B, Pelikan C, Herbold CW et al (2018) Peatland acidobacteria with a dissimilatory sulfur metabolism. ISME J 12:1729–1742. https://doi.org/10.1038/s41396-018-0077-1

Javaid R, Sabir A, Sheikh N, Ferhan M (2019) Recent advances in applications of acidophilic fungi to produce chemicals. Molecules 24:786. https://doi.org/10.3390/molecules24040786

Johnson DB, Quatrini R (2020) Acidophile microbiology in space and time. Curr Issues Mol Biol 63–76:63. https://doi.org/10.21775/cimb.039.063

Jung J, Kim J-S, Taffner J et al (2020) Archaea, tiny helpers of land plants. Comput Struct Biotechnol J 18:2494–2500. https://doi.org/10.1016/j.csbj.2020.09.005

Kanekar PP, Kanekar SP (2022) Acidophilic microorganisms. In: Diversity and biotechnology of Extremophilic microorganisms from India. Singapore, Springer Nature Singapore, pp 155–185

Kielak AM, Cipriano MAP, Kuramae EE (2016) Acidobacteria strains from subdivision 1 act as plant growth-promoting bacteria. Arch Microbiol 198:987–993. https://doi.org/10.1007/s00203-016-1260-2

Leandro IS, Marlon CP, André MXC, Victor HB, Moacir P, Joyce D (2023) Phosphorus-solubilizing microorganisms: a key to sustainable agriculture. Agriculture 13(2):462

Mahmoud K, Leduce L, Ferroni F (2005) Detection of Acidithiobacillus ferrooxidans in acid mine drainage environments using fluorescent in situ hybridization. J Microbiol Method 61:33–45

Mehmet AI, Sena Ö, Duygu A, Bayram K, Elena B, João MFR, Fatih O (2023) The impacts of acidophilic lactic acid bacteria on food and human health: a review of the current knowledge. Foods 12(15):2965

Pinto OHB, Costa FS, Rodrigues GR et al (2020) Soil acidobacteria strain AB23 resistance to oxidative stress through production of carotenoids. Microb Ecol 81:169–179. https://doi.org/10.1007/s00248-020-01548-z

Poomthongdee N, Duangmal K, Pathom-aree W (2015) Acidophilic actinomycetes from rhizosphere soil: diversity and properties beneficial to plants. J Antibiot 68(2):106–114

Rajeev L, Chen A, Kazakov AE et al (2015) Regulation of nitrite stress response in desulfovibrio vulgaris hildenborough, a model sulfate-reducing bacterium. J Bacteriol 197:3400–3408. https://doi.org/10.1128/jb.00319-15

Romoli R, Papaleo MC, de Pascale D et al (2011) Characterization of the volatile profile of Antarctic bacteria by using solid-phase microextraction-gas chromatography-mass spectrometry. J Mass Spectrom 46:1051–1059. https://doi.org/10.1002/jms.1987

Saleem S, Mushtaq NU, Rasool A, Shah WH, Tahir I, Rehman RU (2023) Plant nutrition and soil fertility: physiological and molecular avenues for crop improvement. Academic Press, In Sustainable Plant Nutrition, pp 23–49

Sharma A, Kawarabayasi Y, Satyanarayana T (2012) Acidophilic bacteria and archaea: acid stable biocatalysts and their potential applications. Extremophiles 16:1–19

Tamreihao K, Salam N, Ningthoujam DS (2018) Use of acidophilic or acidotolerant actinobacteria for sustainable agricultural production in acidic soils. Extremophiles in Eurasian Ecosystems: Ecology, Diversity, and Applications, pp 453–464

Vaishnav A, Sahu J, Singh HB (2020) Genomics of extremophiles for sustainable agriculture and biotechnological applications (part I). Curr Genomics 21(2):78

Ward NL, Challacombe JF, Janssen PH et al (2009) Three genomes from the phylum acidobacteria provide insight into the lifestyles of these microorganisms in soils. Appl Environ Microbiol 75:2046–2056. https://doi.org/10.1128/aem.02294-08

X Chen (2021) Thriving at low pH: adaptation mechanisms of acidophiles, vol 1. Chapter Metrics Overview. https://doi.org/10.5772/intechopen.96620

Yang P, Zhou X-F, Wang L-L et al (2018) Effect of phosphate-solubilizing bacteria on the mobility of insoluble cadmium and metabolic analysis. Int J Environ Res Public Health 15:1330. https://doi.org/10.3390/ijerph15071330

Part IV
Multi-omics Approach for Exploring Extremophiles

Chapter 18
Omics Technology in Food and Nutritional Security of Agricultural Crops: Role of Extremophiles

Tamana Khan, Sabba Khan, Diksha Singh, Aaqif Zaffar, Labiba Shah, Rizwan Rashid, Parvaze A. Sofi, Baseerat Afroza, and Sajad Majeed Zargar

18.1 Introduction

Survival of microorganisms under extreme environmental conditions like drought, varied pH, and extremes of temperature has been extensively studied by scientists and researchers over the years to unravel the genetic, physiological, and metabolic changes involved in their survival mechanisms. Such microorganisms harboring under extreme environmental niches are known as extremophiles. They have widespread importance in industry, environment, and agriculture. As far as the agricultural point of view is concerned, the extreme environmental conditions cause the plant soil to become unfavorable, thereby adversely affecting plant growth. Environmental stressors deplete crop productivity and hinder the progression of

T. Khan (✉)
Division of Vegetable Science; Faculty of Horticulture, Sher-e-Kashmir University of
Agricultural Sciences and Technology of Kashmir, Srinagar, Jammu and Kashmir, India

Proteomics Laboratory, Division of Plant Biotechnology, Faculty of Horticulture, Sher-e-
Kashmir University of Agricultural Sciences & Technology of Kashmir,
Srinagar, Jammu and Kashmir, India

S. Khan · L. Shah · R. Rashid · B. Afroza
Division of Vegetable Science; Faculty of Horticulture, Sher-e-Kashmir University of
Agricultural Sciences and Technology of Kashmir (SKUAST- Kashmir), Srinagar, Jammu and
Kashmir, India

D. Singh · A. Zaffar · S. M. Zargar (✉)
Proteomics Laboratory, Division of Plant Biotechnology, Faculty of Horticulture, Sher-e-
Kashmir University of Agricultural Sciences & Technology of Kashmir,
Srinagar, Jammu and Kashmir, India
e-mail: smzargar@skuastkashmir.ac.in

P. A. Sofi
Division of Genetics and Plant Breeding, Sher-e-Kashmir University of Agricultural Sciences
& Technology of Kashmir, Srinagar, Jammu and Kashmir, India

© The Author(s), under exclusive license to Springer Nature Switzerland AG 2024   405
A. Ranjan et al. (eds.), *Extremophiles for Sustainable Agriculture and Soil
Health Improvement*, https://doi.org/10.1007/978-3-031-70203-7_18

attaining sustainable yield. Instead of relying on chemical means to combat the ill effects of adverse soil conditions, the use of microbes isolated from extreme environments as a source of biofertilizers has been recently exploited under stressed soil scenarios (Igiehon and Babalola 2018). Such eco-friendly organisms used as plant growth promotors or biostimulant are generally referred to as plant growth-promoting rhizobacteria/microbes (PGPR/M), and arbuscular mycorrhiza fungi (AMF). There have been reports of plant-associated extremophiles having major roles in enhancing growth and plant adaptation under various abiotic/biotic stresses (Yadav 2017). These "plant-associated extremophilic microbes" or PAEM involve all realms from archaea to eukarya and work efficiently as biocontrol agents and biofertilizers maintaining soil health (Igiehon et al. 2019) and being a low-cost practice to support sustainable agriculture (Yadav et al. 2017). The extremophiles have adopted certain mechanisms to survive under harsh environmental conditions like having thermostable proteins, antioxidants, and osmolytes in higher concentration, ion-specific transporters to maintain pH, and they also maintain the fluidity of cell membranes, thereby protecting their genetic material. They produce some vital metabolites as a function of their unique genes, which remain stable under adverse environmental conditions known as extremozymes.

Detailed investigations on the diversity and survival mechanisms of extremophiles have been made possible through the exploitation of recently advanced omics technologies. Such genome-based studies are in place to decode their survival mechanisms at physiological, metabolic, and genetic levels and elucidate their importance in multiple biotechnological aspects and agriculture (Dumorné et al. 2017). Although the stability at the genomic, proteomic, and molecular levels of extremophiles is yet to be explored in multifaceted aspects; however, some very significant research is underway explaining their adaptive mechanism to varied temperature fluctuations (Orellana et al. 2010). Several factors responsible for the stability of extremophiles have been identified, which include H bond interactions (Vogt et al. 1997; Kumar et al. 2019), amino acid configuration (Zeldovich et al. 2007), composition and folding patterns of tRNA (Dutta and Chaudhuri 2010), the G + C content (Satapathy et al. 2010). Several factors responsible for the thermostability of proteins have been confirmed by some researchers (Perutz and Raidt 1975). Some experimental evidence suggests the role of certain biochemical mechanisms like lipid membrane constitution and expeditious reintegration of heat-inactivated molecules (Hawwa et al. 2009) in imparting survival abilities to extremophiles in extreme environments. Some studies have also confirmed the role of protein and DNA interactions in imparting structural, protein, and cell membrane stability to extremophiles (Hollien and Marqusee 1999; Gerday 2013; Moon et al. 2019). The database of complete genome sequencing of extremophiles (thermophiles and psychrophiles) has also been reported (Kumar et al. 2000) using shotgun sequencing method for meta-transcriptomics and metagenomics that permits assignment of extremophiles to functional groups, specific gene analysis, production of proteins, assignment to taxons, binning of genomes, etc. (Goodwin et al. 2016). With the emergence of advanced bioinformatic tools and well-grounded computational biology, the identification and filtration of contaminants preventing mixing up of genomic contigs and incomplete sequence alignment has been made easy and has simultaneously enhanced the efficiency of shotgun sequencing method (Taffner et al. 2018).

18.2 Concept of Extremophiles

Extremophiles are organisms that can survive under harsh conditions that would normally be inhospitable to most other life forms. They have evolved to be able to live and reproduce in an array of environments, including those with extreme heat or cold, high pressure, acidity or alkalinity, high salt, and even the presence of hazardous compounds. Deep-sea hydrothermal vents, polar ice caps, hot springs, salt flats, and acidic or alkaline lakes are just a few examples of the diverse places on Earth where these ecosystems can be found. Majorly prokaryotic organisms have dominated the evolution of life on Earth, which have developed to occupy almost every environmental niche. Extremophiles can survive in several extremes and are an important subject of study for several disciplines, from the investigation of environmental adaptations to harsh conditions and the elemental cycling in biogeochemistry. The limits of what conditions life can survive in over the past century have been stretched in every direction, embracing greater swathes of radiation, salt, energy, temperature, pH, pressure, and nutrition constraints (Table 18.1). In addition to thriving in such a comprehensive environmental occurrence on Earth, microorganisms are too capable of surviving the hostile state in space, which includes lack of gravity, vacuum pressure, high radiation, and dramatic temperature variations. The "extreme" environment is the norm that permits the organism's metabolic and biochemical processes, thus it is crucial to keep this in mind when thinking about extremophilic species. These organisms are highly suited for the conditions under consideration. The surface of our planet, particularly the subsurface, is home to miscellaneous ecosystems that display extremes in at least one physical or chemical characteristic. As a result, polyextremophiles and other extremophiles in general may be the most numerous life forms on Earth (Capece et al. 2013). Research on (poly)extremophiles, throughout the past few decades, contributed to several breakthroughs in molecular biology and medicine, including the abstraction of culturable (poly)extremophiles and the recognition of extreme microbial communities using a variety of culture-independent techniques (Babu et al. 2014; Coker 2016). Extremophile research has drawn a lot of interest from both the environmental and industrial aspects. In current years, there has been a surge in interest in studying extreme ecosystems that are home to a variety of extremophilic bacteria and their survival strategies. Studying the physiology of these extremophiles can assist in explaining how life evolved because some of them represented very primitive life forms that existed in environments that were very different from those of today.

Table 18.1 Classification of extremophiles based on nomenclature and ranges

pH	Temperature	Salinity	Pressure
Hyperalkaliphile (>pH 11)	Hyperthermophile (>80 °C)	Extreme halophile (>14.6%)	Hyperbarophile or hyperpiezophile (>50 MPa)
Alkaliphile(>pH 9)	Thermophile (45–80 °C)	Halophile (>8.8%)	Barophile or Piezophile (10–50 MPa)
Neutrophile (pH 5–9)	Mesophile (25–45 °C)	Halotolerant (>1.2–2.9%)	Piezotolerant or Barotolerant (0.1–10 MPa)
Acidophile (>pH 5)	Psychrophile (<20 °C)	Non-halophile (<1.2%)	Examples: *Thermococcus piezophilus, Shewanella oneidensis*
Hyperacidophile (>pH 3)	Examples: *Pyrolobus fumarii,* (113 °C), *Synechococcus lividis, Psychrobacter* sp., *Polaromonas vacuolata*	Examples: *Halobacterium salinarum, Danaliella salina*	
Examples: *Natronobacterium, Bacillus firmus OF4, spirulina* spp. (all pH 10.5) *Cyanidium caldarium, Ferroplasma* sp. (both pH 0)			

18.2.1   Thermophiles/Hyperthermophiles

In the early stages of Earth's history, the predominant life forms thrived in extreme conditions, relying on thermophilic anaerobes with chemolithoautotrophic or heterotrophic metabolisms, deriving sustenance from hydrothermal energy sources. These organisms, referred to as thermophiles and hyperthermophiles, thrived in environments characterized by high temperatures, with thermophiles flourishing at around 45 °C and hyperthermophiles thriving in temperatures as high as 80 °C. They could be found in various natural settings, including environments, such as composting facilities and anaerobic reactors, as well as hydrothermal vents, volcanic deposits, and hot springs. Their survival in such extreme environments hinges on their unique protein structure, metabolic activities, and extremophilic enzymes. Such unique proteins retain their functionality even under intense heat due to the resilience of their amino acid structures, preventing misfolding. Adaptation in these organisms involves modifications to essential proteins, enhancing their stability against high temperatures. Thermophilic proteins are characterized by an abundance of alpha-helical residues and shorter amino acid chains. To endure extreme heat, thermophiles employ various strategies, including boosting glycolysis for rapid energy production, activating heat shock proteins (HSPs) for protein folding assistance, advanced DNA repair, cellular structure stabilization with compatible solutes, and membrane fortification with specialized fatty acids and polyamines (Zeldes et al. 2015; Pedone et al. 2020).

18.2.2 Psychrophiles

Psychrophiles are organisms that can survive temperatures as low as $-15°C$, found in permafrost, polar ice, soils, alpine snowpacks, and ocean water. They belong to the bacteria, eukarya, and archaea kingdoms and have unique enzymes called "extremozymes" that help them survive in harsh environments. They also have short-chain fatty acids in their plasma membranes to protect cells from cold stress (Coker 2019), it has also been noted that low-temperature active proteins are produced. The production of antifreeze proteins (AFPs), cold-shock proteins (CSPs) (which protect RNA and protein synthesis), and chaperones, as well as the buildup of mannitol and specific cryoprotectants, are additional mechanisms for survival.

18.2.3 Xerophile

Xerophiles flourish in extremely arid conditions that can also be extremely hot or cold, including the Atacama Desert, the Great Basin, and the Antarctic. Xerophiles, like psychrophilic, can replace water with trehalose, protecting structures, and evading environmental stress, enabling them to thrive in arid environments. Adaptive mechanisms include the production of extracellular polymeric substances, osmoprotectants, DNA-repair proteins, and cell membrane modifications to prevent water loss and increase water retention.

18.2.4 Barophile (Piezophile)

Barophilic organisms are those that flourish at pressures of at least 400 atmospheres. A high-pressure ecosystem includes deep-sea habitats like the bottom of the ocean where pressure is nearly 380 atm). Subsurface rocks with high lithostatic pressures are another example. Those who are barotolerant can survive in less hostile conditions and under high pressures. A Gram-negative proteobacterium, *Halomonas salaria*, is an obligate barophile and requires a pressure of 1000 atm to survive. Phospholipids survive by maintaining their membrane fluid, which balances pressure between cell interior and exterior, and the surrounding environment (Shukla et al. 2020).

18.2.5   Alkaliphile

Microorganisms referred to as alkaliphiles exhibit the remarkable ability to thrive in environments with higher alkalinity, typically ranging from a pH of 8.5 to 11. Alkaliphiles employ electrogenic secondary cation/proton antiporters to regulate cytoplasmic pH and absorb excess H^+ ions. Another defense against alkaline conditions involves the formation of a protective barrier composed of acidic compounds like teichuronopeptide, teichuronic acid, and acidic amino acids surrounding the cell. Additionally, some alkaliphiles have evolved pH-resistant enzymes that sustain hydrophobic interactions and incorporate negatively charged amino acids at essential interfaces, enhancing their alkaline stability. Examples include outer membrane cytochrome C, which facilitates proton transfer across membranes, and phosphoserine aminotransferase (Mamo and Mattiasson 2016; Kulkarni et al. 2019).

18.2.6   Acidophiles

Acidophiles, commonly referred to as acidophilic organisms, can endure under extremely acidic conditions. These species can be found in archaea, bacteria, and eukarya. Microorganisms that are extremely acidophilic flourish at pH 3.0 or lower, whereas moderate acidophiles do the same from pH 3 to 5 (Sharma et al. 2016). To preserve acid-labile cellular components, acidophiles often maintain a cytoplasmic pH close to neutral, which calls for the creation of a sizable pH gradient. They adapt to an acidic environment using three distinct processes or strategies: improved mechanisms for DNA and protein repair, decreased permeability of the cell membrane, and proton flux pumping systems that prevent protons from entering the cytoplasm.

To survive in harsh environments, extremophiles have developed special adaptations and coping mechanisms. By examining these organisms, scientists can learn more about the molecular, cellular, and metabolic processes that enable life to persist and operate in adverse conditions. These modifications have effects on the medical, biotechnological, and astrobiological realms. The types of configurations and conditions that might support life beyond Earth are revealed by extremophiles. Scientists can uncover potential biosignatures and create techniques to look for life on other planets, moons, or exoplanets that experience comparable harsh conditions by researching extremophiles on Earth.

18.3   Plant Growth-Promoting Microbes

Plant growth-promoting microbes (PGPMs), which are commonly found in the rhizosphere (root zone) and non-rhizosphere soil and can be in a symbiotic relationship with plant roots, thereby promoting plant growth and development (Spaepen et al. 2009). PGPMs have garnered global attention as a bioinoculant that holds promise for addressing environmental issues in agroecosystems within the framework of sustainable agriculture. Moreover, they offer the potential to enhance agricultural productivity sustainably, yielding higher yields. Similarly, as a bioinoculant, PGPMs can fulfill diverse functions, including bioremediation of diverse hazardous wastes generated through human activities (Mishra et al. 2017). The rhizosphere, typically rich in nutrients, serves as a prime zone for the colonization of various native fungal and bacterial microbes (Nelson 2004).

There are two main categories of PGPMs: plant growth-promoting fungi (PGPF) and PGPR both representing highly helpful and valuable microbiota for plant health. PGPR specifically refer to soil-borne bacteria that support plant health and growth through diverse mechanisms. Typically, they reside on or within the roots, leaves, or tissues of plants (Glick 2012). Several species of bacteria that benefit plants have been identified, including *Enterobacter, Bacillus, Pseudomonas, Azospirillum, Alcaligenes, Rhizobium, Azotobacter, Klebsiella, Arthrobacter, Serratia*, and *Burkholderia* (Ahemad and Kibret 2014). The microbes mentioned exhibit various properties contributing to plant growth, including nitrogen fixation (Hirel et al. 2011), phosphate solubilization (Bahadir et al. 2018), and the synthesis of phytohormones (Maheshwari et al. 2015). These attributes of a PGPR support the improved growth and development of plants.

PGPF refer to fungi that inhabit the rhizosphere, and exhibit plant growth-promoting activities (Salas-Marina et al. 2011; Murali et al. 2012). *Trichoderma, Piriformospora, Fusarium, Aspergillus, Penicillium, Phoma*, and arbuscular mycorrhizal fungi (AMF) represent significant and important PGPF species prominent for enhancing plant growth and biocontrol of plant diseases (Hossain et al. 2017). PGPF contribute to plant growth through a variety of approaches directly or indirectly, which includes organic matter decomposition, phytohormone production, nutrient solubilization, and protection against both biotic and abiotic stresses (Khan et al. 2010; Olanrewaju and Babalola 2022). Endophytic microbial species, identified in a wide range of host plants, include *Azoarcus, Pantoea, Burkholderia, Gluconoacetobacter, Streptomyces, Enterobacter, Serratia, Nocardioides, Microbiospora, Herbaspirillum, Thermomonospora, Klebsiella, Pseudomonas, Planomonospora, Achromobacter*, and *Micromonospora* (Hallmann et al. 1997; Ryan et al. 2008; Verma et al. 2015). In the case of endophytes like *Rhizobium, Burkholderia*, or *Azospirillum*, which infect the roots, the production of proteases, chitinases, and cellulases has been shown to play a role in PGPMs by degrading the cell walls of plant pathogens (Razie and Anas 2008; Robledo et al. 2018).

PGPMs, like biofertilizers, improve nutrient availability by fixing nitrogen and solubilizing soil minerals like phosphorus and potassium. Some rhizobacteria produce siderophores, aiding iron absorption (Prasad et al. 2019). Moreover, they stimulate plant growth directly by affecting hormone metabolism and reducing ethylene levels (Martinez-Viveros et al. 2010). Additionally, they act indirectly as biopesticides, enhancing plant resistance against pathogens through various mechanisms (Abhilash et al. 2016; Khan et al. 2020).

18.3.1 Contribution of Extremophilic Microbiomes in Promoting Plant Growth Under Stressful Environments

PGPMs that thrive well in adverse conditions include acidophiles, thermophiles, halophiles, psychrophiles, and metal-resistant microorganisms and are primarily introduced to soil, roots, and seed. PGPMs are crucial in enhancing plant health by improving nutrient availability, modulating phytohormones, and boosting plant tolerance to abiotic and biotic stresses. They establish colonization in the plant rhizosphere, resulting in the buildup of osmolyte levels, antioxidants, and the expression of stress-responsive genes, while also bringing changes in root morphology to facilitate improved tolerance under unfavorable environmental conditions (Rajasreelatha and Thippeswamy 2022).

To undergo challenges posed by both biotic and abiotic stress, PGPMs have evolved adaptive characteristics, enabling them to thrive in one or more harsh habitats. Polyextremophiles, however, can grow optimally under multiple challenging conditions (Rajasreelatha and Thippeswamy 2022).

Phytohormones are important to normal plant health and growth and PGPMs have the ability to modulate these hormones. Microbial inoculants can influence the levels of various phytohormones, including gibberellins, auxin, cytokinin, abscisic acid, ACC-deaminase, strigolactones, jasmonates, and brassinosteroids (Saravanakumar 2011; Van Oosten et al. 2017; Arora et al. 2018). Microbial production of auxins is particularly beneficial as it regulates different processes, including cell division, vascular tissue differentiation, shoot growth, root surface area, root elongation, and lateral and adventitious root formation. ACC-deaminase, an enzyme synthesized by microorganisms, plays a critical role in diminishing ethylene levels, consequently alleviating stress in plants. High levels of ethylene can have detrimental effects on plants, leading to issues like leaf chlorosis, compromised photosynthesis, impaired leaf expansion, senescence, decay, decreased fruit yield, and impaired root growth. Certain species, such as *Bacillus* and *Pseudomonas* species, stimulate plant development by generating ACC-deaminase and enhancing auxin levels (Khoshru et al. 2020; Samaddar et al. 2019).

PGPMs possess the ability to generate exopolysaccharides, which actively participate in the creation of a protective biofilm covering the surface of plant roots. This biofilm assumes an essential role in improving water retention within soil particles, thereby sustaining optimal soil moisture levels in the root zone. As a result, it aids in shielding root cells from ionic and osmotic stresses by effectively regulating osmotic balance, even in demanding environmental conditions like fluctuating salinity, pH, drought, and extreme temperatures. Exopolysaccharides produced by PGPMs serve as a mechanism to mitigate stress. They aid in stabilizing the ionic equilibrium of the soil and immobilizing sodium ions (Na^+) during salt stress. The soil's antibacterial activity is enhanced by the exopolysaccharides that *Bacillus* species produce (Hashem et al. 2019). PGPMs, including *Pseudomonas* sp. and *Acinetobacter* sp., produce exopolysaccharides that have demonstrated the competence to impart tolerance to drought in pepper plants. By forming hydrophilic biofilms from EPS surrounding the roots, this is accomplished (Rolli et al. 2014). Additionally, specific strains of *Azospirillum lipoferum* have been identified to produce gibberellins and abscisic acid (ABA), enabling their maize plant hosts to regulate stomatal closure and modulate various stress signal transduction pathways, thereby effectively preventing water loss (Cohen et al. 2007).

Plant development is significantly aided by PGPM inoculants, which also improve resilience to abiotic stressors. This is accomplished by increasing antioxidant levels within cells and reducing the accumulation of ROS (Reactive Oxygen Species)-induced oxidative stress. Abiotic stresses disrupt cellular homeostasis and produce ROS, such as superoxide anion, hydrogen peroxide, singlet oxygen, and hydroxyl radical. These stresses include temperature swings, exposure to heavy metals, pH imbalances, UV-B radiation, and changes in water availability. One of the main effects of abiotic stresses is a rise in the concentration of ROS, which can be extremely harmful and induce oxidative stress in cells. ROS can cause oxidative damage to a variety of cellular constituents, such as proteins and lipids, which can impede enzyme activity and activate programmed cell death (Sharma et al. 2012).

Microbial inoculants play a crucial role in dealing with ROS by enhancing cellular defense mechanisms. They enhance the accumulation of various antioxidants, including catalase (CAT), superoxide dismutase (SOD), ascorbate peroxidase (AsA), glutathione (GSH), phenolics, carotenoids, and tocopherols. These antioxidants collectively contribute to the efficient neutralization and detoxification of ROS, thereby reducing oxidative damage to cellular structures (Gouda et al. 2018). These antioxidants act to protect cellular membranes and biomolecules from oxidative damage, thereby improving plant resilience to abiotic stresses.

In the face of both biotic and abiotic stress conditions, inoculants of these microbes play a significant role in stimulating the production of osmoregulants in plants. These osmoregulants include a range of substances such as amino acids, proteins, carbohydrates, lipids, glycine betaine, trehalose, and proline. Osmoregulation helps to uphold cellular homeostasis by preventing plasmolysis of the cell membrane, increasing the synthesis of HSPs, and regulating enzymatic processes. Microorganisms, especially extremophiles, contribute to enhancing plant resilience by promoting the accumulation of osmolytes within the cytoplasm of plant cells. This accumulation helps in maintaining the turgidity of cells and significantly contributes to heightened stress tolerance in plants. The osmoregulation mechanism emerges as an essential aspect of plant survival and improved resistance under extreme conditions, effectively minimizing the cellular damage induced by abiotic stress (Khoshru et al. 2020; Fernandez et al. 2012).

PGPMs have been extensively studied as biofertilizers, offering potential benefits such as enhanced nutrient supply, improvement in the growth of plants, and reduction in reliance on chemical fertilization (Fig. 18.1). Several microbial species, including *Rhizobium, Azotobacter, Clostridium, Azospirillum, Beijerinckia, Frankia, Nostoc, Bradyrhizobium, Anabaena*, and *Klebsiella* are known as fixers of biological nitrogen by a process called N_2 fixation (Prasad et al. 2019). Phosphorus, which is vital for phospholipid and adenosine triphosphate (ATP) production, is

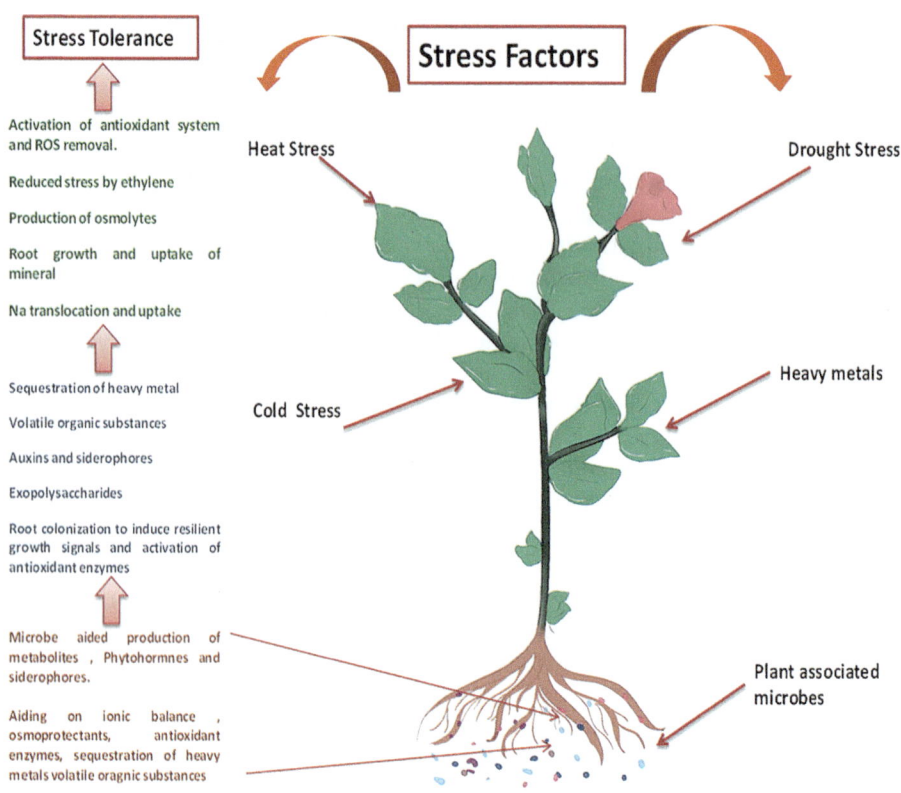

Fig. 18.1 PGPR mediates stress tolerance in plants

significant in enhancing photosynthesis. However, a significant portion of soil phosphorus exists in insoluble forms, making them unavailable to plants, which results in phosphorus deficiency. PGPMs can reduce the pH of soil by releasing organic acids like citrate, succinate, lactate, and gluconate, which aids in solubilization of calcium phosphates ($Ca_3[PO_4]_2$). Also, in acidic soils, these PGPMs help in raising pH through the production of proton at the point of ammonium assimilation, leading to aluminum phosphate ($AlPO_4$) and iron phosphate ($FePO_4$) solubilization (Martínez-Viveros et al. 2010). Sulfur is also a crucial macronutrient present in methionine and cysteine amino acids, which is essential for protein synthesis and enzyme function. Methionine acts as a precursor for ethylene, a hormone influencing fruit ripening. However, cysteine supports cell division (Taiz et al., 2022). The species of *Bacillus* produce some volatile compounds, including dimethyl disulfide, supplying sulfur to plants. Additionally, both *Bacillus* and *Aspergillus* generate inorganic and organic acids, and participate in chelation, exchange reactions, and acidolysis with the ability to solubilize potassium (Prasad et al. 2019).

The increasing global population necessitates a higher level of food production. However, environmental stresses pose significant challenges to the yield and growth of plants, resulting in lower productivity and impacting food security worldwide (Asghari et al. 2020; Scagliola et al. 2021). To address these challenges, the use of PGPMs has gained attention to enhance crop performance, improve food quality, and promote sustainable and environmentally friendly agricultural systems (Etesami and Adl 2020). PGPMs employ direct mechanisms, which include phosphate solubilization and synthesis of plant growth regulators and siderophores, while indirect mechanisms involve controlling phytopathogens and induction of systemic resistance (Yakhin et al. 2017; Olanrewaju and Babalola 2022).

18.3.2 Role of Extremophiles Under Adverse Soil Conditions

The soil is enriched by organisms, viz., bacteria, algae, protozoa, fungi, and underneath plant components. Numerous factors, including nutrient availability, temperature, water, and pH, affect the health of the microbes. Several researchers recently found beneficial microbes that aid plants, including crop plants, in surviving environmental challenges such as nutrient imbalance, salt, drought, and soil acidity and alkalinity. Therefore, based on current knowledge of each physical stress and the part microbes can play in agricultural management of these stresses, there is potential to understand better and go forward with a more sustainable agriculture. In the wild, pH is a crucial element in microbial growth. The dynamics of soil and plant-associated microbial populations have been directly connected to the acidity and alkalinity of soils. Animals, plants, and microbes can find essential natural resources like nutrients in soil. It is a system for maintaining life that offers a variety of essential ecosystem products and services, including carbon storage, water filtration, soil fertility, and agricultural output. Anions and cations must be present for any aqueous solution reaction to take place. In order to maintain biochemical equilibria, proper quantities of proton dissociable groups, and constant cell pH near neutral, a biological system's need for an acceptable pH is essential. Microbes require a certain pH balance to maintain physiological processes, much like any other living cell. The pH scale is used to express the alkalinity or acidity of the soil. The pH of the soil is a crucial factor that has a significant impact on the biological, chemical, and physical processes in the soil that directly affect plant growth and development. It is evident that soil pH affects agricultural productivity. The pH has a significant impact on the solubility, mobility, and bioavailability of trace elements in all soils. Plants face a variety of difficulties in soils that are either acidic or alkaline and fall beyond the range of optimal nutrient availability. Although plants' tolerance for pH extremes varies, most agricultural plants thrive at a pH close to neutral.

(a) Impact of Xerophilic Organisms
   A significant barrier to plant growth and development is water scarcity. A plant often employs its root structure and channels as adaptive mechanisms to fend off drought stress. Numerous secondary roots as well as evenly distributed space between primary and lateral roots make up the root architecture. The addition of

organisms that can withstand drought will enhance the root surface, encourage root and shoot development, and channel secondary roots for simple water and nutrient uptake (Gouda et al. 2018). Microbial exopolysaccharides are also recognized as processes used by extremophiles to enhance drought tolerance in plants (Ma et al. 2016). Numerous environmental studies have investigated the use of microbes as bioinoculants to reduce the issue of water scarcity in plant soil. By releasing trapped soil nutrients, *Bacillus* spp., *Proteus* spp., *Aneurinibacillus aneurinilyticus*, and *Alcaligenes* spp. increase the health of the soil (Patel et al. 2017).

(b) Impact of Temperature Extremophiles

Significant abiotic stress caused by climate change, global warming, and other associated activities is extreme temperature, which prevents plants from sprouting easily. The capacity of psychrotrophs and thermophiles to attribute PGP features either direct or indirect is useful in agricultural applications and has recently gained interest in these organisms. These features and secretions include things like cytokinin, IAA, mitigating environmental stress, and the solubilization of crucial soil nutrients (Verma et al. 2017).

(c) Effect of Halotolerant Organisms

Halophiles and halotolerant organisms actively expend energy to prevent protein aggregation by removing salt from the cytoplasm. A significant proportion of halophilic organisms are adapted and thrive in high-salt environments, enhancing plants' ability to withstand salinity (Ojuederie et al. 2019). Additionally, they produce biofilms and exopolysaccharides, which help regulate soil salinity and mitigate osmotic shock. Halophiles generate polysaccharide molecules that bind rapidly with Na^+ to form chelates, reducing desiccation and ionic toxicity. This process diminishes Na^+ in the rhizosphere, promoting plant growth (Babalola 2010).

(d) Influence of Acidic and Alkaline Tolerant Microbes

Numerous research on soil pH has shown that soil acidity, soil inoculum development, and increased exchangeable sodium percentage viscosity all interact to affect the K-solubilization of rock-forming minerals. Only species capable of tolerating extremely high soil pH levels can overcome this obstacle. The presence of Na_2CO_3 and $NaHCO_3$ can elevate soil pH, disrupting the availability of essential nutrients such as phosphorus, manganese, zinc, and iron. This can lead to osmotic stress and nutrient deficiencies impacting both plant biological processes and the beneficial soil microbiome. Extreme pH ranges are known to support a sizable proportion of beneficial microbes to plants and sources of secondary metabolites. The majority of these organisms are helpful for biotechnology tests and have been used as bioinoculants. Regardless of where these extremophiles are found close to other plants, they always exhibit PGP features in soil with a low or high pH and perform additional tasks, including pathogen biological control. However, P-solubilizing organisms and their capacity to create acid aid in the improvement of plants cultivated under various soil stressors.

18.4 Advancement in Omics Sequencing Techniques for Extreme Soil Microbiome

Over the past few decades, numerous "omics" strategies have evolved as effective technologies for plant systems. Using omics, researchers have been able to identify the ecological functions of organisms in the rhizosphere, endosphere, and phyllosphere, which are close to plants that are cultivated under harsh soil conditions over the past two decades. In order to increase crop yield, metagenomics has also shown the tripartite and synergistic interactions of two or more extremophiles. The integration of functional genomics with other omics highlights the connections between crop genomes and phenotypes under specific physiological and environmental circumstances. Molecular regulatory networks for crop development can be better understood by combining systems biology with multi-omics datasets. Integration of multi-omics and systems biology may be beneficial for crop improvement under environmental stress (Yang et al. 2021).

Utilizing microbes, particularly extremophiles, in the plant biosphere could assist and ease the limitations caused by environmental stress. Additionally, co-inoculating these organisms with cofactors, including proline, micronutrients, and caffeic acid, may aid in improving the control of soil stressors. Although crop cross-breeding and gene engineering have improved plants when compared to biofertilization, they are more costly and time-consuming. To ascertain whether these organisms have the capacity to eliminate environmental soil constraints that limit plant growth will require a focused effort in the future, including the application of omics, to investigate the potential of using extremophiles as bioinoculants.

18.5 Multi-Omics Technologies for the Exploitation of Extremophiles for Crop Improvement

Agriculture has been the basis of human existence since the beginning of the modern era, and science has been constantly seeking to improve agricultural methods and their results. Several factors like harsh environmental factors in the face of biotic and abiotic stressors, limited and sparse space, pollution, and population explosion have put a setback to agricultural progress. To combat such impacts, the exploitation of extremophiles and their vital metabolites can help in improving sustainable agricultural practices for the benefit of the human race. The identified barren lands which are uncultivable with limited resources can be made productive with the help of extremophile application, thereby introducing and enhancing agriculture in such areas.

The novel enzymes involved in extremophilic activity can be identified using "omics" technology. Application of metagenomics (Dávila-Ramos et al. 2019; Gómez-Silva et al. 2019), meta-transcriptomics (Macklaim and Gloor 2018), metabolomics, and meta-proteomics (Li et al. 2014; Hart et al. 2018) in developing

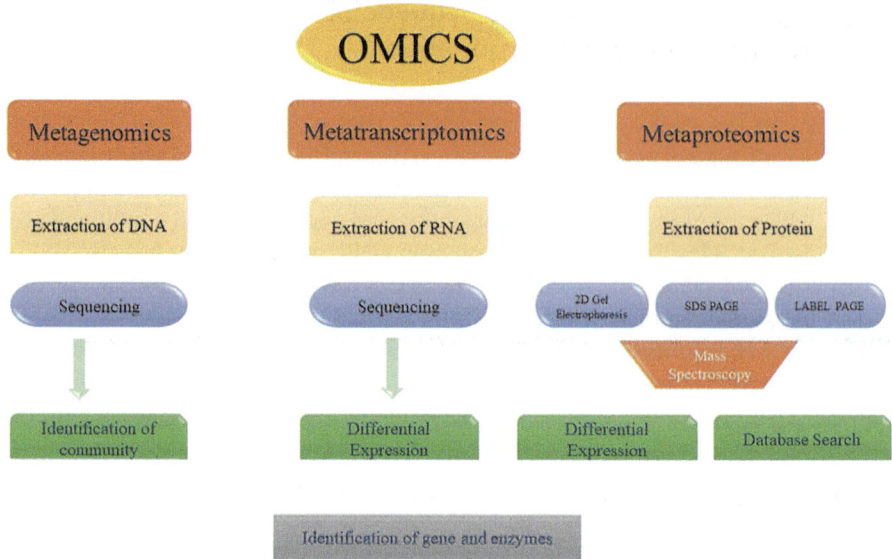

Fig. 18.2 Multi-omics concept and simple operation. Identification of genes and proteins for bio-technological use based on such concept

insights about the applicability of extremophiles in agriculture is indispensable (Zhu et al. 2020) (Fig. 18.2). Application of pertinent genomic approaches to decode the enzymatic and transport mechanisms, and metabolic pathways as a part of adaptive strategies of extremophiles, has been strongly put in place. Protein modifications involve gene shuffling and in situ mutagenesis approaches using bioinformatic tools and certain algorithms, while some recombinant DNA technologies and proteomics approach help in modifying enzymatic activities rendering thermal stability, tolerance to varied pH, etc. (Zhu et al. 2020). The application of the metagenomic to assess the composition of extremophiles using 16S rRNA sequencing markers (Chan et al. 2015; Mardanov et al. 2017) is well documented. However, a better approach is to analyze the whole genome through the shotgun metagenome technique where apart from decoding the functionality of microbiomes, the coding genes having a role in metabolic pathway may also be identified (López-López et al. 2014; Ferrer et al. 2015). The shotgun method of DNA sequencing together with advanced bioinformatic tools has paved the way for reconstitution phylogeny, identification of genomic information, and functions of novel microbiomes. Omics technology has opened new means of decoding the complexity of unculturable microbial taxonomy, which is of great relevance (Nayfach et al. 2015). Omics technology has greatly contributed to identifying the ecological roles of organisms residing in the proximity of the rhizosphere, phyllosphere, and endosphere of adverse soils having plant growth. There have been reports of some novel metabolic compounds being isolated from extremophiles using such technologies, which include the production of siderophore for plant growth-promoting activity (PGPA), antimicrobial activities against certain plant pests, and anti- inflammatory effects (Lubna et al. 2019; Suksaard et al. 2018). The metagenomics approach has significantly contributed to improved plant yield by discovering some synergistic and tripartite interactions between extremophiles (Lubna et al. 2019). Sequencing methods along with the metabolomics approach have highlighted the significant contributions of soil micro-biomes in dealing with abiotic/biotic stresses, fixing N_2 and solubilizing minerals, and other plant growth-promoting traits (Taffner et al. 2018). One classic example of omics application in exploitation of extremophiles as biofertilizer and a source of plant lignocelluloses degradation is *Paenibacillus polymyxa* CR1, which has been isolated from the rhizosphere of maize (Weselowski et al. 2016).

Apart from the sequencing and metagenomics strategy, transcriptomics and pro-teomics approaches have been the source of expressed gene identification and have paved the way for analyzing gene expression under extreme conditions and some physiological responses to such conditions (Schneider et al.; Wilmes and Bond 2006). The metatranscriptome approach reveals transcriptional activity under extremes of the environment (Hassa et al. 2018) but the correlation between the mRNA and the proteins expressed cannot be relied upon due to the post-transcriptional modifications (Pradet-Balade et al. 2001). On account of this, a combined omics approach can be a better alternative to decode the functionality and enzyme expressions under extreme conditions (Hassa et al. 2018). However, the study of extremophilic microbial communities being an emerging field has very scarce information in the literature. Application of multi-omic approaches, including DNA, RNA sequencing, and mass spectrometry, on extremophile *Thermus filiformis* identified some macromolecules, which were thermostable under varied temperature fluctuations and revealed changes in their physiological state (Mandelli et al. 2017).

18.6 Conclusion

Environmental stresses are the major constraints posing a threat to agricultural productivity and food safety. Adoption of eco-friendly means of soil enhancement through the use of extremophiles as bioinoculants and biofertilizers ensures enhanced plant/soil health and increased agricultural production. Integration of plant-associated extremophilic microbes to combat stresses ensures cost-effective cultivation practices and poses no threat to environmental health. Combined omics technologies like genomics, transcriptomics, and proteomics have been a great resource for discovering the potential of novel extremophiles and their extremo-zymes to combat environmental stresses and enhance crop productivity. However, the existing information about extremophiles and their application in defined environments is very scarce. Therefore, a comprehensive gene expression study is needed which can help divulge the microbial diversity and isolation of diversely catalytic extremozymes by screening gene expression libraries with sophisticated detection techniques. A refined study of extremozymes using high-end omics techniques can help in tapping extremophiles at structural, transcriptomic, and proteomic levels, providing a precious resource for sustainability.

References

Abhilash PC, Dubey RK, Tripathi V et al (2016) Plant growth-promoting microorganisms for environmental sustainability. Trends Biotechnol 34:847–850. https://doi.org/10.1016/j. tibtech.2016.05.005

Ahemad M, Kibret M (2014) Mechanisms and applications of plant growth promoting rhizobacteria: current perspective. J King Saud Univ Sci 26:1–20. https://doi.org/10.1016/j. jksus.2013.05.001

Arora NK, Fatima T, Mishra I et al (2018) Environmental sustainability: challenges and viable solutions. Environ Sustain 1:309–340. https://doi.org/10.1007/s42398-018-00038-w

Asghari B, Khademian R, Sedaghati B (2020) Plant growth promoting rhizobacteria (PGPR) confer drought resistance and stimulate biosynthesis of secondary metabolites in pennyroyal (Mentha pulegium L.) under water shortage condition. Sci Hortic 263:109132. https://doi. org/10.1016/j.scienta.2019.109132

Babalola OO (2010) Beneficial bacteria of agricultural importance. Biotechnol Lett 32:1559–1570. https://doi.org/10.1007/s10529-010-0347-0

Babu P, Chandel AK, Singh OV (2014) Survival mechanisms of extremophiles. Springer Briefs in Microbiology, pp 9–23. https://doi.org/10.1007/978-3-319-12808-5_2

Bahadir PS, Liaqat F, Eltem R (2018) Plant growth promoting properties of phosphate solubilizing bacillus species isolated from the Aegean region of Turkey. Turk J Bot 42:183–196. https://doi. org/10.3906/bot-1706-51

Capece MC, Clark E, Saleh JK et al (2013) Polyextremophiles and the constraints for terrestrial habitability. Cellular Origin, Life in Extreme Habitats and Astrobiology 3–59. https://doi. org/10.1007/978-94-007-6488-0_1

Chan CS, Chan K-G, Tay Y-L et al (2015) Diversity of thermophiles in a Malaysian hot spring determined using 16S rRNA and shotgun metagenome sequencing. Front Microbiol 6. https:// doi.org/10.3389/fmicb.2015.00177

Cohen AC, Bottini R, Piccoli PN (2007) Azospirillum brasilense Sp 245 produces ABA in chemically-defined culture medium and increases ABA content in arabidopsis plants. Plant Growth Regul 54:97–103. https://doi.org/10.1007/s10725-007-9232-9

Coker JA (2016) Extremophiles and biotechnology: current uses and prospects. F1000Research 5:396. https://doi.org/10.12688/f1000research.7432.1

Coker JA (2019) Recent advances in understanding extremophiles. F1000Research 8:1917. https:// doi.org/10.12688/f1000research.20765.1

Dávila-Ramos S, Castelán-Sánchez HG, Martínez-Ávila L et al (2019) A review on viral metagenomics in extreme environments. Front Microbiol 10. https://doi.org/10.3389/fmicb.2019.02403

Dumorne K, Cordova DC, Astorga-Elo M, Renganathan P (2017) Extremozymes: a potential source for industrial applications. J Microbiol Biotechnol 27:649–659. https://doi.org/10.4014/ jmb.1611.11006

Dutta A, Chaudhuri K (2010) Analysis of tRNA composition and folding in psychrophilic, mesophilic and thermophilic genomes: indications for thermal adaptation. FEMS Microbiol Lett 305:100–108. https://doi.org/10.1111/j.1574-6968.2010.01922.x

Etesami H, Adl SM (2020) Can interaction between silicon and non–rhizobial bacteria help in improving nodulation and nitrogen fixation in salinity–stressed legumes? A Rev Rhizosphere 15:100229. https://doi.org/10.1016/j.rhisph.2020.100229

Fernandez O, Theocharis A, Bordiec S et al (2012) Burkholderia phytofirmansPsJN acclimates grapevine to cold by modulating carbohydrate metabolism. Mol Plant-Microbe Interact 25:496–504. https://doi.org/10.1094/mpmi-09-11-0245

Ferrer M, Martínez-Martínez M, Bargiela R et al (2015) Estimating the success of enzyme bioprospecting through metagenomics: current status and future trends. Microb Biotechnol 9:22–34. https://doi.org/10.1111/1751-7915.12309

Gerday C (2013) Psychrophily and catalysis. Biology 2:719–741. https://doi.org/10.3390/ biology2020719

Glick BR (2012) Plant growth-promoting bacteria: mechanisms and applications. Scientifica 2012:1–15. https://doi.org/10.6064/2012/963401

Gómez-Silva B, Vilo-Muñoz C, Galetović A et al (2019) Metagenomics of Atacama Lithobiontic extremophile life unveils highlights on fungal communities, biogeochemical cycles and carbohydrate-active enzymes. Microorganisms 7:619. https://doi.org/10.3390/microorganisms7120619

Goodwin S, McPherson JD, McCombie WR (2016) Coming of age: ten years of next-generation sequencing technologies. Nat Rev Genet 17:333–351. https://doi.org/10.1038/nrg.2016.49

Gouda S, Kerry RG, Das G et al (2018) Revitalization of plant growth promoting rhizobacteria for sustainable development in agriculture. Microbiol Res 206:131–140. https://doi.org/10.1016/j.micres.2017.08.016

Hallmann J, Quadt-Hallmann A, Mahaffee WF, Kloepper JW (1997) Bacterial endophytes in agricultural crops. Can J Microbiol 43:895–914. https://doi.org/10.1139/m97-131

Hart EH, Creevey CJ, Hitch T, Kingston-Smith AH (2018) Meta-proteomics of rumen microbiota indicates niche compartmentalisation and functional dominance in a limited number of metabolic pathways between abundant bacteria. Sci Rep 8:10504. https://doi.org/10.1038/s41598-018-28827-7

Hashem A, Tabassum B, Fathi Abd Allah E (2019) Bacillus subtilis: a plant-growth promoting rhizobacterium that also impacts biotic stress. Saudi J Biol Sci 26:1291–1297. https://doi.org/10.1016/j.sjbs.2019.05.004

Hassa J, Maus I, Off S et al (2018) Metagenome, metatranscriptome, and metaproteome approaches unraveled compositions and functional relationships of microbial communities residing in biogas plants. Appl Microbiol Biotechnol 102:5045–5063. https://doi.org/10.1007/s00253-018-8976-7

Hawwa R, Aikens J, Turner RJ et al (2009) Structural basis for thermostability revealed through the identification and characterization of a highly thermostable phosphotriesterase-like lactonase from Geobacillus stearothermophilus. Arch Biochem Biophys 488:109–120. https://doi.org/10.1016/j.abb.2009.06.005

Hirel B, Tétu T, Lea PJ, Dubois F (2011) Improving nitrogen use efficiency in crops for sustainable agriculture. Sustain For 3:1452–1485. https://doi.org/10.3390/su3091452

Hollien J, Marqusee S (1999) Structural distribution of stability in a thermophilic enzyme. Proc Natl Acad Sci 96:13674–13678. https://doi.org/10.1073/pnas.96.24.13674

Hossain MM, Sultana F, Islam S (2017) Plant growth-promoting fungi (PGPF): Phytostimulation and induced systemic resistance. Plant-Microbe Interactions in Agro-Ecological Perspectives 135–191. https://doi.org/10.1007/978-981-10-6593-4_6

Igiehon N, Babalola O (2018) Rhizosphere microbiome modulators: contributions of nitrogen fixing bacteria towards sustainable agriculture. Int J Environ Res Public Health 15:574. https://doi.org/10.3390/ijerph15040574

Igiehon NO, Babalola OO, Aremu BR (2019) Genomic insights into plant growth promoting rhizobia capable of enhancing soybean germination under drought stress. BMC Microbiol 19:159. https://doi.org/10.1186/s12866-019-1536-1

Khan MS, Zaidi A, Ahemad M et al (2010) Plant growth promotion by phosphate solubilizing fungi—current perspective. Arch Agron Soil Sci 56:73–98. https://doi.org/10.1080/03650340902806469

Khan N, Bano A, Ali S, Babar MA (2020) Crosstalk amongst phytohormones from planta and PGPR under biotic and abiotic stresses. Plant Growth Regul 90:189–203. https://doi.org/10.1007/s10725-020-00571-x

Khoshru B, Mitra D, Khoshmanzar E et al (2020) Current scenario and future prospects of plant growth-promoting rhizobacteria: an economic valuable resource for the agriculture revival under stressful conditions. J Plant Nutr 43:3062–3092. https://doi.org/10.1080/01904167.2020.1799004

Kulkarni S, Dhakar K, Joshi A (2019) Alkaliphiles. Microbial Diversity in the Genomic Era, pp 239–263. https://doi.org/10.1016/b978-0-12-814849-5.00015-0

Kumar S, Tsai C-J, Nussinov R (2000) Factors enhancing protein thermostability. Protein Eng Des Sel 13:179–191. https://doi.org/10.1093/protein/13.3.179

Kumar S, Dangi AK, Shukla P et al (2019) Thermozymes: adaptive strategies and tools for their biotechnological applications. Bioresour Technol 278:372–382. https://doi.org/10.1016/j.biortech.2019.01.088

Li W, Xu H, Xiao T et al (2014) MAGeCK enables robust identification of essential genes from genome-scale CRISPR/Cas9 knockout screens. Genome Biol 15:554. https://doi.org/10.1186/s13059-014-0554-4

Lopez-Lopez O, Cerdan M, Siso M (2014) New extremophilic lipases and esterases from metagenomics. Curr Protein Pept Sci 15:445–455. https://doi.org/10.2174/1389203715666140228153801

Lubna AS, Khan AL et al (2019) Growth-promoting bioactivities of Bipolaris sp. CSL-1 isolated from Cannabis sativa suggest a distinctive role in modifying host plant phenotypic plasticity and functions. Acta Physiol Plant 41. https://doi.org/10.1007/s11738-019-2852-7

Ma Y, Rajkumar M, Zhang C, Freitas H (2016) Inoculation of Brassica oxyrrhina with plant growth promoting bacteria for the improvement of heavy metal phytoremediation under drought conditions. J Hazard Mater 320:36–44. https://doi.org/10.1016/j.jhazmat.2016.08.009

Macklaim JM, Gloor GB (2018) From RNA-seq to biological inference: using compositional data analysis in meta-transcriptomics. Methods Mol Biol:193–213. https://doi.org/10.1007/978-1-4939-8728-3_13

Maheshwari DK, Dheeman S, Agarwal M (2015) Phytohormone-producing PGPR for sustainable agriculture. Bacterial Metabolites in Sustainable Agroecosystem, pp 159–182. https://doi.org/10.1007/978-3-319-24654-3_7

Mamo G, Mattiasson B (2016) Alkaliphilic microorganisms in biotechnology. Biotechnology of Extremophiles: 243–272. https://doi.org/10.1007/978-3-319-13521-2_8

Mandelli F, Couger MB, Paixão DAA et al (2017) Thermal adaptation strategies of the extremophile bacterium Thermus filiformis based on multi-omics analysis. Extremophiles 21:775–788. https://doi.org/10.1007/s00792-017-0942-2

Mardanov AV, Gumerov VM, Beletsky AV, Ravin NV (2017) Microbial diversity in acidic thermal pools in the Uzon Caldera, Kamchatka. Antonie Van Leeuwenhoek 111:35–43. https://doi.org/10.1007/s10482-017-0924-5

Martínez-Viveros O, Jorquera MA, Crowley DE et al (2010) Mechanisms and practical considerations involved in plant growth promotion by RHIZOBACTERIA. J Soil Sci Plant Nutr 10. https://doi.org/10.4067/s0718-95162010000100006

Mishra J, Singh R, Arora NK (2017) Plant growth-promoting microbes: diverse roles in agriculture and environmental sustainability. Probiotics and Plant Health 71–111. https://doi.org/10.1007/978-981-10-3473-2_4

Moon S, Kim J, Koo J, Bae E (2019) Structural and mutational analyses of psychrophilic and mesophilic adenylate kinases highlight the role of hydrophobic interactions in protein thermal stability. Structural Dynamics 6:024702. https://doi.org/10.1063/1.5089707

Murali M, Sudisha J, Amruthesh KN et al (2012) Rhizosphere fungus Penicillium chrysogenum promotes growth and induces defence-related genes and downy mildew disease resistance in pearl millet. Plant Biol 15:111–118. https://doi.org/10.1111/j.1438-8677.2012.00617.x

Nayfach S, Bradley PH, Wyman SK et al (2015) Automated and accurate estimation of gene family abundance from shotgun metagenomes. PLoS Comput Biol 11:e1004573. https://doi.org/10.1371/journal.pcbi.1004573

Nelson LM (2004) Plant growth promoting rhizobacteria (PGPR): prospects for new inoculants. Crop Management 3:1–7. https://doi.org/10.1094/cm-2004-0301-05-rv

Ojuederie O, Olanrewaju O, Babalola O (2019) Plant growth promoting rhizobacterial mitigation of drought stress in crop plants: implications for sustainable agriculture. Agronomy 9:712. https://doi.org/10.3390/agronomy9110712

Olanrewaju OS, Babalola OO (2022) Plant growth-promoting rhizobacteria for orphan legume production: focus on yield and disease resistance in Bambara groundnut. Frontiers in Sustainable Food Systems 6. https://doi.org/10.3389/fsufs.2022.922156

Orellana S, Yañez M, Espinoza A et al (2010) The transcription factor SlAREB1 confers drought, salt stress tolerance and regulates biotic and abiotic stress-related genes in tomato. Plant Cell Environ 33:2191–2208. https://doi.org/10.1111/j.1365-3040.2010.02220.x

Patel S, Jinal HN, Amaresan N (2017) Isolation and characterization of drought resistance bacteria for plant growth promoting properties and their effect on chilli (Capsicum annuum) seedling under salt stress. Biocatal Agric Biotechnol 12:85–89. https://doi.org/10.1016/j.bcab.2017.09.002

Pedone E, Fiorentino G, Bartolucci S, Limauro D (2020) Enzymatic antioxidant signatures in Hyperthermophilic archaea. Antioxidants 9:703. https://doi.org/10.3390/antiox9080703

Perutz MF, Raidt H (1975) Stereochemical basis of heat stability in bacterial ferredoxins and in haemoglobin A2. Nature 255:256–259. https://doi.org/10.1038/255256a0

Pradet-Balade B, Boulmé F, Beug H et al (2001) Translation control: bridging the gap between genomics and proteomics? Trends Biochem Sci 26:225–229. https://doi.org/10.1016/s0968-0004(00)01776-x

Prasad M, Srinivasan R, Chaudhary M et al (2019) Plant growth promoting Rhizobacteria (PGPR) for sustainable agriculture. PGPR Amelioration in Sustainable Agriculture 129–157. https://doi.org/10.1016/b978-0-12-815879-1.00007-0

Rajasreelatha V, Thippeswamy M (2022) The role of plant growth promoting extremophilic microbiomes under stressful environments. J Stress Physiol Biochem 18:16–33

Razie F, Anas I (2008) Effect of azotobacter and azospirillum on growth and yield of rice grown on tidal swamp rice field in South Kalimantan. Jurnal Ilmu Tanah dan Lingkungan 10:41–45. https://doi.org/10.29244/jitl.10.2.41-45

Robledo M, Menéndez E, Jiménez-Zurdo JI et al (2018) Heterologous expression of rhizobial CelC2 Cellulase impairs symbiotic signaling and nodulation in medicago truncatula. Mol Plant-Microbe Interact 31:568–575. https://doi.org/10.1094/mpmi-11-17-0265-r

Rolli E, Marasco R, Vigani G et al (2014) Improved plant resistance to drought is promoted by the root-associated microbiome as a water stress-dependent trait. Environ Microbiol 17:316–331. https://doi.org/10.1111/1462-2920.12439

Ryan RP, Germaine K, Franks A et al (2008) Bacterial endophytes: recent developments and applications. FEMS Microbiol Lett 278:1–9. https://doi.org/10.1111/j.1574-6968.2007.00918.x

Salas-Marina MA (2011) The plant growth-promoting fungus aspergillus ustus promotes growth and induces resistance against different lifestyle pathogens in arabidopsis thaliana. J Microbiol Biotechnol 21:686–696. https://doi.org/10.4014/jmb.1101.01012

Samaddar S, Chatterjee P, Roy Choudhury A et al (2019) Interactions between pseudomonas spp. and their role in improving the red pepper plant growth under salinity stress. Microbiol Res 219:66–73. https://doi.org/10.1016/j.micres.2018.11.005

Saravanakumar D (2011) Rhizobacterial ACC deaminase in plant growth and stress amelioration. Bacteria in Agrobiology: Stress Management 187–204. https://doi.org/10.1007/978-3-642-23465-1_9

Satapathy SS, Dutta M, Ray SK (2010) Higher tRNA diversity in thermophilic bacteria: a possible adaptation to growth at high temperature. Microbiol Res 165:609–616. https://doi.org/10.1016/j.micres.2009.12.003

Scagliola M, Valentinuzzi F, Mimmo T et al (2021) Bioinoculants as promising complement of chemical fertilizers for a more sustainable agricultural practice. Front Sustain Food Syst 4. https://doi.org/10.3389/fsufs.2020.622169

Sharma P, Jha AB, Dubey RS, Pessarakli M (2012) Reactive oxygen species, oxidative damage, and antioxidative defense mechanism in plants under stressful conditions. J Bot 2012:1–26. https://doi.org/10.1155/2012/217037

Sharma A, Parashar D, Satyanarayana T (2016) Acidophilic microbes: biology and applications. Biotechnology of Extremophiles: 215–241. https://doi.org/10.1007/978-3-319-13521-2_7

Shukla PJ, Bhatt VD, Suriya J, Mootapally C (2020) Marine Extremophiles. Encyclopedia of Marine Biotechnology 1753–1771. https://doi.org/10.1002/9781119143802.ch74

Spaepen S, Vanderleyden J, Okon Y (2009) Chapter 7 plant growth-promoting actions of rhizobacteria. Adv Bot Res:283–320. https://doi.org/10.1016/s0065-2296(09)51007-5

Suksaard P, Srisuk N, Duangmal K (2018) Saccharopolyspora maritima sp. nov., an actinomycete isolated from mangrove sediment. Int J Syst Evol Microbiol 68:3022–3027. https://doi.org/10.1099/ijsem.0.002941

Taffner J, Erlacher A, Bragina A, et al (2018) What is the role of archaea in plants? New insights from the vegetation of alpine bogs mSphere 3. https://doi.org/10.1128/msphere.00122-18

Taiz L, Møller IM, Murphy A, Zieger E (2022) Plant physiology and development. Sinauer Associates, Incorporated

Van Oosten MJ, Pepe O, De Pascale S et al (2017) The role of biostimulants and bioeffectors as alleviators of abiotic stress in crop plants. Chem Biol Technol Agric 4. https://doi.org/10.1186/s40538-017-0089-5

Verma P, Yadav AN, Khannam KS et al (2015) Assessment of genetic diversity and plant growth promoting attributes of psychrotolerant bacteria allied with wheat (Triticum aestivum) from the northern hills zone of India. Ann Microbiol 65:1885–1899. https://doi.org/10.1007/s13213-014-1027-4

Verma P, Yadav AN, Kumar V et al (2017) Beneficial plant-microbes interactions: biodiversity of microbes from diverse extreme environments and its impact for crop improvement. In: Plant-microbe interactions in agro-ecological perspectives, pp 543–580. https://doi.org/10.1007/978-981-10-6593-4_22

Vogt G, Woell S, Argos P (1997) Protein thermal stability, hydrogen bonds, and ion pairs. J Mol Biol 269:631–643. https://doi.org/10.1006/jmbi.1997.1042

Weselowski B, Nathoo N, Eastman AW et al (2016) Isolation, identification and characterization of Paenibacillus polymyxa CR1 with potentials for biopesticide, biofertilization, biomass degradation and biofuel production. BMC Microbiol 16:244. https://doi.org/10.1186/s12866-016-0860-y

Wilmes P, Bond PL (2006) Metaproteomics: studying functional gene expression in microbial ecosystems. Trends Microbiol 14:92–97. https://doi.org/10.1016/j.tim.2005.12.006

Yadav AN (2017) Agriculturally important micro biomes: biodiversity and multifarious PGP attributes for amelioration of diverse abiotic stresses in crops for sustainable agriculture. Biomed J Sci Techn Res 1. https://doi.org/10.26717/bjstr.2017.01.000321

Yakhin OI, Lubyanov AA, Yakhin IA, Brown PH (2017) Biostimulants in plant science: a global perspective. Front Plant Sci 7. https://doi.org/10.3389/fpls.2016.02049

Yang Y, Saand MA, Huang L et al (2021) Applications of multi-omics technologies for crop improvement. Front Plant Sci 12. https://doi.org/10.3389/fpls.2021.563953

Zeldes BM, Keller MW, Loder AJ et al (2015) Extremely thermophilic microorganisms as metabolic engineering platforms for production of fuels and industrial chemicals. Front Microbiol 6. https://doi.org/10.3389/fmicb.2015.01209

Zeldovich KB, Berezovsky IN, Shakhnovich EI (2007) Protein and DNA sequence determinants of thermophilic adaptation. PLoS Comput Biol 3:e5. https://doi.org/10.1371/journal.pcbi.0030005

Zhu D, Adebisi WA, Ahmad F et al (2020) Recent development of extremophilic bacteria and their application in biorefinery. Front Bioeng Biotechnol 8. https://doi.org/10.3389/fbioe.2020.00483

Chapter 19
Genomic and Metagenomic Prospecting of Extremophiles to Support Sustainable Development

Mohit Gururani, Rishika Malhotra, Abhishek Singh, Raj Kishor Kapardar, and Rajpal Srivastav

19.1 Introduction

There has been a promiscuous increase in the carbon footprint of mankind, especially in developed and developing countries in the last decades. It has posed significant risks to the environment and climate change. The prospecting of extremophiles is important in dealing with these environmental problems. Here, metagenomic exploration and analysis of the extremophilic communities can act as promising tools for solving these issues. The identification of microbial communities in extremophilic environments opens the rapid scope for gene mining associated with different industrial and environmental applications in a sustainable approach. The extremozymes, isolated from extremophiles, have significant industrial importance globally because of their unique molecular and biochemical properties. Extremozymes can be used in harsh conditions (Table 19.1) prevailing in industrial applications and synthesis for the production of enzymes and metabolites. Next-generation sequencing provides high-throughput genomic data and is very useful in whole-genome microbial sequencing, leading to an improved understanding of extremophiles for sustainable environment (Ladoukakis et al. 2014; Suneja and Srivastav 2021a, b).

Numerous enzymes or "extremozymes", produced by extremophiles, like alcohol dehydrogenase, α-amylase, protease, citrate synthase, β-lactamase, subtilisin, and triose phosphate isomerase have specialised activities. "Extremozymes" have

M. Gururani · R. Malhotra · R. Srivastav (✉)
Amity Institute of Biotechnology, Amity University Uttar Pradesh, Noida, UP, India
e-mail: rsrivastav2@amity.edu

A. Singh
Agriculture Biotechnology, Yerevan State University, Yerevan, Armenia

R. K. Kapardar
The Energy and Resources Institute, New Delhi, India

Table 19.1 Summary of a few important extremozymes from various extremophiles for various applications

Extremophile	Habitat	Extremozymes	Species	Application
Thermophiles	Hydrothermal vents, hot springs	Amylase	*Bacillus mojavensis*	Glucose fructose for sweetness
		DNA polymerase	*Thermus aquatics (Taq)*	PCR, diagnostics, molecular biology
		Xylanases	*Bacillus tequilensis*	Biorefinery, food
Halophiles	Salt mines, marine, soil	α-amylase	*Halothermorthrix orenii*	Bio-catalysis
		Protease	*Pseudoalteromonas* sp.	Detergent, peptide synthesis
Psychrophiles	Antarctica soil, deep ocean, Mariana trench	β-Galactosidase	*Arthrobacter species C2*	Biorefinery, ethanol production
		Lipase	*Pyschrobacter okhotskensis*	Food, cosmetics
Alkaliphiles	Salt mines, soil, soda lakes	Protease	*Bacillus firmus*	Detergent, food, and feed
		Cellulase	*Bacillus subtilis*	Fermentation of wine
Acidophiles	Man-made niches, hot springs	Endocellulase	*Thermomonospora (Actinomycetes)*	Leather industries
Piezophiles	Deep seam, Mariana trench	Lipase	*Colwellia hadaliensis BNL-1*	Food processing
		Chymotrypsin	*Shewanella benthica* strains	Pharmaceutical industry
Radiophiles	High ultra-violet radiations (UVR) altitudes, mountains	Mycosporine-like amino acids (MAAs)	*Deinococcus radiodurans*	Diagnostics, bioremediation, and therapeutics

Source: Ali et al. (2023)

comparative functional advantages since they can function in situations where typical "mesophilic enzymes" cannot (Loperena et al. 2012; Ali et al. 2023). The analysis of the adaptation, fitness, and survivability of these organisms can be deciphered for applications and management, associated with a sustainable environment. The physical conditions, including temperature, pH, and pressure, act as driving forces for the emergence of new characteristics in proteins, enzymes, and biomolecules for efficient survival and adaptability as evident in various extremophiles. The application of comparative genomics and metagenomics can provide significant knowledge associated with prospecting the extremophiles for sustainable environments.

19.2   Stability and Flexibility of "Extremozymes" of Extremophiles

The flexibility of the mesophilic and thermophilic enzymes is comparatively different and suggestive of different functionality under different conditions. The thermophilic enzyme is noticeably more rigid when compared to its mesophilic counterpart at room temperature. The results of earlier studies suggest that the thermostability of enzymes is inversely correlated with the values of flexibility indices. Further, some specific regions of the protein structure appear to control protein stability; however, they are different from the thermostabilising domains, while other regions are critical in conferring flexibility, resulting in optimal catalytic efficiency. Therefore, it is speculated that a cold-adapted extremozyme can be endowed with both a high specific activity and good stability at low temperatures (Ali et al. 2023). The conformational changes that should be connected to catalysis are significantly hindered at room temperature. The site-directed mutagenesis performed on enzymes like subtilisin, and a stable thermolysin-like provide evidence in favour of this hypothesis. Limited changes can be introduced to increase the stability of enzymes while also enhancing or maintaining their original catalytic capabilities, which can have applications industrially (Cipolla et al. 2012; Radhakrishnan et al. 2023a). The three-dimensional structure of the enzymes is very critical in determining the novel functionality of the proteins, despite the similar sequence (Nestl and Hauer 2014; Srivastav et al. 2014). The thermostability coming from the stiffness of the molecular structure, which results in a weak specific activity, is thought to weaken the connection between enzyme and substrate. The high specific activity of enzymes like cold-adapted extremozymes is associated with the flexibility of the protein structure (Akanuma et al. 2019). Interestingly, the reduced thermostability observed in all cold-adapted enzymes is a deliberate strategy employed by psychrophiles for better functionality at low-temperature conditions compared to their mesophilic counterparts. On the other side, the mesophilic enzyme isolated exhibits a higher specific activity over the temperature range of 5 °C to 30 °C. It is noteworthy that sometimes recombinant enzymes may not fold properly in mesophilic hosts like *Escherichia coli* when expressed at high temperatures (~30 °C), or they may become partially inactive, dependent on in vitro conditions (Akanuma et al. 2019; Srivastav and Suneja 2019). For example. The threefold less specific activity of triose phosphate isomerase is observed when cells are grown at 37 °C as opposed to 27 °C (Cipolla et al. 2012). Interestingly, there exist various mechanisms through which extremophiles can regulate molecular functioning in extreme environmental conditions, aiding in better adaptations in extremes, the summary of which has been represented in Figs. 19.1 and 19.2.

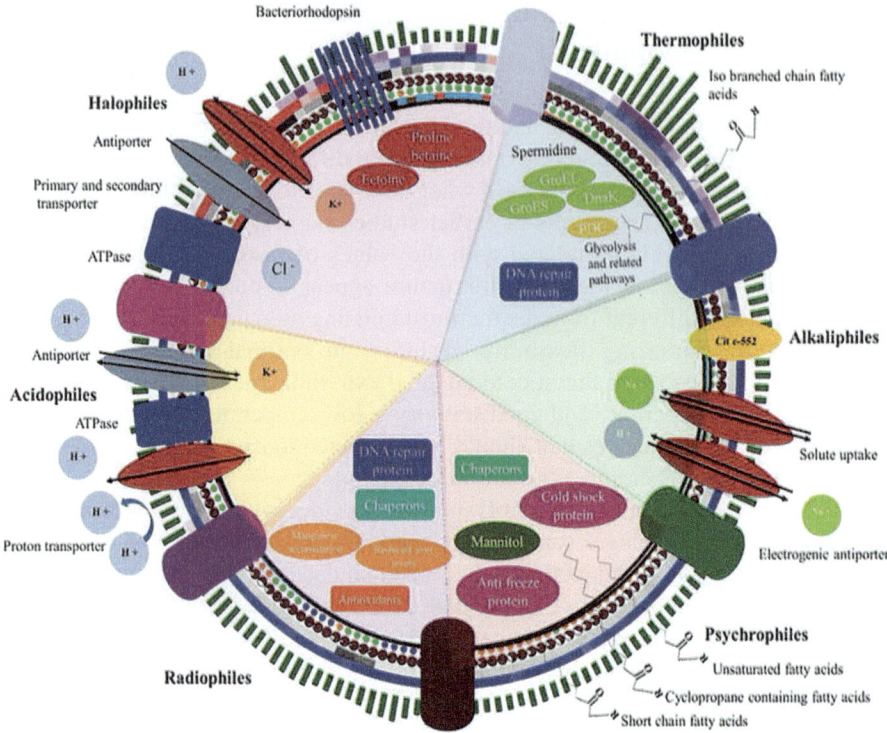

Fig. 19.1 Summary of various adaptations of extremophiles for survival in different extreme environments. (Source: Ali et al. 2023)

19.3 Genomic and Metagenomic Profiling of Extremophiles

It is difficult to create laboratory conditions suitable for cultivation of the extremophiles. Here, metagenomic analysis paves the way for better understanding of these extremophilic organisms. Metagenomic-aided method that involves DNA isolation from microbial communities living in extreme settings and analysing the diversity and functionality of proteins, enzymes, and molecular determinants. Consequently, metagenomic screening allows for the culture-free study of microbial genomes, diversity, and their metabolic pathway (Madhavan et al. 2017; Radhakrishnan et al. 2023c). To understand the functional roles of the microbial proteins, functional metagenomic screening can be performed. The sites like human gut, hydrothermal vents, hot springs, and other extreme places that are sites for extreme and anaerobic growth now can be explored for microbial research because of the advent of metagenomics (Mehta and Satyanarayana 2013; Srivastav and Suneja 2019). There are various geographical sites, which are unexplored for extremophilic microbial diversity for example the Himalayan region is one of India's least explored geographical areas, and little is currently known about the microbial diversity of this area

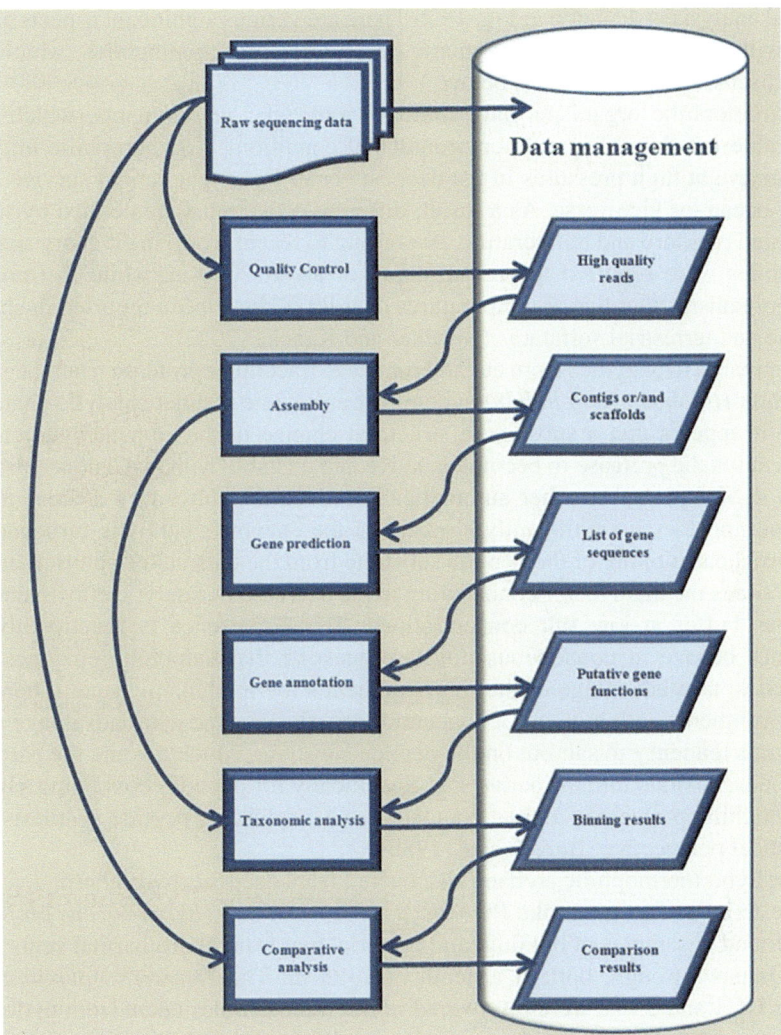

Fig. 19.2 A schematic representation of a typical genomic/metagenomic analysis workflow. (Source: Ladoukakis et al. 2014)

(Venkatachalam et al. 2015). It is a highly diverse and useful site for metagenomic investigations due to the relatively cold regions and some exceptionally high temperatures in hot-spring regions. The biocatalysts from the extremophilic microorganisms have been explored and analysed (Dalmaso et al. 2015; Zarafeta et al. 2016). 16S rRNA genes marker-based techniques are used to identify and categorise the diversity. In contrast, the functional investigation is made possible by shotgun metagenomic sequencing or other approaches as per the analysis requirements (Bella et al. 2013; Patidar and Prakash 2022). A schematic representation of a typical analysis is depicted in Fig. 19.2. There are various significant aspects associated with the usefulness of enzymatic capabilities of extremophiles, which have been discussed in the sections below.

Extremophilic organisms have multiple molecular determinants, which make them able to tolerate extreme environmental conditions. The barophilic microbes can survive at high pressures in the deepest ocean layers but cannot survive in the upper ocean, or vice versa. As a result, different ecosystems are defined by widely separated pressure and temperature. According to recent work, in diversity analysis, barophiles were reported to present depth of about 2000 m, while thermophilic microorganisms that live at temperatures over 80 °C have been reported in shallow marine and terrestrial solfataras (Kanekar and Kanekar 2022).

Compared to enzymes from eubacteria, the extracellular protease from the severe halophile *Halobacterium halobium* demonstrates some distinct catalytic characteristics. It appears that a substantial structural change in the enzyme's structure is what causes the protease to become inactive and unstable with NaCl concentrations below 4 M. Low salt neither automatically reduce the substrate's affinity for the enzyme nor does it significantly slow down the enzyme's catalytic turnover. The decreased partitioning of the peptide substrate from the bulk saline solution into the active site is the main factor contributing to the decrease in catalytic effectiveness of aqueous buffer at low salt concentrations. This occurrence is functionally like enzymes behave in nonaqueous fluids, suggesting that halophilic enzymes have evolved to take advantage of their environment's thermodynamic state. Therefore, in environments with high NaCl concentrations, the enzyme takes advantage of the medium's tendency to salt out on the peptide substrate, which favours the partitioning of the substrate into the active site. Specifically for peptides containing glycine, the halophilic protease is a great contender as a catalyst for peptide synthesis from a practical perspective (Buresh et al. 1980).

The hyperthermophilic archaea, discovered from deep-sea hydrothermal vents in the western Pacific Ocean like *Pyrococcus horikoshii* and *Thermococcus profundus* were found in samples of hot fluid and bacterial mat from hydrothermal vents in the Mid-Okinawa trough, both at a depth of 1395 m. The *Tannerella peptoniphilus* strains OG1 and SM-2 were discovered in hot fluid samples taken from hydrothermal vents in the South Mariana trench at a depth of 1484 m and the Izu Bonin forearc at a depth of 1380 m, respectively. The microbes in these hydrothermal vents likely have very thermostable proteins, SDS-resistant protease that is stable in boiling water (Calderon and Lissi 2013; Jo et al. 2014). These factors are attributed to the unique 3D structure and thermodynamic stability of extremozymes.

An enzyme's long-term stability is aided by such stabilising actions. The stabilising forces, which include hydrogen bonds, hydrophobic bonds, ionic interactions, metal binding, and disulfide bridges, determine the degree of thermostability. Better resistance to most chemical denaturants is also correlated with thermostability (Chapman and Stenzel 2019). Today, isolated enzymes dominate microbes in commercial applications, including thermophiles and thermostable enzymes. A high temperature should be more than the thermophile growth threshold of 55 °C. Since extracellular enzymes cannot be stabilised by cell-specific elements, such as suitable solutes, they often exhibit great thermostability. The intrinsic thermostability of thermostable enzymes is, of course, the primary prerequisite for using them in bioprocessing. This property attributes to extended storage, resistance to organic solvents, and low activity losses during processing (when kept below the Tm of the enzyme), even at high-temperature conditions, frequently used in raw material pretreatments (Umair et al. 2019).

Some of the initial identified barriers (such as restricted access and substrate specificity) to the use of thermostable enzymes in commercial bio-catalysis have been solved using a site-directed mediated evolution approach, further accelerating the process advancements. Many biotechnology enterprises require novel and robust enzymes to process larger-scale reactions with demanding process conditions. By using molecular probing techniques, enzyme prospecting frequently focuses on gene retrieval from nature directly, followed by recombinant creation in a suitable host. The genomic and structural characteristics of the biocatalyst can be exploited both for pharmaceutical and industry settings (Ali et al. 2014).

19.4 Advanced Next-Generation Sequencing Analysis

The throughput of genome sequencing has significantly risen because of next-generation sequencing (NGS) technology. It takes a combination of experimental and computational techniques, as well as the integration of various biological data from various sources, to annotate recently sequenced genomes (Madhavan et al. 2017). A wide range of applications, including molecular cloning, breeding, identifying pathogens, and comparative and evolutionary genetics, could benefit from the use of DNA sequencing technologies. Technologies for DNA sequencing are quick, precise, simple, and affordable (Liu et al. 2012). The genome era, which is characterised by enormous amounts of genome data and subsequently a wide range of research topics and multiple applications, is being driven by the extraordinary development that DNA sequencing technologies in the past decade. In NGS systems, comparative usefulness and drawbacks of platforms are required for the best output. Further, NGS provides diverse applications in genomics, and the recently launched personal genome machines (PGM), is an important development for personalised medicine.

Extremophiles are ecological niche specialists such as hydrothermal vents in the deep ocean, hot springs, geysers, salt flats, deserts, lakes naturally occurring, and so forth (Madigan 2000; Dey et al. 2022; Dutta et al. 2022). Extremophiles are potential sources of fundamental knowledge and valuable enzymes for biotechnological applications for sustainable environments. There are many prospects for enzymes, which are in various phases of development like DNA polymerase, alkali- and heat-resistant β-D-galactosidase, and others. Additionally, numerous CRISPR loci from diverse CRISPR-Cas systems have been discovered in extremophiles. High-throughput sequencing technological advancements led to data generation in a very short period.

In conventional genomic analysis, DNA library preparation and screening are used to consume a significant experimental time. Further, it was labour-intensive process. However, now advanced next-generation sequencing (NGS) can provide you with data in a single day. The widespread use of bacterial whole-genome sequencing has been made possible by NGS technologies like Illumina, MGI, Shenzhen, and Ion Proton. These platforms generate huge data for example millions of paired ends read with a low error rate (0.1%) (Van Dijk et al. 2014). However, it is difficult to completely reconstruct genomic structures of interest due to the short readings of 100–300 bp. The whole chromosome can be assembled, using Oxford Nanopore and SMRT Pacific Biosciences (PacBio) platforms. However, these sequencing methods need to build libraries. Sequencing on two separate platforms is far more expensive and uses larger quantities of high-quality DNA than NGS sequencing (Bigdeli 2012).

It can be difficult to select the most cost-efficient sequencing method while still obtaining high-quality genome sequences when beginning a large-scale bacterial whole-genome sequencing (WGS) project (Nouws et al. 2020). A sequencing strategy must have low cost, high accuracy, and better efficiency for extremophilic genome analysis (Table 19.2). Different platforms capture data with short-read sequencings, such as mate-pair and clonal barcoding methods (e.g. synthetic long read), which have the most promise for bringing the routine long-read capability to second-generation platforms (Henson et al. 2012). There are different approaches and platforms available depending upon the requirement of analysis. The user choice and analysis need should be taken into consideration for deciding the approaches and platforms (Besser et al. 2018).

The use of stLFR technology to accurately sequence complicated extremophile genomes is one of the important advancements. Illumina, ONT, and PacBio data produced from identical DNA extracts were included in the comparison of several methods for hybrid bacterial genome assembly. Radiation-resistant extremophiles that were isolated from the Xinjiang Uygur Autonomous Region of China, genome GC content from 30% to 70%. Additionally, the genomes of extremophilic microorganisms typically range from 4.3 to 6.5 Mb. The goal of this research was to assess and improve the stLFR technology's accuracy when sequencing the genomes of extremophiles with diverse GC contents. Each sequencing platform has various issues on its own. For instance, while being the most popular sequencing method, Illumina produces shorter reads that impede downstream processing. Additionally, PacBio, a third-generation sequencing method, solves Illumina's restriction but has a significant error rate (Mikheyev and Tin 2014). When examining the novel microbial diversity, the read length and mistake frequencies are crucial factors.

Functional genomics may benefit from the use of annotations resulting from information integration as a potent strategy that streamlines subsequent studies. Data warehouses built on integrated information are especially helpful because they enable content exploration from various annotation attributes. InterMine framework

Table 19.2 Characteristics, strengths, and weaknesses of commonly used sequencing platforms

Platforms	Throughput range (Gb)	Read length (bp)	Strength	Weakness
Sanger sequencing				
ABI 3500/3730	0.0003	Up to 1 kb	Read accuracy and length	Cost and throughput
Illumina				
MiniSeq	1.7–7.5	1 × 75 to 2 × 150	Low initial investment	Run and read length
MiSeq	0.3–15	1 × 36 to 2 × 300	Read length, scalability	Run length
NextSeq	10–120	1 × 75 to 2 × 150	Throughput	Run and read length
HiSeq (2500)	10–1000	1 × 50 to 2 × 250	Read accuracy, throughput, low per-sample cost	High initial investment, run length
NovaSeq 5000/6000	2000–6000	2 × 50 to 2 × 150	Read accuracy, throughputlow pre-sample cost	High initial investment, run and read length
IonTorrent				
PGM	0.08–2	Up to 400	Read length, speed	Throughput, homopolymersc
S5	0.6–15	Up to 400	Read length, speed, scalability	Homopolymersc
Proton	10–15	Up to 200	Speed, throughput	Homopolymersc
Pacific biosciences				
PacBio RSII	0.5–1	Up to 60 kb(average 10 kb, N50 20 kb)	Read length, speed	High error rate and initial investment, low throughput
Sequel	5–10	Up to 60 kb(average 10 kb, N50 20 k)	Read length, speed	High error rate
Oxford nanopore				
MInION	0.1–1	Up to 100 kb	Read length, portability	High error rate, run length, low throughput

Source: Besser et al. (2018)

enables the creation of such data warehouses. Before now, it has been used to create model genome data warehouses, leading to resources like FlyMine, modMine, RatMine, and YeastMine supplemental materials for additional information on InterMine's features and a comparison to other systems of a similar nature (Alam et al. 2013). Annotations for microbial genome exploration and analysis can be included in the data warehouse called Included Data Warehouse of Microbial Genomes (INDIGO). It is used for genomic analysis dependent upon the requirements of the research.

19.5   Paradigm Shifts in Genomic Analysis

Despite numerous efforts, however, the expense of the enzymes frequently restricts their use today. However, it is anticipated that the cost will go down as the enzyme market grows and more of them are produced. Additionally, a paradigm change in the industry towards the use of renewable resources is projected to boost the need for microbial catalysts, and there will undoubtedly be an ongoing and rising demand for thermostable selective biocatalysts in the future. The genome mining associated with the extremophilic genome can provide enormous knowledge for genomic and proteomic analysis. Currently, extremophile research is often orientated towards determining what cellular responses are involved in growth and survival, which can be employed for sustainable development. However, lack of nutrients, oxygen, temperature, and osmotic conditions, as well as the absence of growth factors that may be created by other organisms in the community can all contribute to the uncultivability of these microorganisms in the complex community of extremophiles. Even though these organisms may not be the most prevalent or ecologically significant in the environment, cultivating will always favour the organisms, which synchronise their metabolism quickly in response to the conditions used in the laboratory. The sequencing of the 16S rRNA genes of prokaryotes from environmental samples has aided in understanding the uncultured fraction and led to the identification of multiple novel bacterial phyla, of which only a small number are represented by cultivated strains (Srivastav et al. 2013). A promising method for assessing the variety of microorganisms and illuminating metabolic pathways has been identified using next-generation sequencing (NGS) (Mandal and Rath 2014; Vester et al. 2015).

High-throughput, effective, and precise analyses of the entire genome of the microbial community are made possible by NGS advancements (Su et al. 2014). In the recent past, the NGS investigation of hot springs habitat in Central India revealed the existence of a range of hydrocarbon-degrading thermophiles and the pathways needed for bacterial survival under extreme settings (Saxena et al. 2017; Radhakrishnan et al. 2023b). The inclusion of metagenomics makes it easier to analyse the genomics species using advanced NGS platforms (Majumdar et al. 2023). Metagenomics is versatile and relatively simpler in preparation: it applies to any community through which we can extract enough DNA. Metagenomics can explore the phylogenetic and functional aspects of extremophiles and allow us to retrieve the genes encoding for proteins that help adapt in extreme environmental conditions, which can be used as biocatalysts in industries for sustainable environment and other biotechnological aspects (Eloe et al. 2011; Coughlan et al. 2015). The development and commercialisation of novel goods based on biological resources, such as enzymes and bioactive chemicals, can act as important bioprospecting. In general, metagenomic or culture-dependent methods can be used for microorganism bioprospecting. The culture-dependent techniques rely on the cultivation of naturally occurring isolates that may be tested for specific activities, such as enzyme activity, anti-bacterial activity, antibiotic resistance, and novel functionality. The whole-genome analysis provides a better analysis of the novel attributes and a deeper understanding of extremophiles for a sustainable environment.

19.6   Conclusion

The prime reasons for survival capabilities by extremophiles are the presence of unique enzymes referred to as extremozymes, proteins, and biomolecules, which can function in extreme physical conditions. There is constant pressure on industries to produce cost-effective and efficient catalysts to cope with the demands. Various extremozymes have significant industrial, pharmaceutical, and environmental significance for sustainable ecosystems. Next-generation sequencing provides high-throughput genomic data accelerating the pace of gene mining and enzyme discovery useful for industrial applications. The next-generation sequencing, including whole-genome sequencing, is important in comparative genomics analysis and for sustainable agriculture and environments. The research in this area is still in its infancy and only a fraction of microbial diversity has been explored.

References

Akanuma S, Bessho M, Kimura H et al (2019) Establishment of mesophilic-like catalytic properties in a thermophilic enzyme without affecting its thermal stability. Sci Rep 9:1–11

Alam I, Antunes A, Kamau AA et al (2013) INDIGO—INtegrated data warehouse of MIcrobial GenOmes with examples from the red sea extremophiles. PLoS One 8:12

Ali N, Nughman M, Shah SM (2014) Extremophiles and limits of life in a cosmic perspective. In: Extreme environments—diversity, Adaptability and Valuable Resources of Bioactive Molecules. IntechOpen 2023

Ali N, Nughman M, Shah SM (2023) Extremophiles and limits of life in a cosmic perspective. In: Life in extreme environments—diversity, adaptability and valuable resources of bioactive molecules. IntechOpen

Bella D, Julia M, Bao Y et al (2013) High throughput sequencing methods and analysis for microbiome research. J Microbiol Methods 95:401–414

Besser J, Carleton HA, Gerner-Smidt P et al (2018) Next-generation sequencing technologies and their application to the study and control of bacterial infections? Clinical Microbiol Infect 24:335–341

Bigdeli S (2012) Single cell genomics in alginate microspheres. Stanford University

Buresh RJ, Casselman ME, Patrick (1980) Nitrogen fixation in flooded soil systems, a review? Advances Agron 33:149–192

Calderon C, Lissi E (2013) Polyethylene glycol effect on the transient and steady state phases of p-nitrophenyltrimethyl acetate hydrolysis catalyzed by?-chymotrypsin? J Chil Chem Soc 58:2053–2056

Chapman R, Stenzel MH (2019) All wrapped up: stabilization of enzymes within single enzyme nanoparticles. J Am Chem Soc 141:2754–2769

Cipolla A, Delbrassine F, Lage J-LD, Feller G (2012) Temperature adaptations in psychrophilic, mesophilic and thermophilic chloride-dependent alpha-amylases. Biochimie 94:1943–1950

Coughlan LM, Cotter PD, Hill C, Alvarez-Ordez A (2015) Biotechnological applications of functional metagenomics in the food and pharmaceutical industries? Frontiers Microbiol 6

Dalmaso GZL, Ferreira D, Vermelho AB (2015) Marine extremophiles: a source of hydrolases for biotechnological applications. Mar Drugs 13:1925–1965

Dey P, Murthy TPK, Divyashri G et al (2022) Prospects of biofuels, biofertilizers, and therapeutics from extremophiles. Extrem A Parad Nat with Biotechnol Implic 1

Dutta S, Bhattacharjee J, Chakraborty S (2022) Exploring the potential extremophilic microbes for bioremediation. Extrem A Parad Nat with Biotechnol Implic 1

Eloe EA, Fadrosh DW, Novotny M et al (2011) Going deeper: metagenome of a hadopelagic microbial community. PLoS One 6:5

Henson J, Tischler G, Ning Z (2012) Next-generation sequencing and large genome assemblies. Pharmacogenomics 13:901–915

Jo BH, Seo JH, Cha HJ (2014) Bacterial extremo-carbonic anhydrases from deep-sea hydrothermal vents as potential biocatalysts for CO2 sequestration. J Mol Catal B Enzym 109:31–39

Kanekar PP, Kanekar SP (2022) Piezophilic or barophilic microorganisms. In: Diversity and biotechnology of extremophilic microorganisms from India. Singapore, Nature Singapore, pp 269–280

Ladoukakis E, Kolisis FN, AA. C (2014) Integrative workflows for metagenomic analysis. Front Cell Dev BiolNov 19:2. https://doi.org/10.3389/fcell.2014.00070

Liu L, Li Y, Li S et al (2012) Comparison of next-generation sequencing systems. J Biomed Biotechnol

Loperena L, Soria V, Varela H et al (2012) Extracellular enzymes produced by microorganisms isolated from maritime Antarctica? World J Microbiol Biotechnol 28:2249–2256

Madhavan A, Sindhu R, Parameswaran B et al (2017) Metagenome analysis: a powerful tool for enzyme bioprospecting. Appl Biochem Biotechnol 183:636–665

Madigan MT (2000) Extremophilic bacteria and microbial diversity. Ann Missouri Bot Gard:3–12

Majumdar J, Moulik D, Santra SC, Hossain A (2023) Extremophile bacterial and archaebacterial population: metagenomics and novel enzyme reserve. In: Microbial symbionts and plant health: trends and applications for changing climate. Singapore, Nature Singapore, pp 521–544

Mandal S, Rath J (2014) Extremophilic cyanobacteria for novel drug development. Springer

Mehta D, Satyanarayana T (2013) Diversity of hot environments and thermophilic microbes. Thermophilic microbes Environ Ind Biotechnol Biotechnol thermophiles, pp 3–60

Mikheyev AS, Tin MMY (2014) A first look at the Oxford Nanopore MinION sequencer. Mol Ecol Resour 14:1097–1102

Nestl BM, Hauer B (2014) Engineering of flexible loops in enzymes? ACS Catal 4:3201–3211

Nouws S, Bogaerts B, Verhaegen B et al (2020) Impact of DNA extraction on whole genome sequencing analysis for characterization and relatedness of Shiga toxin-producing Escherichia coli isolates. Sci Rep 10:1

Patidar P, Prakash T (2022) Decoding the roles of extremophilic microbes in the anaerobic environments: past, present, and future. Curr Res Microb Sci 100146

Radhakrishnan A, Balaganesh P, Vasudevan M et al (2023a) Bioremediation of hydrocarbon pollutants: recent promising sustainable approaches, scope, and challenges. Sustain For 15. https://doi.org/10.3390/su15075847

Radhakrishnan A, Kapil T, Kapardar RK, Srivastav R (2023b) Microbiome additive therapy for the human health. In: Microbiome Therapeutics. Elsevier, pp 41–61

Radhakrishnan R, Kapil T, Kapardar R, Srivastav R (2023c) Microbiome additive therapy for the human health. In: Chauhan NS, Kumar S (eds) Microbiome therapeutics, vol 9780. Academic Press, pp 41–61

Saxena R, Dhakan DB, Mittal P et al (2017) Metagenomic analysis of hot springs in Central India reveals hydrocarbon degrading thermophiles and pathways essential for survival in extreme environments. Front Microbiol 7

Srivastav R, Suneja G (2019) Recent advances in microbial genome sequencing. In: Kumar P, Tripathi P, Kishore A, Kamle M, Tripathi V (eds) Microbial genomics in sustainable agroecosystems. Ed.1, vol 2. Springer

Srivastav R, Singh A, Jangir PK et al (2013) Genome sequence of Staphylococcus massiliensis strain S46 isolated from surface of healthy human skin. Genome Announc 1:513–553

Srivastav R, Kumar D, Grover A et al (2014) Unique subunit packing in mycobacterial nanoRNase leads to alternate substrate recognitions in DHH phosphodiesterases. Nucleic Acids Res 19:7894–7910

Su X, Pan W, Song B et al (2014) Parallel-META 2.0: enhanced metagenomic data analysis with functional annotation, high performance computing and advanced visualization. PLoS One 9:3

Suneja G, Srivastav R (2021a) Impact of microbial genome sequencing advancements in understanding extremophiles. Book—Extreme Environments, Unique, pp 335–341

Suneja G, Srivastav R (2021b) Impact of microbial genome sequencing advancements in understanding extremophiles. In: Pandey A, Sharma A (eds) Extreme environments1st edn. CRC Press, p 13

Umair MM, Zhang Y, Iqbal K et al (2019) Novel strategies and supporting materials applied to shape-stabilize organic phase change materials for thermal energy storage.A review. Appl Energy 235:846–873

Van Dijk EL, Auger H, Jaszczyszyn Y, Thermes C (2014) Ten years of next-generation sequencing technology. Trends Genet 30:418–426

Venkatachalam S, Gowdaman V, Prabagaran SR (2015) Culturable and culture-independent bacterial diversity and the prevalence of cold-adapted enzymes from the Himalayan mountain ranges of India and Nepal. Microb Ecol 69:472–491

Vester JK, Glaring MA, Stougaard P (2015) Improved cultivation and metagenomics as new tools for bioprospecting in cold environments. Extremophiles 19:17–29

Zarafeta D, Moschidi D, Ladoukakis E et al (2016) Metagenomic mining for thermostable esterolytic enzymes uncovers a new family of bacterial esterases. Sci Rep 6:1–16

Part V
Exploiting Extremophiles by Nano Approach

Chapter 20
Nanotechnology and Extremophiles: Agricultural Applications and Possibilities

Dinoo Gunasekera, Parakkrama Wijerathna, and Disna Ratnasekera

20.1 Introduction

Agriculture plays a major role in the global economy providing food for humans, fodder for farm animals, restoring the environment, and supporting livelihood for most of the people. With the improvement in agriculture, the use of synthetic chemical fertilizers and plastic mulching for better growth, pesticides, herbicides, etc. for protection purposes have been adopted. The long-term application of fertilizer, pesticides, herbicides, use of waste for irrigation, and application of sewage has been leading to soil pollution in agricultural lands. In addition to such direct inputs, some indirect inputs from flooding and atmospheric depositions can contaminate agricultural lands. The agriculture contaminants are broadly categorized into five groups: pesticides, inorganic fertilizers, organic fertilizers, wastewater from irrigation, plastic materials used for mulching, polytunnels, packing materials, and shade netting, drip/sprinkler irrigation tubes, and farmyard waste. Contrary to other pollutions agriculture soil contamination is continued, highly persistent, and chronic due to the aggregative application of agriculture inputs and their toxicity. Unlike other pollutions, agriculture applications contaminated large areas owing to large-scale applications of pesticides, fertilizer, etc., and becoming great public attention. In addition, antibiotics, fungicides, and growth regulators widely used in crops and livestock to promote growth and manage diseases are frequent agricultural soil contaminants. Owing to certain specific nature of pesticide contaminants such as the ability to spread a wide range and largely unknown effects of pesticide mixtures in the

D. Gunasekera
Department of Information and Communication Technology,, Faculty of Technology, University of Ruhuna, Matara, Sri Lanka

P. Wijerathna · D. Ratnasekera (✉)
Department of Agricultural Biology, Faculty of Agriculture, University of Ruhuna, Matara, Sri Lanka

© The Author(s), under exclusive license to Springer Nature Switzerland AG 2024
A. Ranjan et al. (eds.), *Extremophiles for Sustainable Agriculture and Soil Health Improvement*, https://doi.org/10.1007/978-3-031-70203-7_20

environment, make it hard to assess the risk of toxicity in terms of soil ecosystems, human health, and aquatic and terrestrial environments. Thus, understanding the spatial distribution, concentration levels, soil characteristics and temporal movement of contaminants in agricultural areas is vital for designing an efficient decontamination program.

The fate of such contaminants depends on both soil physicochemical properties and characteristics of contaminants, importantly form, concentration, shape, size, and chemical nature of the pollutants. Thus, consideration of relationship between site-specific soil properties and pollutant/s characteristics is vital for employing efficient decontamination program. Our attempt in this chapter is to provide facts on main sources of agriculture soil contaminants, the role of extremophiles in synthesizing bio-nanoparticles for bioremediation, contaminant-soil matrix interaction, and mobility of pollutants.

20.1.1 Sources of Soil Contamination

The agricultural lands are frequently contaminated by excessive application of agrochemicals, including fertilizer, pesticides, contaminated irrigation water, improper usage of plastic materials and sludge, and farm manure. The changed land use approaches along with agronomic practices such as land preparation, tillage and irrigation enhanced toxicant mobility due to introduced cropping patterns that alter soil natural processes (Ondrasek et al. 2019).

(a) Pesticides and Herbicides

Soil is a major part of the ecosystem providing requirements for the persistence of all living organisms while the largest sink for the contaminants. With the development of food demand, adoption of high-input responsive crop varieties and application of pesticides have been increased simultaneously. Any ingredient or mixture of ingredients, including herbicides, fungicides, insecticides, nematicides, antibiotics, rodenticides, molluscicides, and plant growth regulators that have the capability as a repellent, kill, or control pest population or modify plant growth are categorized as pesticides (WHO 2020). Pesticides are incorporated with soil through direct foliar application, accidental releases due to leakages in pipes, damaged containers, etc., poor disposal methods of used pesticide containers, expired pesticides, and inappropriate cleaning methods (Tudi et al. 2021). The intensive application of pesticides for protecting agriculture crops from diseases, insect pests, and weeds has caused a massive accumulation of long-term toxic persistent contaminants in the agricultural soil that can slowly be spread out in soil water, and atmosphere via geochemical cycles. Nearly 90% of sprayed pesticides knock out non-target organisms and remain in crops and soils as residue causing serious health risks and ecological imbalance in agroecosystems. The majority of frequently used pesticides are persistent organic contaminants with the ability of spreading in wider range leading to critical toxicant issues on human health and on ecosystems. The use of vast numbers of various pesticides with different chemical combinations makes it a complicated and hard task to assess the impacts of pesticides on the environment and human health.

In addition, pesticides are also used outside the agriculture sites such as managing organisms that cause damage to food production, processing, storage, and transport, wood and wood products, fibers and other agricultural commodities, vector control for diseases such as dengue, malaria (WHO and FAO 2020). As reported in 2018, more than 1000 pesticide types with more than 800 active compounds were recorded in the market (Zhang 2018). The overall systematic evaluation of spatial distribution of pesticides with temporal trends in cultivated lands is compulsory for designing an effective decontamination process. Among all, organo-chlorine pesticides (OCP) are a large group comprising large number of persistent chemical pesticides. Organochlorine pesticides (OCP) were the most abundantly used pesticides in agricultural lands till they were banned and remaining toxic residues in the environment (Sun et al. 2018a, b).

(b) Organic Fertilizers

The commonly and frequently applied organic fertilizers in agriculture include farmyard manure, compost, urban wastes, sewage sludge and crop residues from large-scale agriculture lands (Khan et al. 2018). The application of organic fertilizer is increasing because of the growing demand for organic foods. The composition of nutrients released from the organic fertilizer solely depends on the composition of the raw materials of the organic dumps used. Thus, urban wastes, sewage sludge, and manure from intensive livestock may contain trace elements that can create soil toxicity during long-term application. Moreover, farmyard manure contained antibiotics, growth regulators, and other microbial-resistant substances that were added to the concentrated feeds of farm animals. The release of such antimicrobials to the soil may lead to development of resistance among microbial communities causing significant human health risks. Use of urban solids and biosolids, septic tank sludge is common in large-scale agriculture, which contains risky microbes that may create serious health issues. Careful analysis of organic fertilizer is recommended before application, as the composition of the fertilizer depends upon the sources of the fertilizer. For example, urban biosolids rich in trace elements, nano- or microplastics, polychlorinated substances, and perfluorino-alkyl substances were reported in biowastes (Srivastava et al. 2016; Weber et al. 2018; Ziajahromi et al. 2016). The composition of the compost relies on sources of raw materials used in compost preparation. The raw materials of composting are green fodder waste, kitchen waste, urban waste, etc., and urban waste may contain various toxic contaminants such as plastics, heavy metals, and other organic pollutants.

Though organic fertilizers are characterized by slow release of nutrients, improving soil's physical and biological properties, they also contribute to soil pollution by mineralization of nitrogen, perfluoro-alkyl materials, and other toxic elements (Gottschall et al. 2017).

(c) Mineral Fertilizers

The over usage of synthetic fertilizer is common among farmer communities, especially in developing countries due to lack of awareness, cheaper fertilizers, readily available (through subsidiary programs), and limited availability of fertilizer recommendations. The application of overdose synthetic fertilizer creates nutrient saturation in the soil, contamination of groundwater due to leaching, eutrophication owing to accumulation of nutrients in freshwater bodies via runoff water, and pollution of drinking water. Primary macro-nutrients are always applied in large amounts. The massive application of various forms of nitrogen was common in most countries but majority of nutrients were lost without absorption to crops. For example, excessive nitrogen fertilizer could be lost by volatilization, leaching, denitrification, and microbial conversion into N_2O, which leads to global warming. The prolonged usage of different nitrogen fertilizers causes soil acidification and eutrophication of water bodies. The acidified soils reduced availability of other nutrients and produced toxic substances, enhancing soil-borne pathogens (Shen et al. 2018).

Application of phosphorus (P) is essential for crop growth and organic or inorganic forms of P are available. The main sources of P, derived from sediment rocks are non-renewable sources that also contain many other trace elements causing air and soil pollution during the mining process (El-Bahi et al. 2017). Thus, long-term application of rock phosphate could lead to accumulation of harmful metallic substances such as lead, mercury, fluorine, arsenic, cadmium, chromium and uranium, radium, and thorium-like radionuclides in soils and plants and circulate via food webs (Jiao et al. 2012).

Some mineral elements such as aluminum (Al), arsenic (As), lead (Pb), mercury (Hg), lithium (Li), cadmium (Cd), and tin (Sn) are classified as toxic elements with no metabolic functions (WHO 1996). The presence of such toxic elements (heavy metals) in very minute concentrations is highly toxic to organisms. Synthetic chemical fertilizers, pesticides and growth regulators may contain some trace amounts of such toxic elements as impurities. The continuous application of pesticides and fertilizers in agricultural lands is the main cause of accumulation of heavy metals, which is toxic to organisms and human beings, posing a serious health risk. The heavy metals may persist in soil in various forms such as dissolved ionic forms, metal-organic complexes, ionic salts, etc. depending upon the pH status of the site. The multi-ionic forms with different electron configurations of heavy metals have a high tendency of forming bioavailable metal-organic compounds with high toxicity.

The use of certain organic supplements such as biochar, compost, peat, etc. to sorb ionic elements is one of the approaches to alleviate metal contaminants. Such amendments can act as metal sorbent by forming soluble and bioactive metal-organic compounds as well as sedating trace elements by chelation with humus components (Egene et al. 2018).

(d) Irrigation Using Wastewater

Clean water is a limited resource for all living organisms and use of wastewater in agriculture has become common due to scarcity of clean water. Irrigation using wastewater has gradually increased due to accumulation of wastewater volumes with the rising population, uplifting living conditions, and growing demand for food and related commodities. Types of wastewater mainly include urban wastewater, industrial wastewater, and septic tanks. Irrigation in croplands using treated wastewater as a low-cost resource is very common worldwide, especially in urban areas. The types of dissolved substances and their concentrations depend on the waste materials and the treatment process adopted. Irrigation using wastewater is cheaper, available throughout the year, contains certain favorable nutrients, and helps to improve soil properties. However, watering using untreated wastewater leads to accumulation of toxic substances, nano and microplastics, disease-causing microbes, and other toxicants (Islam et al. 2018) resulting in serious health issues (Prata 2018). Moreover, prolonged application of treated wastewater also could lead to nutrient imbalance in crop plants, soil salinity, accumulation of heavy metals, and risk of spreading pathogens eventually altering macro and micro fauna, agroecosystems, and aquatic ecosystems.

(e) Plastic Products in Mulching, Netting, Polytunnels, and Packaging

Plastic usage has been drastically increased in recent years because of its easy handling qualities such as durability, lightness, ease of cleaning, occupying less space and low cost, readily available compared to produced using other materials. The plastics and polyethylene are widely used in polythene mulching, shade nets, insect-proof nets, UV-cut polythene for polytunnels, grow bags for vegetative propagation, pots, fertilizer bagging, pesticide packaging, seedling trays and other nursery applications in crop management, while twine, nets, feeders and bailing materials are greatly applied in livestock industry. The proper discarding or recycling mechanisms after usage are not well adopted in most agricultural lands and disposed to field itself with leftover chemicals is a common issue in many places resulting in deadly effects to humans and other organisms, polluting soils and water streams (Vox et al. 2016; Chae and An 2018).

Non-degradable synthetic plastic mulching films are the major source of soil contaminants in agriculture due to their close contact with soil surface and huge accumulation of plastic residues in soils. Features like durability, flexibility, and easy handling increase farmer's preference. Plastics are not completely degrading, instead generating micro and nanoparticles that can absorb plants, ingested by soil organisms and thereby enter food chains (Astner et al. 2019). Plastic mulches retard plant growth, negatively affect soil microbes, emit greenhouse gases, and contain carcinogenic substances (Gao et al. 2019; Boots et al. 2019; Revel et al. 2018). Moreover, greenhouse netting materials, water distributing tubes, seedling trays, fertilizer, and pesticide packing materials are frequently disposed of to open fields that exacerbate the toxication. To overcome the hazardousness of synthetic plastics oxo-degradable and biodegradable plastics have been introduced recently that can totally substitute the non-degradable synthetic plastic mulching films.

20.2 Bio-Remedial Approaches for Contaminated Agriculture Soils

Bioremediation and phytoremediation methods are widely applied in cleaning contaminants in agricultural lands. However, integrated methods are more appropriate and broadly accepted since they overcome most of the drawbacks in bioremediation and phytoremediation methods. For example, plant-microbial associations that showed synergistic connections between plant roots and soil microbes were effective in degrading noxious organic pollutants. The rhizosphere oils are home to a variety of microbes and the ooze of the root zone favors the dissociation of organic pollutants, microbial growth and activity. The combined effects of microbes with surfactants have shown higher decomposition rates, enhanced solubility of organochlorine pesticides, and higher mobilization (Sun et al. 2018a, b).

20.2.1 Use of Nanotechnology in Reclamation of Polluted Agriculture Lands

Use of nanoparticle technology in remediation of environmental issues has become popular and widely adopted due to its efficiency, safety, and low cost. NPs have unique properties such as higher absorbance, strong chemical structure, easy spreading, and diverse composition, which showed enhanced absorption and environmental degradation. NPs are designed to suit specific issues and conditions with various structures, models, and shapes, removing contaminants efficiently. The techniques such as nano-bioremediation, nano-phytoremediation, and use of nano zero-valent iron, nano-Si, nano-TiO$_2$, ZnO-NPs, nano-Fe$_3$O$_4$, nano zeolite (Singh et al. 2020; Tran et al. 2020; Liu et al. 2020; Lv et al. 2020; Sundararaghavan et al. 2020; Wang et al. 2020). In the remediation of contaminants, NPs have shown high efficacy by destroying contaminants via adsorption, oxidation and reduction, ion exchange, and electrostatic and surface reactions (Trujillo-Reyes et al. 2014). Application of NP-associated techniques, including green nanotechnology and NP synthesis using microbes for reclamation of agriculture contaminants has shown promising results (Aliyari Rad et al.2023).

20.2.2 Extremophiles for Nanoparticle Synthesis

Extremophiles are naturally adapted to survive in extreme environments named halophiles, acidophiles according to the prominent extreme event and extremophiles survive in multiple environments known as polyextremophiles. Specific substances such as extremolytes (stabilizing substances), extremozymes (enzymes stable at high temperatures, high pH, etc.), biosurfactants and biomolecules have a wide range of applications in many industries (Kochhar et al. 2022). Extremophiles play key role in agricultural lands by serving as biofertilizers, bioinoculants, and biocontrol means, regulating harmful events such as extreme temperatures, water scarcity, and high salt concentrations. Extremophiles with the capability of nutrient fixation and cycling, mineralization, solubilization, and stabilization act as biofertilizers (Tiwari et al. 2019). Extremophilic-derived biosurfactants are adopted in soil bioremediation and pesticide manufacturing industry for substituting chemical surfactants. For example, rhamnolipids have been used to control Phytophthora spores (Rath and Srivastava 2021) and to enhance hydrophilization toward sustainable agriculture (Markande et al. 2021). The diverse extremophilic microbes have great potential in the different aspects of agriculture as economical, eco-friendly approach to substituting chemical products (Chakraborty and Akhtar 2021).

(a) Nano-phytoremediation

The nano-phytoremediation includes plant-derived NPs having capability of detoxification of pollutants in vast range of environments (Guerra et al. 2018). The hyper-accumulating plant species with absorbents have been identified as ideal sources for remediation of soil contamination. The hyper-accumulating plants have the capability of growing in extreme soils and absorbing huge amounts of contaminants transported to specific organs for storage without causing phytotoxic effects. The hyper-accumulators are characterized by extraordinary intake of contaminants, rapid translocation, and capacity of decontamination or sequestration of contaminants within specific plant organs. With perspective of the plant, accumulation of heavy metals is said to be for defensive purposes (Siyar et al. 2022). Plants with phytoextraction capability can be grown in large extents of agricultural lands to remove metal contaminants successfully. After a certain period, hyper-accumulator vegetation is harvested and removed from the land. The process can be repeated according to the degree of contamination and soil properties (Mocek-Płóciniak et al. 2023). The efficacy of removal of contaminants depends on the phytoextraction capability of the selected plant species and the amount of water-mineral solution moving across the per given time. Certain chemical substances such as ethylene diamino-tetra-acetic acid (EDTA) and citric acid have been identified as accelerators for phytoextraction process associated with plants (Hunt et al. 2014). The plant species, including Sebertia acuminata, Trifolium alexandrinum, Zea mays, Helianthus annuus, Cannabis sativa, and Nicotiana tabacum were reported as heavy metals hyper-accumulators in agricultural lands (Mocek-Płóciniak et al. 2023).

(b) Reclamation by Zero-Valent Iron Nanoparticles (NZVI)

Zero-valent iron nanoparticles (NZVI) readily react with oxygen forming H_2O_2 or react with Fe^{+2} producing OH radicles that can efficiently degrade organic contaminants. Thus, in the presence of oxygenated media, NZVI have efficiently oxidized organic pollutants owing to their strong oxidizing ability. NZVIs are characterized by their ratio of large surface area to volume allowing them to enhance (Gómez-Sagasti et al. 2019). The use of NZVIs in the detoxification of pollutants has been reported, including organic halocarbons and organochlorine pesticides, chlorinated organic solvents and heavy metals (Zhao et al. 2016; Liu et al. 2009).

Decontamination using zero-valent iron particles (ZVINPs) is broadly studied in aquatic environments, and few were reported in soils. The use of NZVIs is limited to aerobic environments as the activity of ZVI relies on redox status. NZVIs are commercially produced by various companies/countries such as Toda Kogyo Ltd. In Japan, Golder Associates Inc. in USA, and nano-iron in the Czech Republic. The rapid aggregation of nano-ZVI and then forming micro- or macro-particles is the biggest drawback in NZVI application. In addition, plain NZVI without any stabilizing will cause agglomeration, weak structural durability, reduced surface area, and reduced redox potential. The use of suitable stabilization materials to form nanocomposite has proven promising results by immobilizing naked NZVI (Sepehri et al. 2023). The formation of porous support using activated carbon, extracted from paddy straw to synthesize "NZVI-paddy straw" composite has successfully engaged in cleaning Pb-contaminated water (Sepehri et al. 2023). Further, bentonite was used as a carrier substance to boost efficacy of ZVI to clean Cr contaminations in soils and in aqueous media (Shi et al. 2011).

(c) Microbe-assisted Nano-remediation

The use of microbe-derived nano-remediation of contaminated agricultural lands has been identified as greater sustainable and eco-friendly approach to substituting chemical nanomaterials that have various disadvantages. Thus, exploring microbe-assisted green nanoparticles has become popular among scientists in reclaiming contaminated soils and water. Plant extracts, enzymes from fungi and bacteria have been identified as promising alternatives, supplemented with metals, proteins or bioactive compounds creating composite NPs. Composite NPs are more stable with higher reducing power leading to greater detoxification ability. The microbe's inhabitants in extreme environments such as mangrove ecosystems that frequently face high salinity and frequent tidal waves possess specific chemical substances. For example, bio-fabricated iron oxide NPs were extracted from *Aspergillus tubingensis* (STSP 25) associated with mangrove species; *Avicennia officinalis* in the largest mangrove ecosystem of the world in Sundarbans, India (Mahanty et al. 2020). The detoxification of trace elements such as Ni, Cu, Pb, and Zn in mangrove estuaries was removed by the naturally synthesized NPs from mangrove sediment-associated microbes by surface binding of heavy metals (Mahanty et al. 2020). Moreover, *Chlorella vulgaris* has been identified as an iron oxide co-precipitator (Govarthanan et al. 2020) and *Escherichia* species, SINT7 was reported to produce biogenic copper NPs, which were capable of degrading azo-dyes in textile industry waste (Noman et al. 2020). These NPs are ideal for the eco-friendly decontamination of toxins in agricultural lands and industrial wastewater.

20.3 Nanoparticles for Agriculture Sustainability

Nano-based fertilizers, pesticides, growth regulators, and herbicides are novel approaches that reduce input amount and wastage, reduce cost, and reclaim pollution (Kottegoda et al. 2011; Aragay et al. 2012; Parisi et al. 2015). The efficacy of absorbance is vital in the process of designing nano-based composites and capsules to prevent wastage and overdosage while minimizing pollution. The application of nano-fertilizers can enhance fertilizer-use-efficiency by increasing absorption while reducing leaching, volatilization, and runoff owing to their slow-eereleasing ability. Nano-fertilizer shows multiple actions such as efficient root penetration and binding with toxic substances immobilizing them (Fig. 20.1). The various materials were reported to be used in producing fertilizer composites, including chemical substances, plant fibers, starch, and plastics, which are used to cover the fertilizer thereby enhancing slow-releasing capacity. For example, Zn and Al $(OH)_3$ were reported as cover in producing nanocomposites and woody encapsulation of urea showed a slow release of nitrogen (Kottegoda et al. 2011). Moreover, composites coated with plastic and starch mixture showed controlled release of urea. Nano-clays, nano-zeolites, and nano-hydrogels were produced and used on a commercial scale for enhancing water holding capacity and nutrient availability (Mahfoudhi and Boufi 2016). Use of green nanotechnology is environmentally friendly and enhances sustainability in agriculture by increasing productivity.

Pesticides encapsulated with Si and Ag/TiO_2 released active ingredients slowly, prevented direct exposure to the environment and prevented degradation by light remaining longer period on target area that enhanced pesticide efficacy (Khot et al. 2012). Antimycotoxin derivatives have been identified as a nano-fungicide that was synthesized by mycosynthesis (Kumari and Khan 2017).

The application of nano-bioplastics as coating materials, biochar for adsorbing toxicants and nano-sensors for soil inspection are some other nano-applications, beneficial for addressing the issues in agriculture for better production.

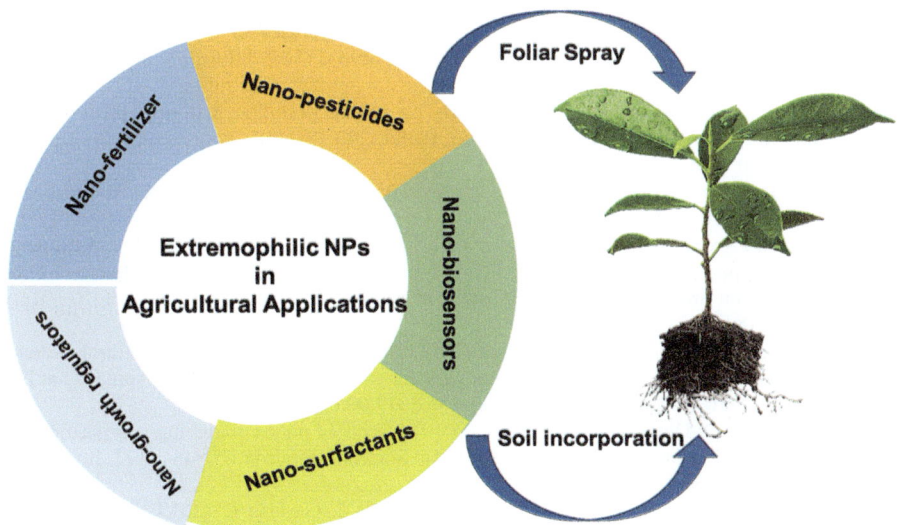

Fig. 20.1 Possible applications of extremophilic nano-products for sustainable agriculture

20.4 Conclusions

The use of extremophiles and the exploration of their secondary metabolites and nano-based bioactive compounds has become a novel eco-friendly and economical trend in agriculture replacing chemical products. Agricultural applications such as nanofertilizers, nanopesticides, nanobiosensors, nanosurfactants, and use as agriculture reclamation agents are some of the key directions that need further studies and attention. The species-wise plant responses and location-specific responses need to be thoroughly studied before commercial applications. The role of NPs, behavior in the environment, interactions, and ecotoxicity evaluations will be helpful in each case of NPs to be introduced. Further, field-scale trials and applications are more acknowledged to understand their efficacy as most trials are performed under controlled environments. However, the full- and long-term understanding of destiny and human and environmental impacts of NPs still a challenge, especially in the agriculture sector.

References

Aliyari Rad S, Nobaharan K, Pashapoor N, Pandey J, Dehghanian Z, Senapathi V, Minkina T, Ren W, Rajput VD, Asgari LB (2023) Nano-microbial remediation of polluted soil: a brief insight. Sustain For 15(1):876. https://doi.org/10.3390/su15010876

Aragay G, Pino F, Merko A (2012) Nanomaterials for sensing and destroying pesticides. Chem 112:5317

Astner AF, Hayes DG, O'Neill H, Evans BR, Pingali SV, Urban VS, Young TM (2019) Mechanical formation of micro- and nano-plastic materials for environmental studies in agricultural ecosystems. Sci Total Environ 685:1097–1106. https://doi.org/10.1016/j.scitotenv.2019.06.241

Boots B, Russell CW, Green DS (2019) Effects of microplastics in soil ecosystems: above and below ground. Environ Sci Technol 53(19):11496–11506. https://doi.org/10.1021/acs.est.9b03304

Chae Y, An Y-J (2018) Current research trends on plastic pollution and ecological impacts on the soil ecosystem: a review. Environ Pollut 240:387–395. https://doi.org/10.1016/j.envpol.2018.05.008

Chakraborty T, Akhtar N (2021) Biofertilizers: prospects and challenges for future. Biofertilizers: Study and Impact, pp 575–590

Egene CE, Van Poucke R, Ok YS, Meers E, Tack FMG (2018) Impact of organic amendments (biochar, compost and peat) on Cd and Zn mobility and solubility in contaminated soil of the Campine region after three years. Sci Total Environ 626:195–202. https://doi.org/10.1016/j.scitotenv.2018.01.054

El-Bahi SM, Sroor A, Mohamed GY, El-Gendy NS (2017) Radiological impact of natural radioactivity in Egyptian phosphate rocks, phosphogypsum and phosphate fertilizers. Appl Radiat Isot 123:121–127. https://doi.org/10.1016/j.apradiso.2017.02.031

Gao H, Yan C, Liu Q, Ding W, Chen B, Li Z (2019) Effects of plastic mulching and plastic residue on agricultural production: a meta-analysis. Sci Total Environ 651:484–492. https://doi.org/10.1016/j.scitotenv.2018.09.105

Gómez-Sagasti MT, Epelde L, Anza M, Urra J, Alkorta I, Garbisu C (2019) The impact of nanoscale zero-valent iron particles on soil microbial communities is soil dependent. J Hazard Mater 364:591–599. https://doi.org/10.1016/j.jhazmat.2018.10.034

Gottschall N, Topp E, Edwards M, Payne M, Kleywegt S, Lapen DR (2017) Brominated flame retardants and perfluoroalkyl acids in groundwater, tile drainage, soil, and crop grain following a high application of municipal biosolids to a field. Sci Total Environ 574:1345–1359. https://doi.org/10.1016/j.scitotenv.2016.08.044

Govarthanan M, Jeon C-H, Jeon Y-H, Kwon J-H, Bae H, Kim W (2020) Non-toxic nano approach for wastewater treatment using *Chlorella vulgaris* exopolysaccharides immobilized in iron-magnetic nanoparticles. Int J Biol Macromol 162:1241–1249

Guerra FD, Attia MF, Whitehead DC, Alexis F (2018) Nanotechnology for environmental remediation: materials and applications. Molecules 23:1760

Hunt AJ, Anderson CWN, Bruce N, García AM, Graedel TE, Hodson M, Meech JA, Nassar NT, Parker HL, Rylott EL et al (2014) Phytoextraction as a tool for green chemistry. Green Process Synth 3:3–22

Islam MA, Romić D, Akber MA, Romić M (2018) Trace metals accumulation in soil irrigated with polluted water and assessment of human health risk from vegetable consumption in Bangladesh. Environ Geochem Health 40(1):59–85. https://doi.org/10.1007/s10653-017-9907-8

Jiao W, Chen W, Chang AC, Page AL (2012) Environmental risks of trace elements associated with long-term phosphate fertilizers applications: a review. Environ Pollut 168:44–53. https://doi.org/10.1016/j.envpol.2012.03.052

Khan MN, Mobin M, Abbas ZK, Alamri SA (2018) Fertilizers and their contaminants in soils, surface and groundwater. In: Encyclopedia of the Anthropocene. Elsevier, pp 225–240. https://linkinghub.elsevier.com/retrieve/pii/B9780128096659098888

Khot LR, Sankaran S, Maja JM et al (2012) Applications of nanomaterials in agricultural production and crop protection: a review. Crop Prot 35:64–70

Kochhar K, Kavya IK, Shrivastava S, Ghosh A, Rawat VS, Sodhi KK, Kumar M (2022) Perspectives on the microorganism of extreme environments and their applications. Curr Res Microb Sci 3:100134. https://doi.org/10.1016/j.crmicr.2022.100134

Kottegoda N, Munaweera I, Madusanka N, Karunaratne V (2011) A green slow-release fertilizer composition based on urea-modified hydroxyapatite nanoparticles encapsulated wood. Curr Sci 101:73–78

Kumari S, Khan S (2017) Synthesis and applications of nanofungicides: a next-generation fungicide. In: Prasad R (ed) Fungal nanotechnology. Fungal biology, Springer, Cham, pp 103–118

Liu W-X, Liu J-W, Wu M-Z, Li Y, Zhao Y, Li S-R (2009) Accumulation and translocation of toxic heavy metals in winter wheat (Triticum aestivum L.) growing in agricultural soil of Zhengzhou. Bull Environ Contam Toxicol 82:343–347

Liu Y, Xu K, Cheng J (2020) Different nanomaterials for soil remediation affect avoidance response and toxicity response in earthworm (Eisenia fetida). Bull Environ Contam Toxicol 104:477–483

Lv Y, Huang S, Huang G, Liu Y, Yang G, Lin C, Xiao G, Wang Y, Liu M (2020) Remediation of organic arsenic contaminants with heterogeneous Fenton process mediated by SiO2-coated nano zero-valent iron. Environ Sci Pollut Res 27:12017–12029

Mahanty S, Chatterjee S, Ghosh S, Tudu P, Gaine T, Bakshi M, Das S, Das P, Bhattacharyya S, Bandyopadhyay S (2020) Synergistic approach towards the sustainable management of heavy metals in wastewater using myco-synthesized iron oxide nanoparticles: bio-fabrication, adsorptive dynamics and chemometric modeling study. J Water Process Eng 37:101426

Mahfoudhi N, Boufi S (2016) Poly (acrylic acid-co-acrylamide)/cellulose nanofibrils nanocomposite hydrogels: effects of CNFs content on the hydrogel properties. Cellulose 23:3691

Markande AR, Patel D, Varjani S (2021) A review on biosurfactants: properties, applications and current developments, vol 330. Bioresour Technol, p 124963

Mocek-Płóciniak A, Mencel J, Zakrzewski W, Roszkowski S (2023) Phytoremediation as an effective remedy for removing trace elements from ecosystems. Plan Theory 12:1653. https://doi.org/10.3390/plants12081653

Noman M, Shahid M, Ahmed T et al (2020) Use of biogenic copper nanoparticles synthesized from a native Escherichia sp. as photocatalysts for azo dye degradation and treatment of textile effluents. Environ Pollut 257:4

Ondrasek G, Bakić Begić H, Zovko M, Filipović L, Meriño-Gergichevich C, Savić R, Rengel Z (2019) Biogeochemistry of soil organic matter in agroecosystems & environmental implications. Sci Total Environ 658:1559–1573. https://doi.org/10.1016/j.scitotenv.2018.12.243

Parisi C, Vigani M, Rodríguez-Cerezo E (2015) Agricultural nanotechnologies: what are the current possibilities? Nano Today 10:124–127

Prata JC (2018) Microplastics in wastewater: state of the knowledge on sources, fate and solutions. Mar Pollut Bull 129(1):262–265. https://doi.org/10.1016/j.marpolbul.2018.02.046

Rath S, Srivastava RK (2021) Biosurfactants production and their commercial importance. Environ Agric Microbiol: Appl Sustain:197–218

Revel M, Châtel A, Mouneyrac C (2018) Micro(nano)plastics: a threat to human health? Curr Opin Environ Sci Health 1:17–23. https://doi.org/10.1016/j.coesh.2017.10.003

Sepehri S, Kanani E, Abdoli S, Rajput VD, Minkina T, Asgari Lajayer B (2023) Pb(II) removal from aqueous solutions by adsorption on stabilized zero-Valent iron nanoparticles—a Green approach. Water 15:222. https://doi.org/10.3390/w15020222

Shen G, Zhang S, Liu X, Jiang Q, Ding W (2018) Soil acidification amendments change the rhizosphere bacterial community of tobacco in a bacterial wilt affected field. Appl Microbiol Biotechnol 102(22):9781–9791. https://doi.org/10.1007/s00253-018-9347-0

Shi L, Zhang X, Chen Z (2011) Removal of chromium (VI) from wastewater using bentonite-supported nanoscale zero-valent iron. Water Res 45:886

Singh R, Behera M, Kumar S (2020) Nano-bioremediation: An innovative remediation technology for treatment and management of contaminated sites. In: Bioremediation of industrial waste for environmental safety. Springer, Singapore, pp 165–182

Siyar R, Doulati Ardejani F, Norouzi P, Maghsoudy S, Yavarzadeh M, Taherdangkoo R, Butscher C (2022) Phytoremediation potential of native hyperaccumulator plants growing on heavy metal-contaminated soil of Khatunabad copper smelter and refinery, Iran. Water 14:3597. https://doi.org/10.3390/w14223597

Srivastava V, de Araujo ASF, Vaish B, Bartelt-Hunt S, Singh P, Singh RP (2016) Biological response of using municipal solid waste compost in agriculture as fertilizer supplement. Rev Environ Sci Biotechnol 15(4):677–696. https://doi.org/10.1007/s11157-016-9407-9

Sun S, Sidhu V, Rong Y et al (2018a) Pesticide pollution in agricultural soils and sustainable remediation methods: a review. Curr Pollution Rep 4:240–250. https://doi.org/10.1007/s40726-018-0092-x

Sun J, Pan L, Tsang DCW, Zhan Y, Zhu L, Li X (2018b) Organic contamination and remediation in the agricultural soils of China: a critical review. Sci Total Environ 615:724–740., ISSN 0048-9697. https://doi.org/10.1016/j.scitotenv.2017.09.271

Sundararaghavan A, Mukherjee A, Suraishkumar GK (2020) Investigating the potential use of an oleaginous bacterium, Rhodococcus opacus PD630, for nano-TiO2 remediation. Environ Sci Pollut Res 27:27394–27406

Tiwari S, Prasad V, Lata C (2019) Bacillus: plant growth promoting bacteria for sustainable agriculture and environment. In: New and future developments in microbial biotechnology and bioengineering. Elsevier, pp 43–55

Tran TD, Dao NT, Sasaki R, Tu MB, Dang GHM, Nguyen HG, Dang HM, Vo CH, Inigaki Y, van Nguyen N (2020) Accelerated remediation of organochlorine pesticide-contaminated soils with phyto-Fenton approach: a field study. Environ Geochem Health 42:3597–3608

Trujillo-Reyes J, Peralta-Videa J, Gardea-Torresdey J (2014) Supported and unsupported nano-materials for water and soil remediation: are they a useful solution for worldwide pollution? J Hazard Mater 280:487–503

Tudi M, Daniel Ruan H, Wang L, Lyu J, Sadler R, Connell D, Chu C, Phung DT (2021) Agriculture development, pesticide application and its impact on the environment. Int J Environ Res Public Health. 18(3):1112. https://doi.org/10.3390/ijerph18031112. PMID: 33513796; PMCID: PMC7908628

Vox G, Loisi RV, Blanco I, Mugnozza GS, Schettini E (2016) Mapping of agriculture plastic waste. Agric Agric Sci Procedia 8:583–591. https://doi.org/10.1016/j.aaspro.2016.02.080

Wang Y, Liu Y, Zhan W, Zheng K, Lian M, Zhang C, Ruan X, Li T (2020) Long-term stabilization of Cd in agricultural soil using mercapto-functionalized nano-silica (MPTS/nano-silica): a three-year field study. Ecotoxicol Environ Saf 197:110600

Weber R, Herold C, Hollert H, Kamphues J, Blepp M, Ballschmiter K (2018) Reviewing the relevance of dioxin and PCB sources for food from animal origin and the need for their inventory, control and management. Environ Sci Eur 30(1):42

WHO (2020). https://www.who.int/news-room/questions-and-answers/item/chemical-safety-pesticides

WHO, ed (1996) Trace elements in human nutrition and health. World Health Organization, Geneva, p 343

Zhang W (2018) Global pesticide use: profile. Trend, cost/benefit and more. Proc Int Acad Ecol Environ Sci 8(1):1

Zhao X, Liu W, Cai Z, Han B, Qian T, Zhao D (2016) An overview of preparation and applications of stabilized zero-valent iron nanoparticles for soil and groundwater remediation. Water Res 100:245–266. https://doi.org/10.1016/j.watres.2016.05.019

Ziajahromi S, Neale PA, Leusch FDL (2016) Wastewater treatment plant effluent as a source of microplastics: review of the fate, chemical interactions and potential risks to aquatic organisms. Water Sci Technol J Int Assoc Water Pollut Res 74(10):2253–2269. https://doi.org/10.2166/wst.2016.414

Chapter 21
Nanoparticles Synthesis Using Extremophilic Microbes and their Potential Agricultural Applications

Girima Nagda, Nitish Rai, Jaya, Shakshi, Chhavi Bhalothia, and Namita Ashish Singh

21.1 Introduction

Nanotechnology (NT), the science and engineering of manipulating matter at the nanoscale, has rapidly advanced in recent years, offering unprecedented opportunities for various industries. Nanotechnology enables the design, synthesis, and manipulation of materials at the nanoscale, ranging between 1 to 100 nanometers (Fendler 1998). At the nanoscale, materials show exclusive physical, chemical, as well as biological properties that differ from their larger scale. These properties have proven advantageous for various applications, including medicine, energy, and agriculture (Prasad et al. 2014). One fascinating avenue of NT research for agricultural applications is the synthesis of NPs using extremophilic microbes (Li et al. 2019). By harnessing the unique properties of extremophiles and the precise manipulation capabilities of NT, researchers can develop innovative solutions that offer sustainable and efficient farming practices. The exploration of extremophilic organisms capable of thriving in extreme environments has captivated researchers due to their remarkable adaptability. In the context of agriculture, the integration of NT and extremophiles presents a compelling avenue for revolutionizing farming practices and addressing the challenges of global food security (Bahrulolum et al. 2021). Extremophiles are microorganisms that thrive in extreme environmental conditions

G. Nagda · C. Bhalothia
Department of Zoology, Mohanlal Sukhadia University, Udaipur, Rajasthan, India

N. Rai
Department of Biotechnology, Mohanlal Sukhadia University, Udaipur, Rajasthan, India

Department of Zoology, University of Lucknow, Lucknow, Uttar Pradesh, India

Jaya · Shakshi · N. A. Singh (✉)
Department of Microbiology, Mohanlal Sukhadia University, Udaipur, Rajasthan, India
e-mail: namita.singh@mlsu.ac.in

that are generally hostile to other life forms. They can be found in diverse habitats such as hot springs, hydrothermal vents, polar regions, and highly acidic or alkaline environments. Extremophiles have progressed remarkable adaptations to survive and function under conditions of high temperature, pressure, salinity, acidity, or aridity (Shu and Huang 2022). In order to thrive under such hostile conditions, these resilient microorganisms have acquired adaptive traits that enable their survival through the ability to produce a variety of bioactive compounds and secondary metabolites as a response to these harsh and extreme conditions (Yadav et al. 2015).

The convergence of NT and extremophiles in the field of agriculture holds significant promise for addressing critical agricultural challenges, such as optimizing crop productivity, improving nutrient uptake, enhancing soil fertility, and mitigating the impact of environmental stresses. The NPs can be tailored to serve various functions, such as enhancing nutrient delivery, improving plant disease resistance, and facilitating targeted delivery of agrochemicals (Adebayo et al. 2021). Moreover, extremophilic microbes possess enzymes capable of functionalizing the nanoparticle surfaces, allowing for the controlled release of nutrients or bioactive compounds. This precise control over nanoparticle synthesis and functionalization enables farmers to optimize nutrient uptake by crops, reduce chemical inputs, and enhance overall agricultural productivity. While the integration of NT and extremophiles presents promising opportunities, several challenges must be addressed. Regulatory frameworks and safety considerations surrounding the use of nanomaterials in agriculture need to be carefully evaluated (Singh and Colonna 2015). Additionally, scalability and cost-effectiveness are important factors to ensure the practical implementation of these technologies on a huge scale.

This chapter explores the intersection of NT and extremophiles in the agricultural context. The chapter provides an overview of the synthesis of different NPs using extremophilic microorganisms like acidophiles, alkalophiles, halophiles, psychrophiles, and thermophiles/hyperthermophiles. It delves further into the distinctive application of synthesized NPs highlighting increasing soil fertility, crop protection, biocontrol potential, and antimicrobial action. Lastly, the chapter discusses the constraints and challenges ahead in the practical implementation of these technologies on a large scale in agriculture.

21.2 Synthesis of Nanoparticles Using Extremophilic Microbes

Biologically synthesized NPs are recognized as an ecologically friendly and cost-effective method. Microbes are valuable sources for the biosynthesis of NPs due to their characteristics such as swift growth rate, easy culture, and ability to grow in different temperatures, pH, as well as pressure. Various types of NPs such as gold (Au), copper (Cu), silver (Ag), zinc (Zn), palladium (Pd), titanium (Ti), as well as nickel (Ni) with precise shapes, sizes, and compositions along with particle distributions, are obtained by microbes (Khan et al. 2018). Microbes can synthesize NPs via enzymatic paths by scavenging metal and exploiting reduction mechanisms (Pal et al. 2019).

Two techniques have emerged for the production of NPs: the top-down method, in which bulk material is broken down, and the bottom-up method, which generates NPs atom-by-atom or molecule-by-molecule. Chemical and biological treatments use the bottom-up strategy, they can be costly and inefficient due to the usage of harmful substances, Therefore, non-toxic, eco-friendly, and biocompatible techniques for producing NPs are being looked for by the scientific world. Plants, bacteria, fungi, and other species have shown promising results in NPs synthesis. Chemical approaches are preferable over biogenic enzymatic processes. Extremophilic bacteria that grow in severe settings provide biotechnology benefits because their potent enzymes permit conversions under hard circumstances that conventional enzymes cannot (Mukherjee et al. 2022). Some NPs synthesized by the extremophilic microbes have been shown in Table 21.1.

21.2.1 Acidophiles

Microorganisms that can survive at pH 3 and below are known as acidophiles. The adaptability of microbes to severe pH values exemplifies their capacity to flourish in harsh settings. Acidophiles and alkaliphiles are typically found in the bacterial and archaeal domains, respectively (Wiegel 2011). Acidophiles are typically found in Fumaroles, acidothermal hot springs, solfataras fields, bioreactors, acid mine drainage channels, and coal spoils. Extremely low levels of pH, temperatures variable between 25 to 90 °C, aerobic or anaerobic conditions, pressures up to 5 MPa, and low oxygen levels characterize these locales (Mukherjee et al. 2022).

(a) Synthesis of Nanoparticles Using Acidophilic Bacteria

An acidophilic aerobe, *Acidithiobacillus ferrooxidans*, produces intracellular electron-dense magnetite (Fe_3O_4) NPs. Besides nanoscale size, good biocompatibility, membrane-bound structure, and ferrimagnetism are its characteristics. Magnetosomes generated from *A. ferrooxidans* have the prospective to be used in a variety of biotechnological and medicinal uses, including the immobilization of bioactive compounds such as enzymes, antibodies, as well as biotin. Because of their distinct features, they are useful instruments for targeted drug administration, enzyme immobilization, and antibody conjugation (Yan et al. 2012). *Lactobacillus acidophilus*, a well-known probiotic bacterium, acts as a capping along with a reducing agent in the room-temperature production of AgNPs. *L. acidophilus* is also used to create NPs of other materials like selenium (Se) and cadmium sulphide (CdS). Its flexibility in the creation of NPs extends beyond Ag to embrace these other materials as well (Mukherjee et al. 2022).

Acidophilic actinomycetes strains *Pilimelia columellifera* SL19 and *P. columellifera* SL24 were separated from the soils of the pine forest with a pH of less than

Table 21.1 Nanoparticles synthesized by the extremophilic microbes

Extremophilic microbe	Class of extremophile	Type of nanoparticle	Size (nm)	References
Verticillium sp.	Acidophiles	Au	20	Mukherjee et al. (2001)
Verticillium sp.		Ag	13–37	Sastry et al. (2003)
Fusarium oxysporum		BaTiO₃	4 ± 1	Bansal et al. (2005)
Aspergillus tamarii		Ag	25–50	Durán et al. (2005)
Bipolaris nodulosa		Ag	10–60	Sahai (2010)
Sulfolobus acidocaldarius		Au	2.5–4	Selenska-Pobell et al. (2011)
Aspergillus terreus		Ag	1–20	Li et al. (2011)
Penicillium waksmanii		Cu	79–179	Honary et al. (2012)
Penicillium nalgiovense		Ag	25.2 ± 2.8	Maliszewska et al. (2014)
Streptacidiphilus sp.		Ag	16	Mohanta and Behera (2014)
Sulfolobus islandicus		Au	50	Kalabegishvili et al. (2014)
Arthrobacter nitroguajacolicus		Au	40	Dehnad et al. (2015)
Acidocella aromatica		Pt	8.5–16.1	Matsumoto et al. (2021)
Pilimelia columellifera		Ag	4–36	Hochvaldová et al. (2022)
Pseudomonas alcaliphila	Alkalophile	Se	50–200	Wang et al. (2010)
Idiomarina sp.		Ag	26	Seshadri et al. (2012)
Spirulina platensis		Ag, au	5–40	Kalabegishvili et al. (2013b)
Bacillus licheniformis		Au	38	Singh et al. (2014)
Bacillus licheniformis		CdS	20–40	Tiquia-Arashiro et al. (2016)
Pseudomonas taiwanensis		Ag	–	Beeler et al. (2020)
Bacillus megaterium	Halophile	Se	200	Mishra et al. (2011)
Navicula atomus		Au	9	Schrofel et al. (2011)
Sargassum wightii		Au	8–12	Oza et al. (2012)
Halococcus salifodinae BK3		Ag	20–12	Srivastava et al. (2013)
Citricoccus sp.		Se	104.46–50.82	Dinc et al. (2022)
Dunaliella salina		Ag	35	Shantkriti et al. (2023)

(continued)

Table 21.1 (continued)

Extremophilic microbe	Class of extremophile	Type of nanoparticle	Size (nm)	References
Pseudomonas proteolytica	Psychrophile	Ag	6.2 ± 2.4	Shivaji et al. (2011)
Arthrobacter gangotriensis		Ag	12.2 ± 5.7	Shivaji et al. (2011)
Euplotes focardii		Ag	20–70	John et al. (2020)
Thermoanaerobacter ethanolicus	Thermophile	Fe_3O_4	100	Roh et al. (2006)
G. Stearothermophilus		Au	5–40	Fayaz et al. (2010)
Thermoanaerobacter sp.		CdS	10	Yeary et al. (2011)
Humicola sp.		Ag	5–25	Syed et al. (2013)
Humicola sp.		CeO_2	12–20	Khan and Ahmad (2013)
Geobacillus sp.		Au	5–50	Correa-Llantén et al. (2013)
Caldicellulosiruptor saccharolyticus		Pd	100	Shen et al. (2015)
Thermoanaerobacter		SiO_2	15 ± 5	Show et al. (2015)
Thermoanaerobacter sp. X513		Cu	70	Jang et al. (2015)
Geobacillus Thermodenitrificans		Ag	1.44–16.3	Youssif et al. (2020)

4.0. The biological production of AgNPs was established by these strains. After treatment with $AgNO_3$ and 24 h of incubation, the colour of the cell filtrate changed from light yellow to dark brown, indicating AgNP production. The existence of spherical and monodispersed NPs was verified by spectroscopy analysis done on a UV-visible spectrophotometer, which produced peaks that were narrow with maximal absorbance at 425 nm and 430 nm. The AgNPs were discovered to be polydispersed, mostly spherical in form, and to exist largely as individual particles, while some aggregates were also seen. *P. columellifera* SL19 and *P. columellifera* SL24 synthesized AgNPs with diameters averaging around 12.7 nm and 15.9 nm, consecutively. For both strains, the AgNP size distribution varied from 4 nm to 36 nm (Hochvaldová et al. 2022).

The external production of AgNPs was reported to be facilitated by acidophilic *Actinobacteria* (SF23, C9). After 24 h of incubation with 1 mM $AgNO_3$, the colour of the cell filtrate changed from colourless to dark brown, suggesting the development of AgNPs. Transmission electron microscopy (TEM) investigation revealed that the AgNPs were nanostructured, equally scattered, and reported to have spherical shape, appearing as separate particles or clusters on occasion. AgNPs synthesized by SF23 varied in size from 4 to 36 nm, whereas those generated by C9 ranged in size from 8 to 60 nm (Anasane et al. 2016). *Acidocella aromatica* and *Acidiphilium crytpum*, two Fe (III)-reducing acidophilic bacteria, were studied for their capacity to create platinum NPs from a solution that was acidic. Platinum NPs were synthesized using *A. aromatic*a cells; these NPs were the smallest having diameters of 16.1 nm and 8.5 nm, consecutively (Matsumoto et al. 2021).

(b) Synthesis of Nanoparticles Using Acidophilic Archaea

The application of cells from the archaeon *Sulfolobus islandicus*, which was thermoacidophilic, to synthesize AuNPs for industrial applications has been known. The procedure involves culturing entire *S. islandicus* cells in an aqueous solution of chloroauric acid ($HAuCl_4$) ranging from 1 to 3 mM at pH 2 and 75 °C with shaking followed by characterization with UV-VIS spectroscopy. Extracellular and intracellular production of gold NPs (AuNPs) that are spherical in shape with diameters in the range of 20 to 80 nm and a usual size of 50 nm occurred inside the biomass of *S. islandicus* (Kalabegishvili et al. 2014). The interaction of *S. islandicus* with an aqueous solution of $AgNO_3$ eventually formed intracellular AgNPs for the first time in only a few hours. The particle size is initially 10 nm, but it progressively grows to a range of 10–50 nm during the process (Kalabegishvili et al. 2015). Gold NPs (AuNPs) may be generated without chemical functionalization on the surface layer (S-layer) of *Sulfolobus acidocaldarius*, which is naturally thiol-containing and proteinaceous. The archaeal AuNPs, which are around 2.5 nm in size, are totally made of metallic Au (0). In comparison, the bacterial S-layer generates 4 nm AuNPs that are a 40:60 combination of Au (0) and Au(III). Interestingly, unlike their bacterial counterparts or bulk gold, archaeal AuNPs have substantial paramagnetic characteristics (Selenska-Pobell et al. 2011).

(c) Synthesis of Nanoparticles Using Acidophilic Fungi

Fungi have received interest in nanoparticle creation as they have added advantages over other microbes. Fungal mycelial mesh is resistant to extreme conditions in bioreactors, making it simple to handle. Their reductive protein extracellular secretions facilitate downstream processing. Metals can be accumulated by fungi through a variety of processes. Enzymatic mechanisms, particularly those involving a reductase which is NADH-dependent, are considered important for the creation of NPs using fungus (Mukherjee et al. 2022).

When exposed to aqueous $AuCl^-_4$ ions, the acidophilic fungus *Verticillium* sp. undergoes a reduction process that results in the creation of AuNPs having a diameter of 20 nm. These NPs are produced on fungi's surface as well as inside its cells, especially on the membrane of cytoplasm. Notably, the metal ions in the surrounding solution are reduced to a minimum. The gold NPs on the cytoplasmic membrane are predominantly spherical, with some triangular and hexagonal particles thrown in for good measure. A huge AuNPs which was quasi-hexagonal was also seen within the cytoplasm (Mukherjee et al. 2001). When exposed to $AgNO_3$, the acidophilic fungus *Verticillium* sp. undergoes a reduction process that results in the buildup of AgNPs inside the fungal biomass i.e., intracellular. The dark brown colour in the fungal biomass depicts that the reduction of metal ions has been successful, and AgNPs were formed in size ranging from 13 to 37 nm (Sastry et al. 2003).

Fusarium oxysporum has been widely explored and used in the synthesis of different NPs. It has the unique capacity to synthesize NPs outside of the cell. Cationic proteins released by *F. oxysporum* promote the formation of zirconia NPs when treated with $ZrF6^{2-}$ anions. The primary component involved in its production has been identified as a protein having a molecular weight range of 24–28 kDa. The zirconia NPs produced are mostly quasi-spherical in form, range 3–11 nm in size (Mukherjee et al. 2002).

At ambient temperature, *F. oxysporum* can also create irregular quasi-spherical barium titanate NPs ($BaTiO_3$) sized average of about 4 ± 1 nm (Bansal et al. 2005). Furthermore, when exposed to a salt combination ($K_3[Fe(CN)_6]$ and $K_4[Fe(CN)_6]$, *F. oxysporum* produces crystalline magnetite NPs with single-domain properties. These magnetite NPs have a quasi-spherical shape as well. These NPs are about 20–30 nm in size (Bharde et al. 2005). *F. oxysporum* has also shown the capacity to create water-soluble quantum dots with strong luminous characteristics when exposed to $CdCl_2$ and $SeCl_4$. At ambient temperature, these quantum dots were made of cadmium selenide (CdSe) and displayed a surface plasmon resonance band at 370 nm. The CdSe quantum dots were stable with a fluorescence half-life of 6–7 ns. They had a polydispersed spherical shape and ranged in size from 9 to 15 nm (Durán et al. 2005). *F. oxysporum* was also discovered to manufacture nanocrystals of optoelectronic material, especially bismuth oxide (Bi_2O_3). These nanocrystals were synthesized extracellularly and have a quasi-spherical form. The nanocrystals varied in size from 5 to 8 nm and exhibited acceptable adjustable characteristics (Uddin et al. 2008).

Using the fungus *Aspergillus tamarii*, researchers investigated an ecologically benign approach for synthesizing AgNPs. Scanning electron microscopy (SEM) imaging indicated the existence of AgNPs, which are spherical in shape ranging in size from 25 to 50 nm (Duran et al. 2005). Copper NPs (CuNPs) were effectively synthesized and stabilized using three distinct fungal strains obtained from soil: *Penicillium aurantiogriseum*, *Penicillium waksmanii* and *Penicillium citrinum*. Scanning electron microscopy revealed that the NPs were homogeneous and in spherical form. The NPs synthesized by *P. aurantiogriseum* were discovered to have a size ranging from 90 to 250 nm by dynamic light scattering. The size range for *P. citrinum* was 85–179 nm, while for *P. waksmanii* was 79–179 nm. These findings reveal that each fungal strain created copper NPs within a defined size range, demonstrating the fungi's potential for nanoparticle creation and stabilization (Honary et al. 2012).

(d) Synthesis of Nanoparticles by Acidophilic Algae

Microalgae have emerged as interesting sources for the production of functionalized NPs due to their availability of biocompatible reductants and environmentally benign character. An aqueous extract of a microalgal isolate termed *Chlorella acidophile*, which flourishes on non-arable terrain, was used to biosynthesize Ag, Au and bimetallic NPs in one research. The monometallic and bimetallic NPs synthesized with the aqueous extract of the microalgal isolate *C. acidophile* were found to have a crystalline structure and to be predominantly spherical in form. These NPs varied in size from 5 to 45 nm (Thangaswamy et al. 2021).

21.2.2 Alkalophiles

The word "alkaliphile" refers to microorganisms that flourish and develop best at pH levels over 9, whereas they grow slowly or not at all at pH levels of 6.5, which is nearly neutral. Alkaliphiles are microorganisms that include prokaryotes, eukaryotes, and archaea. The presence of alkalophiles has been detected in alkali thermal shallow hydrothermal systems, hypersaline soda, alkali thermal hot springs, and sewage lakes like Mono Lake in California and Lake Elementaita in Kenya's Rift Valley. These places have a broad temperature range; however, alkaliphiles are often well-suited to survive and flourish in such environments (Mukherjee et al. 2002; Kanekar et al. 2012).

(a) Synthesis of Nanoparticles Using Bacteria and Alkalophilic Algae

When exposed to aqueous solutions of chloroauric acid ($HAuCl_4$) and $AgNO_3$, the microorganism *Spirulina platensis* produced Au NPs and AgNPs. The NPs produced by algal biomass were largely generated extracellularly and had a crystalline structure. These NPs were usually circular in form having sizes from 5 to 40 nm (Kalabegishvili et al. 2013b). The biological production of AgNPs utilizing *S. platensis* cell-free extract was explored. When the extract was mixed with $AgNO_3$ solution, the resultant solution coloured dark brown, suggesting that Ag ions were reduced and AgNPs were formed. These particles were spherical in form and ranged in size from 30 to 50 nm. Importantly, the NPs were monodispersed, which means they did not form huge agglomerates, suggesting they were stabilized by a capping agent (Sharma et al. 2015).

Using a Se^{2-} resistant strain of *Pseudomonas alcaliphila*, researchers established an effective and ecologically benign technique for synthesizing selenium NPs (SeNPs). The SeNPs expanded in size with time, with spherical particles seen after 6 h and bigger particles ranging from 50 to 200 nm seen after 12 h. After 24 h, many SeNPs with diameters of around 500 nm developed as the reaction continued. The well-known Ostwald ripening mechanism was used in this development phase, in which larger SeNPs devoured smaller ones (Zhang et al. 2011).

Bacillus licheniformis, a bacterial strain, was reported to be competent in synthesizing cadmium sulphide (CdS) NPs. The interaction between cadmium chloride ($CdCl_2$) and sodium sulphide was turned into CdS NPs by using the enzyme sulphate reductase. The precipitation rate was greatest when the cadmium chloride to sodium sulphide ratio was 1:1, and it was lowest when the ratio was 4:1. Surprisingly, the quantity of CdS precipitate formed was found to be inversely related to the amount of nanocrystal production, with maximal nanoparticle synthesis happening during the stationary period of the bacterial cell cycle. The CdS NPs were crystalline and varied in size from 20 to 40 nm, according to the analysis (Tiquia-Arashiro et al. 2016).

Under aerobic circumstances, the bacterium *B. licheniformis* JS2 has been reported to synthesize selenium NPs (SeNPs) by the inner conversion of hazardous selenite ions into non-toxic elemental SeNPs. A cell lysis process utilizing lysozyme and a French press technique was used to collect the intracellular SeNPs, followed by multiple washing with Tris-HCl buffer and extraction using a water-octanol two-phase solvent. When the SeNPs previously purified and cleaned were dispersed on a trypticase soy agar plate, no bacterial growth was seen, showing that the cell lysis procedure was highly competent. The selenium NPs varied in size from 40 to 180 nm (Sonkusre et al. 2014). An alkaliphilic bacteria, *Pseudomonas taiwanensis* SS8, quickly reduced $AgNO_3$ solution into AgNPs. UV-VIS spectrophotometry and electron microscopy were used to characterize the synthesized AgNPs. Polydispersed AgNPs were produced under optimal circumstances of pH 8–9 and 48-hour incubation in nutrient broth growth medium (Beeler et al. 2020).

21.2.3   Halophile

Halophiles are creatures with a liking for salty surroundings and the ability to thrive in them. They are mostly made up of prokaryotic and eukaryotic microorganisms that can maintain the osmotic pressure of their environment and tolerate the potentially harmful effects of excessive salt concentrations. There are around 182 recognized species in the *Archaea* domain that belong to the *Halobacteriaceae* family, which includes aerobic halophiles as well as a few halophilic methanogens and anaerobic ones. In comparison, the bacteria domain contains a broader range of halophiles spread over numerous groups or phyla. *Bacteroidetes, Cyanobacteria, Firmicutes, Proteobacteria*, and *Sulphur-Green bacteria* are all phyla that contain halophilic bacteria. Within the Eucarya domain, halophilic creatures such as fungi, plants, ciliates, and flagellates have been discovered (Oren 2002).

Halophiles have developed a method that requires the regulated ingestion of potassium ions (K^+) into their cytoplasm. This adaptation demands the modification of enzymes and cellular structures for them to operate properly in the presence of high salt concentrations. To function properly, all enzyme systems and structural parts must be appropriately modified to survive the obstacles given by excessive salt levels (Shivanand and Mugeraya 2011).

Halophilic organisms are useful because they include enzymes (extremozymes) that stay stable under severe pH and ionic strength settings. As a result, it is intriguing to investigate these species as biocatalysts when novel nanomaterials are present. Enzymes, halocins (antibiotic proteins from halobacteria), and exopolysaccharides, among other biomolecules generated by these halophilic organisms, demonstrate biological activity even in extreme environments. When these biomolecules are coupled with diverse nanomaterials such as thin layers, nanotubes, and nanospheres, they form unique compounds that have both the biological features of biomolecules and the physicochemical qualities of nanomaterials. Recently, researchers have begun looking into them as prospective sources of metal-tolerant bacteria capable of synthesizing metallic NPs (Metwally et al. 2023; Moopantakath et al. 2023).

(a) Synthesis of Nanoparticles Using Halophilic Bacteria

Most of the research on halophilic bacteria and their byproducts has focused on metallic NPs production. *Halomonas salina, H. maura, Idiomarina* sp.PR-58-8, and *Pseudomonas* sp. were among the halophilic bacteria investigated. *Idiomarina* sp. PR58-8, a halophilic marine bacterium, possesses a high tolerance to Ag and the ability to synthesize intracellular crystalline AgNPs with an average size of about 26 nm (Seshadri et al. 2012).

A newly discovered strain of *Pseudomonas* sp. 591,786 capable of synthesizing internal AgNPs has been identified. These AgNPs come in a variety of sizes, most of which are spherical with some nanotriangles measuring 20 to 100 nm. Smaller particles in the 10 to 20 nm range have also been found. When this bacterial strain is exposed to a 1 mM $AgNO_3$ solution, it quickly produces Ag NPs. However, the generation of intracellular Silver Nanoparticles (SNPs) requires around 6 h of incubation (Rammohan and Balakrishnan 2011). *Halomomas salina*, a kind of halophilic proteobacteria, can create AuNPs of varying forms depending on the pH of the environment. Anisotropic extracellular AuNPs are created under acidic circumstances, whereas spherical NPs are formed in alkaline settings (Shah et al. 2012).

(b) Synthesis of Nanoparticles by Halophilic Archaea.

Haloarchaea can survive at extraordinarily high salinity levels of up to 300 g/L (Zafrilla et al. 2010). In the hypersaline environment, these creatures maintain osmotic balance by collecting potassium ions within their cells (Oren 2010). While haloarchaea are exposed to metals, which are present in their environment, their tolerance to metals has not been thoroughly researched or recorded (Srivastava and Kowshik 2013).

Despite their exposure to metals in their natural environments, there is little known about haloarchaea metal tolerance and nanoparticle creation. Aside from two specific species, *Halococcus salifodinae* BK3 and H. *salifodinae* BK6, no additional cases of metallic nanoparticle formation by haloarchaea have been described. The enzyme NADH-dependent nitrate reductase, which assists in the reduction of Ag ions, is involved in the intracellular manufacture of AgNPs in *H. salifodinae* BK3 and BK6. These organisms can adapt to stress due to metal, as demonstrated by identical growth kinetics characteristics in the presence of $AgNO_3$ compared to those growing in the absence of it (Srivastava et al. 2013, 2014).

(c) Synthesis of Nanoparticle by Halophilic Fungi

A quick approach for producing AgNPs was demonstrated using *Pichia capsulata*, a yeast which is halophilic obtained from mangroves capable of extracellular production of AgNPs at pH 6.0, temp. 5 °C, and 0.3% NaCl concentration. The resultant NPs were predominantly shaped round and 525 nm in size. A partly purified NADH-dependent protein, comparable to nitrate reductase, was discovered and proposed to play a role in the reduction procedure (Subramanian et al. 2010). *Yarrowia lipolytica* (NCIM 3589) isolated from oil-polluted saltwater near Mumbai, synthesized AuNPs in 72 h at 30 °C. TEM examination revealed that both yeast and mycelial forms of this fungus were involved in the synthesis. The NPs size varied according to pH, ranging from huge triangular plates at pH 2.0 to 15 nm structures at pH 7.0 and 9.0. (Agnihotri et al. 2009).

Penicillium fellutanum, a fungus was studied for its capacity to synthesize AgNPs. The maximum nanoparticle production occurred when the filtrate of culture was treated with 1.0 mM AgNO3, 0.3% NaCl concentration, pH 6.0, and incubation at 5 °C for 24 h. The NPs created were spherical in form having a size of 5 to 25 nm. A protein with a molecular weight of around 70 kDa found in cell-free supernatant was involved in zero valence conversion of metal ions (Kathiresan et al. 2009). AgNPs can be synthesized by halophilic fungi viz. *Aspergillus niger* (AUCAS 237), these NPs were globular in form having a diameter from 5 to 35 nm (Seelan et al. 2009). *A. terreus* (MP1) isolated from a marine sponge demonstrated the capacity of synthesis of AgNPs, spherical in shape with a size of around 1520 nm. These NPs inhibited the development of harmful bacterial strains such as *Klebsiella pneumoniae*, *Staphylococcus aureus*, and *Salmonella typhi* (Meenupriya 2011).

The marine yeast *Rhodospiridium diobovatum* may synthesize intracellular lead sulphide (PbS) NPs utilizing non-protein thiols. Characterization studies revealed the existence of spherical PbS NPs ranging in size from 2 to 5 nm. During the stationary phase, the yeast accumulated 90% of the lead in the medium as PbS NPs, with increasing amounts of non-protein thiols. These thiols play an important role in the creation of NPs (Seshadri et al. 2011).

(d) Synthesis of Nanoparticles Using Halophilic Algae

There have been few and most recent discoveries concerning the creation of metallic NPs by salt-loving algae. Algae may be found in both freshwater and saltwater habitats, even those polluted with heavy metals. Several algae species have an established potential to interact with heavy metal ions and select species help to purify and remediate metal waste from water sources (Scarano and Morelli 2003; Pandit et al. 2022). The brown alga *Sargassum wightii* was credited with the first discovery of gold nanoparticle production. This marine species can produce stable AuNPs ranging in size from 30 to 100 nm. Furthermore, when *S. wightii* extract is subjected to AuCl and AgNO$_3$, it exhibits the ability to synthesize AgNPs with sizes between 8 and 12 nm (Oza et al. 2012).

AgNPs with good antifungal activities may be produced by reducing AgNO$_3$ with *S. longifolium* extracts. *S. longifolium* extracts include a variety of active compounds rich in carboxyl or hydroxyl groups, which are involved in the reduction process of the metallic ion. Notably, the AgNPs produced have outstanding antibacterial action against pathogenic fungi such as *Candida albicans*, *Aspergillus*, *Fusarium* sp., and *fumigatus*. To evaluate the antifungal activity, experiments were carried out at various concentrations. It was found that as the quantity of AgNPs grew the zone of inhibition against harmful fungi also increased (Rajesh Kumar et al. 2014).

Under laboratory circumstances, AuNPs were successfully biosynthesized using a mixture of two diatom strains, *Navicula atomus* and *Diadesmis gallica*, as well as an aqueous solution of HAuCl$_4$ containing 500 mg/L of Au. The biosynthesis of

AuNPs utilizing *D. gallica*, on the other hand, resulted in greater mean particle size, averaging approximately 22 nm, and a broader range of size dispersion. The gold NPs synthesized by *N. atomus* had a low average particle size of 9 nm (Schrofel et al. 2011). *Dunaliella salina* aqueous extract was used to decrease $AgNO_3$, resulting in the creation of AgNPs having size of 35 nm. The silver particles encountered the microalgae extract quickly, resulting in the instantaneous formation of AgNPs. The disc-diffusion technique was used to conduct antibacterial experiments, which revealed promising antibacterial effects against *Escherichia coli*, *Bacillus subtilis* etc. (Shantkriti et al. 2023).

The manufacture of selenium NPs utilizing *Citricoccus* sp. was investigated along with the effect of numerous factors, including time, stirring rate, pH, and temperature, on the formation of NPs. The best synthesis conditions were found to be a pH of 8, a reaction period of 24 h, a temperature of 37 °C, and a stirring rate of 150 rpm. The particles were reported to be spherical in form, a size average of around 104.46–50.82 nm (Dinc et al. 2022).

21.2.4 Psychrophiles

Psychrophiles are a subset of extremophiles that live at extremely cold temperatures. These microbes occupy around 75% of the Earth's biosphere, including the polar regions, deep oceans, and air habitats. Surprisingly, 70% of the planet's surface has temperatures ranging from 1 to 5 ° C (Feller 2003). One distinguishing feature of psychrophiles is the synthesis of psychrophilic enzymes. These enzymes prefer temperatures that are low for optimal enzyme activity, demonstrating increased specific activity at cold temperatures while being moderately thermostable. Several psychrophiles have exceptional ability to reduce heavy metal ions, making them appealing candidates for nanoparticle formation (Roulling et al. 2011).

(a) Synthesis of Nanoparticles by Psychrophilic Bacteria

Researchers effectively showed shape anisotropy in AgNPs by manipulating the development kinetics of *Morganella psychrotolerans*, a psychrophilic bacterium with silver resistance, in a ground-breaking work. This discovery is significant since no earlier papers have demonstrated the capacity to regulate the form of AgNPs by changing bacterial growth kinetics throughout the biological production process (Ramanathan et al. 2011). The supernatants of cell-free culture belonging to five bacteria, namely *Pseudomonas meridiana*, *Phaeocystis antarctica*, *Pseudomonas proteolytica*, *Arthrobacter gangotriensis*, and *Arthrobacter kerguelensis* were used in the production of extracellular AgNPs. The generated AgNPs were spherical in form, with average diameters ranging from 6.2 ± 2.4 nm to 12.2 ± 5.7 nm, as measured by transmission electron microscopy (Shivaji et al. 2011).

Researchers highlighted the usage of a new *Pseudomonas* strain acquired from a consortium associated with the Antarctic marine ciliate *Euplotes focardii* to produce AgNPs. The *Pseudomonas* cells were incubated with 1 mM $AgNO_3$ at a temperature of 22 °C for 24 h, resulting in the production of AgNPs. Scanning and transmission electron microscopy indicated the existence of spherical polydispersed AgNPs with sizes 20–70 nm (John et al. 2020).

21.2.5   Thermophiles/Hyperthermophiles

Heat-loving organisms, known as thermophiles, not only tolerate but also rely on high temperatures for development and survival. Thermophiles have the unusual capacity to grow at far greater temperatures than other bacteria, which would be injured or even killed by such severe heat. Their ideal temperature range for growth is from 50 °C to 121 °C, while just a few eukaryotes can grow beyond 50 °C, several fungi have been reported to thrive in the 50–55 °C temperature range (Kristjansson and Stetter 2021). Notably, thermophile categorization has evolved through time, with the idea of a unique division and a border (80 °C and above) for hyperthermophiles, a classification that has received general recognition. As a result, thermophiles are now divided into three types: moderate thermophiles, severe thermophiles, and hyperthermophiles. While the majority of thermophilic bacteria found to date thrive below the hyperthermophilic limit, *Thermotoga* and *Aquifex* are exceptions. Archaea, on the other hand, mostly constitute hyperthermophilic species (Takahata et al. 2001; Cekuolyte et al. 2023). In recent times, scientists have been investigating the utilization of thermophilic microbes for the synthesis of metallic NPs, aiming to achieve precise control over their chemical composition, size, and morphology.

(a) Nanoparticle Synthesis by Thermophilic Bacteria

In earlier work, anaerobic, metal-reducing *Thermoanaerobacter* sp. was used to synthesize cadmium sulphide (CdS) NPs. The extracellular CdS crystallites produced were less than 10 nm in size, and the procedure produced around 3 g/L of growing media each month. The synthesis was highly reproducible and scalable, with a successful output of up to 24 L (Moon et al. 2007). When a cell-free extract of *Geobacillus stearothermophilus* is exposed to metal salts, stable AgNPs and AuNPs develop in the solution. The absorption peaks of these NPs are 423 nm for Ag and 522 nm for Au, respectively. The analysis of TEM pictures indicated that the AgNPs were polydispersed, whereas the AuNPs were monodispersed. The presence of capping proteins released by the bacteria inside the reaction mixture accounts for the great stability found in the nanoparticle solution (Fayaz et al. 2010).

Geobacillus wiegelii GWE1 is an aerobic thermophile; this bacterium may decrease selenite, as evidenced by a notable shift in colour from colourless to red. The generated selenium NPs (SeNPs) have a well-defined spherical form and are elementally selenium. The nanoparticle distributed by supernatants varies from 40 to 160 nm, with roughly 70% of the NPs measuring smaller than 100 nm at pH 4.0. More than 90% of the NPs are smaller than 100 nm in size at pH 6.0 and 8.0. (Correa-Llantén et al. 2014).

Several research projects have been conducted on the bacterial production of AuNPs. One investigation includes the use of the extremophilic actinomycete *Thermomonospora* sp. The researchers discovered that this bacterium has a remarkable capacity to effectively synthesize monodisperse AuNPs. They credited the effective production of these monodisperse AuNPs to the process's severe biological conditions, notably alkaline environments, and somewhat higher temperatures (Ahmad et al. 2003; Sahoo et al. 2022). The production of AgNPs by *Ureibacillus thermosphaericus* using an extracellular route showed great potential. The biosynthetic processes were performed using the culture supernatant at temperatures between 60 and 80 °C and Ag ion concentrations ranging from 0.001 to 0.1 M. At an Ag ion concentration of 0.001 M, the average size of the AgNPs was 57 nm, with diameters of 29 nm and 13 nm at 60 °C, 70 °C, and 80 °C, respectively (Juibari et al. 2011). The cultivable *Thermoactinomycete* sp. 44 was used to generate samples containing Au NPs using a $HAuCl_4$ solution with a concentration of 103 M. During the reaction, AuNPs were formed over a period of 3–4 days with a diameter ranging from 5 to 60 nm (Kalabegishvili et al. 2013a).

The researchers studied the possibility of combining Pd NPs synthesis with H_2 generation by *Caldicellulosiruptor saccharolyticus* for severe thermophilic wastewater treatment. Na_2PdCl_4 was used to obtain a Pd concentration of 50 mg/L. Methyl orange decayed in 30 min in the presence of Pd, whereas diatrizoate degraded in 10 min. Through hydrogen, hydrogenase, and well-dispersed Pd^0 NPs, Pd improved the breakdown of both pollutants. The generated Pd^0 particles were mostly less than 100 nm in size, polyporous, and had a high catalytic activity (Shen et al. 2015). Copper NPs with diameters of 3 and 70 nm were created utilizing anaerobic *Thermoanaerobacter* sp. X513 bacteria in an aqueous solution in a biologically driven metal-reduction method. This method significantly reduced the presence of copper oxide contaminants in the finished product. Notably, this method demonstrated remarkable consistency and scalability, with batch sizes ranging from 0.01 to 1 L (Jang et al. 2015).

21.2.6   Nanoparticle Synthesis by Thermophilic Archaea

The thermophilic archaeon *Sulfolobus islandicus* has been explored for the production of AgNPs within a few hours. AgNPs with sizes between 10 and 50 nm are formed during the first 20-hour interaction between *S. islandicus* biomass and an aqueous solution of $AgNO_3$. Following that, the size of the AgNPs gradually rises over time, averaging 25 nm but varying from 10 to 50 nm. Concurrently, the amount of Ag NPs increases with time (Kalabegishvili et al. 2015).

(a) Nanoparticle Synthesis by Thermophilic Fungi

Scientists have discovered how the thermophilic fungus *Humicola* sp. produces AgNPs. The fungus successfully lowers the precursor solution when exposed to Ag^+ ions, resulting in the creation of extracellular NPs. The researchers were particularly concerned with preserving a size range of 5 to 25 nm (Syed et al. 2013). The bioinspired production of cerium oxide (CeO_2) NPs was reported for the first time utilizing the thermophilic fungus *Humicola* sp. When the fungus *Humicola* sp. is exposed to aqueous solutions containing the oxide precursor cerium (III) nitrate hexahydrate, it produces CeO_2 NPs extracellularly. Notably, the NPs are certainly protected by the proteins generated by the fungus, which prevents them from clumping together. As a result, they have a high degree of stability, water dispersibility, and fluorescence. CeO_2 NPs are spherical and polydispersed, with particle sizes ranging from 12 to 20 nm (Khan and Ahmad 2013).

21.3 Applications of Nanotechnology in Agriculture

Pests and plant diseases are responsible for 20% to 40% annual loss of crops worldwide. Pest treatment can be done by conventional insecticides, fungicides, as well as herbicides. It is vital to generate insecticides that are cost-effective, and less damaging to the environment and soil. Nanotechnology can be the driving force in plant disease management and generating pesticides that are slightly toxic, eco-friendly, and more water-soluble (Mali et al. 2020; Worrall et al. 2018). According to other reports, NT is considered compelling frontier in the present agriculture scenario, which chiefly focuses on sustainable agriculture and production of food to fulfil the requirements of humans and animals (Rehmanullah et al. 2020). Various applications of NT have been depicted in Fig. 21.1.

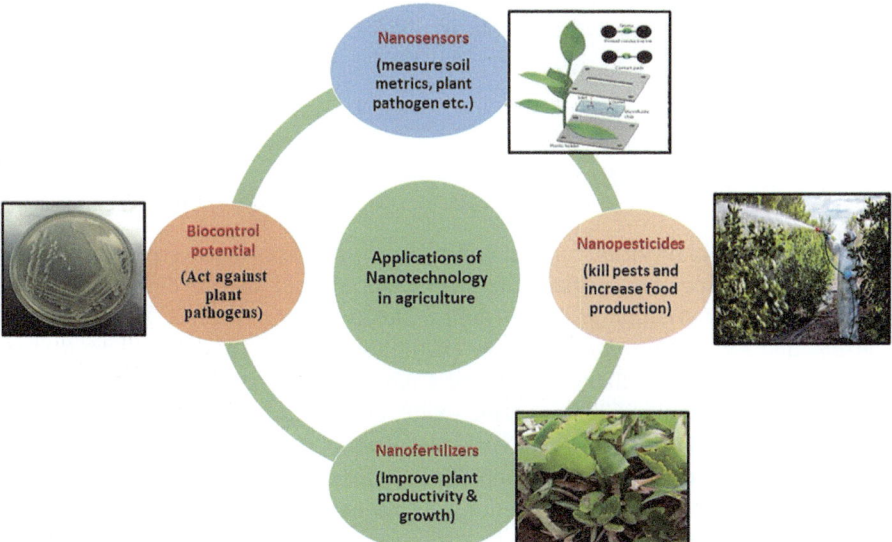

Fig. 21.1 Applications of nanotechnology in the agricultural sector

21.3.1 Increasing Soil Fertility by Nano-Fertilizers

According to Davari et al. (2017), nano fertilizer is a substance that enhances nutrient transfer to plants and manages the slow release of nutrients into the soil in a carefully regulated manner. Extremophile microorganisms can augment the soil with macronutrients viz. phosphorus, nitrogen, and potassium along with micronutrients viz. iron, zinc, and magnesium by living in these soils and adapting to the extreme environment settings. These nutrients can be fixed by a variety of plant-rhizospheric *Bacillus* species, including *B. subtilis*.

By improving the availability of vital nutrients to the plant, nano-fertilizers can boost agricultural output. The use of nano-phosphorus fertilizers in arid circumstances was found to significantly boost the yields of millet and cluster beans. Chitosan nanoparticle suspensions containing sodium, phosphorus, and nitrogen can also increase crop yield. In agricultural research, nanomaterials and nanostructures, including single and multiwalled carbon nanotubes, nanofibers, as well as quantum dots are currently used as biosensors to assess soil quality, plant pathogens as well as fertilizer distribution. Nano-clay, vermiculite, zeolite etc. are also used to increase crop output and fertilizer effectiveness in agriculture (Sivarethinamohan and Sujatha 2021).

Nanoparticles of zinc oxide for chickpeas, colloidal silica, and NPs fertilizer for tomato, silicon dioxide as well as iron slag powder for maize, titanium oxide for spinach, Au, and sulphur fertilizers are just a few examples of the many crops that can be benefited by the growth and enhancement of nano-fertilizers (Sivarethinamohan and Sujatha 2021). In order to deliver nutrients, materials can be coated with thin polymeric films, encapsulated in nanoporous structures, or delivered as nanoscale-sized particles or solutions. The negative effects of high doses of fertilizers are diminished by nanoscale fertilizers because they lessen soil toxicity (Davari et al. 2017).

21.3.2 Crop Protection

Each year, microbial (virus, fungus, and bacteria) illnesses cause significant losses in agriculture. Specific antibacterial nanomaterials aid in preventing microbial infections. *Colletotrichum gloeosporioides, Dematophora necatrix, Fusarium oxysporum*, and *Fusarium solani* are a few examples of pathogenic fungi, which cause infections. Copper and nickel ferrite NPs have compelling antifungal properties against plant pathogens. Chitosan, zinc oxide, and silica NPs are effective in treating viral infections viz. the mosaic virus for tobacco, potatoes, and lucerne. Target-specific herbicide enclosed in a nanoparticle is directed to the roots of the targeted weeds; subsequently, herbicide penetrates the root system. As a result, the particular weed plant will starve to death and die (Chinnamuthu and Kokiladevi 2007). Crops that are resistant to glyphosate have been observed to become vulnerable to it when treated with a nano surfactant made of soybean micelles. NPs not only promote plant growth but also shield it from abiotic stress viz. drought, salinity, temperature variation, metal along with mineral toxicity. Due to the NP's huge surface area and tiny size, toxic metal binds to it, decreasing its availability. Since photosynthesis is a very vulnerable process and a crucial component of plant metabolism, its proper operation could be preserved by reducing oxidative and abiotic stress (Sharifi et al. 2020).

(a) Biocontrol Potential

Utilizing extremophilic microbes to control biological disease is one research conclusion (Mehetre et al. 2021). There are two techniques by which NPs protect plants i.e., as crop protection agents on their own, or as carriers for current pesticides (Worrall et al. 2018). These biocontrol agents restrict diseases as well as a number of nematodes by interfering with their reproductive cycle and competing for nutrition. Examples of bacteria that act as plant pathogen inhibitors include *Bacillus, Clavibacter, Microbacterium*, and *Pseudomonas* (Verma et al. 2017). Some extremophilic microbes live in the plant rhizosphere and protect them from plant pathogens by producing secondary metabolites (antibiotics, siderophores etc.), and biocontrol agents (Torracchi et al. 2020). *Rhizobacteria* protect plants from diseases by creating siderophores (compounds that chelate iron), chitinases, ammonia, hydrogen cyanide, and a number of secondary metabolites (Pandey et al. 2021).

The most researched and used nanoparticle for biosystems is nanosilver. It is well recognized to have a wide range of antibacterial activity as well as potent inhibitory and bactericidal effects. Compared to bulk Ag, AgNPs have a stronger antibacterial impact. Kim et al. (2008) studied the antifungal efficiency of colloidal nanosilver (1.5 nm average diameter) against rose powdery mildew caused by *Sphaerotheca pannosa* var rose. It affects both roses growing outdoors and in greenhouses, and it is widespread. It results in diminished flowering, early defoliation, leaf deformation, and leaf curling. The well-distributed, stabilized Ag NPs solution known as nanosilver colloid is a better fungicide because it sticks to bacteria and fungi more effectively. The biosynthesis of SeNPs using metabolites, i.e., organic acids and biocontrol agents from the supernatant of *Trichoderma harzianum*, has been done. The antifungal activity of developed SeNPs was much higher than that of conventional SeNPs against the pathogens *Fusarium* and *Alternaria* (Hu et al. 2019).

(b) Antimicrobial Activity of Nanoparticles Against Plant Pathogen

Magnesium oxide and zinc oxide NPs were effective against plant pathogenic fungi *Fusarium oxysporum, Rhizopus stolonifera, Alternaria alternata,* and *Mucor pullumbeus* (Wani et al. 2012). Ag, TiO2, carbon, silicon dioxide, and aluminosilicates are also active antifungal agents. Among these, Ag has the uppermost potential to increase seed germination rate and seedling weight (Nikunj et al. 2014). Ag and TiO_2 NPs have been stated to prevent several plant pathogens. It was described that nanosilica increases plant resistance against pathogens by increasing phenolic compounds (Kannan et al. 2014).

Biosynthesized AgNPs inhibit the growth of four plant pathogens: *Alternaria alternata*, *Fusarium oxysporum*, *Pythium ultimum*, and *Aspergillus niger*. Ag/AgClNP biosynthesized by the fungus *Macrophomina phaseolina* can be considered an attractive option for green biopesticides due to its activity against multiple bacterial pathogens. Furthermore, to study the effect of Ag/AgClNP as a soybean protectant, Ag/AgClNP was formulated into water at various concentrations. Despite its antibacterial activity, no oxidative damage has been observed in soybean seed germination (Spagnoletti et al. 2019). Silver NPs (AgNPs) synthesized extracellularly from the supernatants of *Pseudomonas rhodesiae* have been utilized for the treatment of sweet potato stem and root rot disease caused by the bacterium *Dickeya dadantii*. Silver NPs could inhibit the growth of *D. dadantii* by damaging cell membranes, causing oxidative stress by binding to proteins along with DNA, and interfering with the mechanism of respiration as well as DNA replication. AgNPs also depicted effective antifungal activity against *Fusarium verticillioides*, which is a major pest in maize fields and can contaminate the maize seeds before and after harvest. Silver nanoparticles (AgNPs) decrease ergosterol biosynthesis in *F. verticillioides* subsequently delaying conidial germination (Hossain et al. 2019).

Streptomyces sp. 9P is a thermophilic bacterium that is found in soil, and it has many roles in antimicrobial activity against plant pathogens. It hinders *Alternaria brassicae* from infecting plants of *Brassica* spp. and also prevents *Rhizoctonia solani*, a phytopathogen with broader host range. *Streptomyces* sp. 9P hinders *Colletotrichum gloeosporioides* from infecting perennial plants while hindering *Phytophthora capsica* from infecting commercial crops like peppers.

(c) Nano-pesticides

A pest infestation can be noteworthily decreased by the use of insecticides in the soil during the early stages of crop growth. But the persistence of the pesticides or active ingredients pollutes the environment and may even cause bioaccumulation and biomagnification. As a result, we may now think about creating insecticides using NPs or nano-emulsions, these formulations are referred to as nano-pesticides collectively (Kitherian 2017). Due to their small size, nano-pesticides are more easily absorbed by the bugs they are intended to kill, making them more potent than traditional pesticides. As a result, both the amount of formulation needed and the frequency of application of nano-pesticides are 10–15 times smaller than for chemical pesticides. NPs are used to decrease the amount of pesticides dispersed, decrease the number of nutrients lost during fertilization, and improve quality as well as production by using the right nutrients (Sangeetha et al. 2021). Additionally, because less pesticides are used, fewer people are exposed to dangerous pesticides (Nuruzzaman et al. 2016). Therefore, the goal of NT in agriculture is to develop safe and effective nano-pesticides to boost global food production while minimizing adverse environmental effects on ecosystems.

Deka et al. (2021) explained that NPs are explored to deliver pesticides by releasing active chemicals to the target site at a given time also minimizes the negative effects on non-target organisms. For the control of crop pests, several metal NPs produced from salts of Ag, Au, Cu, Al, and Ti are being developed (Kitherian 2017).

Alumina, silica, silver, and lead NPs have all been used to suppress the wheat, rice, and maize pest *Sitophilus oryzae* (Debnath et al. 2011; Sankar and Abideen 2015).

Methods for the preparation of nano-biopesticides include mixing NPs with active pesticides formed in biosystems, reduction of metal salts etc. In both cases, polymers are used in nano-biopesticide preparations for precise release and long-standing activity of pesticides. These decomposable polymers also aid the pesticide formulation to dissolve in water and soil (Ragaei and Sabry 2014). Zinc oxide NPs biosynthesized from the brown alga *Turbinaria ornata* are helpful in improving rice seed quality and crop yield as well as better quality rice agronomic traits (Itroutwar et al. 2020).

(d) Nanoherbicides and Nanofungicides

Weeds are considered to be the greatest threat because they use nutrients that are naturally available to plants and wreak havoc on crops. Traditional methods such as manual weeding are time-consuming and labour-intensive. Many herbicides are now available on the market that can kill weeds in fields, but they can also damage crops and make soil infertile. Nanoherbicides are a better alternative for real weed control without damaging the soil (Perez-de-Luque 2017). Chitosan, alginates, starch, and polyesters are examples of the kinds of polysaccharides that have been considered for manufacturing nano insecticides (Mali et al. 2020).

Long-term utilization of the same herbicide can weaken weed resistance. Poly(ε-caprolactone) nanocapsules were explored as a carrier for the conventional herbicide atrazine, which exhibited stronger herbicidal activity against post-emergent mustard plants compared to conventional atrazine formulations (Oliveira et al. 2015). Chitosan NPs were prepared by cross-linking diuron via disulphide bonds to control release based on glutathione concentration (Yu et al. 2015). In this process, fungal-mediated metal NP biosynthesis is favoured over bacteria because fungi can produce more enzymes such as nitrate reductase which are helpful in metal reduction. Antifungal activity was observed against powdery mildew infection by spraying Ag-silica NPs on pumpkin leaves for 3 days (Park et al. 2006). Copper also exhibits potent antifungal activity in nanocomposites with polymers (Ciofi et al. 2005).

Biosynthesized AgNPs have a good potential to protect wheat from fungal infections i.e. *Fusarium graminearum*. *Fusarium* causes head blight in wheat, resulting in irreparable damage to wheat crops. The AgNPs inhibited hyphal growth and deformed hyphal morphology resulting in the leakage of DNA and proteins from the cells. Silver NPs can hinder spore germination, germ tube length, as well as production of mycotoxin from *Fusarium* (Ibrahim et al. 2020).

(e) Nano-sensors

In agriculture, nano-biosensors have empowered farmers to effortlessly identify plant pathogens, pesticides, heavy metals and fertilizers as well as monitor soil fertility using precision agriculture (War et al. 2020). Precision agriculture is the utilization of computer systems, drones, and the internet of nano-things to estimate environmental surroundings such as nutrient deficiencies, irrigation problems, and parasite infestations via nano-sensors to improve crop yields. Accurately measuring soil metrics (pH, nutrients, pesticide residues, and soil moisture), and estimating nitrogen uptake using nano-sensors benefits farmers to use their fields more competently (Bellingham 2011; Srivastava et al. 2017). These nano-sensors can be used to estimate the duration, amount of watering, and pesticide treatments based on the needs of desired crops. Critical factors for this progress are water retention capacity, the ability to retain water locally, the supply of water near the roots, the plant's ability to consume water, and the trapped water that is released according to plant needs.

Monreal et al. (2015) reported that there are interactions among roots and microbes in the rhizosphere of wheat and other crops play a significant role in the chemical signalling network. As a result, microorganisms and plant metabolites enter the soil, this compound concept is used to develop an excellent nano fertilizer delivery platform for micro-nutrients such as Zn and Fe. It has been suggested that good nano-fertilizers have nano-biosensors suspended in biopolymers that form a layer around the fertilizer flakes. The process of nutrient release relies on the identification and binding of specific plant distress signals by nano-biosensors encapsulated in membranes overlying zinc fertilizer NPs. After attachment, fertilizer zinc oxide NPs are released in response to root signals. After the development of such a micro-nutrient nano fertilizer delivery system, its evaluation should be performed in different crops, climates, and operating conditions to improve micro-nutrient utilization efficiency and reduce fertilizer waste to the environment (Kaushal & Wani 2017). Agricultural drones are an interesting technique for analysing crop health using nano-biosensors. The drone sensor can indicate health status at an initial phase and estimate the vegetation index while the multispectral sensor records the area affected by the microbes. Consequently, we can determine the precise quantity of chemicals needed to eliminate the epidemic (Sajoy 2021). Liposomal biosensors can detect very low concentrations of organophosphate pesticides such as dichlorvos and paraoxon (Rupesh et al. 2021). The application of NPs in agriculture has been listed in Table 21.2.

21.4 Constraints and Challenges Ahead

The integration of NT with extremophiles presents a compelling frontier in agricultural research, offering innovative solutions to address global food security challenges. The nanoparticle synthesis using extremophilic microbes and their use in agriculture hold immense promise for enhancing crop productivity, nutrient management, pest control, and soil health. The potential benefits of NPs in agriculture are substantial but challenges remain. Diffusion and bioaccumulation of NPs in soil, water, as well as food, the indiscriminate practice of nano-pesticides may have an adverse effect on water quality and human health.

Major constraints linked by the synthesis of NPs via thermophiles as well as psychrophiles include monotonous purification steps and the unapproachability of an efficient method for the recovery of intracellular NPs (Shakibaie et al. 2010),

Table 21.2 Application of nanoparticle synthesized using extremophiles in Agriculture

Extremophilic microbe	Class of extremophile	Source	Application	References
Bacillus subtilis	Thermophile	Soil and plant roots	Antimicrobial	Khan et al. (2011)
Bacillus subtilis	Mesophile	Leaves of *Catharanthus roseus*	Enlargement of plant height	Morrison and Mark (2016)
Trichoderma asperellum	Alkaliphile	–	Antifungal activity against plant pathogens	Guo et al. (2018)
B. subtilis, B. licheniformis	Thermophile	Hot spring	Plant growth promotion	Verma et al. (2018)
Thermomyces lanuginosus	Thermophile	–	Biocontrol potential	Okongo et al. (2019)
Bacillus methylotrophicus	Halophiles	–	Biocontrol potential	Castro et al. (2020) and Torres et al. (2020)
Staphylococcus aureus	Mesophile	–	Antimicrobial activity	Win et al. (2020)
Pseudomonas aeruginosa	Psychrophile	–	Antimicrobial activity	Ghosh et al. (2021)
Rhizobacteria	Drought resistance	Root nodules of leguminous plant	Biocontrol potential	Pandey et al. (2021)
B. mycoides	Psychrophilic	Soil	Increasing soil fertility	Roy et al. (2022)

scientists are not able to fully explore these extremely competent extremophiles and mostly rely on extracellular nanoparticle-synthesizing microbes. To release NPs that are produced intracellularly, further steps are required viz. ultrasonication, reaction with detergents, and enzymes (Sonkusre et al. 2014).

Nanoparticle-based techniques have been developed at the lab level but nano-toxicology studies for adverse health effects due to extended exposure at different concentration levels in humans as well as in the environment have not yet been studied thoroughly before commercialization. Further research is necessary to optimize nanoparticle synthesis, ensure their stability and safety, and understand their long-term effects on ecosystems.

21.5   Conclusion

The synthesis of NPs using extremophilic microbes offers a natural and sustainable alternative to conventional chemical methods. These NPs can be tailored to meet specific agricultural needs, such as targeted delivery of nutrients and agrochemicals, improving plant disease resistance, and enhancing nutrient uptake. Additionally, extremophilic microbes contribute to soil health by improving soil structure, water-holding capacity, as well as nutrient cycling, thus promoting overall crop resilience. The application of synthesized NPs in agriculture presents exciting possibilities. NPs can enhance seed germination, root development, and nutrient uptake, ultimately leading to increased crop yields. Furthermore, NPs can act as antimicrobial agents, combating plant diseases and reducing reliance on conventional pesticides.

As per our study, we found that majorly acidophilic microbes have been used for the synthesis of NPs followed by thermophiles, halophiles, and alkalophiles. Few psychrophilic microbes have been explored for nanoparticle synthesis while barophiles and xerophiles have not been explored yet. To fully utilize the potential of NT and extremophiles in agriculture, collaboration among scientists, policymakers, and industry stakeholders is crucial. Robust regulatory frameworks should be established to ensure the safe and responsible deployment of nanomaterials in agricultural practices. Furthermore, knowledge dissemination and capacity-building initiatives are vital to enable farmers to adopt these technologies effectively.

In conclusion, the integration of NT and extremophiles opens new possibilities for sustainable and efficient agricultural practices. The synthesis of NPs using extremophilic microbes, coupled with their application in agriculture, offers transformative solutions to enhance crop productivity, reduce environmental impact, and address global food security challenges. By embracing this interdisciplinary field, we can pave the way for a resilient and sustainable agricultural future, ensuring food availability and nutritional security for generations to come.

Acknowledgement The authors are grateful to the authorities of Mohanlal Sukhadia University, Udaipur, for supporting this work.

References

Adebayo EA, Azeez MA, Alao MB, Oke AM, Aina DA (2021) Fungi as veritable tool in current advances in nanobiotechnology. Heliyon 7:e08480

Agnihotri M, Joshi S, Kumar AR, Zinjarde S, Kulkarni S (2009) Biosynthesis of gold nanoparticles by the tropical marine yeast *Yarrowia lipolytica* NCIM 3589. Mater Lett 63:1231–1234

Ahmad A, Senapati S, Khan MI, Kumar R, Sastry M (2003) Extracellular biosynthesis of monodisperse gold nanoparticles by a novel extremophilic actinomycete, Thermomonospora sp. Langmuir 19:3550–3553

Anasane N, Golińska P, Wypij M, Rathod D, Dahm H, Rai M (2016) Acidophilic actinobacteria synthesised silver nanoparticles showed remarkable activity against fungi-causing superficial mycoses in humans. Mycoses 59:157–166

Bahrulolum H, Nooraei S, Javanshir N, Tarrahimofrad H, Mirbagheri VS, Easton AJ, Ahmadian G (2021) Green synthesis of metal nanoparticles using microorganisms and their application in the agrifood sector. J Nanobiotechnol 19:1–26

Bansal V, Rautaray D, Bharde A, Ahire K, Sanyal A, Ahmad A, Sastry M (2005) Fungus-mediated biosynthesis of silica and titania particles. J Mater Chem 15:2583–2589

Beeler E, Choy N, Franks J, Mulcahy F, Singh OV (2020) Extracellular synthesis and characterization of silver nanoparticles from alkaliphilic pseudomonas sp. J Nanosci Nanotechnol 20:1567–1577

Bellingham BK (2011) Proximal soil sensing. Vadose Zone J 10:1342–1342. https://doi. org/10.2136/vzj2011.0105br

Bharde A, Wani A, Shouche Y, Joy PA, Prasad BLV, Sastry M (2005) Bacterial aerobic synthesis of nanocrystalline magnetite. J Am Chem Soc 127:9326–9327

Castro D, Torres M, Sampedro I, Martínez-Checa F, Torres B, Bejar V (2020) Biological control of Verticillium wilt on olive trees by the salt-tolerant strain Bacillus velezensis XT1. Microorganisms 8(7):1080. https://doi.org/10.3390/microorganisms8071080

Cekuolyte K, Gudiukaite R, Klimkevicius V, Mazrimaite V, Maneikis A, Lastauskiene E (2023) Biosynthesis of silver nanoparticles produced using Geobacillus spp. bacteria. Nanomaterials 13:702

Chinnamuthu CR, Kokiladevi E (2007) Weed management through nanoherbicides. In: Application of nanotechnology in agriculture. Tamil Nadu Agricultural University, Coimbatore, pp 23–36

Cioffi N, Ditaranto N, Torsi L, Picca RA, Sabbatini L, Valentini A, Novello G, Tantillo T, Bleve-Zacheo T, Zambonin PG (2005) Analytical characterization of bioactive fluoropolymer ultra-thin coatings modified by copper nanoparticles. Anal Bioanal Chem 381:607–616

Correa-Llantén DN, Muñoz-Ibacache SA, Castro ME, Muñoz PA, Blamey JM (2013) Gold nanoparticles synthesized by Geobacillus sp. strain ID17 a thermophilic bacterium isolated from Deception Island, Antarctica. Microb Cell Fact 12:1–6

Correa-Llantén DN, Muñoz-Ibacache SA, Maire M, Blamey JM (2014) Enzyme involvement in the biosynthesis of selenium nanoparticles by Geobacillus wiegelii strain GWE1 isolated from a drying oven. Int J Bioeng Life Sci 8:637–641

Davari MR, Kazazi SB, Pivehzhani OA (2017) Nanomaterials: implications on agroecosystem. In: Nanotechnology. Springer, pp 59–71

Debnath N, Das S, Seth D, Chandra R, Bhattacharya SC, Goswami A (2011) Entomotoxic effect of silica nanoparticles against Sitophilus oryzae. J Pest Sci 84(1):99–105

Dehnad A, Hamedi J, Derakhshan-Khadivi F, Abuşov R (2015) Green synthesis of gold nanoparticles by a metal resistant Arthrobacter nitroguajacolicus isolated from gold mine. IEEE Trans Nanobioscience 14:393–396

Deka B, Babu B, Baruah C, Barthakur M (2021) Nanopesticides: a systematic review of their prospects with special reference to Tea pest management. Front Nutr 6:1–16

Dinc SK, Vural OA, Kayhan FE, San Keskin NO (2022) Facile biogenic selenium nanoparticle synthesis, characterization and effects on UV-generated oxidative stress in microalgae. Particuology 70:30–42. https://doi.org/10.1016/j.partic.2021.12.005

Duran N, Marcato PD, Alves OL, De Souza GIH, Esposito E (2005) Mechanistic aspects of biosynthesis of silver nanoparticles by several Fusarium oxysporum strains. J Nanobiotechnol 3:1–7

Fayaz AM, Balaji K, Girilal M, Yadav R, Kalaichelvan PT, Venketesan R (2010) Biogenic synthesis of silver nanoparticles and their synergistic effect with antibiotics: a study against gram-positive and gram-negative bacteria. Nanomedicine 6:103–109

Feller G (2003) Molecular adaptations to cold in psychrophilic enzymes. Cell Mol Life Sci 60:648–662

Fendler JH (1998) Nanoparticles and nanostructured films: preparation, characterization and applications. Wiley, New York, p 488

Ghosh S, Ahmad R, Banerjee K, Alajmi MF (2021) Mechanistic aspects of microbe-mediated nanoparticle synthesis. Front Microb 12(1):1–10

Guo R, Wang Z, Huang Y, Fan H, Liu Z (2018) Biocontrol potential of saline-or alkaline-tolerant Trichoderma asperellum mutants against three pathogenic fungi under saline or alkaline stress conditions. Braz J Microbiol 49:236–245

Hochvaldová L, Večeřová R, Kolář M, Prucek R, Kvítek L, Lapčík L, Panáček A (2022) Antibacterial nanomaterials: upcoming hope to overcome antibiotic resistance crisis. Nanotechnol Rev 11:1115–1142

Honary S, Barabadi H, Gharaei-Fathabad E, Naghibi F (2012) Green synthesis of copper oxide nanoparticles using Penicillium aurantiogriseum, Penicillium citrinum and Penicillium waksmanii. Dig J Nanomater Bios 7:999–1005

Hossain A, Hong X, Ibrahim E, Li B, Sun G, Meng Y, Wang Y, An Q (2019) Green synthesis of silver nanoparticles with culture supernatant of a bacterium *Pseudomonas rhodesiae* and their antibacterial activity against soft rot pathogen *Dickeya dadantii*. Molecules 24(12):2303. https://doi.org/10.3390/molecules24122303

Hu D, Yu S, Yu D et al (2019) Biogenic Trichoderma harzianum-derived selenium nanoparticles with control functionalities originating from diverse recognition metabolites against phytopathogens and mycotoxins. Food Control 106:106748. https://doi.org/10.1016/j.foodcont.2019.106748

Ibrahim E, Zhang M, Zhang Y, Hossain A, Qiu W, Chen Y, Wang Y, Wu W, Sun G, Bin L (2020) Green-synthesization of silver nanoparticles using endophytic bacteria isolated from garlic and its antifungal activity against wheat Fusarium head blight pathogen *Fusarium graminearum*. Nano 10:219

Itroutwar PD, Govindaraju K, Tamilselvan S et al (2020) Seaweed-based biogenic ZnO nanoparticles for improving agro-morphological characteristics of rice (*Oryza sativa* l.). J Plant Grow Reg 39:717–728

Jang GG, Jacobs CB, Gresback RG, Ivanov IN, Meyer III HM, Kidder M, Joshi PC, Jellison GE, Phelps TJ, Graham DE (2015) Size tunable elemental copper nanoparticles: extracellular synthesis by thermoanaerobic bacteria and capping molecules. J Mater Chem C Mater 3:644–650

John MS, Nagoth JA, Ramasamy KP, Mancini A, Giuli G, Natalello A, Ballarini P, Miceli C, Pucciarelli S (2020) Synthesis of bioactive silver nanoparticles by a pseudomonas strain associated with the antarctic psychrophilic protozoon Euplotes focardii. Mar Drugs 18:38

Juibari MM, Abbasalizadeh S, Jouzani GS, Noruzi M (2011) Intensified biosynthesis of silver nanoparticles using a native extremophilic Ureibacillus thermosphaericus strain. Mater Lett 65:1014–1017

Kalabegishvili T, Kirkesali E, Ginturi E, Rcheulishvili A, Murusidze I, Pataraya D, Gurielidze M, Bagdavadze N, Kuchava N, Gvarjaladze D (2013a) Synthesis of gold nanoparticles by new strains of thermophilic actinomycetes. Nano Stud 7:255–260

Kalabegishvili T, Murusidze I, Kirkesali E, Rcheulishvili A, Ginturi E, Kuchava N, Bagdavadze N, Gelagutashvili E, Frontasyeva MV, Zinicovscaia I (2013b) Gold and silver nanoparticles in Spirulina platensis biomass for medical application. Ecol Chem Eng S 20:621–631

Kalabegishvili TL, Murusidze IG, Prangishvili DA, Kvachadze LI, Kirkesali EI, Rcheulishvili AN, Ginturi EN, Janjalia MB, Tsertsvadze GI, Gabunia VM (2014) Gold nanoparticles in Sulfolobus islandicus biomass for technological applications. Adv Sci Eng Med 6:1302–1308

Kalabegishvili TL, Murusidze IG, Prangishvili DA, Kvachadze LI, Kirkesali EI, Rcheulishvili AN, Ginturi EN, Janjalia MB, Tsertsvadze GI, Gabunia VM (2015) Silver nanoparticles in Sulfolobus islandicus biomass for technological applications. Adv Sci Eng Med 7:797–804

Kanekar PP, Kanekar SP, Kelkar AS, Dhakephalkar PK (2012) Halophiles–taxonomy, diversity, physiology and applications. Microorganisms Environ Managet:1–34

Kannan N, Rajendran V, Yuvakkumar R, Karunakaran G, Kavitha K, Suriyaprabha R (2014) Application of silica nanoparticles in maize to enhance fungal resistance. IET Nanobiotechnol 8:133–137

Kathiresan K, Manivannan S, Nabeel MA, Dhivya B (2009) Studies on silver nanoparticles synthesized by a marine fungus, Penicillium fellutanum isolated from coastal mangrove sediment. Colloids Surf B Biointerfaces 71:133–137

Kaushal M, Wani S (2017) Nanosensors: frontiers in precision agriculture. https://doi.org/10.1007/978-981-10-4573-8_13

Khan SA, Ahmad A (2013) Fungus mediated synthesis of biomedically important cerium oxide nanoparticles. Mater Res Bull 48:4134–4138

Khan OF, Kowalski PS, Doloff JC et al (2018) Endothelial siRNA delivery in nonhuman primates using ionizable low–molecular weight polymeric nanoparticles. Sci Adv 4(6):eaar8409. https://doi.org/10.1126/sciadv.aar8409

Kim H, Kang H, Chu G, Byun H (2008) Antifungal effectiveness of nanosilver colloid against rose powdery mildew in greenhouses. Solid State Pheno 135:15–18

Kitherian S (2017) Nano and bio-nanoparticles for insect control. Res J Nanosci Nanotechnol 7:1–9

Kristjansson JK, Stetter KO (2021) Thermophilic bacteria. In: Thermophilic bacteria. CRC Press, pp 1–18

Li G, He D, Qian Y, Guan B, Gao S, Cui Y, Yokoyama K, Wang L (2011) Fungus-mediated green synthesis of silver nanoparticles using *Aspergillus terreus*. Int J Mol Sci 13:466

Li J, Webster TJ, Tian B (2019) Functionalized nanomaterial assembling and biosynthesis using the extremophile *Deinococcus radiodurans* for multifunctional applications. Small 15:1900600

Mali SC, Raj S, Trivedi R (2020) Nanotechnology a novel approach to enhance crop productivity. Biochem Biophys Rep 24:100821

Maliszewska I, Juraszek A, Bielska K (2014) Green synthesis and characterization of silver nanoparticles using ascomycota fungi *Penicillium nalgiovense* AJ12. J Clust Sci 25:989–1004

Matsumoto T, Phann I, Okibe N (2021) Biogenic platinum nanoparticles' production by extremely acidophilic Fe (III)-reducing bacteria. Fortschr Mineral 11:1175

Meenupriya J (2011) Biogenic silver nanoparticles by *Aspergillus terreus* MP1 and its promising antimicrobial activity

Mehetre G, Leo VV, Singh G, Dhawre P, Maksimov I, Yadav M, Upadhyaya K, Singh BP (2021) Biocontrol potential and applications of extremophiles for sustainable agriculture. In: Microbiomes of extreme environments. CRC Press, pp 230–242

Metwally RA, El Nady J, Ebrahim S, El Sikaily A, El-Sersy NA, Sabry SA, Ghozlan HA (2023) Biosynthesis, characterization and optimization of TiO_2 nanoparticles by novel marine halophilic Halomonas sp. RAM2: application of natural dye-sensitized solar cells. Microb Cell Factories 22:1–17

Mishra RR, Prajapati S, Das J, Dangar TK, Das N, Thatoi H (2011) Reduction of selenite to red elemental selenium by moderately halotolerant *Bacillus megaterium* strains isolated from Bhitarkanika mangrove soil and characterization of reduced product. Chemosphere 84:1231–1237

Mohanta YK, Behera SK (2014) Biosynthesis, characterization and antimicrobial activity of silver nanoparticles by Streptomyces sp. SS2. Bioprocess Biosyst Eng 37:2263–2269

Monreal CM, DeRosa M, Mallubhotla SC, Bindraban PS, Dimkpa C (2015) The application of nanotechnology for micronutrients in soil-plant systems, VFRC report 2015/3. Virtual Fertilizer Research Center, Washington, DC, p 44

Moon J-W, Roh Y, Yeary LW, Lauf RJ, Rawn CJ, Love LJ, Phelps TJ (2007) Microbial formation of lanthanide-substituted magnetites by Thermoanaerobacter sp. TOR-39. Extremophiles 11:859–867

Moopantakath J, Imchen M, Anju VT, Busi S, Dyavaiah M, Martínez-Espinosa RM, Kumavath R (2023) Bioactive molecules from haloarchaea: scope and prospects for industrial and therapeutic applications. Front Microbiol 14

Morrison TJ, Mark (2016) Nanotechnology in agriculture and food. Institute of Nanotechnology, Chicago

Mukherjee P, Ahmad A, Mandal D, Senapati S, Sainkar SR, Khan MI, Ramani R, Parischa R, Ajayakumar PV, Alam M (2001) Bioreduction of AuCl4−ions by the fungus, Verticillium sp. and surface trapping of the gold nanoparticles formed. Angew Chem Int Ed 40:3585–3588

Mukherjee P, Senapati S, Mandal D, Ahmad A, Khan MI, Kumar R, Sastry M (2002) Extracellular synthesis of gold nanoparticles by the fungus *Fusarium oxysporum*. Chem Bio Chem 3:461–463

Mukherjee S, Atique U, Mukherjee R, Chatterjee S, Altaf M, Sinha D, Dey R, Dutta SR, Mondal A, Chowdhury S (2022) Potential of extremophiles: a review of current research in nanoparticle synthesis. Extremophiles:289–314

Nikunj P, Purvi D, Niti P et al (2014) Agronanotechnology for plant fungal disease management: a review. Int J Curr Microbiol App Sci 3:71–84

Nuruzzaman MD, Rahman MM, Liu Y et al (2016) Nanoencapsulation, nano-guard for pesticides: a new window for safe application. J Agricul Food Chem 64(7):1447–1483

Okongo RN, Puri AK, Wang Z et al (2019) Comparative biocontrol ability of chitinases from bacteria and recombinant chitinases from the thermophilic fungus Thermomyces lanuginosus. J Biosci Bioeng 127(6):663–671

Oliveira HC, Stolf-Moreira R, Martinez CB, Grillo R, de Jesus MB, Fraceto LF (2015) Nanoencapsulation enhances the post-emergence herbicidal activity of atrazine against mustard plants. PLoS One 10:e0132971. https://doi.org/10.1371/journal.pone.0132971

Oren A (2002) Halophilic microorganisms and their environments. Kluwer Academic Publishers, Dordrecht, Boston

Oren A (2010) Industrial and environmental applications of halophilic microorganisms. Environ Technol 31:825–834

Oza G, Pandey S, Shah R, Sharon M, Phata J, Ambernath W, Sharon M (2012) A mechanistic approach for biological fabrication of crystalline gold nanoparticles using marine algae, *Sargassum wightii*. Eur J Exp Biol 2:505–512

Pal G, Rai P, Pandey A (2019) Green synthesis of nano-particles: a greener approach for a cleaner future. In: Characterization and applications of nanoparticles SBT-GS micro and nano technologies. Elsevier, Amsterdam, pp 1–2

Pandey KD, Patel AK, Singh M, Kumari A (2021) Secondary metabolites from bacteria and viruses. Natural Bioact Compd:19–40

Pandit C, Roy A, Ghotekar S, Khusro A, Islam MN, Bin ET, Lam SE, Khandaker MU, Bradley DA (2022) Biological agents for synthesis of nanoparticles and their applications. J King Saud Univ-Sci 34:101869

Park HJ, Kim SH, Kim HJ et al (2006) A new composition of nanosized silica-silver for control of various plant diseases. Plant Pathol J 22:295–302. https://doi.org/10.5423/PPJ.2006.22.3.295

Perez-de-Luque A (2017) Interaction of nanomaterials with plants: what do we need for real applications in agriculture. Frontier Environment Sci 5:12. https://doi.org/10.3389/fenvs.2017.00012

Prasad R, Kumar V, Prasad KS (2014) Nanotechnology in sustainable agriculture: present concerns and future aspects. African J Biotech 13:705–713

Ragaei M, Sabry AH (2014) Nanotechnology for insect pest control. Int J Sci Environ Technol 3:528–545

Rajeshkumar S, Malarkodi C, Paulkumar K, Vanaja M, Gnanajobitha G, Annadurai G (2014) Algae mediated green fabrication of silver nanoparticles and examination of its antifungal activity against clinical pathogens. Int J Met 2014:1–8

Ramanathan RO, Mullane AP, Parikh RY, Smooker PM, Bhargava SK, Bansal V (2011) Bacterial kinetics-controlled shape-directed biosynthesis of silver nanoplates using *Morganella psychrotolerans*. Langmuir 27:714–719

Rammohan M, Balakrishnan K (2011) Rapid synthesis and characterization of silver nano particles by novel pseudomonas sp. "ram bt-1". J Ecobiotechnol 3(1):24–28

Rehmanullah, Muhammad, Z, Inayat, N, Majeed, A (2020). Application of Nanoparticles in Agriculture as Fertilizers and Pesticides: Challenges and Opportunities. In: Rakshit, A., Singh, H., Singh, A., Singh, U., Fraceto, L. (eds) New Frontiers in Stress Management for Durable Agriculture. Springer, Singapore. https://doi.org/10.1007/978-981-15-1322-0_17

Roh Y, Vali H, Phelps TJ, Moon J-W (2006) Extracellular synthesis of magnetite and metal-substituted magnetite nanoparticles. J Nanosci Nanotechnol 6:3517–3520

Roulling F, Piette F, Cipolla A, Struvay C, Feller G (2011) Psychrophilic enzymes: cool responses to chilly problems. In: Extremophiles handbook. Springer, Tokyo, pp 891–913

Roy B, Maitra D, Ghosh J, Mitra AK (2022) Extremophilic *Bacillus*: their application in plant growth promotion and sustainable agriculture. https://doi.org/10.1016/B978-0-323-90452-0.00021-9

Rupesh K, Sinha R, Ramesh KV (2021) Introduction of nano-biosensor in agriculture sector. Int J Modern Agric 10(2):1063–1070

Sahai S (2010) Production of silver nanoparticles by a phytopathologic fungus Bipolaris nodulosa and its antimicrobial activity. Dig J Nanomater Biostruct 5:887–895

Sahoo A, Satapathy KB, Sahoo SK, Panigrahi GK (2022) Microbased biorefinery for gold nanoparticle production: recent advancements, applications and future aspects. Prep Biochem Biotechnol, pp 1–12

Sajoy PB (2021) Emerging trends in the use of IoT in agriculture and food supply chain management: a theoretical analysis. Turkish J Comp Maths Education 12(3):3293–3297

Sangeetha J, Hospet R, Thangadurai D, Adetunji CO, Islam S, Pujari N, Al-Tawaha ARMS (2021) Nanopesticides, nanoherbicides, and nanofertilizers: the greener aspects of agrochemical synthesis using nanotools and nanoprocesses toward sustainable agriculture. In: Handbook of nanomaterials and nanocomposites for energy and environmental applications. Springer. https://doi.org/10.1007/978-3-030-36268-3_44

Sankar MV, Abideen S (2015) Pesticidal effect of green synthesized silver and lead nanoparticles using Avicennia marina against grain storage pest *Sitophilus oryzae*. Int J Nanomater Biostruct 5(3):32–39

Sastry M, Ahmad A, Khan MI, Kumar R (2003) Biosynthesis of metal nanoparticles using fungi and actinomycete. Curr Sci:162–170

Scarano G, Morelli E (2003) Properties of phytochelatin-coated CdS nanocrystallites formed in a marine phytoplanktonic alga (*Phaeodactylum tricornutum*, Bohlin) in response to Cd. Plant Sci 165:803–810

Schrofel A, Kratosova G, Bohunicka M, Dobrocka E, Vavra I (2011) Biosynthesis of gold nanoparticles using diatoms—silica-gold and EPS-gold bionanocomposite formation. J Nanopart Res 13:3207–3216

Seelan JSS, Ali AAKF, Muid S (2009) Aspergillus species isolated from mangrove forests in Borneo Island, Sarawak, Malaysia. J Threat Taxa:344–346

Selenska-Pobell S, Reitz T, Schonemann R, Herrmansdörfer T, Merroun M, Geißler A, Bartolomé J, Bartolomé F, García LM, Wilhelm F (2011) Magnetic Au nanoparticles on archaeal S-layer ghosts as templates. Nanomater and Nanotechnol 1:13

Seshadri S, Saranya K, Kowshik M (2011) Green synthesis of lead sulfide nanoparticles by the lead resistant marine yeast, *Rhodosporidium diobovatum*. Biotechnol Prog 27:1464–1469

Seshadri S, Prakash A, Kowshik M (2012) Biosynthesis of silver nanoparticles by marine bacterium, Idiomarina sp. PR58-8. Bull Mater Sci 35:1201–1205

Shah R, Oza G, Pandey S, Sharon M (2012) Biogenic fabrication of gold nanoparticles using Halomonas Salina. J Microbiol Biotechnol Res 2:485–492

Shakibaie M, Khorramizadeh MR, Faramarzi MA, Sabzevari O, Shahverdi AR (2010) Biosynthesis and recovery of selenium nanoparticles and the effects on matrix metalloproteinase-2 expression. Biotechnol Appl Biochem 56:7–15

Shantkriti S, Pradeep M, Unish KK, Viji Das MS, Nidhin S, Gugan K, Murugan A (2023) Biosynthesis of silver nanoparticles using Dunaliella salina and its antibacterial applications. Appl Surf Sci Advan 13:100377. https://doi.org/10.1016/j.apsadv.2023.100377

Sharifi M, Faryabi K, Talaei AJ, Shekha MS, Ale-Ebrahim M, Salihi A (2020) Antioxidant properties of gold nanozyme: a review. J Mol Liq 297:112004. https://doi.org/10.1016/j.molliq.2019.112004

Sharma G, Jasuja ND, Kumar M, Ali MI (2015) Biological synthesis of silver nanoparticles by cell-free extract of *Spirulina platensis*. J Nanotechnol 2015:1–6

Shen N, Xia X-Y, Chen Y, Zheng H, Zhong Y-C, Zeng RJ (2015) Palladium nanoparticles produced and dispersed by *Caldicellulosiruptor saccharolyticus* enhance the degradation of contaminants in water. RSC Adv 5:15559–15565

Shivaji S, Madhu S, Singh S (2011) Extracellular synthesis of antibacterial silver nanoparticles using psychrophilic bacteria. Process Biochem 46:1800–1807

Shivanand P, Mugeraya G (2011) Halophilic bacteria and their compatible solutes–osmoregulation and potential applications. Curr Sci:1516–1521

Show S, Tamang A, Chowdhury T, Mandal D, Chattopadhyay B (2015) Bacterial (BKH1) assisted silica nanoparticles from silica rich substrates: a facile and green approach for biotechnological applications. Colloids Surf B Biointerfaces 126:245–250

Shu WS, Huang LN (2022) Microbial diversity in extreme environments. Nat Rev Microbiol 20:219–235

Singh OV, Colonna T (2015) Nanotechnology: overview of regulations and implementations. In: bio-nanoparticles: biosynthesis and sustainable biotechnological implications. Wiley, pp 303–329

Singh S, Vidyarthi AS, Nigam VK, Dev A (2014) Extracellular facile biosynthesis, characterization and stability of gold nanoparticles by *Bacillus licheniformis*. Artif Cells Nanomed Biotechnol 42:6–12

Sivarethinamohan R, Sujatha S (2021) Unlocking the potentials of using nanotechnology to stabilize agriculture and food production. AIP Publishing LLC 2327(1):20022

Sonkusre P, Nanduri R, Gupta P, Cameotra SS (2014) Improved extraction of intracellular biogenic selenium nanoparticles and their specificity for cancer chemoprevention. J Nanomed Nanotechnol 5:1–9

Spagnoletti FN, Spedalieri C, Kronberg F, Giacometti R (2019) Extracellular biosynthesis of bactericidal Ag/AgCl nanoparticles for crop protection using the fungus Macrophomina phaseolina. J Environ Sci. https://doi.org/10.1016/j.jenvman.2018.10.081

Srivastava P, Kowshik M (2013) Mechanisms of metal resistance and homeostasis in Haloarchaea. Archaea 732864. https://doi.org/10.1155/2013/732864

Srivastava P, Bragança J, Ramanan SR, Kowshik M (2013) Synthesis of silver nanoparticles using haloarchaeal isolate *Halococcus salifodinae* BK 3. Extremophiles 17:821–831

Srivastava P, Braganca J, Ramanan SR, Kowshik M (2014) Green synthesis of silver nanoparticles by haloarchaeon *Halococcus salifodinae* BK6. In: Advanced materials research. Trans Tech Publ, pp 236–241

Srivastava K, Srivastava AD, Karmakar S (2017) Nanosensors and nanobiosensors in food and agriculture. Environ Chem Letters 16(1):161–182

Subramanian M, Alikunhi NM, Kandasamy K (2010) In vitro synthesis of silver nanoparticles by marine yeasts from coastal mangrove sediment. Adv Sci Lett 3:428–433

Syed A, Saraswati S, Kundu GC, Ahmad A (2013) Biological synthesis of silver nanoparticles using the fungus *Humicola sp.* and evaluation of their cytotoxicity using normal and cancer cell lines. Spectrochim Acta A Mol Biomol Spectrosc 114:144–147

Takahata Y, Nishijima M, Hoaki T, Maruyama T (2001) Thermotoga petrophila sp. nov. and Thermotoga naphthophila sp. nov., two hyperthermophilic bacteria from the Kubiki oil reservoir in Niigata. Japan Int J Syst Evol Microbiol 51:1901–1909

Thangaswamy SJK, Mir MA, Muthu A (2021) Green synthesis of mono and bimetallic alloy nanoparticles of gold and silver using aqueous extract of chlorella acidophile for potential applications in sensors. Prep Biochem Biotechnol 51:1026–1035

Tiquia-Arashiro S, Rodrigues DF, Tiquia-Arashiro S, Rodrigues D (2016) Alkaliphiles and acidophiles in nanotechnology. Extremophiles: Applications in Nanotechnology, pp 129–162

Torracchi JE, Morel MA, Tapia-Vázquez I, Castro-Sowinski S, Batista-Garcia RA et al (2020) Fighting plant pathogens with cold-active microorganisms: biopesticide development and agriculture intensification in cold climates. Appl Microbiol Biotechnol 104(19):8243–8256

Torres M, Llamas I, Torres B, Toral L, Sampedro I, Bejar V (2020) Growth promotion on horticultural crops and antifungal activity of *Bacillus velezensis* XT1. Appl Soil Ecology 150:103453. https://doi.org/10.1016/j.apsoil.2019.103453

Uddin I, Adyanthaya S, Syed A, Selvaraj K, Ahmad A, Poddar P (2008) Structure and microbial synthesis of sub-10 nm Bi2O3 nanocrystals. J Nanosci Nanotechnol 8:3909–3913

Verma P, Yadav AN, Kumar V, Singh DP, Saxena AK (2017) Beneficial plant-microbes interactions: biodiversity of microbes from diverse extreme environments and its impact for crop improvement. In: Plant-microbe interactions in agro-ecological perspectives. Springer, pp 543–580

Verma JP, Jaiswal DK, Krishna R, Prakash S, Yadav J, Singh V (2018) Characterization and screening of thermophilic *Bacillus* strains for developing plant growth promoting consortium from hot spring of Leh and Ladakh region of India. Front Microbiol 9:1293

Wang T, Yang L, Zhang B, Liu J (2010) Extracellular biosynthesis and transformation of selenium nanoparticles and application in H2O2 biosensor. Colloids Surf B Biointerfaces 80:94–102

Wani AH, Shah MA (2012) A unique and profound effect of MgO and ZnO nanoparticles on some plant pathogenic fungi. J Appl Pharma Sci 2-4:18671981

War JM, Fazili MA, Mushtaq W et al (2020) Role of nanotechnology in crop improvement. In: Nanobiotechnology in agriculture. Springer, pp 63–97

Wiegel J (2011) Anaerobic alkaliphiles and alkaliphilic polyextremophiles. Extremophiles Handbook. Springer, Tokyo, pp 81–97

Win TT, Khan S, Fu PC (2020) Fungus- (*Alternaria sp.*) mediated silver nanoparticles synthesis, characterization, and screening of antifungal activity against some phytopathogens. J Nanotech. Article ID 8828878

Worrall EA, Hamid A, Mody KT, Mitter N, Pappu HR (2018) Nanotechnology for plant disease management. Agronomy 8(12):285

Yadav AN, Verma P, Kumar M, Pal KK, Dey R, Gupta A, Padaria JC, Gujar GT, Kumar S, Suman A, Prasanna R, Saxena AK (2015) Diversity and phylogenetic profiling of niche-specific Bacilli from extreme environments of India. Ann Microbiol 65:611–629

Yan L, Yue X, Zhang S, Chen P, Xu Z, Li Y, Li H (2012) Biocompatibility evaluation of magnetosomes formed by *Acidithiobacillus ferrooxidans*. Mater Sci Eng C 32:1802–1807

Yeary LW, Moon J-W, Rawn CJ, Love LJ, Rondinone AJ, Thompson JR, Chakoumakos BC, Phelps TJ (2011) Magnetic properties of bio-synthesized zinc ferrite nanoparticles. J Magn Magn Mater 323:3043–3048

Youssif AM, Soliman NA, Sabry SA, Ghozlan HA (2020) Biosynthesis, characterization and application of silver nanoparticles by *Geobacillus thermodenitrificans* Az1 as antimicrobial, antibiofilm and dye catalyist. Asian J Microbiol Biotech Environ Sci 22:50–56

Yu Z, Sun X, Song H, Wang W, Ye Z, Shi L et al (2015) Glutathione-responsive carboxymethyl chitosan nanoparticles for controlled release of herbicides. Mater Sci Appl 6591. https://doi.org/10.4236/msa.2015.66062

Zafrilla B, Martínez-Espinosa RM, Alonso MA, Bonete MJ (2010) Biodiversity of Archaea and floral of two inland saltern ecosystems in the Alto Vinalopó Valley, Spain. Saline Syst 6:1–12

Zhang W, Chen Z, Liu H, Zhang L, Gao P, Li D (2011) Biosynthesis and structural characteristics of selenium nanoparticles by *Pseudomonas alcaliphila*. Colloids Surf B Biointerfaces 88:196–201

Correction to: Biotechnology of Promising Genes from Extremophiles to Produce Stress-Resilient Plants and Microbes for Sustainable Agriculture

Manmeet Kaur, Diksha Singla, Kamal Kapoor, Gautam Chhabra, Sezai Ercisli, Mehmet Ramazan Bozhuyuk, Shiv K. Yadav, and Ravish Choudhary

Correction to:
Chapter 16 in: A. Ranjan et al. (eds.), *Extremophiles for Sustainable Agriculture and Soil Health Improvement*,
https://doi.org/10.1007/978-3-031-70203-7_16

The original version of this chapter "Biotechnology of Promising Genes from Extremophiles to Produce Stress-Resilient Plants and Microbes for Sustainable Agriculture" was inadvertently published with incorrect surname for author Mehmet Ramazan Bozhuyuk. The correct surname has been updated in the chapter.

The updated version of this chapter can be found at
https://doi.org/10.1007/978-3-031-70203-7_16

Index

A

Abiotic stresses, 63–81, 222, 239, 260–262,
264–266, 282, 286, 308, 337, 338, 351,
374, 375, 411–413, 416, 471
Acidophilus microbes, 389–399
Agricultural applications, 96, 124, 146–151,
222, 392, 416, 441–450, 455–476
Agriculture, 4, 23, 55, 65, 95, 125, 138, 163,
200, 222, 252, 281, 303, 321, 405, 435,
441, 455

B

Bacillus spp., 22, 27, 126, 153, 237, 307, 318,
349, 416
Bioactive, 23, 111, 115, 121–133, 137–154,
203, 207, 251–268, 281, 434, 444
Bioactive compounds, 63, 80, 99, 105,
121–133, 138, 140, 141, 146, 149–151,
153, 154, 228, 308, 309, 448, 450,
456, 457
Biocontrol agents, 107–109, 123–125, 127,
129, 130, 133, 252, 264, 307, 308, 347,
351, 379, 380, 406, 471
Bio-nanoparticles, 442
Bioremediations, 10, 12–14, 24, 29, 33, 37,
53, 55, 57–58, 78, 80, 94, 108, 109,
151–152, 164, 165, 169–174, 178–186,
199–201, 204–213, 231, 255, 259, 261,
264, 303, 322, 328, 340, 341, 380, 411,
426, 442, 446, 447
Biotechnological applications, 12, 59, 94, 98,
281, 309–310, 317–330, 400, 431
Biotic stresses, 39, 64, 81, 126, 131, 252,
262–263, 337–352, 389, 406, 412, 419

C

Cold stresses, 66, 71, 80, 230, 257, 261,
305–306, 309–311, 366, 378, 409
Compounds, 3, 23, 48, 66, 96, 121, 137, 168,
201, 222, 262, 283, 308, 322, 390, 407,
443, 456
Crop health, 397, 474
Crop plants, 65, 201, 234–236, 241, 259, 261,
292, 306, 338, 347, 349–351, 415, 445

E

Ecosystems, 9, 21, 25, 50, 54, 59, 64, 65, 67,
77–80, 95, 99–101, 103, 104, 110, 111,
114, 138, 164, 169, 203, 224, 228, 235,
252, 254, 257, 263–264, 302–304, 347,
390, 396, 398, 399, 407, 409, 415, 430,
435, 442, 445, 448, 472, 475
Enhancing plant growth, 79, 114, 147, 411
Enzymes, 5, 7–11, 14, 23–29, 33–39, 41, 51,
53, 56, 58, 67, 72, 75, 76, 94–97,
99–101, 105, 106, 108, 109, 116,
121–123, 125, 128, 131, 132, 141, 143,
150–154, 164, 166–168, 173, 178, 184,
202, 204, 207, 208, 210–213, 229,
239–242, 254, 256, 257, 261, 262, 266,
283–286, 293, 304, 305, 317, 318, 320,
322, 325–329, 340, 342, 344–347, 349,
350, 361–368, 372, 376, 377, 394, 399,
400, 408–410, 412–414, 417, 419,
425–428, 430, 431, 434, 435, 446, 448,
456, 457, 462–464, 466, 473, 475
Evolution, 6, 39, 49, 50, 54, 137, 164, 265,
330, 407, 431
Extreme adaptation, 222

Extreme environments, 21, 22, 24, 35, 48–53, 58, 64, 66, 67, 72, 76, 78, 79, 92, 109, 111, 112, 114–116, 121, 129, 130, 138, 139, 151, 154, 199–213, 222, 242, 252, 253, 256, 263, 281, 339, 364–368, 390–392, 406, 408, 428, 446, 448, 455, 463, 470

Extremophiles, 3–15, 21–41, 48–50, 52, 53, 56–58, 63–81, 91–116, 121–133, 137, 138, 151, 206, 207, 213, 221–242, 251–268, 281, 318, 323, 328, 329, 339, 340, 361–381, 393, 405–419, 425–435, 441–450, 455, 456, 458, 466, 470, 474–476

Extremozymes, 5, 23, 27, 38, 39, 58, 105, 123, 222, 266, 281, 328–329, 344, 361, 406, 409, 419, 425–428, 430, 435, 446, 463

G

Genomics, 6, 26, 71, 76, 77, 98, 138, 210, 264, 281–294, 329, 330, 341, 345, 346, 371, 406, 407, 417–419, 425–435

H

Halotolerant, 22, 23, 50, 77, 139, 240, 241, 254, 265, 266, 281–294, 408, 416

Harsh conditions, 3, 14, 21, 23, 24, 53, 109, 138, 255, 259, 262, 281, 308, 348, 407, 410, 425

Heavy metals (HMs), 3, 5, 8, 9, 11, 37, 48, 56, 57, 59, 65, 66, 77–78, 80, 104, 116, 148, 164–186, 199, 205, 222, 236, 239, 255, 256, 261, 266, 293, 311, 322, 324, 338, 339, 347, 361, 391, 413, 443–445, 447, 448, 465, 466, 473

I

Industrial applications, 10, 11, 23, 53, 58, 105, 107–116, 121, 124, 128, 131, 146, 153, 210, 325, 329, 394, 425, 435, 460

M

Marine, 10, 25, 27, 50, 68, 128, 138–142, 145, 146, 150, 153, 184, 262, 367, 426, 430, 464–466

Membrane fluidity, 9, 13, 25, 27, 71, 72, 75, 94, 231, 256, 304, 305, 309, 310, 365

Membrane permeability, 8, 72, 175, 258, 318, 319, 349, 372

Mesophiles, 8, 12, 22, 39, 70, 238, 320, 322, 325, 327, 329, 330, 345, 346, 408, 475

Metallotolerant microbes, 165, 176, 178–185

Microbe–metal interactions, 163–186

Microbial adaptation, 47–59, 64

Microbial synthesis, 293

Microorganisms, 3–5, 7–13, 15, 21–25, 27, 29, 31, 32, 35–37, 39, 47–57, 59, 64, 66, 67, 69–81, 93–111, 113–116, 121–123, 125, 127, 129, 130, 137, 164–171, 173, 174, 177, 178, 185, 199, 204, 208, 227, 228, 232, 234, 236, 237, 239–240, 252, 254, 258–261, 263–266, 268, 281, 283, 284, 287, 292, 301–311, 317–319, 322, 324–326, 330, 339–342, 346–350, 352, 361–363, 367, 371, 380, 390–392, 394, 398, 399, 405, 407, 410, 412, 413, 429, 430, 432, 434, 455–457, 462, 463, 470, 474

Multi omics approaches, 328–329

N

Nano-fertilizers, 449, 470, 474

Nanoparticles (NPs), 154, 178, 208, 288, 446–449, 455–476

Next-generation sequencing (NGS), 6, 64, 81, 285–287, 294, 425, 431–435

Nutrient availability, 48, 54, 97, 98, 100, 102, 121, 169, 211, 222, 238, 283, 288, 293, 395, 398, 411, 412, 415, 449

O

Organic pollutants (OPs), 59, 163, 184, 199–213, 231, 262, 311, 443, 446, 447

P

Plant-associated extremophilic microbes (PAEM), 252, 406, 419

Plant disease management, 121–133, 469

Plant growth-promoting rhizobacteria (PGPR), 58, 59, 127, 129, 228, 231, 260, 282–288, 292–294, 349, 351, 406, 411, 414

Plant growth promotion, 55, 58, 129, 222, 224, 228, 229, 261, 303, 347–349, 397, 398, 475

Plant health, 178, 233, 241, 242, 251–268, 282, 283, 288, 292, 294, 301–311, 391, 396, 411, 412

Plant protection, 203

Potential microbes, 252

Psychrophiles, 3, 6, 8–10, 22, 24–27, 38, 49, 50, 55, 64, 70, 72, 93–95, 110, 122,

124, 138, 153, 222, 224–231, 242, 253, 256–257, 259, 261, 302, 339, 366, 369, 372, 393, 406, 408, 409, 412, 426, 427, 456, 459, 466–467, 474, 475
Psychrotrophs, 257, 416

S
Saline soils, 4, 5, 100, 260, 281–294
Soil fertilities, 97, 104, 108, 115, 128, 164, 228, 234, 242, 251–268, 282–285, 389, 392, 394–396, 399, 415, 456, 470, 473, 475
Soil improvement, 116, 231, 301–311, 396
Soil nutrients, 236, 283, 309, 389–399, 416
Soil remediation, 55, 104, 164
Stabilities, 23–28, 33, 39, 48, 52, 53, 58, 71, 72, 93, 96, 97, 102, 112, 122, 131, 151, 174, 201, 207, 209, 213, 229, 255, 256, 286, 317–320, 325, 328–330, 344–346, 349, 362, 368, 372, 373, 399, 406, 408, 410, 418, 427–428, 430, 467, 469, 475
Stress adaptation, 67–78, 302, 303, 406

Stress tolerance, 67, 99, 241, 260, 265, 287, 362, 378–379, 381, 413, 414
Survival mechanisms, 24, 31, 39–40, 49, 76, 80, 371, 405, 406
Sustainability, 41, 53, 59, 81, 96, 97, 99, 103, 104, 111, 254, 262, 283, 341, 352, 390, 399, 419, 448–449
Sustainable agriculture, 4, 24, 41, 54, 67, 93, 99–102, 104–106, 116, 123, 128, 133, 185, 233, 252, 264, 311, 322, 341, 342, 347, 348, 361–381, 390, 393, 406, 411, 415, 435, 447, 449, 469
Symbiosis, 105, 230

T
Thermophiles, 3, 6, 8–9, 22, 27–29, 36, 39, 40, 64, 70–73, 78, 80, 95–96, 110, 113, 122, 124, 138, 222, 237–239, 242, 253, 255–256, 317–330, 339–346, 368, 393, 406, 408–409, 412, 416, 426, 430, 434, 456, 459, 467–468, 474–476
Thermophilic microbes, 239, 337–352, 467
Thermozymes, 317, 318